T0142360

Advances in Intelligent Systems and Computing

Volume 673

Series editor

Janusz Kacprzyk, Polish Academy of Sciences, Warsaw, Poland
e-mail: kacprzyk@ibspan.waw.pl

The series "Advances in Intelligent Systems and Computing" contains publications on theory, applications, and design methods of Intelligent Systems and Intelligent Computing. Virtually all disciplines such as engineering, natural sciences, computer and information science, ICT, economics, business, e-commerce, environment, healthcare, life science are covered. The list of topics spans all the areas of modern intelligent systems and computing.

The publications within "Advances in Intelligent Systems and Computing" are primarily textbooks and proceedings of important conferences, symposia and congresses. They cover significant recent developments in the field, both of a foundational and applicable character. An important characteristic feature of the series is the short publication time and world-wide distribution. This permits a rapid and broad dissemination of research results.

More information about this series at http://www.springer.com/series/11156

Subhash Bhalla · Vikrant Bhateja
Anjali A. Chandavale · Anil S. Hiwale
Suresh Chandra Satapathy
Editors

Intelligent Computing and Information and Communication

Proceedings of 2nd International Conference, ICICC 2017

 Springer

Editors
Subhash Bhalla
Department of Computer Software
University of Aizu
Aizuwakamatsu, Fukushima
Japan

Anil S. Hiwale
Department of Information Technology
MIT College of Engineering
Pune, Maharashtra
India

Vikrant Bhateja
Department of Electronics and
  Communication Engineering
Shri Ramswaroop Memorial Group of
  Professional Colleges
Lucknow, Uttar Pradesh
India

Suresh Chandra Satapathy
Department of Computer Science
  and Engineering
Anil Neerukonda Institute
  of Technology and Sciences
Visakhapatnam, Andhra Pradesh
India

Anjali A. Chandavale
Department of Information Technology
MIT College of Engineering
Pune, Maharashtra
India

ISSN 2194-5357   ISSN 2194-5365   (electronic)
Advances in Intelligent Systems and Computing
ISBN 978-981-10-7244-4   ISBN 978-981-10-7245-1   (eBook)
https://doi.org/10.1007/978-981-10-7245-1

Library of Congress Control Number: 2017959316

Printed on acid-free paper

This Springer imprint is published by Springer Nature
The registered company is Springer Nature Singapore Pte Ltd.
The registered company address is: 152 Beach Road, #21-01/04 Gateway East, Singapore 189721, Singapore

Preface

The second International Conference on Intelligent Computing and Communication (ICICC 2017) was successfully organized by Dr. Vishwanath Karad, MIT World Peace University, Pune, during August 2–4, 2017, at MIT College of Engineering, Pune, India. The conference had technical collaboration with Computer Society of India and College of Engineering Pune and was supported by All India Council for Technical Education (AICTE) and Council of Scientific and Industrial Research (CSIR). The objective of this international conference was to provide a platform for academicians, researchers, scientists, professionals, and students to share their knowledge and expertise in the field of intelligent computing, communication, and convergence and address various issues to increase awareness of technological innovations and to identify challenges and opportunities for the development of smart cities using multidisciplinary research techniques. Research submissions in various advanced technology areas were received, and after a rigorous peer-review process with the help of program committee members and external reviewers, only quality papers were accepted. The conference featured eight special sessions on various cutting-edge technologies which were conducted by eminent professors and many distinguished academicians like Dr. Vijay Bhatkar, Padma Bhushan, Chancellor of Nalanda University, Bihar, India; Dr. Hemant Darbari, Executive Director, CDAC, Pune, India; Dr. Bipin Indurkhya, Jagiellonian University, Kraków, Poland; and Dr. Subhash Bhalla, University of Aizu, Japan.

Our sincere thanks to all special session chairs Prof. Dr. Priti Rege, Prof. Dr. R. A. Patil, Dr. Anagha Kulkarni and distinguished reviewers for their timely technical support. We would like to extend our special thanks here to our publication chairs for doing a great job in making the conference widely visible. Thanks to dynamic team members for organizing the event in a smooth manner. Our sincere thanks to all sponsors, press, print, and electronic media for their excellent coverage of this conference.

Aizuwakamatsu, Japan   Subhash Bhalla
Lucknow, India   Vikrant Bhateja
Pune, India   Anjali A. Chandavale
Pune, India   Anil S. Hiwale
Visakhapatnam, India   Suresh Chandra Satapathy

Organizing Committee

Chief Patrons
Prof. Dr. Vishwanath Karad, India
Prof. Dr. Vijay Bhatkar, India

Patrons
Prof. Dr. Mangesh Karad, India
Prof. Dr. Rahul Karad, India

Organizing Chair
Dr. Ramchandra Pujeri, India

Organizing Co-chair
Dr. Anil Hiwale, India

Advisory Committee
Dr. Seeram Ramakrishna, Singapore
Dr. Subhash Bhalla, Japan
Dr. Shirish Sane, India
Dr. Suresh Borkar, USA
Dr. Shabnam Ivkovic, Canada

TPC Chair
Dr. Anjali A. Chandavale, India

Publication Chair
Dr. Suresh Chandra Satapathy, Vijayawada, India

Contents

About the Editors

Prof. Subhash Bhalla joined as a Faculty School of Computer and Systems Sciences, Jawaharlal Nehru University (JNU), New Delhi, in 1986. He was a Visiting Scientist at Sloan School of Management, Massachusetts Institute of Technology (MIT), Cambridge, Massachusetts, USA (1987–1988). He is a Member of the Computer Society of IEEE and SIGMOD of ACM. He is associated with the Department of Computer Software at the University of Aizu, Japan. He has also toured and lectured at many industries for conducting feasibility studies and for the adoption of modern techniques. He has received several grants for research projects. He currently participates in the Intelligent Dictionary System Project. He is exploring database designs to support models for Information Interchange through the World Wide Web. He is working with a study team on creating user interfaces for Web users and transaction management system for mobile computing. He is studying transaction management and algorithmic designs for distributed real-time systems. He is also pursuing performance evaluation and modeling of distributed algorithms. His research interests include managing components and application services; distributed client/middleware/service-oriented computing; XML, e-commerce; mobile database management systems; Web query and Web data mining; synchronization and crash recovery; and integration of technologies.

Prof. Vikrant Bhateja is Associate Professor, Department of Electronics and Communication Engineering, Shri Ramswaroop Memorial Group of Professional Colleges (SRMGPC), Lucknow, and also the Head (Academics and Quality Control) in the same college. His areas of research include digital image and video processing, computer vision, medical imaging, machine learning, pattern analysis and recognition, neural networks, soft computing, and bio-inspired computing techniques. He has more than 90 quality publications in various international journals and conference proceedings. He has been on TPC and chaired various sessions from the above domain in the international conferences of IEEE and Springer. He has been the track chair and served in the core technical/editorial teams for the international conferences: FICTA 2014, CSI 2014 and INDIA 2015 under Springer-AISC Series, and INDIACom-2015 and ICACCI-2015 under IEEE.

He is Associate Editor in International Journal of Convergence Computing (IJConvC) and also serving in the editorial board of International Journal of Image Mining (IJIM) under Inderscience Publishers. At present, he is guest editor for two special issues floated in International Journal of Rough Sets and Data Analysis (IJRSDA) and International Journal of System Dynamics Applications (IJSDA) under IGI Global Publications.

Dr. Anjali A. Chandavale is currently working as Associate Professor (Information Technology) in Maharashtra Academy of Engineering and Educational Research's Maharashtra Institute of Technology College of Engineering, Pune, India. She has over 18 years of teaching and over 6 years of industrial experience in the field of process instrumentation. Her research interests include Internet security, image processing, and embedded systems. She has worked as Research and Development Engineer in Lectrotek Systems (Pune) Pvt Ltd, Pune, for 4 years and also as a Software Engineer (Design and Development) in Ajay Electronics Pvt Ltd (1997–1998), Pune. She has several publications in journals as well as in conference proceedings of international repute.

Prof. Anil S. Hiwale is currently working as Professor and Head of Department of Information Technology, MIT College of Engineering, Pune. He completed his Ph.D. from SGB Amravati University, Amravati. Prof. Hiwale has 28 years of teaching experience and published over 50 research papers in National/International Journals and conferences. His areas of research include signal processing, digital communications, multi-antenna systems and satellite communications. He is presently supervising four Ph.D. scholars. He is a Fellow of Institution of Engineers (India) and a member of IET, CSI, ISTE and Broadcast Engineering Society of India.

Prof. Suresh Chandra Satapathy is currently working as Professor and Head, Department of Computer Science and Engineering, PVP Siddhartha Institute of Technology, Andhra Pradesh, India. He obtained his Ph.D. in Computer Science and Engineering from JNTU Hyderabad and M.Tech. in CSE from NIT Rourkela, Odisha, India. He has 26 years of teaching experience. His research interests include data mining, machine intelligence, and swarm intelligence. He has acted as program chair of many international conferences and edited six volumes of proceedings from Springer LNCS and AISC series. He is currently guiding eight scholars for Ph.D. He is also a Senior Member of IEEE.

35 W GaN Solid-State Driver Power Amplifier for L-Band Radar Applications

Vivek Ratnaparkhi and Anil Hiwale

Abstract In this paper, 35 W driver power amplifier was designed and simulated using GaN HEMT for L-band radar. GaN HEMT is used because it can provide high output power and high gain as compared to other semiconductor technologies. The 35 W output power is generated using CGHV40030 GaN HEMT which is sufficient to drive further stages of power amplifier. The driver amplifier is designed at 1.3 GHz of center frequency. This amplifier is designed in class AB and 60.5% of PAE is achieved.

Keywords Class AB · GaN HEMT · L-band radar · PAE

1 Introduction

Rapid development in the field of gallium nitride (GaN) semiconductor devices since last two decades is changing the rules of power amplifier design for wireless communications. To generate and amplify high microwave power, people were using microwave tubes which are having certain limitations. It is all dependent on particular application where microwave tube is used for signal generation and amplification. For microwave applications, traveling wave tube tubes (TWT), magnetrons, and klystrons were conventionally used. TWT amplifiers have certain limitations such as high noise, shorter lifespan, and wide bandwidth which results in interference with other wireless communication system operating in an adjacent band. Consequently, there is a strong desire for solid-state power amplifiers (SSPA) that are superior in long-term reliability and signal noise to replace conventional TWT amplifiers at microwave frequencies [1–3]. GaN devices are becoming a

V. Ratnaparkhi (✉)
Department of EXTC, SSGMCE, Shegaon, Maharashtra, India
e-mail: ratnaparkhi_vivek@yahoo.co.in

A. Hiwale
Department of IT, MITCOE, Pune, Maharashtra, India
e-mail: anil.hiwale@mitcoe.edu.in

© Springer Nature Singapore Pte Ltd. 2018
S. Bhalla et al. (eds.), *Intelligent Computing and Information and Communication*,
Advances in Intelligent Systems and Computing 673,
https://doi.org/10.1007/978-981-10-7245-1_1

1

Table 1 Comparative analysis of different device types

Device type	Pmax (kW)	Efficiency (%)	Gain (dB)	Bias voltage (kV)	Operation time
GaN devices	0.8	50–80	10–20	0.025–0.1	High
Power grid tubes	0.5–10	50–60	10–15	0.5–10	Less
Electron beam devices	0.1–2000	25–60	20–45	25–100	Moderate

promising choice for SSPA design for microwave applications. Most of the transistor manufacturers are now developing GaN high-power transistors with attractive performance characteristics. Table 1 shows the comparison of power devices family on the basis of device characteristics. It is apparent that new GaN semiconductor devices are capable of producing almost 1 KW of output power with the highest efficiency of 65%. If several stages of solid-state power amplifiers are combined properly, very high output power in the range of few kilowatts can be generated. It is also important to note that lifetime of power amplifiers using SSPDs is very high when compared with the power grid tubes (PGTs) and electron beam devices (EBDs) such as traveling wave tubes (TWTs) and klystrons. From the reliability point of view, SSPDs are more reliable than PGTs and EBDs.

GaN solid-state transistors have the potential to disrupt very large vacuum tube market and can replace some conventional vacuum tubes. There is high demand for GaN SSPAs for microwave applications.

2   GaN Device Technology

For radar applications, traveling wave tubes (TWT) such as magnetrons and klystrons were conventionally used because of the required power level as high as 1 kw [4]. GaN transistors were first demonstrated in the 1990s and have started to become commercially available in last decade [5]. GaN transistors have many advantages when it is compared with other semiconductor materials like Si, GaAs, and SiC. Table 2 compares material characteristics of Si, GaAs, SiC, and GaN. It is apparent that GaN has higher breakdown voltage which allows GaN HEMTs to

Table 2 Material parameters comparison

Property	Si	GaAs	SiC	GaN
Band gap energy (eV)	1.11	1.43	3.2	3.4
Critical breakdown field (MV/cm)	0.3	0.4	3.0	3.0
Thermal conductance (W/cm K)	1.5	0.5	4.9	1.5
Mobility(cm^2/V s)	1300	6000	600	1500
Saturated velocity ($\times 10^7$ cm/s)	1.0	1.3	2.0	2.7
JFOM versus Si	1.0	1.7	20	27

operate at biasing voltages about 50VDC. Large drain voltages lead to high output impedance per watt of RF power and result in easier matching circuit design.

Since GaN HEMT devices have higher impedance than other semiconductor devices, this can help to enhance the PAE and bandwidth of the required SSPA. High-saturated drift velocity results in higher saturation current densities and watts per unit periphery. This ensures GaN HEMTs suitability for switched-mode power amplifiers [6].

It is evident that high-power amplifiers made by Gallium arsenide produce high efficiency than silicon high-power amplifiers. GaN devices even provide higher efficiencies than GaAs power amplifiers. Due to such high-efficiency capabilities, GaN high-power SSPAs are being used extensively in microwave applications. GaN devices can operate at higher voltage and have higher saturated velocity which allows GaN transistors to generate higher power in smaller space which results in high power density. Consequently, smaller size high-power amplifiers can be fabricated with GaN transistors.

There are many researchers and manufacturers working in the field of GaN SSPA design for microwave applications. Recently, solid-state power amplifiers designed using GaN HEMT devices have replaced many conventional vacuum tubes used for radar and space applications. It is expected that GaN technology will certainly grow and will be promising choice for high output power applications at microwave frequencies with high reliability and lesser noise when compared to conventional vacuum tubes.

3 GaN SSPA Design

Over the last decade, GaN-based high electron mobility transistors (HEMTs) have emerged as excellent devices for a number of applications [7]. GaN HMTs technology has been shown to provide high power and high efficiency making it the perfect candidate for this next-generation radar system [8]. George Solomon [9] and his group from Communications and Power Industries, LLC Beverly Microwave Division have reported the VSX3622, a 1.5 kW X-B, and GaN power amplifier for radar applications. In this section, a step-by-step design of GaN HEMT driver power amplifier for L-band radar applications is explained.

Figure 1 shows topology of proposed solid-state power amplifier for L-band radar applications. Using proposed topology, at least 500 W output power can be achieved. In this paper, design and simulation of GaN solid-state driver power amplifier are discussed.

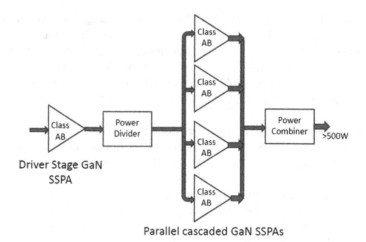

Fig. 1 Topology of proposed GaN SSPA

4 Results

In this section, simulation results of GaN driver power amplifier are presented. Keysigth's ADS software is used for design and simulation of GaN driver power amplifier. Since our requirement is to generate at least 30 W output power from driver stage, CGHV40030 GaN HEMT is selected.

4.1 DC Analysis

Cree's CGHV40030 GaN HEMT is used for designing driver power amplifier [10]. First stage of PA design is to carry out DC analysis and find out the operating point of proposed amplifier. Since maximum output power is expected, class AB is the best choice for proposed driver power amplifier. Upon performing DC analysis in ADS, VDS = 48 V and VGS = −3 V selected as operating points, which ensures IDS = 20 mA. The device power consumption at this bias is 88 mW.

4.2 Stability Analysis

It is very important to perform and ensure stability of the device before proceeding for further design. Any instabilities of the device may generate unwanted oscillations and amplifier performance will affect significantly. Device may have unconditionally stable or conditionally stable. There are two stability checks available to check stability of the device: stability factor and stability measure.

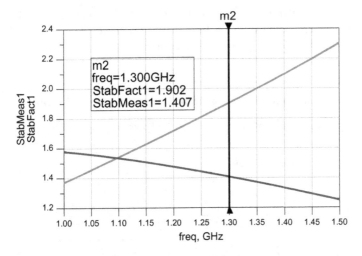

Fig. 2 Stability factor and stability measure

Figure 2 shows stability factor greater than one and stability measure greater than zero which ensures that device is unconditionally stable. This stability analysis is performed after proper DC biasing to the device.

4.3 Load Pull Analysis

Load pull analysis is the important step in PA design. Having a good nonlinear model is essential for starting PA design. We have obtained nonlinear model of CGHV40030 from device manufacturer. Load pull analysis is carried out using nonlinear models of the device to find out optimum impedance to be presented to device for specific output power, PAE, IMD, etc. Once optimum load and source impedances are found using load pull, impedance matching network is designed. For CGHV40030 GaN HEMT, load pull analysis is performed and results are shown in Fig. 3.

4.4 Impedance Matching Network Design and Complete Amplifier Performance

Once load pull analysis is completed and obtained values for source and load impedances, input and output impedance matching networks can be designed. Results of complete amplifier with input and output matching networks after some optimization are shown in this section. Figure 4a shows output power of designed GaN SSPA, which is almost 35 W. Figure 4b shows PAE of designed GaN SSPA and it is acceptable for our proposed driver amplifier.

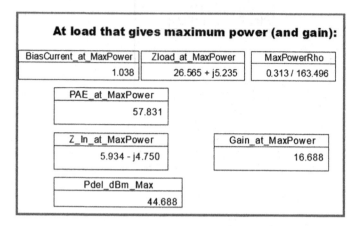

At load that gives maximum power (and gain):

BiasCurrent_at_MaxPower	Zload_at_MaxPower	MaxPowerRho
1.038	26.565 + j5.235	0.313 / 163.496

PAE_at_MaxPower
57.831

Z_In_at_MaxPower	Gain_at_MaxPower
5.934 - j4.750	16.688

Pdel_dBm_Max
44.688

Fig. 3 Results of load pull simulation at maximum power and gain

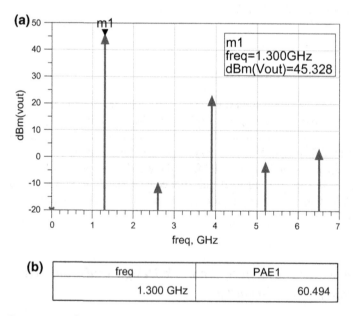

(a)

m1
freq=1.300GHz
dBm(Vout)=45.328

(b)

freq	PAE1
1.300 GHz	60.494

Fig. 4 **a** Output power, **b** power added efficiency

5   Conclusion

Solid-state driver power amplifier using CGHV40030 GaN HEMT is designed at 1.3 GHz for L-band radar applications. Output power of 45 dBm and PAE of 60.494% are achieved. Load pull analysis is carried out to determine device output and input impedances. The nonlinear model of the device is obtained from device manufacturer. Complete GaN SSPA with four parallel cascaded stages with power divider and combiner can be designed for generating 500 W of output power.

References

1. Allen Katz and Marc Franco "GaN comes of Age" in IEEE microwave magazine, December 2005. pp. 524–534.
2. David Schnaufer "GaN: The Technology of the Future http://www.rfglobalnet.com/doc/gan-the-technology-of-the-future-0001.
3. A. Katz et al. "A linear GaN UHF SSPA with record high efficiency," in IEEE MTT-S Int. Microwave Symp. Dig., Boston, MA, June 7–12, 2009, pp. 769–772.
4. Ken Kikkuchi et al., "An X-Band 300-Watt Class High Power GaN HEMT Amplifier for Radar Applications", SEI Technical Review, Number 81, October 2015, pp. 40–44.
5. Francesco Fornetti "Characterisation and performance optimization of GaN HEMTs and amplifiers for Radar applications" December 2010.
6. Andrew Moore and Jose Jimenez "GaN RF Technology For Dummies", TriQuint Special Edition Published by John Wiley & Sons, Inc. 111 River St. Hoboken, NJ 07030-5774,2014.
7. Guest Editorial Special Issue on GaN Electronic Devices "IEEE Transactions on Electron Devices," vol. 60, no. 10, October 2013, pp. 2975–2981.
8. T. Thrivikraman and J. Hoffman, "Design of an ultrahigh efficiency GaN high-power amplifier for SAR remote sensing," in IEEE Aerospace Conference, pp. 1–6, 2013.
9. George Solomon et al., "The VSX3622, a 1.5 kW X-band GaN Power Amplifier for Radar Application", Communications & Power Industries LLC, Beverly Microwave Division.
10. Cree's Datasheet Online available at: http://www.wolfspeed.com/cghv40030.

Validation of Open Core Protocol by Exploiting Design Framework Using System Verilog and UVM

Gopika Rani Alekhya Pamarthy, M. Durga Prakash
and Avinash Yadlapati

Abstract Today's scenario of semiconductor technology is a tremendous innovation; it includes a large number of intellectual property (IP) cores, interconnects, or buses in system on chip (SOC) design and based upon the necessity its complexity keeps on increasing. Hence, for the communication between these IP cores, a standard protocol is developed. The necessity of IP reuse, abridging the design time and the complexity makes large-scale SOC more challenging in order to endorse IP core reusability for SOC designs. An efficient non-proprietary protocol for communication between IP cores is open core protocol (OCP). OCP comes under socket-based interface and openly licensed core concentric protocol. This paper addresses on the verification of implemented design of OCP. The proposed paper is to verify the implemented design by using System Verilog and Universal Verification Methodology (UVM) in SimVision tool.

Keywords Open core protocol (OCP) · Intellectual property (IP)
System on chip (SOC) · Socket-based interface · Core concentric

1 Introduction

Open core protocol (OCP) is a competent protocol for communication on SoC. OCP [1] compliance IP cores can be reversed by the designer, depending on system integration and verification approach in multiple designs without reinstallation,

G. R. A. Pamarthy (✉) · M. Durga Prakash
Department of Electronics and Communication Engineering, K L University, Guntur,
Andhra Pradesh, India
e-mail: alekhya.rani.gopika@gmail.com

M. Durga Prakash
e-mail: mdprakash82@gmail.com

A. Yadlapati
CYIENT Ltd, Hyderabad, India
e-mail: avinash.amd@gmail.com

© Springer Nature Singapore Pte Ltd. 2018
S. Bhalla et al. (eds.), *Intelligent Computing and Information and Communication*,
Advances in Intelligent Systems and Computing 673,
https://doi.org/10.1007/978-981-10-7245-1_2

9

reducing the development time, cost, and design risk. OCP is an interface for communication between IP cores on an SOC. OCP defines a bus-independent configurable interface.

OCP renovates IP cores making them independent of the architecture and design of the systems in which they are used and shortens system verification and testing by providing a secure boundary around each IP core. OCP is simple, synchronous, point-to-point, highly scalable, and configurable to match the communication requirements associated with different IP cores. Even complex high-performance cores can be accommodated capably with OCP extensions. Cores with OCP interfaces enable true plug-and-play [2] approach and automated design processes, thus allowing the system integrator to choose the best cores and best interconnect system.

Unlike bus approach, with reference to the standard communication approach, there are mainly two protocols: VCI (Virtual Component Interface) and OCP. OCP which is scalable and bus independent is a superset of VCI which reports only data flow aspects; moreover, OCP supports sideband control signaling and tests harness signals which are configurable. OCP is the only protocol which unifies all the inter-core communication. OCP establishes a point-to-point interface between two IP cores [3]. One of them acts as the master who is the controller and generates the commands and other as the slave responding to commands generated by the master, either by accepting or giving data to the master.

The OCP defines a point-to-point interface between two communicating entities such as IP cores and bus interface modules (bus wrappers). One entity acts as the master of the OCP instance and the other as the slave. Only the master can present commands and is the controlling entity. The slave responds to commands presented to it, either by accepting data from the master or presenting data to the master. For two entities to communicate in a peer-to-peer fashion, there need to be two instances of the OCP connecting them—one where the first entity is a master, and one where the first entity is a slave.

2 Project Scope

Implementation is carried out using behavioral Verilog HDL [4] simulation environment. The design implements a simple memory read and write operations and Burst transactions between two IP cores. The design complies with subset of OCPIP handshake signals. The implemented design, i.e., DUT, is verified using system Verilog and UVM. As the system Verilog is superset of Verilog and is based upon OOPS [5] (Object Oriented Programming) concepts, the environment can be extended without modifying the intention of the original existing code by adding all the required new features.

3   OCP Protocol

OCP is a configurable protocol that defines one of the communicating entities as master and other as slave. Master initiates the operation by generating a request signal to the slave, and in turn slave responds by sending the acknowledgement to the master. Once master receives the acknowledgement [5] from slave, it transmits the data to the slave; this phenomenon illustrates the typical handshaking process of communication as shown in Fig. 1.

3.1   OCP Signal Description

The dataflow signals consist of a small set of required signals and a number of uncompelled signals that can be configured to support additional requirements [1]. The dataflow signals are grouped into basic signals, burst extensions (support for bursting). The naming conventions for dataflow signals use the prefix M for signals driven by the OCP master and S for signals driven by the OCP slave [3].

- **Clk**: Input clock signal for the OCP clock.
- **Maddr**: Input address to the master in which data has to be written into the corresponding memory location and is to be accessible later during read operation. Maddr width is configurable.
- **MCmd**: Mode assignment command of width 3. This signal specifies the mode of OCP which is requested by the master. Usually, depending upon the command generated by the master, slave responds and performs that operation. If OCP is in idle state, it does not perform any mode of transfer, whereas during non-idle state depending on the direction of data flow, OCP [2] performs either read/write operation.
- **Mdata**: The data sent by the master that has to be written into the prescribed memory location of slave which is a configurable one.
- **MDatavalid** [6]: This is the indication to the slave that the data sent by the master is valid only. When it is set, it specifies that the Mdata field is valid.
- **MRespAccept**: Acknowledgment from master that the response from slave was accepted by it when MRespAccept is set.

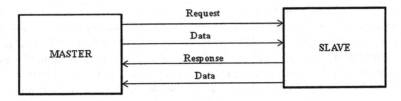

Fig. 1 Signal transmission between master and slave

- **SCmdAccept**: Acknowledgment from slave that the response from master was accepted by it when SCmdAccept is set.
- **Sresp**: Response from the slave to a transfer request from the master. It is of width 2.
- **SData**: It transfers the requested data by master as read data from the slave to the master.
- **MBurstLength**: This illustrates configurable number of transfers in a burst. MBurstLength of value 0 is illegal.
- **MBurstPrecise**: This specifies whether the precise burst length is known at the start of the burst or not.
- **MBurstSeq**: This 3-bit-width field indicates the sequence of addresses for requests in a burst.
- **SRespLast**: Last response from the slave in a burst.
- **MReqlast**: Last request from the master in a burst.

4 Functional Description

The data transfer between master and slave is communicated non-serially. Commands are generated by the master to the slave. Based on the command given by the master, slave responds and decides the mode of operation need to be performed that is requested by the master. Data is either written or read based upon the command from master into the particular memory location which is specified by the master [6].

4.1 Master

Master is the one who starts the transactions by providing data and address to the slave. It is the commander and controller of the entire design. It makes the slave to function on what it needs. The basic block diagram of master with all the input and output signal specifications is shown in the figure.

Master activates the OCP by sending command, address, and data to the slave when clock is active. Slave responds to the master's command and sends an acknowledgment to the master indicating the acceptance of command from master as SCmdAccept [7]. Write/read mode of transmission is performed based on the master's request. In order to indicate that the data sent by master is valid, it sends a valid bit on MDatavalid to the slave; on receiving it, slave starts writing the data in the corresponding memory location. Similarly, master reads the data from slave by activating data to be read on SData by specifying the exact memory location.

Master block with all its inputs and outputs driving to slave are shown in Fig. 2.

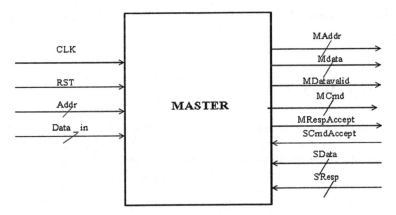

Fig. 2 OCP master block

4.2 *Slave Block*

Slave simply performs the operation depending on the command sent by the master. It activates itself to respond on to the master request. Data handshaking process of communication is adopted to have a proper and efficient communication [8]. Slave acknowledges for each and every signal from master as a correspondence or acceptance of request from the master. The detailed slave block is shown in Fig. 3.

Slave responds to the request sent by the master [9]. It performs read/write mode of transfer by responding to the signals from master. Slave acknowledges the master for each and every transfer at enabled clock. During read mode of operation, it acknowledges the master that the data sent by it is a valid one by sending its

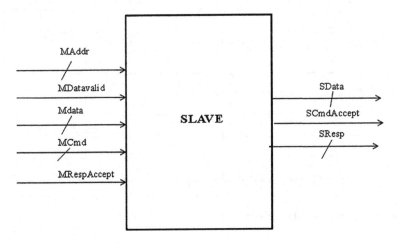

Fig. 3 OCP slave block

response by means of Sresp signal. Both master and slave are controlled by the same clock [1].

5 Timing Analysis

To connect two IP cores together by allowing them to communicate over an OCP interface, the protocols, signals, and pin-level timing must be compatible. OCP establishes the communication based on the applied clock to the master [3]. Timing plays a vital role while sending the data to the slave or from the slave.

5.1 Basic Mode of OCP

The timing analysis for simple write and simple read of OCP is illustrated in Fig. 4. The diagram shows a write with no response enabled on the write [5].

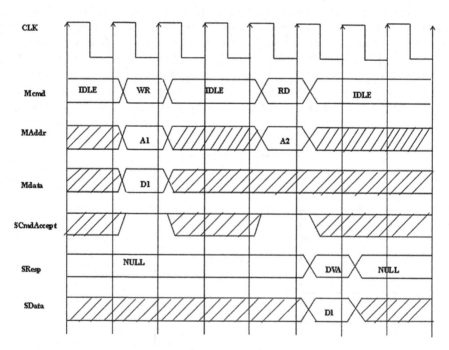

Fig. 4 Timing analysis of basic mode OCP

5.2 Burst Mode of Operation

OCP burst mode for a burst of four 32-bit words, incrementing precise burst write, with optional burst framing information (MReqlast) is illustrated [4]. As the burst is precise (with no response on write), the MBurstLength signal is constant during the whole burst. MReqlast flags the last request of the burst, and SRespLast flags the last response of the burst. The slave monitors MReqlast for the end of burst. The timing diagram is shown in Fig. 5.

6 Simulation Results

The implemented design of OCP is verified by creating the UVM verification environment by using SimVision, cadence tool. SimVision Debug can be used to debug digital, analog, or mixed-signal designs. These can be written in Verilog, SystemVerilog, e, VHDL, and SystemC® languages. It supports all IEEE standard designs of signal-level and transaction-based flows, testbench, and assertion languages. SimVision Debug provides a unified simulation and debug environment. It allows Incisive Enterprise Simulator to manage multiple simulation runs easily [8]. To analyze both design and testbench at any point in the verification process is quiet easy.

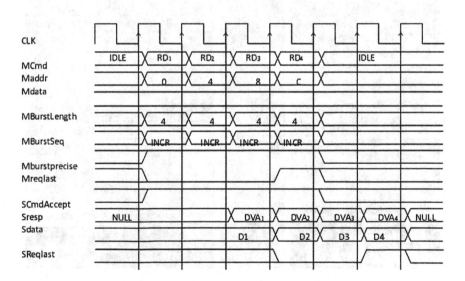

Fig. 5 Timing analysis of burst mode OCP

6.1   Simple Mode of Operation (Both Write and Read)

Basic mode of OCP with simple write and read is implemented and verified [2].
Master initiates the transfer by sending the address and data to the slave on Maddr
and MCmd, respectively. Slave acknowledges the master by asserting the
SCmdAccept to the master. It performs either read or write based upon the command generated by the master. The simulation results obtained for the applied
transaction between master and slave when RTL is verified by using UVM
Methodology on SimVision tool are shown in Fig. 6.

6.2   Burst Mode of Operation (Both Write and Read)

Precise burst mode was implemented and verified. MBurstLength,
MBurstSequence, and MBurstPrecise are considered constant throughout the
transaction. Master initiates precise burst write operation by sending the request to
slave (MCmd), address (Maddr), and data (Mdata). Mreqlast = 1 indicates the end
of request from master. Slave acknowledges the master by enabling SCmdAccept.
A high on SRespLast represents the end of precise burst read [6]. The simulation
result obtained for the applied transaction between master and slave when RTL is
verified by using UVM methodology on SimVision tool is shown in Fig. 7.

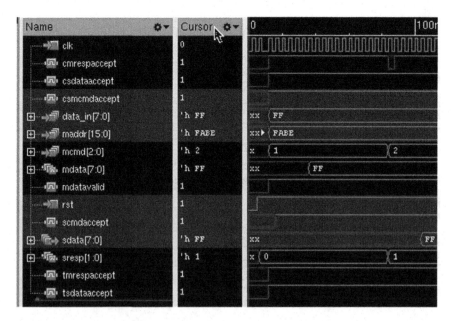

Fig. 6 Simulation results of basic mode OCP

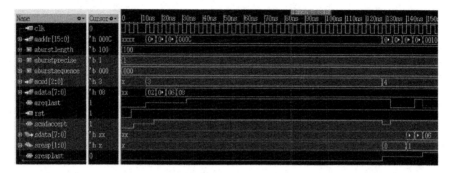

Fig. 7 Simulation results of burst mode OCP

7 Conclusion

This paper is mainly about the implementation and verification of basic modes of OCP. This work delivers that OCP is an efficient protocol for secured data core communication. OCP is capable enough in reducing the design time and risk by simultaneously designing the cores and working in the system. Based upon the real-time applications, IP cores are supposed to be designed such that they can be redesigned by cores which are not having inbuilt system logic that can be reused with no additional time for cores to be re-created.

References

1. Technical Information on Open Core Protocol, [online] Available: http://www.ocpip.org/, Open Core Protocol - International Partnership (OCP-IP).
2. Shihua Zhang, Asif Iqbal Ahmed and Otmane Ait Mohamed, "A Re-Usable Verification Framework of Open Core Protocol (OCP)". OCP-IP, 2008.
3. Elina Rajan Varughese and Rony Antony, "Implementation of extended open core protocol interface memory system using Verilog HDL," P. 978-1-4673- 6126-2/13/$31.00 c 2013 IEEE.
4. Open Core protocol (OCP) Specifications 2.2, Revision 1.0, Accellera 2004, [online] Available: http://www.ocpip.org/home.
5. OCP-IP, The Importance of Sockets in SoC Design, [online] Available: http://www.ocpip.org/ white papers.php.
6. Chin-Yao Chang, Yi-Jiun Chang, Kuen-Jong Lee, Jen-Chieh Yeh, Shih-Yin Lin and Jui-liangma "Design of On-Chip Bus with OCP Interface,".
7. R. Usselmann, "Implementation of Re-configurable Open Core Protocol Compliant Memory System using VHDL 2010 5th International Conference on Industrial and Information Systems," ICIIS 2010, Jul 29–Aug 01, 2010, India.
8. Chris Spear, "System Verilog for verification: A Guide to Learning the test bench for Language Features": Springer, second edition.
9. S. Palnitkar, "Verilog HDL: A Guide to Digital Design and Synthesis", Upper Saddle River, New Jersey: Prentice Hall, Jan. 1996.

Cellular Automata Logic Block Observer Based Testing for Network-on-Chip Architecture

Shaik Mohammed Waseem and Afroz Fatima

Abstract Necessity to test the logic circuits has increased rapidly due to the increase in number of applications being hosted on a single chip. This in turn has demanded the design of testing architectures which are capable of providing high fault coverage with less resource utilization and minimal power usage. Cellular Automata Logic Block Observer (CALBO), a technique homologous to Built-In Logic Block Observer (BILBO), has been considered in this paper to test the routers which are considered as important components of a Network-on-Chip (NoC) architecture. The resource utilization and power report of the design have been successfully generated to list out the advantages of the CALBO in comparison to BILBO for the architecture considered.

Keywords CALBO · BILBO · Network-on-chip · 7-port router
Cellular automata · Testing

1 Introduction

Increase in demand for miniaturization of the electronic devices across the globe has triggered a challenge for design engineers to look at various alternatives and in such an attempt to design the low-power devices with reduced chip area, the probability for occurrence of faults has proportionately increased [1]. Different techniques and architectures have been under practice for providing testability to the SoCs and one among such is the CALBO which is a technique similar to BILBO and comes with the advantages of cellular automata-based rules being used for test pattern generation and response compaction [2]. CALBO has its presence in

S. M. Waseem (✉)
IF&S, Kadapa, Andhra Pradesh, India
e-mail: waseem.vlsi@gmail.com

A. Fatima
Mosaic, London, ON, Canada
e-mail: afatima.es@gmail.com

© Springer Nature Singapore Pte Ltd. 2018   19
S. Bhalla et al. (eds.), *Intelligent Computing and Information and Communication*,
Advances in Intelligent Systems and Computing 673,
https://doi.org/10.1007/978-981-10-7245-1_3

testing for quite a long time and in [3], small ternary circuits has been tested using CALBO to demonstrate its ability as a good technique for providing testability. The ability of CALBO to act as a simple register, test pattern generator, and response compactor depending upon the different modes of operation forced on it through control signals has made it a peculiar technique when compared to others [4].

2 Background

G. Jervan et al. in [5] analyzed the impact of NoC parameters on the test scheduling through quantization methods. In [6], C. L. Hsu et al. proposed a low-power BILBO structure based on the LPTM approach, which can functionally reduce the switching activity during test operation. K. Namba et al. in [7] proposed a construction of FF that can effectively correct soft errors during system operations and can work as a BILBO FF on manufacturing testing. In [8], E. Sadredini et al. proposed IP-BILBO that generates test data locally and makes use of two techniques, namely direct and indirect reseedings of internal registers for generating pseudorandom test patterns. L. Gao et al. in [9] proposed a linear hybrid cellular automata using Cellular Automata (CA) Rule 90 and 150 and used it successfully as test pattern generator and signature analyzer for implementing built-in self-test capability. In [10], S. M. Waseem et al. observed the behavior of a 7-port router architecture for three-dimensional NoC, by implementing the same for a mesh topology and compared the results of its behavior for algorithms like Hamiltonian routing and conventional routing.

3 Cellular Automata Logic Block Observer

CALBO is homologous to BILBO register [11] in its use but slightly differs in its mechanism as it hosts cellular automata-based test pattern generator and multiple-input signature register. Among the 256 rules of CA and their usage proposed by Stephen Wolfram in [2], Rule 45 of CA is considered in this paper due to its peculiar cyclic nature and good account of randomness in patterns generated that could be applied to Circuit Under Test (CUT), here being the router architecture for NoC. The uniqueness of CALBO is due to its ability to act as a Cellular Automata Test Pattern Generator (CATPG) in one test cycle and Cellular Automata Multiple-Input Signature Register (CAMISR) in another. In normal operation, it can also be used as a simple register to store data, if required. The different modes of operation of CALBO are shown in Table 1.

Table 1 Modes of operation for CALBO

Control signal-1 (B1)	Control signal-2 (B2)	Mode of operation
0	1	Cellular automata test pattern generator
0	0	Serial scan chain
1	1	Cellular automata MISR
1	0	Normal D flip-flop

3.1 CALBO as Test Pattern Generator

The ease to detect the presence of faults often depends on the extent of randomness in the generated patterns to be applied to CUT in this criterion, CA-based test pattern generator has overcome the conventional LFSR-based TPGs. Rule 45 of CA has been considered in this paper, for test pattern generation of CALBO due to its higher degree of randomness [12, 13]. Rule 45 of CA is represented through a logical correlation of its neighbors as shown in Eq. (1):

$$C^n = C_{i-1}^p \text{XOR} \left(C_i^p \text{OR} \left(\text{NOT } C_{i+1}^p \right) \right), \tag{1}$$

where

C^n Succeeding stage output state of the cell, which is under consideration
C_{i-1}^p Existing state of the adjacent (left) cell
C_i^p Existing state of the cell which is under consideration
C_{i+1}^p Existing state of the adjacent (right) cell

The pictorial representation of Eq. (1) could be best described with the help of Fig. 1, where the black cell can be considered as "logical 1" and white cell as "logical 0".

Numerical representation of the same is shown as below:

$\{\{1, 1, 1\} \rightarrow 0, \{1, 1, 0\} \rightarrow 0, \{1, 0, 1\} \rightarrow 1, \{1, 0, 0\} \rightarrow 0, \{0, 1, 1\} \rightarrow 1,$ $\{0, 1, 0\} \rightarrow 1, \{0, 0, 1\} \rightarrow 0, \{0, 0, 0\} \rightarrow 1\}$

Numerical correspondence can, respectively, be evaluated as $\{C_{i-1}^p, C_i^p,$ and $C_{i+1}^p\} \rightarrow C^n$

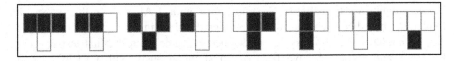

Fig. 1 Pictorial cell representation for rule 45 of cellular automata

3.2   CALBO as Multiple-Input Signature Register

In this paper, to implement CALBO, in order to act as CA-based MISR (CAMISR), CA-based MISR from [14] has been considered in which the signature is formed by first doing XOR of each cell with the corresponding circuit output and then incrementing the cellular automaton. This method can be mathematically represented as in (2):

$$C(t+1) = (C(t)\text{XOR } O(t))' \qquad (2)$$

where

$C(t)$ Contents of cellular automaton at time "t"
$O(t)$ Output of the circuit at time "t"
$C(t+1)$ Contents of cellular automaton at time "$t+1$" and symbol ($\prime$) indicates incremented value at respective time

4   NoC Architecture Testing with CALBO

4.1   NoC Architecture

Router is often considered as the main component responsible to route data on a chip and hence is of greater importance to successfully implement the network-on-chip architecture. The router designed in this paper consists of seven ports and hence could possibly route data in seven different directions with a scope for three-dimensional NoC implementation. The arbitration mechanism is of round robin in nature and the crossbar switch being the multiplexer based, one which drives the data to related FIFO buffer. The data is routed to the concerned port with the information from LUT (Look Up Table) which is responsible for hosting information related to routing logic. The router architecture can be best described as shown in Fig. 2.

4.2   CALBO-Based Testing for NoC

In an attempt to test the NoC architecture with CALBO, in this paper, a four-router-based NoC architecture is considered and testing has been accomplished in a scheduled three test sessions as detailed in Table 2. In the first test session, Router 1 and Router 4 have been tested simultaneously with CALBO 1 and CALBO 3 made to act as CAMISR and CATPG, respectively, by forcing appropriate control signals as stated in Table 1. Test session-2 was aimed at successfully

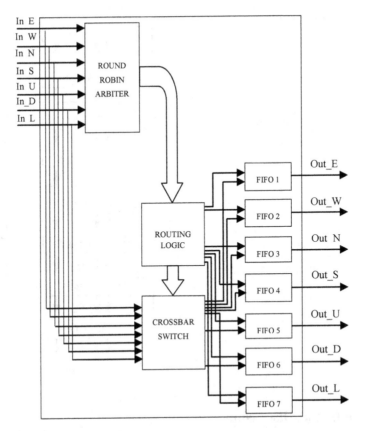

Fig. 2 7-port router architecture for NoC implementation

Table 2 Test session for routers in NoC architecture for CALBO-based testing

Test session-1(Router-1 and Router-4)
CALBO1-CAMISR
CALBO3-CATPG
Test Session-2(Router-2)
CALBO1-CATPG
CALBO2-CAMISR
Test Session-3(Router-3)
CALBO2-CATPG
CALBO3-CAMISR

testing Router 2 by forcing CALBO 1 and CALBO 2 to work under modes of CATPG and CAMISR respectively. Router 3 in third test session has been tested by making CALBO 2 to operate in CATPG mode and CALBO 3 in CAMISR mode as shown in Table 2 and Fig. 3.

Fig. 3 Testing of four routers
in an NoC architecture with
CALBO

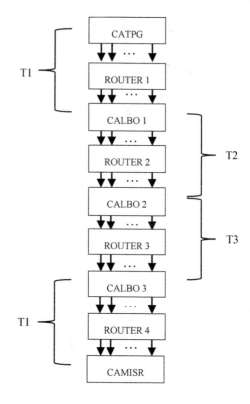

5   Results and Discussion

Four-router-based NoC architecture has been considered for CALBO-based testing
and the design has been implemented on Kintex Ultrascale device of Xilinx Inc.
with the help of Xilinx Vivado 2016.2 V suite. The generated power and resource
utilization reports for the design have been tabulated in Tables 3 and 4, respectively, along with that of the BILBO-based testing statistics. From the graphical
representation of the statistics in Figs. 4 and 5, it can be noticed that the
CALBO-based testing is beneficial in both the important criteria, i.e., power and
resource utilization which are rigorously being pursued by researchers working on
System-on-Chip (SoC) architectures. Even though the difference in the numerical
for some of the factors of CALBO in power and resource utilization report is almost
same or is less statistically comparable to that of BILBO, still the difference in total
is considerably acceptable and it is sought to be increased with increase in the size
of the target application (i.e., the number of routers being taken into consideration
due to increase in the size of the NoC architecture).

Table 3 Power report for CALBO- and BILBO-based testing

Type of testing	Dynamic power (W)				
	Clocks (W)	Signals (W)	Logic (W)	I/O (W)	Total (W)
CALBO-based testing	0.017	0.014	0.004	0.322	0.356
BILBO-based testing	0.018	0.016	0.004	0.349	0.387

Table 4 Resource utilization report for CALBO- and BILBO-based testing

Type of testing	Resource utilization			
	LUT	FF	I/O	BUFG
CALBO-based testing	1253	2021	482	1
BILBO-based testing	1460	2656	483	1

Fig. 4 Graphical representation of power report for CALBO and BILBO testings

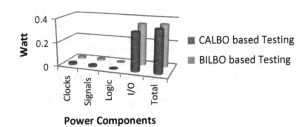

Fig. 5 Graphical representation of utilization report for CALBO and BILBO testings

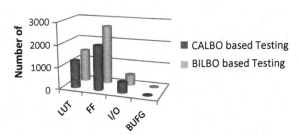

6 Conclusion

The CALBO-based testing though demands higher level of complexity in the design, the degree of randomness that it provides in order to detect the faults, is often beneficial in comparison to BILBO. Further, the statistical and graphical reports of power and resource utilization tend to prove the advantages of CALBO-based testing over BILBO, for the considered application.

References

1. G. Strawn, C. Strawn, "Moores law at fifty," IT Professional, vol. 17, no. 6, pp. 69–72, 2015.
2. Stephen Wolfram, "A New Kind of Science", Wolfram Media Inc., 2001.
3. C. Rozon and H. T. Mouftah, "Pseudo-random testing of CMOS ternary logic circuits," *[1988] Proceedings. The Eighteenth International Symposium on Multiple-Valued Logic*, Palma de Mallorca, Spain, 1988, pp. 316–320.
4. P. D. Hortensius, R. D. McLeod and B. W. Podaima, "Cellular automata circuits for built-in self-test," in *IBM Journal of Research and Development*, vol. 34, no. 2.3, pp. 389–405, March 1990.
5. G. Jervan, T. Shchenova and R. Ubar, "Hybrid BIST Scheduling for NoC-Based SoCs," *2006 NORCHIP*, Linkoping, 2006, pp. 141–144.
6. Chun-Lung Hsu and Chang-Hsin Cheng, "Low-Power built-in logic block observer realization for BIST applications," *2007 7th International Conference on ASIC*, Guilin, 2007, pp. 994–997.
7. K. Namba and H. Ito, "Soft Error Tolerant BILBO FF," *2010 IEEE 25th International Symposium on Defect and Fault Tolerance in VLSI Systems*, Kyoto, 2010, pp. 73–81.
8. E. Sadredini, M. Najafi, M. Fathy and Z. Navabi, "BILBO-friendly hybrid BIST architecture with asymmetric polynomial reseeding," *The 16th CSI International Symposium on Computer Architecture and Digital Systems (CADS 2012)*, Shiraz, Fars, 2012, pp. 145–149.
9. L. Gao, Y. Zhang and J. Zhao, "BIST using Cellular Automata as test pattern generator and response compaction", *Consumer Electronics, Communications and Networks (CECNet), 2012 2nd International Conference on*, Yichang, 2012, pp. 200–203.
10. S. M. Waseem and A. Fatima, "Pursuance measures of CRA & HRA for 3D networks on a 7-port router schema," *2015 International Conference on Communications and Signal Processing (ICCSP)*, Melmaruvathur, 2015, pp. 0055–0061.
11. M. L. Bushnell and V. D. Agrawal, Essentials of Electronic Testing for Digital, Memory, and Mixed-Signal VLSI Circuits. Springer, 2000.
12. Stephen Wolfram, "Random sequence generation by cellular automata", Advances in Applied Mathematics, vol. 7, Issue 2, 1986, pp. 123–169.
13. A. Fatima and S. M. Waseem, "Cellular Automata based Built-In-Self Test implementation for Star Topology NoC," *2017 11th International Conference on Intelligent Systems and Control (ISCO)*, Coimbatore, 2017, pp. 45–48.
14. P. D. Hortensius, R. D. McLeod and H. C. Card, "Cellular automata-based signature analysis for built-in self-test," in *IEEE Transactions on Computers*, vol. 39, no. 10, pp. 1273–1283, Oct 1990.

Structural Strength Recognizing System with Efficient Clustering Technique

Sumedha Sirsikar and Manoj Chandak

Abstract Internet of Things (IoT) visualizes future, in which the objects of everyday life are equipped with sensor technology for digital communication. IoT supports the concept of smart city, which aims to provide different services for the administration of the city and for the citizens. The important application of IoT is Structural Strength Recognition (SSR). This approach is becoming popular to increase the safety of buildings and human life. Proper maintenance of historical buildings requires continuous monitoring and current conditions of it. Sensor nodes are used to collect data of these historical buildings or large structures. Structural strength recognition covers huge geographical area and it requires continuous monitoring of it. It involves more energy consumption during these activities. Hence, there is need for efficient energy management technique. Clustering is one of the important techniques for energy management in Wireless Sensor Networks (WSN). It helps in reducing the energy consumed in wireless data transmission. In this paper, SSR system is designed with efficient clustering algorithm for wide network and also finds out optimum number of clusters.

Keywords Structural strength recognition · Energy management
Optimal clustering · Energy efficient WSN

1 Introduction

The idea of Internet of Things was developed with combination of Internet and wireless sensor networks. The term Internet of things was devised by Kevin Ashton in 1999 [1]. It refers to identify objects and its virtual representations in

S. Sirsikar (✉)
Sant Gadge Baba Amaravati University, Amravati, Maharashtra, India
e-mail: sumedha.sirsikar@mitpune.edu.in

M. Chandak
Shri Ramdeobaba College of Engineering and Management, Nagpur, Maharashtra, India
e-mail: chandakmb@gmail.com

© Springer Nature Singapore Pte Ltd. 2018
S. Bhalla et al. (eds.), *Intelligent Computing and Information and Communication*,
Advances in Intelligent Systems and Computing 673,
https://doi.org/10.1007/978-981-10-7245-1_4

"Internet-like" structure uniquely. These objects can be anything like huge buildings, parts of a large system, industrial plants, planes, cars, machines, any kind of goods and animals, etc. Wireless communication technologies play a major role in IoT. Major advancements in WSNs will be carried out by integrating sensors with objects. IoT is developed based on this principle.

One of the aims of IoT is to design and develop cities smartly. It will offer better services for the administration of the city and citizens. It is used in variety of applications such as structural strength recognition, agriculture, military, biodiversity mapping, disaster relief operations, and medical health care.

1.1 Structural Strength Recognition

Structural strength recognition is a system used to evaluate real-time condition of buildings. This application has enormous importance in safety of human life. The data collected from sensor devices is used to recover the operation. This maintenance operation of the system is to repair and replace the structures. Detection of current damages can be used to distinguish eccentricities from the design performance. Monitored data needs to be integrated in structural management systems. It will help to take superior decisions by providing reliable and accurate information. The defective houses and multistorey buildings have serious implications. Sometimes, buildings get collapse just because of irregular monitoring and no maintenance which is very dangerous for human life.

The structural strength recognizing system design requires focus on the particular requirements. The first thing is the probable reasons for degradation and the related risks, second the expected requirements for this degradation, and finally, a suitable structural strength recognizing system to detect such conditions. In SSR, the appropriate sensors are required for appropriate monitoring. If the process of data collection, updation, and data analysis is correct, then recognizing strength of structures becomes really easy.

If the selection of the sensors for specific risks associated with a given structure is done properly, then it is necessary to combine different parameters to analyze the building. It requires maintenance system to insure proper data collection process, integrated data acquisition, and management system.

Multistorey buildings have complex constructions with a combination of different elements. Generally, building structure differs in size, operational system, geometry, construction material, and foundation characteristics. Due to the stress of natural actions, these elements create impact on a building performance.

Few years before, implementation of sensor nodes in large civil structures was not suitable, in terms of cost and power requirements [1]. Today, due to the tremendous progress of low-power semiconductor devices and high functionality microcontroller units results in the great development of sensor nodes which are used in SSR [2]. A large amount of energy is consumed for data transmission over the sensor network. Sensor nodes are powered through small batteries that possess short life.

The network lifetime can be improved using efficient energy management technique. This can be done using efficient clustering techniques [2] that serves an important role in large SSR systems. The proposed model would be useful to carry out survey of large geographical area. But large area consumes more energy [3]. Hence, efficient clustering technique with optimum number of clusters [4, 5] is required for proper utilization of energy because it is very difficult to recharge or replace nodes in sensor network as it is complex in nature [6–8].

1.2 Structural Damage Detection

Building damages structural damage detection can be analyzed from different parameters of building such as columns, beams, and flooring. Nonstructural damage detection can be analyzed by observing damage or partition in ceilings, walls, and glasses.

2 Related Work

2.1 SSR Techniques

Civil structural strength recognizing system consists of various popular techniques such as acoustic emission (AE) analysis, vibration study, and optical sensing. The most popular technology is optical sensing, accurate in detecting small distortions in structures. Variety of fiber optic strain sensors are available. Due to their characteristic as small physical size as compared to the civil structures and multiplexing capability, the fiber optic sensors have become very widespread [9]. Moreover, it is a perfect choice for civil structures of any shape to embed such cost- and size-effective sensors. Passive Radio Frequency Identifier (RFID) system was developed by Ikemoto et al. which does not require to care about battery lifetime problems of sensor nodes [9]. Each sensor node holds a passive RFID tag that is embedded into the structure in duration of construction or can be attached later on surface. The main aim of the development of this device is zero battery power of sensor node.

2.2 Strength Recognizing Components

There are different components to recognize structural strength. These are classified into several categories such as **Scale** (Local scale, Global scale, and Network scale),

Periodicity (Periodic, Semi-continuous, and Continuous), **Parameter** (Mechanical, Chemical, Physical, and Environmental actions), and **Data collection** (Manual, None, Online, Off-line, and Real time).

2.3 Advantages of SSR

The benefits of a structural strength recognizing systems are increased in security, concealed structural assets determination, increase in safety, and long-term quality and structural management preservation.

2.4 System Integration

The SSR system is designed as an integrated system. In the system integration, all data flow into a single database from buildings and represent through a user interface. Various sensors can be attached to the same data logger single data management system. As shown in Fig. 1, different types of sensors are positioned into buildings that collect actual data of buildings. This data will be stored in the database for further analysis. After analysis, results are displayed either on site or at a remote location.

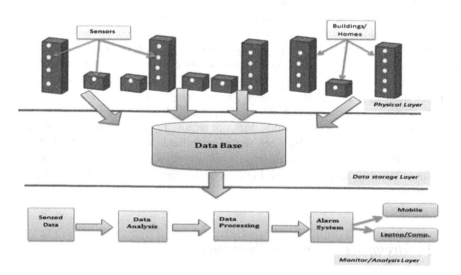

Fig. 1 SSR system integration

3 Problem Statement

Technical monitoring of buildings and its structures are the basic needs nowadays. The bulk of buildings are growing every year. SSR covers large geographical area. The data collection from sensor nodes to database requires large amount of energy. The main objective is to determine optimum number of clusters for a large sensor network so that energy of sensor network will be saved and thus it increases the network lifetime with the help of efficient clustering techniques.

4 Proposed Solution

Energy management is an important factor in WSN. Transmission of data from one location to other requires energy consumption. There are different techniques to manage the network energy efficiently. Clustering is one of the techniques to decrease the energy depletion and increase network lifetime. SSR requires large area for monitoring and analysis. This large area needs huge amount of energy to collect information and to transmit it to the base station. Clustering technique is used to solve issues related to scalability, energy, and lifetime of sensor networks. At the same time, clustering process has problem of determining the optimum number of clusters that will minimize energy consumption. Number of clusters are directly proportional to the amount of energy consumed by the network. The solution is proposed to design SSR system and find out the optimum number of clusters for wide network as shown in Fig. 2.

5 Proposed Model

This section describes mathematical model for finding Optimum Number of Clusters (ONC) and average energy consumption.

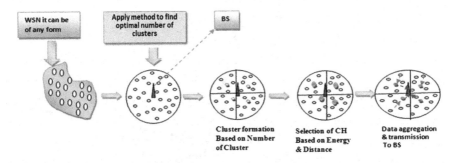

Fig. 2 Proposed system architecture

$$ONC = \{ \, N, C, NC, R \}$$

N = *Set of all nodes present in the network* = *{n1, n2, n3, ...}*
C = *Cluster which contains group of sensor nodes*
NC = *Number of cluster* = *{nc1, nc2, nc3, ...}*
R = *Range*

Optimum number of clusters is obtained from the intersection of R and NC as shown in Fig. 3.

Ei = Initial energy of each node

Energy consumed by node *i* initially requires some amount of energy for certain events, for example, send, receive, and drop.

Ici = Energy consumed by ith node = Initial Energy − Residual Energy

Total Consumed Energy (TCE)

$$TCE = \sum_{i=1}^{N} (\text{Initial Energy} - \text{Residual Energy}) \text{ of node}$$

Average Energy Consumption (AvgEC) = Total energy consumed by nodes in the network

$$AvgEC = \sum_{i=1}^{Nt} Ici/N$$

Fig. 3 Intersection of number of clusters and range

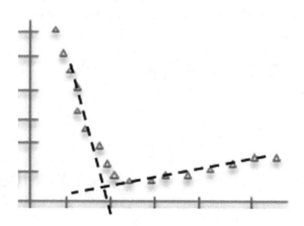

6   Proposed Methodology

One of the important constraints of WSN is lifetime of network and energy. In this research work, this problem is kept in mind and a solution for finding optimum number of clusters and efficient clustering is described. The proposed solution contains the following steps:

    I. Calculation of optimum number of clusters
   II. Formation of clusters and selection of cluster head
  III. Reformation of cluster

Step I: Calculation Optimum Number of Clusters
Optimum number of clusters for a sensor network is calculated using proposed method by plotting graph, where x-axis represents the number of clusters and y-axis represents the range. Two lines are drawn that covers the maximum points on the curve. Each line must contain at least two points and must start at either end of the data. The intersection point of these two lines is used as optimum number of clusters as shown in Fig. 3.

Step II: Formation of Clusters and Selection of Cluster Head
After getting exact number for clusters, divide the circular network or make partitions of circular network according to number of clusters.

$$Degree \ of \ partition = 360/Optimum \ Number \ of \ Clusters$$

Degree of partition = 360/Optimum Number of Clusters
    Cluster Head (CH) is selected after formation of clusters. The cluster head is selected by Base Station (BS). Here, it is assumed that the BS knows all the information of sensor nodes. BS will select CH based on this information. Node which has high energy level and suitable geographical location is selected as CH. BS then broadcasts this information to all nodes [10].

Step III: Reformation of Cluster
During the continuous operation of sensor network, the battery level of sensor nodes as well as cluster heads decreases. Due to this, sensor nodes and cluster heads become inactive. If the node energy is below the threshold value, then the sensor network should be re-organized into new clusters. In this process, the node with highest residual energy and degree will become new cluster head [10].

6.1   SSR Application Execution Based on Proposed Methodology

As per the phases, working of application of SSR is described in Fig. 4. Here, circular geographical area is considered, which contains certain number of houses and buildings. In the houses, single node is deployed but for buildings single sensor

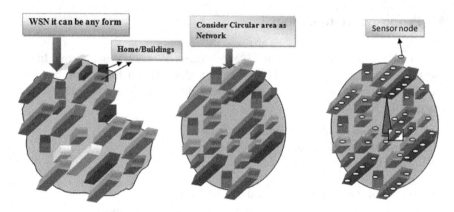

Fig. 4 Deployment of nodes in houses

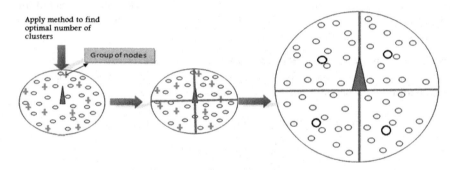

Fig. 5 Formation of clusters

node is not sufficient. Building contains multiple sensor nodes as shown in Fig. 4. Now the number of required clusters for that circular area is calculated as shown in Fig. 5. Here, yellow circles represent single sensor node for house and blue star represents group of sensor nodes of a particular building.

Initially, calculate optimum number of clusters for network and partition the circular network into that much number of clusters. Example is shown in Fig. 5. After obtaining intersection point as per Fig. 3, divide network into four clusters for example.

Now the cluster head is selected for data collection process. Finally, the CH will gather data from sensor nodes and send it to base station. Later, the data acquisition system collects data from BS. Data processing system will monitor that data for further analysis. Alarm system generates notifications and alert alarms on computer system and mobile devices as shown in Fig. 6.

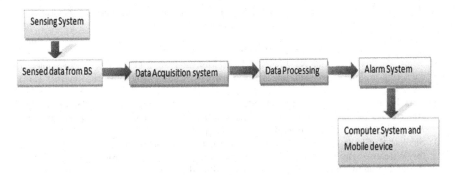

Fig. 6 Data aggregation and sensing system

7 Conclusion

The lifetime of a sensor network is one of the most important parameters in WSNs. It depends on energy level of battery-powered sensor nodes. The network is left unattended to perform monitoring and reporting functions. Energy of sensors in WSN is depleted due to direct transmission of data between sensors and a base station, which can be avoided by clustering the nodes in WSN. It further enhances scalability of WSN in real-world applications while conserving limited energy. Thus, proposed algorithm aims to minimize the number of clusters and in turn energy consumption per cluster. Hence, lifetime of WSN can be increased with efficient clustering technique. Our proposed algorithm gives optimum number of clusters that certainly help to improve the performance of network. Next step is to implement the proposed solution in NS2 and to observe the performance in terms of energy and network lifetime.

References

1. Zanella, A., Bui, N., Castellani, A., Vangelista, L., & Zorzi, M. (2014), Internet of things for smart cities. IEEE Internet of Things Journal, 1(1), pp. 22–32.
2. Sasikumar, P., & Khara, S. (2012), K-means clustering in wireless sensor networks. Fourth international conference on Computational Intelligence and Communication Networks (CICN), IEEE, pp. 140–144.
3. Navjot Kaur Jassi, Sandeep Singh Wraich, (2014), A Review: An Improved K-means Clustering Technique in WSN, Proceedings of the International Conference on Advances in Engineering and Technology (ICAET).
4. Bholowalia, P., & Kumar, A. (2014). EBK-means: A clustering technique based on elbow method and k-means in Wireless Sensor Network, International Journal of Computer Applications, 105(9).
5. Alrabea, A., Senthilkumar, A. V., Al-Shalabi, H., & Bader, A. (2013), Enhancing k-means algorithm with initial cluster centers derived from data partitioning along the data axis with PCA. Journal of Advances in Computer Networks, 1(2), pp. 137–142.

6. Bhawna, Pathak, T., & Ranga, V., (2014), A Comprehensive Survey of Clustering Approaches in Wireless Sensor Networks, Elsevier Publications.
7. Solaiman, B. F., & Sheta, A. F. Energy optimization in wireless sensor networks using a hybrid K-means PSO clustering algorithm.
8. Tong, W., Jiyi, W., He, X., Jinghua, Z., & Munyabugingo, C. (2013), A cross unequal clustering routing algorithm for sensor network. Measurement Science Review, 13(4), pp. 200–205.
9. Alahakoon, S., Preethichandra, D. M., & Ekanayake, E. M. (2009), Sensor network applications in structures–a survey. EJSE Special Issue: Sensor Network on Building Monitoring: From Theory to Real Application, pp. 1–10.
10. Sirsikar, S., Chunawale, A., & Chandak, M. (2014), Self-organization Architecture and Model for Wireless Sensor Networks. International Conference on Electronic Systems, Signal Processing and Computing Technologies (ICESC), IEEE, pp. 204–208.

Trajectory Outlier Detection for Traffic Events: A Survey

Kiran Bhowmick and Meera Narvekar

Abstract With the advent of Global Positioning System (GPS) and extensive use of smartphones, trajectory data for moving objects is available easily and at cheaper price. Moreover, the use of GPS devices in vehicles is now possible to keep a track of moving vehicles on the road. It is also possible to identify anomalous behavior of vehicle with this trajectory data. In the field of trajectory mining, outlier detection of trajectories has become one of the important topics that can be used to detect anomalies in the trajectories. In this paper, certain existing issues and challenges of trajectory data are identified and a future research direction is discussed. This paper proposes a potential use of outlier detection to identify irregular events that cause traffic congestion.

Keywords Trajectory data · Map matching · Trajectory outlier detection
GPS data · Similarity measures

1 Introduction

Trajectory data is the data about moving objects like vehicles, animals, and people. With the advent of technologies of mobile computing and location-aware services, there is a massive generation of this trajectory data. Trajectory data analysis helps in identifying several real-world phenomena like, for example, identifying the buying patterns of humans in shopping malls, understanding the migratory patterns of animals and birds, and detecting the travel path of hurricanes and tornadoes. Vehicle trajectory analysis is used in various traffic dynamics like traffic jam prediction, roadmap construction, and real-time traffic flow; in route navigation like suggest optimum route, suggest places to visit, and identify frequently visited

K. Bhowmick (✉) · M. Narvekar
Dwarkadas J. Sanghvi College of Engineering, Vile Parle, Mumbai, Maharashtra, India
e-mail: kiran.bhowmick@djsce.ac.in

M. Narvekar
e-mail: meera.narvekar@djsce.ac.in

© Springer Nature Singapore Pte Ltd. 2018
S. Bhalla et al. (eds.), *Intelligent Computing and Information and Communication*,
Advances in Intelligent Systems and Computing 673,
https://doi.org/10.1007/978-981-10-7245-1_5

locations; in itinerary navigation like discover popular routes, predict next location; and in other applications like predicting human flow, suggesting alternate transport means, etc. [1–4].

In recent years, there has been an increase in the research of outlier analysis of trajectory data. This outlier analysis is used to identify anomaly in the regular data. Outlier detection can also be used to identify events that occur in regular traffic. This paper discusses different outlier detection techniques, the research literature available, and suggests the use of trajectory outlier detection for detecting irregular events such as traffic congestion due to road accidents or roadblocks or any other reason.

2 Related Concepts

2.1 Basic Concepts

This section describes the concepts related to trajectory mining and outlier detection basics.

2.1.1 Trajectory data

Definition 1: A trajectory T is an ordered list of spatiotemporal samples $p1, p2, p3, ..., pn$. Each $pi = (x_i, y_i, t_i)$, where x_i, y_i are the spatial coordinates of the sampled point and t_i is the timestamp at which the position of the point is sampled, with $t1 < t2 < ... < tn$ [5].

In general, a trajectory depicts the time and location of a moving object where it can be tracked. A single trajectory can include many trajectory points and can have any length for a single moving object. Basically, the length depicts the places that the object has traveled from point A to point B before it comes to a halt. A sub-trajectory depicts a partition or a segment of a full trajectory.

2.1.2 Trajectory outlier

Definition 6: An outlier in a trajectory data is an item that is significantly different from other items in terms of some form of similarity [1].

Outliers are objects that are inconsistent with remaining data items. Outliers are also sometimes considered to be noise. But the noise for someone can be someone else's data [6]. For example, in the following, figures t4 and t5 are outliers in the path of the regions R1 and R2.

However, t4 can be considered as an alternate path, while t5 can be a long detour (Fig. 1).

Fig. 1 Outliers between regions

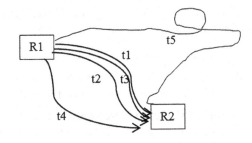

2.2 *Similarity Measures*

The distance between two trajectories is usually measured by some kind of aggregation of distances between trajectory points. The various similarity measures are defined in [1, 7]. These include closest pair distance (CPD), sum of pairs distance (SPD), Euclidean distance (ED), dynamic time warping (DTW and PDTW), edit distance measures (ERP and EDR), and LCSS. The CPD, SPD, and ED require equal length of the trajectories and calculate the minimum distance between two trajectories. Edit distances, DTW and LCSS, assume variable length trajectories.

2.3 *Map Matching*

Map matching is a preprocessing technique in the trajectory data mining used to convert the sequence of GPS points to a sequence of road segments. There are various categories of map matching algorithms depending upon the type of information considered and the range of sampling [8]. The category with the type of information includes geometrical, topological, weight-based, probabilistic, and advanced, whereas the range of sampling has two types local/incremental and global methods.

2.4 *Outlier Detection Techniques*

This section introduces different outlier techniques.

Distance-based outlier detection techniques These techniques consider the distance between trajectory and its neighbors. If the trajectory does not have neighbors above a particular threshold, it is an outlying trajectory [9, 10].

Density-based outlier detection techniques This means a trajectory is an outlier if its density is relatively much lower than that of its neighbors [9, 11].

The distance-based outlier techniques detect global outliers due to the way the distance is measured between the points. It also requires uniform distribution of data. With different density distributions of data, the distance-based techniques cannot be used. The density-based outlier techniques detect local outliers and can do analysis in different density distributions but require a large number of k-nearest neighborhood queries and are computationally very intensive.

Historical similarity-based outlier detection techniques This technique suggests the use of historical similarity to identify trends between data points. At each time step, a road segment is compared with its historical recorded temporal neighborhood values to identify similarity values. A drastic change in the values results in the outlier being detected.

Motifs-based outlier detection techniques A motion classifier to detect trajectory outliers using an objects motion feature is called as motifs. These motifs are a sequence of motion features with values related to time and location. The classifier works on this high-dimensional feature space to distinguish an anomalous trajectory with a normal one.

3 Related Work

Although it is an upcoming field, a lot of work have already been done in the outlier detection and overall trajectory mining. With respect to the different similarity measures, H. Wang et al. in their paper [7] have done a comparative study about various similarity measures under different circumstances.

Map matching helps overcome the problem of uncertainty, but the existing map matching algorithms also require high sampling rate of the data which is not always the case. Yin Lou et al. in [12] have used spatiotemporal matching and analysis to find the candidate graph which helps to map low sampling digital data to a roadmap. K. Zheng et al. [13] proposed the use of similarity between query data points and historical data to infer path for low sampling data. The inference is done using traversal graph and nearest-neighbor-based heuristics approach. In [14], the authors propose a technique to consider the complex nature of the urban roads and address the problem of identifying a data point on elevated roads. The problem of identifying the candidate segment selection for the first GPS point is addressed in [15], where the authors employ a heuristic A* algorithm to find the shortest path between a previous point and the candidate segment. Paolo Cintiaa and Mirco Nannia in [16, 17] argue that the shortest path should not be the measure to find the best matching point on the road network when the data samples are low and that a time-aware Dijkstra's shortest path is suggested to match the points. G. Hu et al. in [18] suggested the use of an information fusion of data to accurately perform map matching as well as consider the complexity of road networks. And lastly, the authors in [19] suggest using LCSS and clustering algorithm totally avoiding the use of any map matching algorithm. They conclude with their experimental results that none of the measures are better than others. In all, each similarity measure is affected by the decrease in the sampling rate. None of the related techniques consider time-aware heuristics methods together with the complexity of road networks like multilayer and parallel roads, complex interchanges, and elevated roads.

Outlier detection has also been researched in different scenarios. Next, paragraph review of literature on the research is been discussed.

Lee et al. first proposed a trajectory outlier detection algorithm TRAOD [19] to detect outlying sub-trajectories using a partition and detect framework. It uses density as well as distance-based measures and does not suffer from local density problem and has minimal overhead for density measures. However, it has a complexity of $O(n^2)$ and considers only spatial data. Piciarelli et al. in [20] used a single-class SVM classifier on fixed-dimensional vectors of spatial data to identify outliers. But the algorithm uses a training data with outliers. Outliers not seen in training data will not be detected. In [21], Li et al. proposed a temporal outlier detection TOD algorithm that considers time dimension and historical data to detect outliers. The authors suggest the use of exponential function to detect outliers. This exponential function, however, cannot be used for spatial dimension. Yong Ge et al. in [22] proposed TOP-EYE to detect an evolving trajectory using a decay function. If a trajectory is going to be an outlier, then it will be detected at an early stage by this method. Wei Liu et al. in [23] suggested the use of frequent pattern tree to identify the causal interactions between the detected outliers. They have considered temporal information about current and historical time frames trajectory to compare using Euclidean distance for detecting outliers. Daqing Zhang et al. proposed their algorithm iBAT [24] for detecting outliers based on the isolation techniques. An iTree based on the data-induced random partition is generated for all the trajectories. If a trajectory is an outlier, then it will have comparatively shorter paths than those of a normal trajectory. Alvares et al. in [25] detected an outlier between surveillance camera and inbound trajectories by analyzing the behavior of trajectories. If a trajectory seems to avoid the object in certain patterns, then it is detected as outlier. But it fails to identify if the avoidance was forced or intentional. Chawla et al. in [26] used PCA analysis to identify anomalies and then L1 optimization to infer possible routes whose flow traffic caused the outlier. Zhu et al. in [27, 28] used time-dependent transfer graphs to identify outliers online.

From the above literature review of outlier, it is observed that, while considering trajectory data, its inherent characteristics affect the outlier detection method. Preprocessing of the data can reduce most of the challenges. But the fact that the trajectory data is not equally distributed on all roads, it is important to consider historical data. The choice of hybrid outlier detection technique and suitable similarity measure can improve the accuracy majorly.

4 Issues and Challenges Identified

Trajectory data can be efficiently used in traffic management. Considering the characteristics of trajectory data and the literature review, the following issues and challenges related to trajectory data, map matching, and outlier detection are identified.

4.1 Trajectory Length and Sampling

Every trajectory have different lengths, i.e., the number of data points is different for each, i.e., T1 = (p1, p2, ..., pn) and T2 = (q1, q2, ..., qm) and n and m can be different. It is important to identify the region of interest to consider the start and end of each trajectory.

4.2 Low and Uneven Sampling Rates

The GPS data of the moving vehicles usually have a low sampling rate to avoid the overhead of communication cost, data storage, and battery life of devices. So the time interval between two consecutive GPS points can be very large (average sampling rate 2 min). This leads to an uncertainty of the path taken by a trajectory between these time gaps. Also, the data sampling rates are different across all trajectories, i.e., T1 may be sampled for every 1 m and T2 is sampled every 2 m.

4.3 Trajectory Directions, Regions, and Road Networks

Trajectories moving in different directions and different regions should be considered different. Trajectories moving in different directions but close proximity and those moving in the same direction but different regions should be considered as different. Also for vehicles, the trajectories are bound by the underlying road networks, so the data points should match the road networks map.

4.4 Similarity Measures

There are different types of similarity measures as mentioned in Sect. 2.2. However, with different lengths, low samplings, uncertainties, and noises, it is difficult to identify a suitable similarity measure to compare trajectories together. All the measures described in above section are sensitive to decrease in sampling rate. It is a challenge to process data with low sampling rate.

4.5   *Map Matching Challenges*

Most of the existing map matching algorithms assume high sampling of data as they usually perform local or incremental approach. Also, they consider only spatial data for mapping to road networks. However, there is a lot of information available that can improve the map matching considerably like time, speed, and direction. The complexity of urban road networks is another issue. The topology of road networks in urban areas has many challenges like parallel roads, multilayer roads, complex interchanges, and elevated roads. Considering all these to perform map matching increases the complexity, but will be realistic and accurate.

4.6   *Outlier Detection Challenges*

Different types of outlier detection techniques exist, viz., distance-based and density-based. The existing trajectory outlier detection algorithms have different techniques and each tries to solve the outlier detection with a different objective. The issue is to devise an algorithm that can overcome the problem of uncertainty, low sampling, and uneven length data with mixed distribution of trajectories and considering spatial as well as temporal similarity constraints efficiently identifies outliers even with large amount of data. A data structure that can help to deal with the variety of information and the length of the data is needed. The algorithm should be able to identify sparse as well as dense trajectories for online data. Also, there needs to be a way to propagate the effects of the outlier to linked traffic nodes.

5   Research Problem

The outlier detection is still an upcoming research due to its application in traffic management. A huge number of trajectory data are generated for a single moving object and there are thousands of moving objects in the form of GPS enabled vehicles and human beings carrying smartphones. The large number of sampling points and the uneven length requires a suitable data structure like a modified R-tree

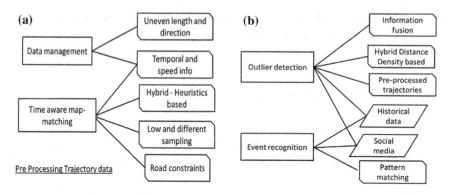

Fig. 2 a Preprocessing **b** Outliers between regions

for fast retrieval and indexing of the trajectory data. The R-tree should be able to store temporal and related information (Fig. 2).

The trajectory data has uneven rate of sampling and an inherent uncertainty about the locations between two sampled points. This uncertainty between sampling points can be reduced by using map matching algorithms. But the existing map matching algorithms consider only spatial data and work well with high sampling. Also, it is important to consider the urban road complexity like multi-layer, complex interchanges, and elevated roads. With low sampling, this makes it more challenging. The use of shortest path to find the actual location of the trajectory is not helpful. There has to be a consideration of hybrid technique like combining the global method with weight-based technique or global method with advanced or probabilistic method. A time-aware heuristic method can improve the accuracy of map matching algorithms in low sampling data. Since the distribution of trajectory data is skewed, using only density-based or distance-based method of outlier detection is not the solution. A hybrid approach is used by many existing algorithms but temporal information is not considered. Using time-dependent popular routes graph based on historical data to deal with the changing nature of outliers is beneficial. The detected outliers can be used to identify events by performing pattern matching and comparing it with similar patterns that were seen earlier when an irregular event occurred.

6 Conclusion

This paper is a part of an on-going research on identification of traffic events using vehicle trajectory. This paper mainly focusses on the identifying problems in trajectory data and outlier detection. A detailed literature survey on existing map matching techniques and outlier detection technique is discussed. The paper also discusses the proposed research problem for identifying traffic events using outlier detection techniques.

7   Future Scope

In the future, we plan first to enhance our research by designing a novel outlier detection technique focussing on the issue of low sampling and uncertainty using taxi trajectory data and second to design pattern matching algorithms to identify traffic events.

References

1. Zheng Y. 2015. "Trajectory Data Mining: An Overview", ACM Transactions on Intelligent Systems and Technology (TIST), Volume 6 Issue 3, May 2015 Article No. 29
2. Zaiben Chen et. al., "Discovering Popular Routes from Trajectories", *ICDE Conference 2011*, IEEE 2011, DOI 978-1-4244-8960-2/11
3. Lin, Miao and Wen-Jing Hsu. "Mining GPS Data for Mobility Patterns: A Survey". *Pervasive and Mobile Computing* 12, Elsevier, pp. 1–16, July 2013
4. Li Z, et. al., MoveMine: Mining Moving Object Databases. *SIGMOD '10 ACM*, pp. 1203–1206, June 2011
5. Wagner et. al., "Mob-Warehouse: A Semantic Approach for Mobility Analysis with a Trajectory Data Warehouse", Springer, pp. 127–136, Nov 2013
6. J. Han and M. Kamber, Data Mining: Concepts and Techniques, 2nd ed. Morgan Kaufmann, 2006
7. H. Wang et.al., "An Effectiveness Study on Trajectory Similarity Measures", Proceedings of the Twenty-Fourth Australasian Database Conference (ADC 2013), Adelaide, Australia, vol. 137, 2013
8. Mahdi Hashemi and Hassan A. Karimi, "A critical review of real time map matching algorithms: Current issues and future directions", Computers, Environment and Urban Systems, Elseveir, vol 48, pp. 153–165, Aug 2014
9. Gupta, Manish et al. "Outlier Detection for Temporal Data: A Survey". *IEEE Transactions on Knowledge and Data Engineering* 26.9, vol. 25, no. 1, 2014: 2250–2267
10. Knorr, Edwin M., Raymond T. Ng, and Vladimir Tucakov. "Distance-Based Outliers: Algorithms And Applications". The VLDB Journal, 8.3–4 (2000): 237–253
11. Liu, Zhipeng, Dechang Pi, and Jinfeng Jiang. "Density-Based Trajectory Outlier Detection Algorithm". Journal of Systems Engineering and Electronics 24.2 (2013): 335–340
12. Y. Lou, C. Zhang, Y. Zheng, X. Xie, W. Wang and Y. Huang, "Map-matching for low-sampling-rate GPS trajectories", Proceedings of the 17th ACM SIGSPATIAL—GIS '09
13. K. Zheng, Y. Zheng, X. Xie, and X. Zhou, "Reducing uncertainty of low-sampling-rate trajectories", 28th IEEE International Conference on Data Engineering. IEEE, 2012, pp:1144–1155
14. Z. He, S. Xi-wei, P. Nie and L. Zhuang, "On-line map-matching framework for floating car data with low sampling rate in urban road networks", IET Intelligent Transport Systems, vol. 7, no. 4, pp. 404–414, 2013
15. M. Quddus and S. Washington, "Shortest path and vehicle trajectory aided map-matching for low frequency GPS data", Transportation Research Part C: Emerging Technologies, Elsevier, vol. 55, pp. 328–339, March 2015
16. Paolo Cintiaa, and Mirco Nannia "An effective Time-Aware Map Matching process for low sampling GPS data", Elsevier, March 2016
17. G. Hu, J. Shao, F. Liu, Y. Wang and H. Shen, "IF-Matching: Towards Accurate Map-Matching with Information Fusion", IEEE Transactions on Knowledge and Data Engineering, vol. 29, no. 1, pp. 114–127, Oct 2016

18. J. Kim and H. Mahmassani, "Spatial and Temporal Characterization of Travel Patterns in a Traffic Network Using Vehicle Trajectories", Transportation Research Procedia, vol. 9, pp. 164–184, July 2015

19. J. G. Lee, J. W. Han and X. L. Li. "Trajectory outlier detection: a partition and detect framework". 24th International Conference on Data Engineering ICDE, IEEE, pages 140–149, 2008

20. Piciarelli, C., C. Micheloni, and G.L. Foresti. "Trajectory-Based Anomalous Event Detection". IEEE Trans. Circuits Syst. Video Technol. 18.11 (2008): 1544–1554

21. X. L. Li, Z. H. Li, J. W. Han and J. G. Lee. "Temporal outlier detection in vehicle traffic data". 25th International Conference on Data Engineering, pages 1319–1322, 2009

22. Yong Ge, et. al., "TOP-EYE: Top-k Evolving Trajectory Outlier Detection", CIKM'10 ACM, 2010

23. Wei Liu et. al., "Discovering Spatio-Temporal Causal Interactions in Traffic Data Streams", KDD'11, ACM, August 2011, California, USA

24. Daqing Zhang et. al., "iBAT: Detecting anomalous taxi trajectories from GPS traces", UbiComp'11, ACM 978-1-4503-0603-0/11/09, September 2011, Beijing, China

25. L. Alvares, A. Loy, C. Renso and V. Bogorny, "An algorithm to identify avoidance behavior in moving object trajectories", Journal of the Brazilian Computer Society vol. 17, no. 3, pp. 193–203, 2011

26. S. Chawla, Y. Zheng, and J. Hu, Inferring the root cause in road traffic anomalies, 12th IEEE International Conference on Data Mining. IEEE, 141–150, 2012

27. Zhu, Jie et al. "Time-Dependent Popular Routes Based Trajectory Outlier Detection". Springer International Publishing (2015): pp. 16–30, Switzerland, https://doi.org/10.007/978-3-319-26190-4_2

28. Zhu, Jie et al. "Effective And Efficient Trajectory Outlier Detection Based On Time-Dependent Popular Route". *Springer Science* (2016), NY, https://doi.org/10.007/s11280-016-000-6

A Privacy-Preserving Approach to Secure Location-Based Data

Jyoti Rao, Rasika Pattewar and Rajul Chhallani

Abstract There are number of sites that provide location-based services. Those sites use current location of user through the web applications or from the Wi-Fi devices. Sometimes, these sites will get permission to the user private information and resource on the web. These sites access user data without providing clear detail policies and disclosure of strategies. This will be used by the malicious sites or server or adversaries breaches the sensitive data and confidentiality of the user. User shares original context of the location. An adversary learns through the user's original context. Due to the lack of secure privacy-preserving policies, it has shifted them to specific goals for various hazards. In order to secure or preserve privacy of user, new privacy-preserving technique called FakeIt is proposed. In FakeIt, system works around privacy, security to satisfy privacy requirements and the user decides context before sharing. If the current location context is sensitive to the user, then user decides to share the fake location context to location-based services instead of original. System restricts the adversaries to learn from the shared sensitive location context of the user.

Keywords Privacy · Protection · Location-based service · Security
Sensitive context

J. Rao (✉) · R. Pattewar · R. Chhallani
Department of Computer Engineering, Dr. D.Y. Patil Institute of Technology,
Pimpri, Pune, India
e-mail: jyoti.aswale@gmail.com

R. Pattewar
e-mail: pattewar.rb@gmail.com

R. Chhallani
e-mail: rajuldchhallani@gmail.com

© Springer Nature Singapore Pte Ltd. 2018
S. Bhalla et al. (eds.), *Intelligent Computing and Information and Communication*,
Advances in Intelligent Systems and Computing 673,
https://doi.org/10.1007/978-981-10-7245-1_6

1 Introduction

The increasing popularity of the location-based service provides new opportunities to users. They provide personalized services to the user based on their location details provided to the LBSN sites. For example, some sites provide remainder if user is at a particular location. Due to large number of location-sharing applications, the large amount of privacy issues are raised. User trusts on an application location-sharing web application. But sometimes these types of service may use users personal data for other purposes like to sell third parties or may use adversaries for malicious purpose, etc., privacy leakage without consent of user or without user's permissions. Sometimes, this concerns to threat user privacy and location sensitivity. To prevent such type of privacy leakage, user has privacy-preserving approach to restrict the adversaries. To restrict the adversaries from getting the personal information of user, sensitive details are disclosed to adversaries harmful to the user. The need of policies in user decides when or where to share location or share other random location to the location-sharing sites [1].

Current system aims to limit the location-sharing site from accessing sensitive data, which would prevent adversaries from inferring sensitive details of user. Location context control policy gets privileges depending on the user. To restrict adversaries on policy used called as deception policy, this policy is used in location sharing to decide whether to release fake or not; if location is sensitive, then shared location will be fake location. Adversaries may get know from previous user context; for these temporal correlations, we use MaskIt policy. MaskIt is privacy checking policy that decides releasing context of user, to share or to conceal current location. FakeMask is a new technique to control the location sharing in web application. Decide while sharing location with the application or fake location context shared to web applications [1].

In existing privacy-preserving techniques, a simple technique is used. In this technique, all the private information is secreted and other is transmitted. Though the private information not transmitted, an attacker can infer some secreted private information and resources from the transmitted information. For applying those preventions related to information that distinguishes among nearly situated subareas, the main reason first is that the public and private locations may unexpectedly release their secrete context and hazarded their individual security for locations, concerning on preventing adversary from accessing sensitive information and resources. But many times it shortfall approaches security [1].

2 Review of Literature

Cai et al. [1] presented FakeMask technique. FakeMask chooses whether to spread the present information or release fake information. FakeMask is useful for protecting secrecy of locality. The information distributed under FakeMask creates trouble to attacker to collect original information. Deception technique is acquired inside FakeMask because of non-secret information releasing like secret information. Semi-Morkov model and secrecy examining methods are used for modeling. Privacy of user is protected from dangerous software which behaves like mediator. Present information should not be revealed to the adversary. Handling with the FakeMask and understanding information spread from user are not simple for the attacker. Attacker is not capable of understanding this kind of information.

Shen et al. [2] studied structural design, communication progression, protection, and privacy of network. Authors studied three groups of smartphone applications accompanied by a center of attention on two independent smartphone approaches. Business card plus assistance assessment are approaches that have learned. They search practical method to relate with the connected security and confidentiality challenge. In previous approach, there are deficiencies of the technique, so author presents various capable research guidelines.

Jedari et al. [3] presented mechanism which mainly focuses to give logical group on safety challenge and an enormous study on some present way in mobile sensor networks. This exertion decreases safety troubles and resolution techniques throughout opportunistic networks and stoppage tolerable networks to mobile sensor networks containing anticipation of wrapping all effort started about protection, privacy, and confidence in mobile networks. Correlation among social belief and reputation system is unstated. One problem is energy custom of points and humanity outcome on points. Process of applying matchmaking procedures by innovative people to realize additional concerning usage of mobile sensor network approaches is not determined.

Kelley et al. [4] observed users' risk and advantage perception associated with practice of these apparatuses and privacy restrict of current location-sharing approaches. They studied an online survey and observed that although the great figure of their responders had heard of location-sharing approaches, they do not yet understand the probable value of mentioned approaches; also they are worry regarding sharing current location context online. Next to evaluating present commercial locality-sharing approaches' secrecy controls, author analyzed that location-sharing approaches never offer their users different group of rules to manage disclosure of user's place; they offer only modicum of privacy.

Gao et al. [5] presented a technique connected toward locality susceptible hashing toward partition consumer area in groups each consisting at minimum number of K users, also designed an efficient algorithm to reply kNN queries for any point in the spatial cloaks of arbitrary polygonal shape. Great replication study shows that two algorithms provide superior effect having normal measuring complexity. Two drawbacks associated with mentioned technique are present. First, it fragments dataset; it is not locality preserving. Next, time complexity is more.

Ying et al. [6] proposed a motivational technique having privacy conservation within smartphone crowdsourcing systems. Combining profits of offline incentive techniques and online incentive techniques, a motivation apparatus presents that choose the worker public continually, and then energetically prefer defenders after bidding. The proposed motivation technique includes two algorithms. Algorithms are improved two-stage auction algorithm also straightforward online status upgrading algorithm. By simulation, author validates capacity and benefits of the represented incentive technique. Represented motivational technique can determine freeriding problem and also enhance the capability and usefulness of mobile crowdsourcing system sufficiently.

Gehrke et al. [7] presented MaskIt approach. This approach keeps secrecy which is helpful to choose a user information series. The chosen text information could be passed to approaches. It may be useful to feedback requirements from these kinds of approaches. User explained privateness which is associated with a group of secret context mentioned. Although adversaries are powerful and understand context about choosing system and time-related cooperations in the context sequence, information realized by adversaries based on filtered sequence about to user being in sensitive data is limited by MaskIt. Inside MaskIt, privacy inspect determines whether to deliver or suppress the present user's context. This technique gives two novel privacy inspects and explains how to choose the check with more utility for a user.

Zhang et al. [8] at the beginning acknowledged that leak privacy issue has a belief of data dynamics and hazardous third parties who have potentials of tuning their rushing policies, and then represent the correlative contest within users, adversaries as a zero-sum stochastic game. As well, the author presents well-planned least learning algorithm to obtain best protection strategy.

Weiss et al. [9] defined BlurSense, a tool which allow secure, customized permission to all sensors in smartphones, tablets, and same type of gadgets. The present permission restrict to mobile tools, like sensor information, is consistent and coarse-grained. BlurSense is not static, fine-grained, feasible permission restrict approach, behaving as a line of protection that permits users to describe and join secrecy filters. As an output, the user may reveal filtered sensor information to untrusted applications, and scientists can gather data in a manner that guards users' privacy.

Oluwatimi et al. [10] presented an access control approach. In mentioned approach, benefits would be dynamically allowed or disallowed to mechanisms related to particular context of user. This application of context differentiates between closely located subareas in a similar location. After a change in operating system, context-based access control restrictions can be defined and enforced.

Ghafoor et al. [11, 12] presented a scheme. It is accuracy-forced privacy-preserving access control scheme. The scheme is a composition of access reticence and privacy protection methods. The access reticence method grants only allowed query predicates on secrete information. The privacy-preserving unit anonymizes information to reach privacy needs and imprecision restrictions on premise identified by the permission restrict method. The permission restrict policies describe that chosen premises were able to obtain the roles, whereas the privacy requirement is to fulfill k-anonymity or l-diversity. An additional restriction which wants it should be fulfilled by Privacy Protection Mechanism (PPM) is the imprecision bound for each chosen premise.

3 System Overview

Figure 1 shows that users share location to Location-Based Service Network (LBSN) through application. After obtaining the current location of user, FakeIt runs and location given by FakeIt is shared. Adversary can take information when interaction between user and LBSN is performing.

Whereas the third party server is not trustworthy, this type of server directly attacks the user's privacy or sells user private information to individual. An adversary refers to malicious parties, which aim at user location details or hack data and loose privacy. System is to expand the complication of extrapolate private information; very difficult policies are acquired in secrecy protecting. To provide high level of user security and manageability, it also reduces the transmission risk and enhances security and privacy preserving.

4 System Architecture

In this system, a technique is proposed to preserve privacy of user-sensitive location. Figure 2 shows proposed system architecture.

Using Google, API's user can share his current location to the application through the GPS or Wi-Fi. User can decide whether to release his location or not. FakeIt decides before location shared with application. User can share current location but if user's current location is sensitive, then user decided that not to release current real location. On such situations, system releases fake location. With the FakeIt approach, we use to release fake location where KNN algorithm and ranking algorithm are used. The use of KNN algorithm adds more privacy to the

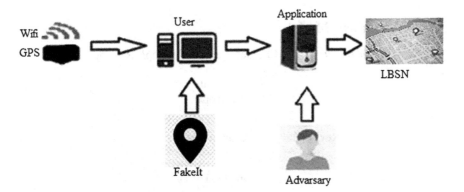

Fig. 1 FakeIt approach

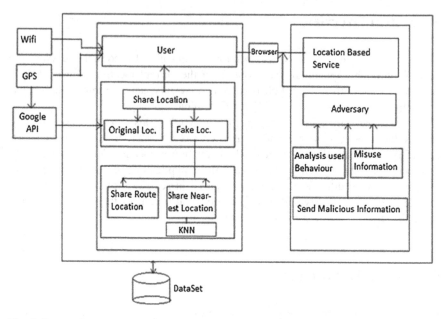

Fig. 2 Proposed system architecture

over user-sensitive location. KNN algorithm used to find nearest locations from the current location of the user. By using KNN algorithm, system finds n number of nearest location. These n locations then reversed by the ranking algorithm. From reversed ranked n locations, select top location and share it to the location-based service. This has great difficulty to adversary to find users sensitive context from release user context.

5 Mathematical Model

$$S = \{\{u_1, u_2, \ldots u_n\}\text{CL}, f(\text{FL}), \text{lat}, \text{lng}, \text{FI}\{\text{FL}, \text{Top}k, \text{RevTop}k, \text{PC}\}\}$$

$$
\begin{aligned}
\text{CL} &= \text{lat} + \text{ln} \\
\text{FL} &\in \text{RevTop}k
\end{aligned}
\tag{1}
$$

$$\text{RevTop}k = \int_0^k \text{ReverseOrder}(\text{CL} \in \text{Top}k) \tag{2}$$

$$\text{FL} = \text{Top}(\text{RevTop}k) \tag{3}$$

$$f(\text{FL}) = \sum_{i=0}^{k} \text{TopReverse}k \qquad (4)$$

$$\text{FK} = \text{PC}\{\text{FL}\}, \qquad (5)$$

where

S = Start of the program, $f(\text{FL})$ = Fake Location, CL = Current Location,
lat = Lattitude, lng = Longitude, Top$k$ = Top k Results,
RevTop$k$ = Reverse order of Top k Results,
FK = Fake Location, FI = FakeIt, and PC = Privacy check.

Equation (1) gives current location of user where user presents at the stage. Location is the combination of (lat) latitude and (lng) longitude on the geospatial data. Here, consider latitude as the X-axis and longitude as the Y-axis. By using a KNN algorithm, find the nearest neighbor to current location. Consider Top$k$ results are found from the current location. Equation (2) gives reverse order of result and from these results, Eq. (3) gives fake location where Eq. (4) gives function for obtaining finally Eq. (5) that checks and gives the fake location.

6 Expected Results

There are privacy utility approaches for the efficiency of the different privacy checks. Shown in table the average times for making the decision to privacy checks and filter the current instances of traces. The existing approaches do not prevent adversaries from inferring sensitive context.

The current context data may be breached to this exceed the belief than certain level. This has a normal privacy guarantee.

Table 1 shows average processing time for existing approaches. Mask sensitive shows lowest initialization time but it breaches privacy. The FakeIt provides maximum utility than mask sensitive in which approach conceals data without breaching privacy for more sensitive context.

Table 1 Avarage processing time

Algorithm	Initialization time on PC (min)	Release time on web (ms)
Mask sensitive	0	<1
MaskIt	18	<128
Naive fake	>60	<1
Efficient fake	3	<1
FakeIt	<3	<1

7 Conclusion

In this, the problem of user's privacy in location-based web application is discussed. Proposed work is related to the privacy preserving of user's location. Proposed work conceals the user-sensitive context being leaked from the location-based services. The fake location context will be released for sensitive context. An attacker has great trouble to learn original information from these shared contexts.

Acknowledgements Success is never achieved single-handed. Apart from our humble efforts, this paper is outcome of the help, co-operation, and guidance from various corners. I would like to add a few heartfelt words for the people who were part of this in numerous ways and the people who gave unending support right from the stage of ideas.

References

1. Lichen Zhang, Zhipeng Cai, Xiaoming Wang, "FakeMask: A Novel Privacy Preserving Approach for Smartphones" IEEE Transactions on Network and Service Management, 2016. J. Clerk Maxwell, A Treatise on Electricity and Magnetism, 3rd ed., vol. 2. Oxford: Clarendon, 1892, pp. 68–73.
2. X. Liang, K. Zhang, X. Shen, and X. Lin, "Security and privacy in mobile social networks: challenges and solutions," IEEE Wireless Communications, vol. 21, no. 1, pp. 33–41, 2014.
3. Y. Najaflou, B. Jedari, F. Xia, L. T. Yang, and M. S. Obaidat, "Safety challenges and solutions in mobile social networks," IEEE Systems Journal, vol. 9, no. 3, pp. 834–854, 2015.
4. J. Tsai, P. G. Kelley, L. F. Cranor, and N. Sadeh, "Location sharing technologies: Privacy risks and controls," I/S: A Journal of Law and Policy for the Information Societ, vol. 6, no. 2, pp. 119–317, 2010.
5. K. Vu, R. Zheng, and J. Gao, "Efficient algorithms for k-anonymous location privacy in participatory sensing," in Proceedings of the 31st Annual IEEE International Conference on Computer Communications (INFOCOM'12), Orlando, FL, USA, March 25–30 2012, pp. 2399–2407.
6. Y. Wang, Z. Cai, G. Ying, Y. Gao, X. Tong, and G. Wu, "An incentive mechanism with privacy protection in mobile crowdsourcing systems," ComputerNetwork., p. In Press, 2016.
7. M. Gotz, S. Nath, and J. Gehrke, "Maskit: Privately releasing user context streams for personalized mobile applications," in Proceedings of the 2012 ACM SIGMOD International Conference on Management of Data (SIGMOD'12), Scottsdale, Arizona, USA, May 20–24 2012, pp. 289–300.
8. W. Wang and Q. Zhang, "A stochastic game for privacy preserving context sensing on mobile phone," in Proceedings of the 33rd Annual IEEE International Conference on Computer Communications (INFOCOM'14), Toronto, Canada, April 27–May 2 2014, pp. 2328–2336.
9. J. Cappos, L. Wang, R. Weiss, Y. Yang, and Y. Zhuang, "Blursense: Dynamic fine-grained access control for smartphone privacy," in Proceedings of the IEEE Sensors Applications Symposium (SAS'14), Queenstown, New Zealand, Febrary 18–20 2014, pp. 329–332.
10. B. Shebaro, O. Oluwatimi, and E. Bertino, "Context-based access control systems for mobile devices," IEEE Transactions on Dependable and Secure Computing, vol. 12, no. 2, pp. 150–163, 2015.

11. Z. Pervaiz, W. G. Aref, A. Ghafoor, and N. Prabhu, "Accuracy constrained privacy preserving access control mechanism for relational data," IEEE Transactions on Knowledge and Data Engineering, vol. 26, no. 4, pp. 795–807, 2014.
12. Rasika Pattewar and Jyoti Rao, "A Survey on Privacy Preserving Approaches for Location Based Data", International Journal of Advanced Research in Computer and Communication Engineering, 240–243, Vol. 5, Issue 12, December 2016.

Analysis of Blind Image Watermarking Algorithms

Chhaya S. Gosavi and Suresh N. Mali

Abstract This paper presents an overview of blind image watermarking algorithms. In this paper, we analyzed these algorithms for different criteria like robustness, security, and imperceptibility. We also compared pros and cons of using blind method for embedding and extraction of watermark. Most of these algorithms are implemented using MATLAB 2011 and tested on the standard image dataset. We used true color images of size 256×256 and binary watermarks of size 32×32 for testing. This paper will help watermarking researcher to choose the particular algorithms depending on their need for the application they are working on.

Keywords Blind · Watermark · DCT · DWT · SVD

1 Introduction

A watermark is irremovable, imperceptibility, and robustly embedded in the multimedia data. The process of embedding watermark is called watermarking. A watermark is embedded for protection of digital data from illegally used. The main applications of watermarking are copyright protected, authentication, piracy detection, etc. Many researchers have been working in this area since two decades. Image watermarking is explored by many researchers as it is easy compared with audio and video data and also having a lot of redundant information which can be used for embedding.

Watermarking process is mainly divided into three parts. First, watermark is added into an image. In the next part, it is extracted from the watermarked image and lastly authentication is done by comparing original and extracted watermarks. Watermark insertion can be done in two ways, first is spatial domain and the second

C. S. Gosavi (✉) · S. N. Mali
Department of Computer Engineering, DYPIET, Pimpri, Pune, Maharashtra, India
e-mail: chhaya.gosavi@gmail.com

S. N. Mali
e-mail: snmali@rediffmail.com

© Springer Nature Singapore Pte Ltd. 2018

S. Bhalla et al. (eds.), *Intelligent Computing and Information and Communication*,
Advances in Intelligent Systems and Computing 673,
https://doi.org/10.1007/978-981-10-7245-1_7

is transform domain. Watermarked image can be tampered with using different attacks like common image processing attacks, geometric attacks, or compression attacks. The watermark should be robust to such attacks. Spatial domain embedding is fragile and cannot withstand even simple manipulations. Transform domain technique is generally used for robust embedding.

The data is needed to remove the watermark partition; the watermarking algorithm is divided into two types, blind algorithms and non-blind algorithms. At the time of watermark removal if we need original image, then the algorithms are called non-blind algorithms otherwise blind algorithm. In non-blind algorithms, it is easy to compare original images and watermarked image and generated watermark data. The sender has to transfer original image along with watermarked image to the receiver. But practically, it is not practicable and this does not assist the purpose of watermarking. Non-blind algorithms can be more robust compared with blind algorithms but not good.

Hence, in this paper, we compare different methods for blind insertion and detection of watermarks.

2   Literature Review

Let us consider I1 is the image, W1 is the watermark, I2 is watermarked image, and W2 is extracted watermark. Here, we discuss algorithms to embed W1 in I1 and extract W2 from I2.

Algorithm 1 In [1–3], authors have inserted watermark in third level Discrete Wavelet Transform (DWT) of an image, that is, in LH_3 sub-band. Maximum value of pixel's eight neighbors is "t". Every coefficient is replaced by the following equation:

$$
\begin{aligned}
&\text{New\_LH}_3 = LH_3 + W_1 * LH_3; \\
&W_1 = 1 \text{ if } W = 1 \text{ and } LH_3 > t \\
&\text{or } W = -1 \text{ and } LH3 < t \\
&W_1 = -1 \text{ if } W = 1 \text{ and } LH_3 < t \\
&\text{or } W = -1 \text{ and } LH_3 > t
\end{aligned}
\tag{1}
$$

for extraction, authors have used W as follows:

$$
\begin{aligned}
&W_2 = 1 \text{ if } W = 1 \text{ and } LH_3 > t \\
&\text{or } W = -1 \text{ and } LH_3 < t \\
&= -1 \text{ otherwise.}
\end{aligned}
\tag{2}
$$

The algorithm that uses original watermark for extraction is known as non-blind algorithm.

Algorithm 2 In [4], DWT and SVD techniques are used for embedding. First, DWT of image is obtained by the SVD of the low-frequency part that is calculated and then a watermark is embedded into highest singular value. These steps are explained by following equations:

$$[LL, LH, HL, HH] = DWT(I1)$$
$$[U S V] = svd(LL)$$
$$if (W_1 == 1)$$
$$if ((S\%Q) <= Q/4)$$
$$S = S - (S\%Q) - Q/4;$$
$$else$$
$$\tag{3}$$

$$S = S - (S\%Q) + 3 * Q/4;$$
$$end$$
$$else$$
$$if ((S\%Q) > = 3 * Q/4)$$
$$\tag{4}$$

$$S = S - (S\%Q) + 5 * Q/4;$$
$$else$$
$$\tag{5}$$

$$S = S - (S\%Q) + Q/4;$$
$$end$$
$$end$$
$$\tag{6}$$

Here, Q is the embedding strength.

Extraction Procedure:

$$NewLL = DWT(I2)$$
$$[U1\ S1\ V1] = svd(NewLL)$$
$$W_2(i,j) = 1\ if\ ((S1\%Q) > Q/2)$$
$$= 0\ otherwise$$
$$\tag{7}$$

Q is needed for extraction. The algorithm is semi-blind.

Algorithm 3 In [5], watermark is embedded by the following method:

$$C = DWT(I1);$$
$$if\ W1[j] = 1,$$
$$interchange\ C[i]\ with\ max\ of\ (C[i], C[i + 1]....C[i + 5])$$
$$Else$$
$$\tag{8}$$

Interchange $C[i]$ with min of $(C[i], C[i+1], \ldots \ldots C[i+5])$

For the extraction of watermark, they have used   (9)

$$W2 = DWT(I1)$$

if $W2 > \text{median}(WC[i], WC[i+1], \ldots \ldots WC[i+5]$,

$$W2 = 1$$

Else   (10)

$$W2 = 0$$

This algorithm is a pure blind algorithm.

Algorithm 4 In [6], watermark is embedded using

$$
\begin{aligned}
&\text{if } W1(k) == 1\\
&I2\,(4, 4, 1) = I1(4,\ 4,\ 1) + k;\\
&\text{end}
\end{aligned}
\tag{11}
$$

$$
\begin{aligned}
&\text{if } W1(k) == 0\\
&I2\,(4, 4, 1) = I1\,(4, 4, 1) - k;\\
&\text{End}
\end{aligned}
\tag{12}
$$

for extraction

$$
\begin{aligned}
&\text{If } ((I2(5, 5, 1))) > \; = I2(4, 4, 1)\\
&W2(i) = 0;\\
&\text{else}\\
&W2(i) = 1;\\
&\text{end}
\end{aligned}
\tag{13}
$$

This algorithm is also pure blind algorithm.

Algorithm 5 In [7], embedding in HH band of DWT

$$
\begin{aligned}
&\text{If } (W1(i) == 1) \text{ then}\\
&\text{blocHH} = \text{blocHH} + K \times 1\\
&\text{Else}\\
&\text{blocHH} = \text{blocHH} + K \times -1\\
&\text{End If}
\end{aligned}
$$

Here, K is the robustness factor = 20
for extraction,

find gradient of each HH block, i.e., GHH

> If (Sum_Of_GHH_Coefficients > 0) then
>
> W2(i) = 1
>
> Else
>
> If (Sum_Of_GHH_Coefficients < 0)
>
> W2(i) = 0
>
> End If
>
> End If

Algorithm 6 In [8, 9], singular value decomposition is used for embedding the watermark:

$$[s1, s2, s3 \ldots] = \text{svd}(I1);$$
$$s1, s2, s3 \ldots \text{ are singular values}$$

Watermark is embedded by the following method:

$$Si = 0.5[(Si - 1 + Si + 1) + W1(i) * (Si - 1 - Si + 1)] \quad (14)$$

Extraction process is as follows:

$$W2(i) = 0 \text{ if } Si > 0.5(Si - 1 + Si + 1)$$
$$= 1 \text{ otherwise}$$

In [10], they are using the same key for embedding and extraction of watermark it makes the algorithm semi-blind.

In [11], embedding procedure is not explained here. They have applied genetic algorithms and extracted watermark by doing fitness test.

In [12], image I1 is decomposed into three-level DWT. SVD is applied on HH band to get singular values. These are replaced by singular values of watermark. Reverse method is applied for extraction.

Algorithm 7 Rajab [9] uses LSB method to embed watermark as follows:

$$HL = \text{DWT}(I1)$$
$$[U \, SHL \, V] = \text{svd}(HL)$$
$$\text{LSB} (SHL (i, i)) = W1(i)$$

Extraction method

$$HL1 = DWT(I2)$$
$$[U\,SHL1\,V] = svd(HL1)$$
$$W2(i) = LSB\,(SHL1\,(i,i))$$

Algorithm 8 Karmakar [13, 14] uses the following algorithm for embedding:

$$
\begin{aligned}
&if(W1(k) == 1)\\
&if\,((I1)\%d) < a)\\
&I2 = I1 - (I1\%d) - a\\
&else
\end{aligned}
\tag{16}
$$

$$
\begin{aligned}
&I2 = I1 - (I1\%d) + c\\
&endif\\
&elseif\,(W1(k) == 0)\\
&if\,((I1\%d) > c)
\end{aligned}
\tag{17}
$$

$$
\begin{aligned}
&I2 = I1 - (I1\%d) + e\\
&else
\end{aligned}
\tag{18}
$$

$$
\begin{aligned}
&I2 = I1 - (I1\%d) + a\\
&endif
\end{aligned}
\tag{19}
$$

Extraction process is

$$
\begin{aligned}
&if((I2\%d) > b)\\
&W2(b) = 1\\
&else\\
&W2(b) = 0\\
&endif
\end{aligned}
\tag{20}
$$

This algorithm is also semi-blind as it uses "d" for embedding as well as extraction.

As seen from the literature survey, three things are important while embedding a watermark into the image.

1. Watermark embedding position should be known at extraction process.
2. Embedded bit should survive after different manipulations and attacks.
3. Embedding watermark should not affect quality of the cover.

3 Proposed Algorithm

The proposed algorithm for embedding and extraction is explained below:
Embedding algorithm:
Input—Image I1 (256 × 256), watermark W1 (32 × 32);
Output—Embedded Image I2 (256 × 256).

1. Divide I1 into 8 × 8 blocks Iblk;
2. Calculate [U T] = SCHUR(Iblk);
3. Max = Maximum(T);
4. Min = Minimum (T);
5. if(W1(i) == 1)
   T(0,0) = Max
   else
   T(0,0) = Min;
6. Continue till all watermark bits get embedded in the image; and
7. Save I2 as watermarked image.

Extraction algorithm:
Input—Watermarked Image I2 (256 × 256);
Output—Extracted watermark W2 (32 × 32).

1. Divide I2 into 8 × 8 blocks Iblk;
2. Calculate [U T] = SCHUR(Iblk);
3. Max = Maximum(T);
4. Min = Minimum (T);
5. if T(1,1) = Max
   W2(i) = 1
   else
   W2(i) = 0;
6. Continue till all watermark bits are extracted from the image; and
7. Save W2 as extracted watermark.

Table 1 summarizes the results.
PSNR is calculated to check imperceptibility of watermarked image by the following equation:

$$MSE = \frac{1}{MN} \sum_{i=1}^{M} \sum_{j=1}^{N} (x(i,j) - y(i,j))^2 \tag{21}$$

$$PSNR = 10 \log_{10} \frac{(2^n - 1)^2}{\sqrt{MSE}} \tag{22}$$

Normalized cross-correlation (NK) is calculated to test robustness of extracted watermark by the following equation:

Table 1 Comparison of blind watermarking algorithms

Algorithm	Algo1	Algo2	Algo3	Algo4	Algo5	Algo6	Algo7	Algo8	Proposed
Blindness	Non	Semi	Pure	Pure	Pure	Pure	Pure	Semi	Pure
PSNR (dB)	45.44	43.21	47.32	41.53	54.43	41.03	49.7	40.6	60.20
NK—No attack	0.9593	0.9231	0.9421	0.9133	0.9741	0.9265	0.9434	0.9121	0.9982
Median filtering (3 × 3)	0.9382	0.9072	0.9112	0.8913	0.9512	0.9051	0.9211	0.8943	0.9895
Smoothing	0.9322	0.9245	0.9176	0.9063	0.9476	0.9143	0.9212	0.8987	0.9863
Cropping (10%)	0.9002	0.8910	0.8941	0.8921	0.9384	0.8965	0.9138	0.8912	0.9754
Compression	0.8893	0.8928	0.8942	0.8933	0.8974	0.8926	0.8944	0.8912	0.9482

$$\mathrm{NK} = \frac{\sum\limits_{i=1}^{M}\sum\limits_{j=1}^{N} \left(x(i,j) \times y(i,j) \right)}{\sum\limits_{i=1}^{M}\sum\limits_{j=1}^{N} \left(x(i,j) \right)^2} \tag{23}$$

4 Conclusion

The algorithm presented here is a pure blind algorithm. It is a block-based algorithm so one bit of watermark is embedded in 8 × 8 block of cover image which helped to maintain quality of watermarked image. This method is robust for all signal processing attacks like median filtering, smoothing, cropping, and compression.

References

1. A. Essaouabi and F. Regragui: A Blind Wavelet-Based Digital Watermarking for Video, In: International Journal of Video & Image Processing and Network Security, Vol 09, No 09, (2009), pp. 37–41
2. Majid Masoumi and Shervin Amiri: A Blind Video Watermarking Scheme Based on 3D Discrete Wavelet Transform In: International Journal of Innovation, Management and Technology, Vol. 3, No. 4, (2012) pp. 487–490
3. DING Hai-yang, ZHOU Ya-jian, YANG Yi-xian, ZHANG Ru: Robust Blind Video Watermark Algorithm in Transform Domain Combining with 3D Video Correlation In: Journal of Multimedia, Vol. 08, NO. 2 (2013) pp. 161–166
4. Chetan K. R Raghavendra K.: A Blind Video Watermarking Algorithm Resisting to Rotation Attack. In: IEEE International Conference on Computer and Communications Security, (2009) pp. 111–114

5. Chetan K. R Raghavendra K: DWT Based Blind Digital Video Watermarking Scheme for Video Authentication, In: International Journal of Computer Applications (0975—8887), Volume 4– No. 10, (2010), pp. 19–26

6. D. Niranjan babu, D. Jagadeesh: A Blind and Robust Video Water Marking Technique in DCT Domain, In: International Journal of Engineering and Innovative Technology (IJEIT), Volume 2, Issue 2, (2012) pp. 128–132

7. Henri Bruno Razafindradina and Attoumani Mohamed Karim: Blind and Robust Image Watermarking based on Wavelet and Edge insertion In: International Journal on Cryptography and Information Security (IJCIS), Vol. 3, No. 3, (2013) pp. 23–30

8. S. Naveen kumar, U. Nageswar Rao: Robust & Blind Video Watermarking Algorithm Using SVD Transform, In: International Journal of Engineering Research & Technology (IJERT), ISSN: 2278-0181, Vol. 3 Issue 5, (2014) pp. 1359–1364

9. Lama Rajab, Tahani Al-Khatib, Ali Al-Haj: A Blind DWT-SCHUR Based Digital Video Watermarking Technique, In: Journal of Software Engineering and Applications, 8, (2015), pp. 224–233

10. Md. Asikuzzaman, Md. Jahangir Alam, Andrew J. Lambert, Mark Richard Pickering,: Imperceptible and Robust Blind Video Watermarking Using Chrominance Embedding: A Set of Approaches in the DT CWT Domain, In: IEEE Transactions of Information Forensics and Security, Vol. 09, No. 09, (2014), pp. 1502–1517

11. Nitin A. Shelke, Dr. P. N. Chatur: Blind Robust Digital Video Watermarking Scheme using Hybrid Based Approach, In: International Journal of Computer Science and Information Technologies, Vol. 5 (3), (2014) pp. 3619–3625

12. Sakshi Batra, Rajneesh Talwar: Blind Video Watermarking based on SVD and Multilevel DWT, In: European Journal of Advances in Engineering and Technology, 2(1), (2015), pp. 80–85

13. Amlan Karmakar, Amit Phadikar, Baisakhi Sur Phadikar, Goutam Kr. Maity: A blind video watermarking scheme resistant to rotation and collusion attacks, In: Elsevier, Journal of King Saud University—Computer and Information Sciences, 28, (2016) pp. 199–210

14. X. Y. Wang, Q. L. Shi, S. M. Wang, H. Y. Yang: A blind robust digital watermarking using invariant exponent moments, In: Elsevier, Int. J. Electron. Commun. (AEÜ), 70, (2016) pp. 416–426

ClustMap: A Topology-Aware MPI Process Placement Algorithm for Multi-core Clusters

K. B. Manwade and D. B. Kulkarni

Abstract Many high-performance computing applications are using MPI (Message Passing Interface) for communication. The performance of MPI library affects the performance of MPI applications. Various techniques like communication latency reduction, increasing bandwidth, and increasing scalability are available for improving the performance of message passing. In multi-core cluster environment, the communication latency can be further reduced by topology-aware process placement. This technique involves three steps: finding communication pattern of MPI application (application topology), finding architecture details of underlying multi-core cluster (system topology), and mapping processes to cores. In this paper, we have proposed novel "ClustMap" algorithm for the third step. By using this algorithm, both system and application topologies are mapped. The experimental results show that the proposed algorithm outperforms over existing process placement techniques.

Keywords High-performance computing · Topology-aware process placement
System topology · MPI application topology

K. B. Manwade (✉)
Department of Computer Science and Engineering, Ph.D. Research Center, Walchand
College of Engineering, Sangli, Maharashtra, India
e-mail: mkarveer@gmail.com

K. B. Manwade
Shivaji University, Kolhapur, Maharashtra, India

D. B. Kulkarni
Department of Information Technology, Walchand College of Engineering, Sangli,
Maharashtra, India
e-mail: d_b_kulkarni@yahoo.com

© Springer Nature Singapore Pte Ltd. 2018   67
S. Bhalla et al. (eds.), *Intelligent Computing and Information and Communication*,
Advances in Intelligent Systems and Computing 673,
https://doi.org/10.1007/978-981-10-7245-1_8

1 Introduction

MPI is a dominant message passing library used in high-performance clusters. Therefore, improving its performance is very much essential. Computer architectures are evolved from uni-core to multi-core as illustrated in [1]. Also, modern interconnect technologies provide bandwidth in gigabytes [2]. The development in these two areas gives an opportunity to take advantages of shared memory, processor locality, and high communication bandwidth which are available in multi-core clusters. Current MPI implementations map processes on processors or cores without considering topology of application or topology of the underlying system. While placing processes in multi-core clusters, the affinity between cores and processes should be known. This leads to the development of new topology-aware process placement technique as suggested in [3]. In this paper, we suggest new topology-aware process placement algorithm.

The rest of this paper is organized as follows. In Sect. 2, concepts of topology-aware process placement, the data structure for system, and application topology are illustrated. The topology-aware process placement algorithm, its working, and time complexity are given in Sect. 3. In Sect. 4, works related to topology-aware process placement are listed. The results of the "ClustMap" algorithm for benchmark applications are given in Sect. 5. The summary of our work, conclusion, and future work are given in Sect. 6.

2 Topology-Aware Process Placement

2.1 Process Placement Techniques

As shown in Fig. 1, there are two approaches for process placement: static and dynamic. Static approach places the process before the start of the execution, whereas dynamic approach reschedules the process at run time. There are two sub-techniques of static approach. The first technique, core binding [4], determines physical core on which a specific MPI process should be placed so that the communication latency will be reduced. In this technique, modification of MPI application is not required. So the performance of legacy as well as new MPI application can be improved by using this technique. But before process placement, both system and application topologies should be known. Second technique, rank reordering [5] maps processes with default policy and then by using communication pattern of processes the ranks of processes are reordered under the communicators. The constraint for rank reordering is that it should be performed before application data is loaded and for that purpose few new lines need to be added to MPI code.

In the second approach, processes are remapped to another core or processor during runtime. Two techniques are available in this approach; process migration and rank reordering. Process migration is more complicated and time-consuming

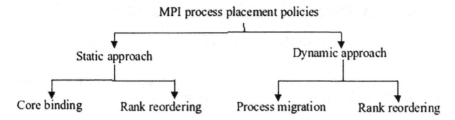

Fig. 1 Taxonomy of topology-aware MPI process placement techniques

technique. Implementation of rank reordering is easy but in time slice-based rank, reordering handling of intermediate result is difficult. We have used core binding technique for process placement.

2.2 Data Structure for System and Application Topology

Topology-aware process placement is an NP-complete problem as mentioned in [6]. The mapping time depends on data structure used for the representation of the system topology and application topology. We have used a data structure called as "dendrogram" [7] to represent both system and application topologies. The key benefits of this data structure are as follows:

1. It explores the hierarchical nature of the system.
2. It forms clusters of cores based on the various nodes as well as network parameters and hence finds the nearest cores.
3. Its tree-like structure reduces the mapping time.
4. It is compatible with existing work as it works on data stored in matrix format.

The sample network-based system is shown in Fig. 2a and its topology in "dendrogram" format is shown in Fig. 2b.

Figure 2a shows that four nodes are connected in a network. Each node has eight cores that are situated on two sockets. In this case, cores can be grouped hierarchically according to their location. This grouping is socket wise (groups G1–G8), node wise (groups G9–G12), and cluster-wise (group G13) as shown in Fig. 2b.

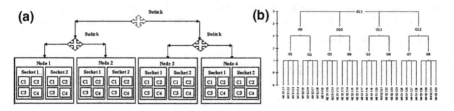

Fig. 2 a Sample network-based system **b** Sample network-based system using "dendrogram"

3 Topology-Aware Process Placement Algorithm

3.1 The Process Mapping Problem

We have treated process to core mapping as cluster mapping problem. Let P and C are the set of clusters of processes and cores, respectively.

$$P=\{p_1, p_2, p_3, \ldots, p_n\} \tag{1}$$

$$C=\{c_1, c_2, c_3, \ldots, c_m\} \tag{2}$$

The mapping function "F" is defined as

$$F : P \rightarrow C \tag{3}$$

The mapping function maps processes from set P to cores in set C, if the following conditions are satisfied:

$$|P_i| \leq |C_k| \tag{4}$$

where $1 < i < n$, $1 < k < m$, and number of processes in cluster P_i and number of cores in cluster C_k are same. The first condition from Eq. (4) states that if the size of sub-cluster C_k is greater than or equal to sub-cluster P_i, then one-to-one mapping of the process to the core is possible. Otherwise, more processes need to be scheduled on a core instead of one process per core:

$$E(P_i) \geq \frac{\sum_{j=1}^{n} E(P_j)}{(n-1)}, \tag{5}$$

where $i \neq j$, E is the affinity function, and process from P_i have more communication affinity than process from P_j. The second condition in Eq. (5) indicates that processes from P_i are chosen for mapping if they have more affinity between them than processes from P_j.

$$E(C_k) \geq \frac{\sum_{k=1}^{n} E(C_1)}{(m-1)}, \tag{6}$$

where $k \neq 1$, E is the affinity function, and cores from C_k have more affinity than cores from C_1. The third condition from Eq. (6) indicates that cores in C_k have more affinity than cores from C_1

$$P_i \text{ and } C_k \text{ are unmapped clusters of processes and cores respectively} \tag{7}$$

According to the fourth condition from Eq. (7), only unmapped clusters, P_i and C_k, are chosen for mapping.

3.2   The *"ClustMap"* Algorithm

```
-------------------------------------------------------------------------------
Algorithm1: ClustMap
-------------------------------------------------------------------------------
1.  Input:
2.  S //Hierarchical cluster of cores
3.  A // Hierarchical cluster of processes
4.  Sᵢ and Aᵢ //Sub-clusters of S & A respectively
5.  Output:
6.  M //Process to core mapping
7.  Procedure: ClustMap(S,A)
8.  If(Size of Cluster(S) = Size of Cluster(A))
9.  MapClusters(S,A); //Perform mapping
10. ElsIf(Size of Cluster(S) < Size of Cluster(A))
11. Remap(S, A); //Perform mapping with more processes on each node
12. ElsIf(Size of Cluster(S) > Size of Cluster(A))
13. For (Each sub cluster Sᵢ of S)
14. ClustMap(Si, A); //Recursively check for the mapping
15. EndFor
16. EndIf
17. EndProcedure
-------------------------------------------------------------------------------
18. Procedure: ReMap(S,A)
19. For (Each process Pᵢ from A and core Cⱼ from S)
20. M[0][i]=Pᵢ and M[1][i]=Cⱼ //Map Pᵢ to Cⱼ as 1:1
21. EndFor
22. For (Each remaining process Pⱼ from A)
23. If (Affinity(M[0][i] , Pⱼ) is more)
24. M[1][i]= Pⱼ //Map remaining process
25. EndIf
26. EndFor
27. EndProcedure
-------------------------------------------------------------------------------
28. Procedure: MapClusters(S,A)
29. For (Each leaf node of A)
30. M[0][j]=Sⱼ
31. M[1][j]=Aⱼ
32. EndFor
33. EndProcedure
-------------------------------------------------------------------------------
```

The algorithm 1 takes "dendrogram" of system and application topology as inputs and produces a process to core mapping as output. If the size of both system and application "dendrograms" is equal, then process to core mapping is performed (Lines 8–9). If the size of system "dendrogram" is less than application "dendrogram", then map processes to cores with more processes on each node based on process affinity (Lines 10–11). If the size of system "dendrogram" is greater than application "dendrogram", then it recursively goes to next level of "dendrogram" until the application "dendrogram" maps to system "dendrogram" (Lines 12–14).

The "ClustMap" algorithm maps processes to cores hierarchically, and therefore the complexity varies linearly with a number of cores in the system and number of processes in the application. If M be a number of cores in the system and N be a

number of processes in the application, then the complexity of this algorithm is
$O(M + N)$.

4 Related Work

In literature, various topology-aware process placement algorithms are given. The
process placement is treated as graph mapping or tree mapping problem. MPI
standard [8] itself provides set of routines to create and manipulate topologies, but
all MPI implementations do not support these features. Also, MPI standard does not
support mapping of application topology on system topology.

A recursive bi-partitioning algorithm [9] is used to map application topology on
system topology. In [10], graph-based mapping algorithm is used. For process
mapping, lists of subgraphs of both application and system topologies are created
and by using permutation, topologies are mapped. In [11], author treated
topology-aware process placement as graph embedding problem. The virtual
topology of processes is mapped on physical topology of cores. To represent the
physical topology of system, the tree data structure is used.

In [12], authors treated the mapping as graph isomorphism problem. They have
developed a "mapper" algorithm for processor array and mapped processes in
pair-wise manner. In [13], authors treated process mapping as graph mapping
problem. They have considered weighted graph to indicate the volume of data
exchanged between processes and distance between processors. Initially, processes
are mapped using greedy approach and then pair-wise exchange is used. Similarly,
in [14], authors treated mapping as graph mapping problem. They developed
GRASP heuristic algorithm for mapping. To evaluate the quality of mapping,
dilation metric is used.

In [15, 16], authors proposed different process mapping strategies for Blue Gene
system. This implementation is specific to Blue Gene system.

In [17], authors constructed application topology using MPI Graph create
function and to construct system topology, they have used hwloc tool and network
discovery tool. For process mapping, rank reordering technique is used.

In [18], authors have used TreeMatch algorithm for process mapping. To optimize mapping function, two parameters, network contention and load balance,
among processors are used. To collect hardware topology, hwloc tool is used and
for application topology, a profiling tool is used.

5 Results

5.1 Experimental Setup

For performance evaluation, we have used "SimGrid" [19] simulator. The MPI programs are compiled using "smpicc" compiler and are executed using "smpirun" command. A platform file of four nodes with 64 cores each is defined. The simulation system contains totally 1024 cores. The "ClustMap" algorithm gives topology-aware mapping of MPI processes in host file and it is used as one of the parameters for "smpirun" command. For testing purpose, programs from OSU micro-benchmarks version 5.3.2 [20] are used. This benchmark contains programs for point-to-point and collective communications.

The performance of topology-aware process placement technique is measured as the extent of latency reduced through effective process placement. The performance of the "ClustMap" algorithm is measured as the time required for mapping processes to cores.

5.2 Latency Reduction for MPI Communication

To compare the performance of topology-aware process placement with default process placement, the benchmark programs are used. The communication latency of these programs is measured by varying data size from "0" to "1024" bytes. As shown in Fig. 3a, b topology-aware placement significantly reduces latency of MPI programs having point-to-point communication.

In collective communication, all processes are involved; therefore, topology-aware process placement does not reduce communication latency. Figure 4a shows that latency of the benchmark program containing only collective communication is not reduced. Thus, topology-aware process placement alone is not sufficient to improve performance collective communication.

5.3 Performance Evaluation of "ClustMap" Algorithm

The time complexity of "ClustMap" algorithm is $O(M + N)$. Its performance is measured by varying number of processes. The process of core mapping time is measured.

As shown in Fig. 4b, the "ClustMap" algorithm is scalable as it shows linear performance for increasing number of processes in the application.

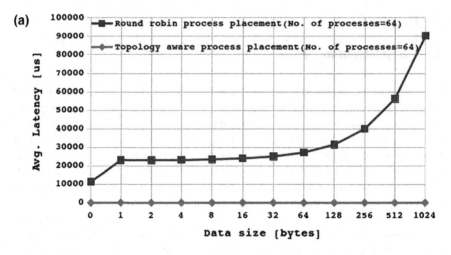

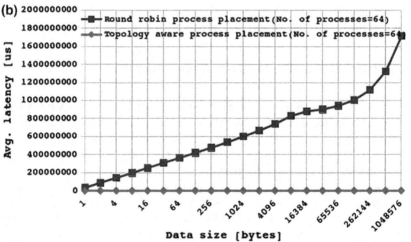

Fig. 3 a Latency reduction for point-to-point communication **b** Latency reduction for point-to-point and collective communication

6 Conclusion

To reduce the latency of MPI program and improve its performance by using topology-aware process placement, three steps are required: finding application topology, finding system topology, and topology-aware process placement. This paper contributes to the third step. We have proposed "ClustMap" algorithm which places processes efficiently.

It is observed that topology-aware process placement approach gives better performance than default process placement. The extent of improvement depends

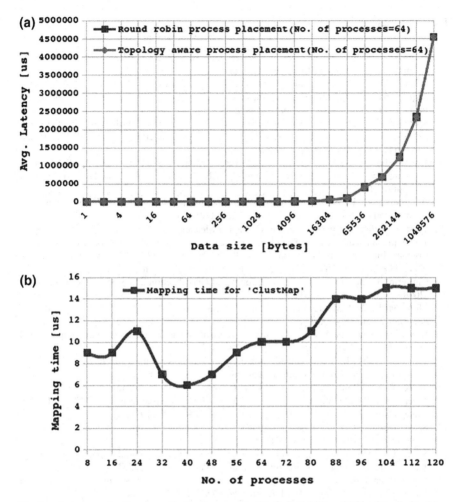

Fig. 4 **a** Latency reduction for collective communication **b** Scalability of "ClustMap" algorithm

on the affinity between the processes to be mapped. Thus, the performance improvement of topology-aware process placement is proportional to the affinity between the processes. The complexity of "ClustMap" varies linearly with number of processes. By using "ClustMap" algorithm, latency of only point-to-point communication can be reduced. In collective communication, all processes are involved therefore for collective communication latency reduction though topology-aware placement is not possible.

We have planned to improve the performance of collective communication by dividing it into inter-node and intra-node communications. For inter-node communication, point-to-point communication functions and for intra-node communication, shared memory technique will be used.

References

1. Balaji Venu, "Multi core processors: An overview", Department of Electrical Engineering and Electronics, University of Liverpool, Liverpool, UK, (2010).
2. Dennis Abts and John Kim, "High Performance Datacenter Networks: Architectures, Algorithms, and Opportunities", Synthesis Lectures on Computer Architecture, Book series, Pages 1–115, Vol. 6, No. 1, Morgan Claypool Publishers, (2011).
3. Mohammad Javad Rashti, Jonathan Green, Pavan Balaji, Ahmad Afsahi and William Gropp, "Multi-core and Network Aware MPI Topology Functions, EuroMPI'11, ISBN: 978-3-642-24448-3 (2011), pp. 50–60.
4. Joshua Hursey, Jerey M. Squyres and Terry Dontje, "Locality-Aware Parallel Process Mapping for Multi-Core HPC Systems", IEEE International Conference on Cluster Computing, (2011).
5. B. Brandfass, T. Alrutz and T. Gerhold, "Rank reordering for MPI communication optimization", 23rd Int. Conf. on Parallel Fluid Dynamics, (2013), pp. 372–380.
6. Torsten Hoefler, Emmanuel Jeannot, Guillaume Mercier, "An Overview of Process Mapping Techniques and Algorithms in High-Performance Computing", High Performance Computing on Complex Environments, Wiley, pp. 75–94, 2014, 978-1-118-71205-4 <hal-00921626>, (2014)
7. Alberto Fernndez and Sergio Gmez, "Solving Non-Uniqueness in Agglomerative Hierarchical Clustering Using Multi-dendrograms", Journal of Classification 25:43-65, ISSN: 0176–4268, (2008).
8. MPI, "MPI 2.2: A Message-Passing Interface Standard", MPI Forum, www.mpi-forum.org/docs/mpi-2.2/mpi22-report.pdf (2009).
9. B. W. Kernighan and S. Lin, "An efficient heuristic procedure for partitioning graphs", Bell System Technical Journal, 59 (1970), pp. 291–307.
10. T. Hoeer and M. Snir, "Generic topology mapping strategies for large-scale parallel architectures", Int. Conf. on Supercomputing, (2011), pp. 75–84.
11. T. Hatazaki, "Rank reordering strategy for MPI topology creation functions", Recent Advances in Parallel Virtual Machine and Message Passing Interface, EuroPVM/MPI 1998, Lecture Notes in Computer Science, vol 1497, pp. 188–195, (1998).
12. S. H. Bokhari, "On the mapping problem", IEEE Transactions on Computers, Volume 30 Issue 3, Pages 207–214, ISSN: 0018-9340, (1981).
13. S. Y. Lee and J. K. Aggarwal, "A mapping strategy for parallel processing", IEEE Transactions on Computers, Volume 36 Issue 4, Pages 433-442, ISSN: 0018–9340, (1987).
14. C. Sudheer and A. Srinivasan, "Optimization of the hop-byte metric for effective topology-aware mapping", 19th Int. Conf. on HPC, (2012), pp. 19.
15. P. Balaji, R. Gupta, A. Vishnu, and P. H. Beckman, "Mapping communication layouts to network hardware characteristics on massive-scale Blue Gene systems", Journal of Computer Science - Research and Development, Volume 26 Issue 3-4, Pages 247–256, ISSN:1865-2034 (2011).
16. B. E. Smith and B. Bode, "Performance effects of node mappings on the IBM Blue-Gene/L machine", Euro-Par 2005. Lecture Notes in Computer Science, vol 3648, Pages 1005–1013, ISBN: 978-3-540-28700-1, (2005).
17. J. L. Traff, "Implementing the MPI process topology mechanism", Proceedings of the 2002 ACM/IEEE conf. on Supercomputing, IEEE Computer Society Press, (2002), pp. 114.
18. Guillaume Mercier, Jerome Clet-Ortega, "Towards an efficient process placement policy for MPI applications in multi-core environments", Recent Advances in Parallel Virtual Machine and Message Passing Interface. EuroPVM/MPI 2009, pp. 104–115, ISBN: 978-3-642-03769-6 (2009).
19. Henri Casanova, Arnaud Giersch, Arnaud Legrand, Martin Quinson, Frederic Suter, "Versatile, Scalable, and Accurate Simulation of Distributed Applications and Platforms", Journal of Parallel and Distributed Computing, Volume 74, Issue 10, pp. 2899–2917, (2014).
20. http://mvapich.cse.ohio-state.edu/benchmarks/.

Mobile Agent-Based Frequent Pattern Mining for Distributed Databases

Yashaswini Joshi, Shashikumar G. Totad, R. B. Geeta and
P. V. G. D. Prasad Reddy

Abstract In today's world of globalization, business organizations produce information from many branch offices of their business while operating across the globe and hence lead to large chunk of distributed databases. There is an innate need to look at this distributed information that leverages the past, monitors the present, and predicts the future with accuracy. Mining large distributed databases using client–server model is time-consuming and sometimes impractical because it requires huge databases to be transferred over very long distances. Mobile agent technology is a promising alternative that addresses the issues of client–server computing model. In this paper, we have proposed an algorithm called MADFPM for frequent pattern mining of distributed databases that use mobile agents. We have shown that the performance of MADFPM is better compared to the conventional client–server approach.

Keywords Business intelligence · Distributed database · Data mining
Mobile agents · Frequent patterns · FP-tree · Parallel computing
Association rules · Decision-making

Y. Joshi (✉) · S. G. Totad
Department of Computer Science and Engineering, BVBCET, Hubballi, India
e-mail: yashaswinijoshi@gmail.com

S. G. Totad
e-mail: totad@bvb.edu

R. B. Geeta
Department of Information Science, KLEIT, Hubballi, India
e-mail: geetatotad@yahoo.co.in

P. V. G. D. Prasad Reddy
Department of CS & SE, Andhra University, Visakhapatnam, India
e-mail: prasadreddy.vizag@gmail.com

© Springer Nature Singapore Pte Ltd. 2018   77
S. Bhalla et al. (eds.), *Intelligent Computing and Information and Communication*,
Advances in Intelligent Systems and Computing 673,
https://doi.org/10.1007/978-981-10-7245-1_9

1 Introduction

Business intelligence is all about making right decision at right time. A lot of business intelligence is hidden in the enterprise database that helps in making wise decisions. Association rules ([1, 2]) are very useful for business promotions.

Furthermore, in the today's world of globalization and internet, related data is more often distributed and shared among multiple parties across remote locations. In case of large distributed databases, the conventional centralized client–server model of computation is time-consuming and sometimes impractical, because it requires huge databases to be transmitted over very long distances. Chattratichat et al. [3], in their work, explained how the distributed systems support parallel computation while describing the demands of the distribute databases for data mining. Mobile agents are a promising alternative technology attracting interest of researchers from the fields of distributed systems that address the issues of centralized client—server computing model. In this paper, we propose a frequent pattern mining approach for distributed databases that uses mobile agents. The proposed approach not only minimizes the data transmission but also reduces time required to compute frequent patterns by the way of inherent parallel execution.

2 Related Works

The notations given by Ruan et al. [4] are as follows: *LFI* denotes the set of local frequent itemsets, *GFI* is the set of global frequent itemsets, and *CGFI* denotes the set of candidate *GFI*. The works proposed by Kulkarni et al. [5] and Ruan et al. [4] obtain frequent patterns by computing *LFIs, GFIs, and CGFI* and performing parallel computations. The parallel algorithm for Mining Frequent Itemsets (PMFI) proposed by You-Lin Ruan et al. constructs FP-tree at each of the distributed sites DB_i. It computes *LFI* and sends it to the central site. The central site receives *LFIs* from all distributed sites and computes *GFI* and *CGFI*. It sends *CGFI* back to each of the sites. An itemset $X \subseteq CGFI$ may be frequent or infrequent at a distributed site. Each of the distributed sites computes support count of all the itemsets from *CGFI* that are infrequent at that site and sends it back to the central site for consideration to include into the *GFI*.

Realizing the negative effect of heavy network traffic on the performance of distributed/parallel data mining applications, in their work proposed a programmable data parallel to primitive D-DOALL, a generic mechanism to schedule a set of independent tasks on a network of workstations. A framework called MAD-ARM is proposed by Saleem Raja and George Dharma Prakash Raj [6], which attempts to reduce the communication overhead. Research work published by Liu [7] demonstrates how set of mobile agents having specialized expertise can be used to gather relevant information and how they cooperate with each other to arrive at timely decisions, in dealing with various supply chain finance–business analysis scenarios. The technique proposed by Kulkarni et al. [5] also computes *LFIs* concurrently like PMFI algorithm, but using mobile agents. Also, Folorunso et al. [8] proposed an improved cost model for mining association rule for distributed databases.

3 Proposed MADFPM Algorithm

The two approaches are proposed by Kulkarni et al. [5] and Ruan et al. [4] and discussed above in Sect. 2, indulge thrice in transfers of itemsets between central site and distributed sites or between distributed sites themselves. This transfer of itemsets, particularly the transfer of *LFIs* between the distributed sites, results in heavy network traffic leading to network congestion. This in turn reduces the overall computation speed of the algorithm, as the distributed sites will have to wait long for the receipt of the *LFIs* from all other neighbors. The technique that we present in this section transfers itemsets only once and avoids the multiple transfers of itemsets.

The technique proposed has a main component called *Parent* that performs operations of central site. It is responsible for creating mobile agents, dispatching them to remote distributed sites, receiving results from retracted mobile agents, and processing the received results. The *Parent* creates and dispatches mobile agents twice in its course of action. First time, it dispatches mobile agents called SupportCount_MobileAgent to distributed sites to get Local Support Count (*LSC*) of items. It computes Global Support Count (*GSC*) by adding the *LSCs* of corresponding items received from distributed sites. Then, it computes L-order from the computed *GSCs* of the items. L-order is an ordering of items with support count $\geq$ minSup that is arranged in the decreasing order of item support count. Second time, the *Parent* dispatches mobile agents called FP-tree_MobileAgents to each of the distributed sites for getting frequent itemsets. FP-tree_MobileAgents carry the L-order computed by the *Parent* while migrating. Each mobile agent migrates to its respective distributed site and constructs FP-tree by accessing the distributed database located there. Mobile agents explore the constructed FP-tree [9–11] and retract back to the *Parent* with *LocalDisjoint Matrix* (*LDM*) containing "*Composite Itemsets*". A *Composite Itemset* contains multiple itemsets hidden within it. It represents a path from a child of the root node to a leaf node of the FP-tree. Below-mentioned property and the theorem, defined by Tanbeer et al. [12], are equally applicable and hold good here as well.

Property *The total frequency count of any node in the FP-tree is greater than or equal to the sum of the total frequency counts of its children.*

Theorem *If $C = \{C_1 \cup C_2, \ldots, C_x\}$ is a pattern where x is a positive integer and C_j, $j \in [1, x]$ is a sub-pattern of C, then*

$\sup(C) \leq \sum_i \min\{\sup_i(C_1), \sup_i(C_2), \ldots, \sup_i(C_x)\}$,
where $i \in [1, n]$ and n = number of sites.

Transmission of the *LDM* containing *Composite Itemsets* reduces the network traffic (in terms of number of itemsets), as the composite itemset contains multiple itemsets hidden within it. Finally, it computes *Global Disjoint Matrix* (*GDM*) by merging all the *LDMs* received from FP-tree_MobileAgents. This *GDM* that contains composite itemsets can be further used for obtaining frequent itemsets.

Further, here in our approach, as we avoid computation of *CGFI*, we also avoid transfer of *CGFI* back to distributed sites. Instead, we sent mobile agents to distributed sites to get back *LSC* of the items. Here again, much of the network traffic is reduced, as support count of the items only is transferred instead of *CGFI* which is a huge set of itemsets. Algorithm given below explains the sequence of operations performed by the Parent.

The FP-tree_MobileAgent is a mobile agent that migrates to the specified remote distributed site carrying with it the information required and the code to be executed at the remote location. After migrating, FP-tree_MobileAgent connects to the remote database, accesses the transactions and constructs disjoint matrix representation of FP-tree of the remote database.

FP-tree_MobileAgent may store its *LDM* into a file and further transfer it to the parent by the way of FTP, if *LDM* is too large. Algorithm given below explains the sequence of operations performed by an FP-tree_MobileAgent.

4   Performance Analysis

The proposed MADFPM algorithm mines distributed database by performing independent and parallel mining of the partitions DB_1, $DB_2, \ldots$, DB_n at distributed sites S_1, $S_2, \ldots$, S_n, respectively, and thus adopts divide-and-conquer technique. It also avoids computation and hence transfer of *CGFI*, thereby reducing the cost of computation and communication significantly. Tests have been carried out by running the algorithms on the IBM Almaden Quest research group's dataset.

4.1   Comparison with Client–Server Approach

Client–server approach for distributed frequent pattern mining involves transfer of all the distributed database partitions DB_1, $DB_2, \ldots$, DB_n located at sites S_1, $S_2, \ldots$, S_n to the central site and then building FP-tree/disjoint matrix at the central site for obtaining *GFI*. Therefore, time, T_{CS}, required to construct FP-tree/disjoint matrix in client–server approach is

$$T_{\text{CS}} = \sum_{i=1}^{n} \text{Time to transfer each database DB}_i + T_{\text{B-GDM}}$$

where $T_{\text{B-GDM}}$—Time to build global disjoint matrix.

All transactions take approximately the same time to transfer:

$$T_{\text{CS}} = \sum_{i=1}^{n} \sum_{j=1}^{p} \text{Time to transfer each transaction} + T_{\text{B-GDM}}$$

Let, $T_{\text{T-Tranx}}$ be the time to transfer each transaction. Then,

$$T_{\text{CS}} = n \times p \times T_{\text{T-Tranx}} + T_{\text{B-GDM}} \qquad (1)$$

(Note: $n \times p = D$, Total distributed database)

Let T_{MADFPM} be the time required to construct disjoint matrix for the given distributed database DB, using MADFPM approach:

$$\therefore T_{\text{MADFPM}} = n \times T_{\text{S-MA-LSC}} + T_{\text{C-LSC}} + n \times T_{\text{R-LSC}} + n \times T_{\text{S-MA-LDM}}$$
$$+ T_{\text{C-LDM}} + n \times T_{\text{R-LDM}} + T_{\text{M-GDM}}, \qquad (2)$$

where $T_{\text{S-MA-LSC}}$—Time to send mobile agent to a distributed site to compute LSC, $T_{\text{C-LSC}}$—Time to compute LSC for all items at a distributed site, $T_{\text{R-LSC}}$—Time to receive LSC from a distributed site, by the central site, $T_{\text{S-MA-LDM}}$—Time to send mobile agent to a distributed site to compute LDM, $T_{\text{C-LDM}}$—Time to compute LDM at a distributed site, $T_{\text{R-LDM}}$—Time to receive LDM from a distributed site, by the central site, and $T_{\text{M-GDM}}$—Time to merge n $LDMs$ at central site.

Computations of LSC and LDM are done carried out simultaneously at all sites. Rewriting the equation by moving the related terms in Eq. (2) toward one end, we get

$$T_{\text{MADFPM}} = n \times T_{\text{S-MA-LSC}} + n \times T_{\text{R-LSC}} + n \times T_{\text{S-MA-LDM}} + n \times T_{\text{R-LDM}}$$
$$+ T_{\text{C-LSC}} + T_{\text{C-LDM}} + T_{\text{M-GDM}}$$

But $T_{\text{R-LSC}} = m \times$ Time to receive support count of one item.

Assuming that it takes 1 unit of time to receive support count of one item, then, $T_{\text{R-LSC}} = m$.

But $T_{\text{C-LSC}} + T_{\text{C-LDM}} + T_{\text{M-GDM}} =$ Time to build GDM $= T_{\text{B-GDM}}$

$$\therefore T_{\text{MADFPM}} = n + (n \times m) + n + (n \times T_{\text{R-LDM}}) + T_{\text{B-GDM}}$$
$$\text{i.e., } T_{\text{MADFPM}} = 2n + (n \times m) + n \times T_{\text{R-LDM}} + T_{\text{B-GDM}} \qquad (3)$$

But

$$T_{\text{R-LDM}} = q \times \text{Time to transfer a Composite Itemset}$$
$$= q \times T_{\text{T-CI}},$$

where q is the number of composite itemsets.

As *Composite Itemsets* are formed by merging transactions, we may suitably assume $T_{\text{T-CI}} \cong T_{\text{T-Tranx}}$:

$$\therefore T_{\text{R-LDM}} = q \times T_{\text{T-Tranx}}$$

Substituting this in Eq. (2), we get

$$T_{\text{MADFPM}} = (2 \times n) + (n \times m) + (n \times q \times T_{\text{T-Tranx}}) + T_{\text{B-GDM}} \qquad (4)$$

where m and n are the constants for a given distributed database system.

$$\therefore T_{\text{MADFPM}} \cong n \times q \times T_{\text{T-Tranx}} + T_{\text{B-GDM}} \qquad (5)$$

Comparing Eqs. (1) and (5), it is evident that the run times of client–server and distributed systems for frequent patterns mining are governed by the parameters p and q, respectively. p is the number of transactions (i.e., database size) and q is the number of *Composite Itemsets*. It is to be noted that q is always less than p. Hence, as the new transactions are added, most of the time only frequencies of existing itemsets increase and not the number of itemsets. Figure 1 below represents a graph of number of transactions of databases versus number of itemsets of the FP-tree of transaction databases. It may be observed from the graph that, as the database size increases, the number of itemsets also increases. But increase in number of itemsets, which is little less compared to the increase in database size initially, drops further as the database size increases which eventually reduces to half of the size of the database. Hence, as *Composite Itemset* contains multiple itemsets hidden within it, as the number of transactions increases, number of *Composite Itemsets* also. Therefore, MADFPM always takes much less time than its client–server version.

Fig. 1 Number of itemsets versus number of transactions

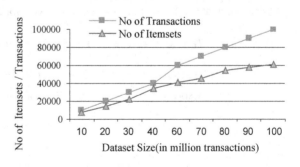

4.2 Comparison with PMFI-A Approach

As discussed in Sect. 2, PMFI-A, while distributed sites send their *LFIs* to the neighbors, they also receive *LFIs* from the neighbors at the same time, that is, distributed sites perform overlapped sending and receiving operations. Further, each distributed site starts computation of *PGFI* only after it completes sending and receiving of *LFIs*, whereas the central site only receives *LFIs* (and does not send). Hence, probability of central site completing its computations of GFI and *CGFI* before distributed sites complete their computation of *PGFI* is very high. This fact is ignored while measuring performance. Therefore, sending of *PGFI* is sequential operation done after the distributed sites complete computation of their *PGFI* and not a parallel operation mkl;';l.,;'. Hence, we have considered the time required to receive *PGFIs* by the central site (from all distributed sites) for calculating overall time required to compute *GFI*. Let $T_{\text{PMFI-A}}$ be the time required to construct *GFI* for the given distributed database DB, using PMFI-A approach.

$$\therefore T_{\text{PMFI-A}} = n \times T_{\text{S-MA-LFI}} + T_{\text{C-LFI}} + n \times T_{\text{R-LFI}} + T_{\text{C-C/GFI}} + n \times T_{\text{R-PGFI}} + \qquad (6)$$
$$n \times T_{\text{PGFI->GFI}}$$

where $T_{\text{S-MA-LFI}}$—Time to send mobile agent to a distributed site to compute *LFI*, $T_{\text{C-LFI}}$—Time to compute *LFI* at a distributed site, $T_{\text{R-LFI}}$—Time to receive *LFI* from a distributed site, by the central site, $T_{\text{C-C/GFI}}$—Time to compute *GFI and CGFI* at central site, $T_{\text{R-PGFI}}$—Time to receive PGFI from a distributed site, by the central site, and $T_{\text{PGFI->GFI}}$—Time to compute final GFI using PGFI count.

Comparing Eqs. (6) and (2), we have

$$T_{\text{S-MA-LFI}} \cong T_{\text{S-MA-LSC}}, T_{\text{C-LFI}} \cong T_{\text{C-LSC}} + T_{\text{C-LDM}},$$
$$T_{\text{R-LFI}} \cong T_{\text{R-LDM}}, T_{\text{PGFI->GFI}} \cong T_{\text{M-GDM}}.$$

Furthermore, $T_{\text{C-C/GFI}} + n \times T_{\text{R-PGFI}} \gg n \times T_{\text{R-LSC}} + n \times T_{\text{S-MA-LDM}}$, because *LSC* is an array of items' support count that depends on the number of items "m", which is very small compared to size of *PGFI*. *PGFI* is a set of itemsets where each itemset contains many items. Hence, transfer of *PGFI* (i.e., $T_{\text{R-PGFI}}$) takes much more time compared to that of *LSC*. It needs no explanation to say that the computation of *CGFI and GFI*, using all of the *LFIs*, received from distributed sites, at the central site taking more time compared to the time required to dispatch "n" mobile agents [13]. Further, though we have assumed that $T_{\text{R-LFI}} \cong T_{\text{R-LDM}}$, considering the fact that the *Composite Itemset* contains multiple itemsets hidden within it, size of the *LDM* is much less compared to that of *LFI*.

$$\therefore T_{\text{MADFPM}} \ll T_{\text{PMFI-A}}.$$

In PMFI-A, every distributed site transfers set of itemsets twice to the central site in the course of execution, namely *LFI* first time and then *PGFI* later. This transfer of itemsets twice contributes to network traffic apart from contributing to the computational time. Whereas, in case of MADFPM, each distributed site transfers LDM, which is a set of *Composite Itemsets*, only once in the course of execution. Hence, the performance of MADFPM is superior to PMFI-A.

5 Conclusion

Our proposed algorithm MADFPM for frequent pattern mining of distributed databases uses mobile agents and found to be better in terms of performance. Performance of MADFPM is better than conventional client–server approach as preprocessed compact data in the form of disjoint matrix is transferred to the central site rather than moving the entire data to the central site and then processing it there. Performance of MADFPM is also better compared to the other mobile agent-based approaches, PMFI and PMFI-A, as it reduces the computational cost and network traffic. Computational cost is reduced because the distributed sites compute local itemsets in the form of LDM only once, unlike PMFI-A which computes local itemsets (at the distributed sites) twice, in the form of *LFI* and *PGFI*. Network traffic is reduced because only the *LDMs* are transferred from distributed sites to the central site once. Moreover, in case of PMFI-A, each distributed site transfers *LFI* to all neighbors along with the central site contributing to the network traffic further.

References

1. Agrawal R., Imielinski, T., and Swami, A. (1993), "Mining association rules between sets of items in large databases". In Proc. of ACM-SIGMOD, (SIGMOD'93), pp. 207–216.
2. Paul S. Bradley, J. E. Gehrke, Raghu Ramakrishnan and Ramakrishnan Srikant (2002), 'Philosophies and Advances in Scaling Mining Algorithms to Large Databases". *Communications of the ACM.*
3. Chattratichat, J., Darlington, J, et al. (1999), "An Architecture for Distributed Enterprise Data Mining", 7th Intl. Conf. on High Performance Computing and Networking.
4. You-Lin Ruan, Gan Liu, Quin-Hua Li (2005), "Parallel Algorithm for Mining Frequent Items", Proceedings of the Fourth International Conference on Machine Learning and Cybernetics, Guangzhou, pp-18–21.
5. U.P. Kulkarni, P.D. Desai, Tanveer Ahmed, J.V. Vadavi, A.R. Yardi (2007), "Mobile Agent Based Distributed Data Mining". International Conference on Computational Intelligence and Multimedia Applications, pp. 18–24.
6. Saleem Raja, George Dharma Prakash Raj, (2013), "Mobile Agent based Distributed Association Rule Mining", International Conference on Computer Communication and Informatics (ICCCI), 2013.

7. LIU Xiang (2008), "An Agent-based Architecture for Supply Chain Finance Cooperative Context-aware Distributed Data Mining Systems". 3rd International Conference on Internet and Web Applications and Services.

8. Ogunda A.O., Folorunso O., Ogunleye G.O., (2011), "Improved cost models for agent-based association rule mining in distributed databases, Anale. SeriaInformatică. Vol. IX fasc. 1 – 2011.

9. J. Han, J. Pei, Y. Yin and R. Mao (2004), "Mining Frequent Patterns without Candidate Generation: A Frequent-Pattern Tree Approach". Data Mining and Knowledge Discovery, 8 (1), pp. 53–87.

10. Keshavamurthy B.N., Mitesh Sharma and DurgaToshniwal (2010), "Efficient Support Coupled Frequent Pattern Mining Over Progressive Databases", International Journal of Database Systems, Vol.-2, No-2, pp-73–82.

11. Mengling Feng, Jinyan Li, Guozhu Dong, Limsoon Wong (2009), "Maintenance of Frequent Patterns: A Survey", published in IGI Global, XIV Chapter, pp-275–295.

12. Syed K. Tanbeer, C. F. Ahmed, B-S Jeong (2009), "Parallel and Distributed Algorithms for Frequent PatternMining in Large Databases". IETE Technical Review, Vol. 26, Issue 1, pp-55–66.

13. Raquel Trillo, Sergio Ilarri, Eduardo Mena (2007), "Comparison and Performance Evaluation of Mobile Agent Platforms", Third International Conference on Autonomic and Autonomous Systems (ICAS'07), pp. 41.

A Hybrid Approach for Preprocessing of Imbalanced Data in Credit Scoring Systems

Uma R. Salunkhe and Suresh N. Mali

Abstract During the last few years, classification task in machine learning is commonly used by various real-life applications. One of the common applications is credit scoring systems where the ability to accurately predict creditworthy or non-creditworthy applicants is critically important because incorrect predictions can cause major financial loss. In this paper, we aim to focus on skewed data distribution issue faced by credit scoring system. To reduce the imbalance between the classes, we apply preprocessing on the dataset which makes combined use of random re-sampling and dimensionality reduction. Experimental results on Australian and German credit datasets with the presented preprocessing technique has shown significant performance improvement in terms of AUC and F-measure.

Keywords Credit scoring · Imbalanced data · Preprocessing · Classifier ensemble

1 Introduction

Classification techniques in data mining are applied to diverse data in order to predict the category to which the given instance belongs. With the rapid growth of data, usage of classification techniques by various applications has also increased. One of the most common applications of classification is credit scoring systems where applicants are classified as creditworthy or non-creditworthy based on the analysis of their characteristics. Credit risk management is the major concern for the financial organizations so that the probable risk associated with repayments can be decreased. Credit scoring systems are designed to handle this issue by categorizing

U. R. Salunkhe (✉)
Department of Computer Engineering, Smt. Kashibai Navale College
of Engineering, Pune, Maharashtra, India
e-mail: umasalunkhe@yahoo.com

S. N. Mali
Sinhgad Institute of Technology and Science, Pune, Maharashtra, India
e-mail: snmali@rediffmail.com

© Springer Nature Singapore Pte Ltd. 2018 87
S. Bhalla et al. (eds.), *Intelligent Computing and Information and Communication*,
Advances in Intelligent Systems and Computing 673,
https://doi.org/10.1007/978-981-10-7245-1_10

the applicants into good and bad applicants [1]. Traditionally, human experts used to do this categorization by applying their experience and following the guidelines given by the organization. Basel Committee on Banking Supervision has also specified some guidelines for this purpose [2]. Five characteristics namely character, capacity, capital, collateral, and conditions were considered for making the decision of whether applicant is good or bad [3]. But this could lead to incorrect decisions and decision was expert dependent. Hence, those traditional models are replaced by credit scoring models that have the following advantages:

1. Credit scoring models are cheaper than traditional methods.
2. The decision-making is faster.
3. Those models are consistent in decision-making.
4. Those models are adaptive to policy changes.

2 Related Work

Marques et al. [1] presented two-level classifier ensembles for credit scoring systems in order to improve performance in terms of AUC compared to single-level ensemble. The proposed method introduces diversity using different training datasets as well as different feature sets. Authors conclude that two-level ensemble that combines Bagging and Rotation Forest gives the best performance. Wu et al. [4] proposed a prediction model based on hybrid re-sampling approach that combines undersampling with SMOTE and AdaBoost. Presented method shows performance improvement in terms of evaluation measures like AUC, accuracy, and G_mean.

Xiao et al. [5] proposed a classifier ensemble technique to improve the performance of credit scoring system by diversifying the base classifiers. K-means clustering divides instances of each class into different clusters. Clusters from two classes are combined to form various diverse training subsets which are then used to train base classifiers. Abellan et al. [6] presented an experimental study wherein different classifier ensembles are tested with different base classifiers in order to select the best base classifier. Authors suggest that credal decision tree (CDT) which is very unstable classifier is a good choice for classifier ensembles.

Pozzolo et al. [7] made an experimental comparison of many existing classification techniques on fraud detection datasets. Their focus is on three important issues: unbalancedness, nonstationary, and assessment. Study recommends average precision, AUC, and precision rank as preferable performance measures. Oreski et al. [8] introduced hybrid genetic algorithm with neural network that proves to increase the accuracy and scalability of credit scoring system. Presented method first removes redundant attributes by dimensionality reduction which is then further analyzed for refinement and applied with incremental learning.

Han et al. [9] addressed the issue of curse of dimension faced by logistic regression and support vector machine. The orthogonal transform process is introduced for the feature extraction. Grid search method is used and then dimensionality reduction is done. Experimental results show better results in terms of sensitivity, specificity, and prediction accuracy. Kim et al. [10] resolved multi-co linearity problem using genetic algorithm-based technique.

Xiao et al. [11] proposed a dynamic classifier ensemble method for imbalanced data (DCEID) that makes combined use of classifier ensemble and cost-sensitive learning approaches. Dynamic classifier ensembles use either dynamic classifier selection (DCS) or dynamic ensemble selection (DES). The DCEID adaptively chooses the best of these two for each test customer. Also, a new evaluation criterion called cost-sensitive symmetric regularity criterion (CS-SRC) is constructed.

3 Proposed Model

Credit scoring systems suffer from the issue of skewed distribution of dataset, i.e., the available training dataset has imbalanced distribution of data. This is because customers that are likely to be defaulter for making repayments are very less compared to the customers that are creditworthy. The classification model that is trained with such imbalanced data is likely to make biased decision toward the creditworthy class. This is due to unavailability of sufficient data that belongs to "bad" (non-creditworthy) applicant's class. Thus, the applicant who is likely to be defaulter in repayments is classified as "good" (creditworthy) and will be granted the credit. This can lead to major financial loss to the organization.

In order to handle above issue, classification model should be designed such that it tries to focus on the accuracy of the class with less number of samples, i.e., minority class. Traditional classifiers are not effective when applied to imbalanced dataset. Recent studies present different ways to handle imbalanced dataset which can be categorized into classes, namely data-level approaches, algorithm-level approaches, cost-sensitive algorithms, and classifier ensemble approaches.

In this paper, we present an approach that can handle the imbalance between two classes of credit scoring system by making use of data-level approach. To achieve this, initially noisy instances and instances with missing values are identified. Then, re-sampling of training data is done and imbalance between them is reduced to some extent. The term "Imbalance Ratio" is used to indicate the imbalance between classes and is defined as ratio of number of samples of majority class to that of minority class.

Figure 1 represents the experimental setup in which initially original credit scoring dataset is provided as input to the preprocessing stage. It involves handling instances with missing values and noisy instances. The instances which are having more than half nearest neighbors of different classes are known as noisy instances and are removed from the dataset. Then, the dimension reduction is done in order to remove the redundant attributes. The aim is to select the appropriate subset of

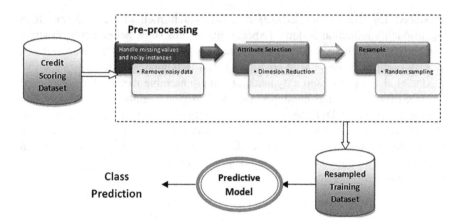

Fig. 1 Proposed method

attributes that is sufficient to train the model [8]. This is done to overcome the issues raised by the existence of redundant attributes as listed below [9].

1. If model is trained with redundant attributes, it may predict incorrectly as predictions are skewed by those attributes.
2. Redundant attributes may lead to over-fitting issue.
3. Removal of such attributes reduces the size of the training dataset which in turn reduce the time required to train model.

The output of previous step is applied with random sampling with replacement. This generates the re-sampled credit scoring dataset which is then provided as input to train the classifier model.

4 Experimental Design

This section discusses the experimental design of the work that is presented in this paper.

4.1 Experimental Setup

The experiments were carried out using Weka environment with its default parameters. In this work, we implemented a preprocessing technique that selects suitable subset of attributes of the training dataset and then applies re-sampling to it. Preprocessed data is provided as input to the classification algorithm. Different individual classifiers such as J48, Naïve Bayes, Multilayer Perceptron (MLP), and

SVM are used for the experimentation. Also, ensemble techniques namely AdaBoostM1, Bagging, random subspace, and rotation forest are tested. All experiments were carried out using tenfold cross-validation. Results of the different classifiers when trained with preprocessed data are compared with the results when classifier is trained on original imbalanced training dataset.

4.2 Datasets

For experimentation, we have chosen two publicly available credit scoring datasets, namely Australian and German datasets from UCI repository. Details of those datasets are given in Table 1.

4.3 Evaluation Parameters

Classification algorithms can be evaluated by using different evaluation parameters such as accuracy, G-mean, F-measure, AUC, Type-I error, etc. [12]. To derive these parameters, we need to use confusion matrix for binary classification problem which is shown in Table 2.

Previous studies [13] suggest that AUC is more appropriate measure than accuracy for the datasets that have imbalanced distribution of data. In this paper, we have used AUC and F-measure as our evaluation measures.

AUC (Area under ROC curve)
AUC can be defined as arithmetic average of the mean predictions for each class [1]. AUC is represented in Eq. (1):

$$AUC = \frac{\text{sensitivity} + \text{specificity}}{2}, \tag{1}$$

where sensitivity and specificity are computed using Eqs. (2) and (3),

$$\text{Sensitivity} = \frac{TP}{TP + FN} \tag{2}$$

$$\text{Specificity} = \frac{TN}{FP + TN} \tag{3}$$

Table 1 Characteristics of the datasets used in the experiments

Dataset	# Examples	# Minority	# Majority	# Attr.
Credit-a	690	307	383	14
Credit-g	1000	300	700	21

Table 2 Confusion matrix
for a two-class problem

	Predicted as positive	Predicted as negative
Positive class	True positive (TP)	False negative (FN)
Negative class	False positive (FP)	True negative (TN)

F-measure F-measure represents accuracy by using combination of two measures, precision and recall. It is computed by using [14] Eq. (4),

$$F\text{ - measure} = \frac{2 \times \text{precision} \times \text{recall}}{\text{precision} + \text{recall}}, \tag{4}$$

where precision and recall are computed using Eqs. (5) and (6),

$$\text{Precision} = \frac{TP}{TP + FP} \tag{5}$$

$$\text{Recall} = \frac{TP}{TP + FN} \tag{6}$$

5   Results and Discussion

The experimental results in this paper are presented in two subsections. The first subsection describes the performance in terms of AUC with different classifier models. Results of the classifiers are compared between the one that is trained with re-sampled credit scoring dataset and another trained with dataset without re-sampling. The next section shows performance in terms of F-measure. Table 3 summarizes the AUC performance of different classification models on two credit scoring datasets.

Figure 2 plots the performance of the different classifier models in terms of AUC when they are trained with original credit scoring dataset and then with preprocessed dataset.

Table 4 summarizes the performance of different classification models using F-measure as evaluation measure on two credit scoring datasets.

Figure 3 plots the performance of the different classifier models in terms of F-measure when they are trained with original imbalanced credit scoring dataset and then with preprocessed dataset.

Analysis of the graphs shown in Figs. 2 and 3 clearly shows that preprocessing has enhanced the performance of most of the tested classifiers. Although the

Table 3 Performance evaluation using AUC on the credit scoring datasets

Classification algorithm	AUC			
	Credit-a		Credit-g	
	Original dataset	Re-sampled dataset	Original dataset	Re-sampled dataset
J48	0.867	0.879	0.639	0.716
Naïve Bayes	0.892	0.904	0.787	0.737
Multilayer perceptron	0.89	0.936	0.733	0.741
Logistic	0.906	0.935	0.715	0.743
SVM	0.853	0.853	0.671	0.681
Rotation forest	0.906	0.954	0.777	0.786
Bagging	0.919	0.936	0.752	0.761
AdaBoostM1	0.902	0.963	0.724	0.742
Random subspace	0.912	0.952	0.754	0.744

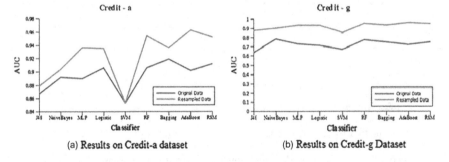

(a) Results on Credit-a dataset   (b) Results on Credit-g Dataset

Fig. 2 Performance evaluation using AUC on the credit scoring datasets

improvement in AUC or F-measure values seems smaller, it is beneficial for credit scoring systems where correct identification of the customer who is likely to be defaulter is critically important and may save huge financial loss. While experimenting, we also recorded the training time for those classifiers in different scenarios. It helped us to conclude that classifier ensemble techniques require significantly higher training time compared to individual classifiers. When we applied proposed preprocessing to the dataset, we could achieve considerable performance improvements in individual classifiers. Hence, it could be beneficial to save the time and achieve good performance with the help of combined use of the proposed preprocessing and individual classifier method. Multilayer preceptron classifier model is exception to this as it requires very high time for training than that required for the classifier ensemble technique.

Table 4 Performance evaluation using F-measure on the credit scoring datasets

Classification algorithm	F-measure			
	Credit-a		Credit-g	
	Original dataset	Re-sampled dataset	Original dataset	Re-sampled dataset
J48	0.861	0.883	0.692	0.73
Naïve Bayes	0.767	0.77	0.746	0.753
Multilayer perceptron	0.831	0.874	0.715	0.75
Logistic	0.854	0.858	0.744	0.751
SVM	0.847	0.847	0.741	0.738
Rotation forest	0.862	0.917	0.737	0.746
Bagging	0.864	0.89	0.731	0.741
AdaBoostM1	0.843	0.91	0.694	0.748
Random subspace	0.863	0.901	0.696	0.673

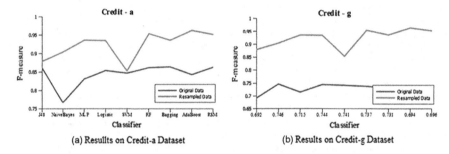

(a) Results on Credit-a Dataset   (b) Results on Credit-g Dataset

Fig. 3 Performance evaluation using F-measure on the credit scoring datasets

6 Conclusion

We have presented an approach to enhance performance of credit scoring systems which face challenge of skewed dataset. In order to resolve this issue, we initially applied preprocessing to the imbalanced credit scoring dataset and tried to convert it to balanced form to some extent. While doing this, redundant attributes were removed from the feature space which helped to improve performance. Then, it was given as input to the classification algorithm and has shown significant improvements in classifying the customers correctly. Another concern of this work was to test this model on different classifiers and compare their behavior. Comparison of performance improved and amount of time to train the model indicates that it is better to achieve considerable improvement with less training time for individual classifiers.

References

1. Marqués, A. I., Vicente García, and Javier Salvador Sánchez. "Two-level classifier ensembles for credit risk assessment." Expert Systems with Applications 39.12 (2012): 10916–10922.
2. BIS. Basel III: a global regulatory framework for more resilient banks and banking systems. (2011). Basel Committee on Banking Supervision, Bank for International Settlements, Basel. ISBN print: 92-9131-859-0. <http://www.bis.org/publ/bcbs189.pdf>.
3. Marqués, A. I., Vicente García, and Javier Salvador Sánchez. "Exploring the behaviour of base classifiers in credit scoring ensembles." *Expert Systems with Applications* 39.11 (2012): 10244–10250.
4. Wu, Xiaojun, and SufangMeng. "E-commerce customer churn prediction based on improved SMOTE and AdaBoost." *Service Systems and Service Management (ICSSSM), 2016 13th International Conference on*. IEEE, 2016.
5. Xiao, Hongshan, Zhi Xiao, and Yu Wang. "Ensemble classification based on supervised clustering for credit scoring." *Applied Soft Computing* 43 (2016): 73–86.
6. Abellán, Joaquín, and Javier G. Castellano. "A comparative study on base classifiers in ensemble methods for credit scoring." *Expert Systems with Applications* 73 (2017): 1–10.
7. Dal Pozzolo, Andrea, et al. "Learned lessons in credit card fraud detection from a practitioner perspective." *Expert systems with applications* 41.10 (2014): 4915–4928.
8. Oreski, Stjepan, and Goran Oreski. "Genetic algorithm-based heuristic for feature selection in credit risk assessment." *Expert systems with applications* 41.4 (2014): 2052–2064.
9. Han, Lu, Liyan Han, and Hongwei Zhao. "Orthogonal support vector machine for credit scoring." *Engineering Applications of Artificial Intelligence* 26.2 (2013): 848–862.
10. Kim, Myoung-Jong, and Dae-Ki Kang. "Classifiers selection in ensembles using genetic algorithms for bankruptcy prediction." *Expert Systems with applications* 39.10 (2012): 9308–9314.
11. Xiao, Jin, et al. "Dynamic classifier ensemble model for customer classification with imbalanced class distribution." *Expert Systems with Applications* 39.3 (2012): 3668–3675.
12. Salunkhe, Uma R., and Suresh N. Mali. "Classifier Ensemble Design for Imbalanced Data Classification: A Hybrid Approach." *Procedia Computer Science* 85 (2016): 725–732.
13. Liu, Xu-Ying, Jianxin Wu, and Zhi-Hua Zhou. "Exploratory under-sampling for class-imbalance learning." Systems, Man, and Cybernetics, Part B: Cybernetics, IEEE Transactions on 39.2 (2009): 539–550.
14. Kamalloo, Ehsan, and Mohammad SanieeAbadeh. "An artificial immune system for extracting fuzzy rules in credit scoring." *Evolutionary Computation (CEC), 2010 IEEE Congress on*. IEEE, 2010.

Vision-Based Target Tracking Intelligent Robot Using NI myRIO with LabVIEW

Anita Gade and Yogesh Angal

Abstract Robots are worked to do tasks that are risky to people, for example, defusing bombs and discovering survivors in unsteady environments and investigation. The rising exploration field on scaled-down programmed target following robot is given significance in hazardous, military and industrial areas for navigation, observation, safety, and goal acknowledgment by image processing. The real-time vision-based technique is used for target tracking. The objective of this system is to develop the proto robot capable of following a target using image processing. Normally, the way of a mobile robot is controlled by a pre-identified data about objects in nature. In this developed system, the robot tracks the desired destination goal in a systemized way.

Keywords Camera · LabVIEW · myRIO · Target tracking · Vision

1 Introduction

The world we live in is a tridimensional one. That is the reason it is critical for artificial vision framework to see the world in three dimensions. Ordinary sensors, utilized as a part of video cams, are just bidimensional. In spite of the fact that, by consolidating at least two cameras in a stereoscopic design, the tridimensional parts of the world might be evaluated, the development speaks to on its hand a critical wellspring of data over a succession of images. Consolidating the information separated from tridimensional space utilizing the stereovision with those in regards to development in bidimensional arrangement permits acquiring some performing arrangement of tracking objects [1]. The video stream is handled by

A. Gade (✉) · Y. Angal
Department of E&TC Engineering, JSPM's Bhivarabai Sawant Institute of Technology & Research, SPPU, Pune, India
e-mail: anita.gade@yahoo.co.in

Y. Angal
e-mail: yogeshangal@yahoo.co.in

© Springer Nature Singapore Pte Ltd. 2018
S. Bhalla et al. (eds.), *Intelligent Computing and Information and Communication*,
Advances in Intelligent Systems and Computing 673,
https://doi.org/10.1007/978-981-10-7245-1_11

LabVIEW-VDM to decide the way that the robot must take after to achieve the objective. Depending upon the adjustment in position of target, LabVIEW recalculates the way and commands are sent to the robot to move toward the new area of target. The elements of the moving target like shape, size, and color can be observed for tracking it. In this venture, color and shape of the target were utilized to track the moving target. In this paper, we will present an incorporated vision-based strategy for adaptable robots that are able to track the target using monocular camera images.

2 Problem Definition

In a substitute method, a robot is furnished with sequence of images within space. By differentiating these pre-recorded pictures and the camera pictures, the robot is able to decide its area [2]. Previous methodologies have identified ultrasonic sensors or IR sensors for target tracking but by using these sensors, we cannot estimate color, size, and shape of object. Considering this problem, we have implemented this system using stereovision camera for target detection and tracking.

3 System Overview

Target recognition and tracking are carried out in LabVIEW using vision assistant module. Two cameras are used to acquire real target image. NI myRIO device is control and processing element of this system [3]. Four-wheel drive robot is used to track the object. Opt-interrupter encoder is used to count the pulses. Camera is calibrated to acquire image of the entire arena (Fig. 1).

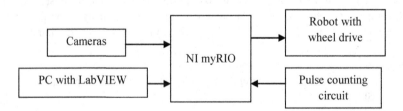

Fig. 1 System overview

4 Algorithm

Two cameras (Left and Right) have been configured for continuous acquisition of image. The hypothesis made for ideal stereovision may not be done for the application of stereovision in the real world. Even the best cameras and lenses will present certain distortion level of acquired image, so in order to compensate this distortion stereovision system needs to be calibrated [4]. Calibration processes the use of matrix of alignment with the assistance of sets of images which are acquired from various angles to figure the distortion of the image. The calibration grill contains in vision development module. Disparity information provides relative depth information. The disparity information is made by using semi-block matching algorithm from NI vision because it provides more details and works in lesser region in case of proposed system. We have used two servomotors for pan/tilt camera support: The development of the cameras in horizontal arrangement and the development of the cameras in vertical arrangement.

- System Execution Flow:

  1. Adjustment of camera is done in camera calibration setup, i.e., Focus.
  2. Image is acquired using left and right cameras. Two images of the same scene are captured at the same time by two different cameras.
  3. Quality of the image is improved using image preprocessing.

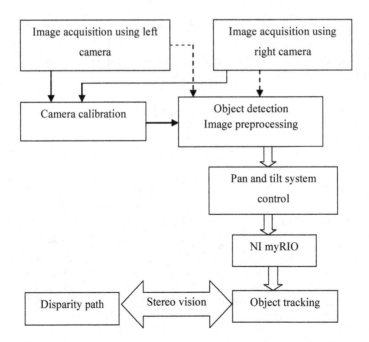

Fig. 2 Algorithm

4. Particle analysis of image is done.
5. Analyzed output of image is given to myRIO controller.
6. Control signals are generated for hardware to track the object (Fig. 2).

- Binocular Stereovision Principle

We have used binocular stereovision algorithm to calculate the position of objects within the frame of the scene and in particular, the depth of the elements contained in the two images [5]. The principle of binocular stereo vision is illustrated in Fig. 3, where f is the central length of left and right cameras, b is the baseline, $P(x, y, z)$ is a real-world point, uQ is the projection of this present reality point P in an image got by the left camera, and uR is the projection of this present reality point P in an image got by the right camera. Since the two cameras are separated by "d", both cameras see a same real-world point P. The x-directions of point's uQ and uR are given by relations (1) and (2):

$$uQ = f * \frac{x}{z} \qquad (1)$$

$$uR = f * \frac{x - d}{z} \qquad (2)$$

Distance between the two projected points is known as "disparity" and it is computed using relation (3):

$$\text{Disparity} = uQ - uR = f * \frac{d}{z} \qquad (3)$$

The depth information and the distance between real-world point "P" and the stereovision system, is given by the relation (4):

$$\text{Depth} = f * \frac{d}{\text{disparity}} \qquad (4)$$

Fig. 3 Stereovision system principle

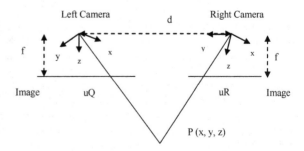

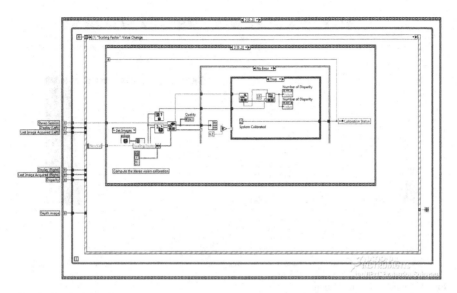

Fig. 4 Stereovision VI

The coordinates of focus of the object (*x*1, *y*1) and (*x*2, *y*2) obtained from left and right cameras, respectively, are changed relatively to the center of the images of sizes 640 × 480.

5 Experimental Results

1. Calibrate stereovision system (Fig. 4)

Disparity image indicates the disparity between left and right rectified images. Depth image provides comprehensive real-world depth information about the scene. Error map gives the inherent error in measurement for each pixel in the depth image.

2. Stereovision front panel (Fig. 5)

A calibration grid of dots was introduced to both cameras and images are acquired simultaneously. Brighter pixels indicate points that are closer to the camera and darker pixels indicate points that are farther away from the camera.

3. Target tracking VI
4. Robot and target tracking by image processing

Figure 6 shows the block diagram VI in LabVIEW for stereovision. Figure 7 shows images captured from left and right camera which are taken precisely at same time for stereovision. At the point when target is detected by image processing

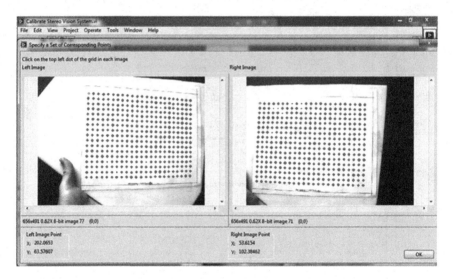

Fig. 5 Stereovision front panel

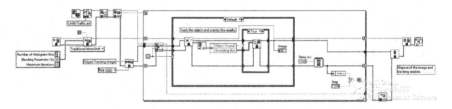

Fig. 6 Target tracking VI

techniques, we have determined coordinates of mass focal point of it in the two images [6].

6 Conclusion

We have implemented target tracking robot using image processing for developed proto system which detects the target by vision-based method. This system is implemented with wireless communication using LabVIEW-VDM image processing techniques to reach target destination. We have developed the application with remote computer in which the user gives a command to capture an image, process, and control remotely.

Left Camera Images   Right Camera Images

Fig. 7 Target tracking robot

Acknowledgements Authors put on record and warmly recognize to Savitribai Phule Pune University for partially funded by BCUD Research Grant vide, BCUD Sanction Letter, OSD/ BCUD/113/48. I take it a deemed privilege to express thanks to JSPM'S Bhivarabai Sawant Institute of Technology and Research, Wagholi, to complete this research work.

References

1. Francois Guerin, Simon G. Fabri, Marvin K. Bugeja "Double Exponential Smoothing for Predictive Vision Based Target Tracking of a Wheeled Mobile Robot", IEEE 2013.
2. Chung-Hao Chen, David Page, Chang Cheng, Andreas Koschan, and Mongi Abidi "A Moving Target Tracked by A Mobile Robot with Real-Time Obstacles Avoidance Capacity" IEEE, 0-7695-2521-0/06, 2006.
3. You-Wei Lin and Rong-Jong Wai, "Design of Adaptive Moving-Target Tracking Control for Vision-Based Mobile Robot, IEEE 2013.
4. Anita Gade, Yogesh Angal, "Automation in Library Management Using LabVIEW", International Conference on Computing Communication Control and automation, IEEE 2016. https://doi.org/10.1109/ICCUBEA.2016.7860133.
5. Robert Bichsel and Paulo V K Borges, "low–Obstacle Detection Using Stereo Vision", International Conference on Intelligent Robots and Systems, IEEE 2016.
6. Prajin Palungsuntikul, Wichian Premchaiswadi, "Object Detection and Keep on a Mobile Robot by using a Low Cost Embedded Color Vision System", IEEE 2010.

Detection of Misbehaviors Nodes in Wireless Network with Help of Pool Manager

Asha Chaudhary and Pournima More

Abstract As in recent time, wireless networks were developed in various extreme ends but then also facing from the problem of malicious nodes in the network. In this case, networks drop the packets at the receiver side and create various routing attacks. By the help of Dijkstra's algorithm, network can escape from the problem of dropping packets. In that, nodes can calculate shortest path on the basis of Fuzzy logic. Biggest downsides of existing propose are that it is unable to handle occurring of blackhole and grayhole in the network and nodes can easily befool with each other. To avoid occurring of attacks, network can use a scheme in which it creates pool manager at the starting of communication by which load of the nodes get reduces and routing attacks can easily detect. And at the time of further communication, that node is banned and does not engage for delivery packets from source to destination. By ejection of malicious nodes from the network with the help of pool manager that can iterative more established data by comparing hash function.

Keywords Blackhole · Grayhole · Pool manager · MANETs · Titlemapping DOS

1 Introduction

Mobile adhoc networks (MANET) is a scenario in which a group of mobile nodes comes together to give communication between them through wireless connection. In this case, there is no need of any centralized administration; it can be easily resemble in the situations where it is not possible to set up any infrastructure.

A. Chaudhary (✉) · P. More
Computer Network Department, G.H. Raisoni College of Engineering and Management,
Wagholi, Pune, India
e-mail: Asha92chaudhary@gmail.com

P. More
e-mail: Pournima.more1@gmail.com

© Springer Nature Singapore Pte Ltd. 2018
S. Bhalla et al. (eds.), *Intelligent Computing and Information and Communication*,
Advances in Intelligent Systems and Computing 673,
https://doi.org/10.1007/978-981-10-7245-1_12

Advantages of MANET are dynamic in nature, by which different nodes of the network can participate or exit at any time.

In MANETs, each mobile node operates not only as a host but also acts like router which forward packets from one node to other mobile nodes that require cooperation of nodes in the network. As described earlier, it is dynamic in nature and there is a high risk of security. By this way, DSR (Dynamic Source Routing) protocol is used instead of using routing table at very hop; its divides the operation into route discovery and route maintenance. One of the biggest drawbacks faced by the network is when nodes of the network start misbehaving to each other due to that a concept of malicious nodes arises. By this effectiveness, approaches become weak and nodes collude with each other that results in more devastating damages of the network [1]. A weakness of infrastructure leads to highly susceptibility to routing attacks (blackhole and grayhole) in the network. Blackhole attack is one kind of Denial of Service (DOS) attacks. In this type of attacks, shortest path and sequence number of the nodes, the message is intercepting to its destination, but when the malicious (fake) node made a use of the vulnerabilities of the route discovery packets of routing protocols at that time, there is a arise of blackhole attack in the network. A node which acts as malicious node has a tendency to attract packet through false route reply packet (RREP) to falsely claim fake shortest route from source to destination; instead of forwarding to destination, it discards the packets in between the routes [1] (Fig. 1).

But detection of grayhole attack was very difficult to identify because in this there is a continuous dropping of the packet data. In this case, packet may drop data for specific IP address; meanwhile, other nodes of the network receive correct information. Due to selectively drops of the data it is difficult to identify. To avoid this, nodes have to take more responsibility and detect grayhole attack by their own due to that proper data transmission can be held. All this attack mainly affects the routing speed and performance of the network itself (Fig. 2).

To avoid creation of malicious nodes in the network, concept of pool manager was that condition network creates pools that consist of number of mobile nodes. In this type of network, pool manager is all time in active state whenever transmission

Fig. 1 Blackhole attack

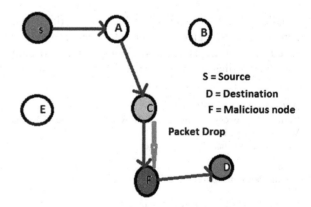

S = Source
D = Destination
F = Malicious node

Packet Drop

Fig. 2 Grayhole attack

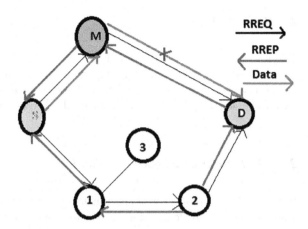

process is going on first; all the information pass through pools of the network, due to that network can easily identify all nodes that are misbehaving to each other due to the creation of attacks. DTN (Delay Tolerant Network) also plays an important role in the detection of blackhole and grayhole in the network. It provides a solution of resistant network that faces the problem of intermittent connectivity, long delay, or frequency disrupt. It also makes use of hop-to-hop routing and store-and-forward model to overcome the problem of end-to-end paths [2]. Now, the question arises how DTN helps in network to escape that nodes which cause attacks; in that case, there is a concept of intermediate nodes. These nodes receive, store, and forward packets to other destination which sometimes works as intermediate nodes. Suppose intermediate nodes are not in the path, then those node store entire packets. By the help of probabilistic behavior of nodes, it can predict and measure all information which is stored by the node in contact of its potential forwarder in terms of deliver message to the destination node. Most of the methodologies in wireless paradigm are to identify the vulnerability issues like grayhole and blackholes. Most of those are assigned this task to the routing nodes which can create the more havoc at the node end to decrease the routing performance of the nodes. By keeping this issue as of priority, proposed system assigns this task to a node's head of the pool and called it as "pool manager".

This paper is segmented into various sections in which Sect. 2 tell about related work of how another scheme helps to find malicious nodes. Section 3 details about our proposed methodology of how efficiently network can remove all the malicious nodes from the network by the creation of pool manager with help of title mapping. The consequence of our methodology is performed in Sect. 4. Section 5 concludes the paper with some scope of future.

2 Related Work

Various numbers of methods were discovered to detect black- and grayhole nodes in the network, in which some of them work during route discovery process while others at the time of data transmission. Those who work during route discovery have more advantage, because in that case they do not lose the data.

2.1 Detection of Blackhole Attacks

In this, author proposed a method in which it can design a network which can use destination unique ID to detect malicious node of blackhole attack and then it keeps apart the node between source and destination [3]. In this route, node has to check and analyze all neighbors node by comparing response time of sender by the help of AODV route request message. Author describes a detection method that is known as cooperative bait detection Scheme (CBDS). In this scheme, source node has the tendency to select an adjacent node to whom it has to cooperate, and then after this, address is used for bait destination address to bait malicious nodes to send a reply RREP message. With the help of the reverse tracing technique, network can detect all malicious nodes and then prevent them from participating in the routing operation of the network for the transmission of data packets [1]. The main advantage of this scheme is that it can work for both proactive detection and reactive detection.

2.2 Detection of Grayhole Attacks

In this paper, author uses a concept of sequence number. For detection of grayhole node, use sequence number for normal nodes by which it eliminates itself to enter in the blacklist. Algorithm works to check the peak value of the sequence number whether it is greater or less than reply packet [4]. Author describes an algorithm named channel aware detection (CAD) which consists of two techniques: one is called hop-by-hop observation and other is called traffic overhearing due to that network can easily detect malicious nodes which belong to that particular path, where transfer of packet is going on. By this, mesh node of the network can be easily detected by appropriate subject of attack [5]. Describe a scheme which is called as course-based detection method. Whatever nodes involved in the routing route keep total focus on the neighboring node. In this, all neighbors do not include monitoring [6]. Here, only neighboring node of the route gets the packet which was sent by source node, which consists of itself as a copy of packets in the form of buffer named Fwd packet buffer. By this source, node can analyze overhear rate and get percentage of data packets transferred from source and received by destination.

2.3   Use of DTN for Detection of Routing Attacks

Author ensures a mechanism for the contrast analysis of blackhole attack in DTN that is known as Combined Faith Value (CFV) [7]. By checking the node history separately, CFV value can be calculated for each node of the network. Node history can be stored in the form of node buffers. By this history of nodes information gets regarding to the data transmission, number of packets which was created, how much packets received, dropped and sent numbers of packets. In this blackhole, nodes have maximum packets to drop and deliver wrong message based on CFV mechanism. Author proffers how DTN performs store–carry–forward method. DTN only works when both sender and receiver nodes come in contact with each other. Till they come in contact, nodes store the packet in buffer and then carries to next node when it gets the range for transmission. Distributed scheme is introduced for detection of packet dropping in that node select contains in the form of signed contact record of its previous contact. At the time of collecting record of nodes, contact can easily detect which node dropped the packet [8].

2.4   Detection of Blackhole and Grayhole Attacks on Same Network

Author recommends a cryptographic system which consists of secure data transfer mechanism that protects the data from encryption and hashing during the time of attackers attacking the system. In this system, if the receiver is unable to decrypt the cipher data correctly, then taint propagation detects it, and immediately sends to sender that particular node start acts like malicious node [9]. Author represents a probabilistic misbehavior detection scheme (iTrust) that can reduce the detection overhead more effectively. For this, iTrust creates Trusted Authority (TA) to determine the node's behavior based on collecting routing evidences and probabilistically checking. It works on the basis of two phases: one is called evidence generation phase and other is called routing evidence phase [10]. Author suggests a scheme that is known as statistical-based detection of blackhole and grayhole attackers (SDBG) by which network can address both individual and collusion attacks. In this scheme, nodes are required to record their nodes histories by which it helps to evelute forwarding behaviors. To aviod continous dropping of messages, scheme promotes metrices at the same time by which attackers have to create fake encounter records frequently [11]. Robust algorithm helps to detect colluding attacks for that it exploits abnormal pattern for appearance frequency and number of message sent in fake encounter [12].

3 Proposed Methodology

In this section, it is discussed how pool title system works regarding providence of perfect network in which there is no chance of occurrence blackhole and grayhole in the network by which network detects all nodes that behave like malicious nodes and for further communication it escapes from the network (Fig. 3).

Following are the steps which were used in the pool title method for the wireless communication by that our system is more effective for its working.

In the below diagram, it is easily understood that it shows the relation between user and admin. For appropriate system, there should be a healthy relationship between admin and its users. In this system, all the parameter of the system should be considered as perfect organizer, due to which information of system should reach its proper destination.

3.1 Create Pool Manager

In the initial step of the proposed system, first all the nodes of the network should register to wireless router and make scenario in which we have to make our own user ID and passwords by which selective nodes of the system can only take part in the communication and no further nodes take part. Create pool manager and all the nodes which were taking part for communication have to register to it. For that,

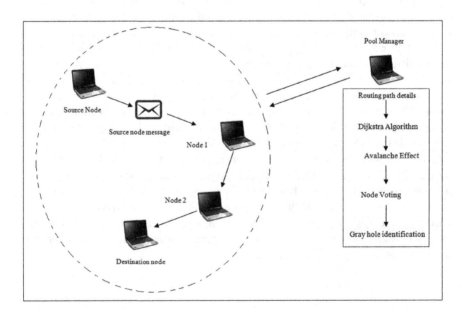

Fig. 3 System overview

nodes have to send their node names and IP address. To check the IP address, first go to command prompt and put IP configure; then the system generates its IP address and Mac Address. After this, whatever information used for communication is stored in pool manager for further transactions.

3.2 Select Destination Node

For this step, source node of the network can view the entire registered node which was required for communication in our system. After source node is selected, destination node information is sent. After the destination node is selected by the source node, then a routing path is requested to the pool manager to update its profile. In this, systems have option to select all the nearby nodes due to which communication can be easy and short. What nodes select by source node to send data those nodes only receive the data rest all are also active but at that time not used for communication.

3.3 Shortest Path Estimation

After receiving the routing path request from the source node, pool manager assigns the random integer value for each node of the network, which involves the communication of data. Network uses Dijkstra's algorithm for calculating the shortest path for routing process, where each node assigned a random weight and then based on the random weights, node distances are been evaluated for instance nodes. As Dijkstra's algorithm not only work for shortest path between nodes but it can put common variant fixes a single nodes as the source node and finds the shortest paths from all source to all the other nodes in the graph and produce shortest path tree.

3.4 Use of Hash Key

As early as shortest path is calculated by the help of Dijkstra's algorithm with help of prior step in which data is start routing based on the route map provided by the pool manager. And for every node, a signature of the data is sent in the form of hash code that will be generated by using MD5 algorithm. Hash function is used to convert the data of arbitrary length to a fixed length and it provides much faster than a symmetric encryption.

3.5 Identification of Malicious Nodes

Here, in the last step, a strict analysis of the hash signature is carried out for the pattern of the blackhole and grayhole attacks though avalanche effect for a fix time slot is called as "tile". By the help of avalanche effect, system can describe cryptography in which there is a slight change in the input value (message) that causes a significant change in the output (hash value). If a node is behaving as the grayhole, then it partially drops the data or it malfunctions the data, whereas for the blackhole it completely drops the data so that only one hash signature will be received by the pool manager.

The whole process of blackhole and grayhole detection process can be shown in Algorithm 1

Algorithm 1: Gray- and Blackhole Identification Using Pool Tile Method

//Input: Sender Data bytes **B**
//Destination Node N_{st}
//Output: Successful identification of Gray- and Blackhole node
Step 0: **Start**
Step 1: Activate pool managers T_m
Step 2: Register Node N_i with pool manager T_m
Step 3: Set Tile **T** for Pool managers T_m(Where tile is Time in Seconds)
Step 4: Choose Data bytes **B** by source node S_{rc}
Step 5: Set Destination node N_{st}
Step 6: Identify the shortest path P_{th}
Step 7: **WHILE B** $\notin B_n$
Step 8: for each tile **T**
Step 9: $B_d \rightarrow S_{nt} \in B$ (P_d = previous data in Hash and S_{nt} = Run time Source node)
Step 10: $C_d \rightarrow T_{nt} \in B$($C_d$ = Current data in hash and T_{nt} = Run time Destination node)
Step 11: check **IF**$C_d \neq P_d$(Avalanche Effect)
Step 12: Label Current instance nodeC_n as Grayhole
Step 13: **END IF**
Step 14: IF C_d = NULL
Step 15: **Label** Current instance node C_n as Blackhole
Step 16: **END IF**
Step 17: **END WHILE**
Step 18: **Stop**

Table 1 Comparison between pool title method and cluster analysis method

Number of hops	Number of packets sent	Average transmission time(ms) Tile bitmap process	Average transmission time(ms) DSDV protocol
1	5334	675.64	1000
2	5334	728.98	1000
3	5334	1268.38	1000
4	5334	1369.06	1000
5	5334	1102.36	1000
6	5334	693.42	1000
7	5334	728.98	1000
8	5334	746.76	1000
9	5334	656.86	1000
10	5334	497.84	1000

4 Results and Discussions

An explicit experiment is conducted for the performance evaluation of the system based on the time taken to identify the blackhole and grayhole in the wireless network. System is set to detect the blackhole and grayhole for much number of runs to record the timings of performance, which are tabled in Table 1.

When it was compared with the pool tile and cluster analysis method, following difference occurs due to which it can easily analyze pool tile mapping that is far better than any other method. Cluster analysis method works for the destination sequence distance vector routing protocol in which time requires more to transmit and analysis blackhole and grayhole of the network (Fig. 4).

5 Conclusion and Future scope

Due to increased use of wireless networks, there is problem which faces a huge entry point to the attacker in the network, in which blackhole and grayhole attacks are most common and highly dangerous forms of threats to the wireless network. In this paper, it introduced pool tile method to identify the attacker node which efficiently works than that of most of the other methods. This is because proposed system eradicates the burden of detection of malicious nodes from the routing nodes which come under shortest path sequence, thereby assigning the entire detection job to high configured pool manager. In the future, this method can work for more efficiently incorporated by assigning the detection of blackhole and grayhole jobs to the multiple pool manager on increasing of pool size.

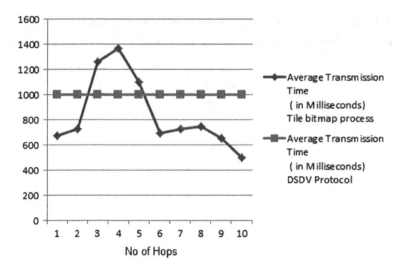

Fig. 4 Performance comparison of different methods

References

1. Jian-Ming Chang, Po-Chun Tsou, Isaac Woungang, Han-Chieh and Chin-Feng Lai, "Defending Against Collaborative Attacks by Malicious Nodes in MANETs: A Cooperative Bait Detection Approach" IEEE System Journal, Vol. 9, No. 1, March 2015
2. Pham Thi Ngoc Diep, Chai kiat Yeo, "Detecting Colluding Blackhole and Greyhole Attacks in Delay Tolerant Network", IEEE Transactions on Mobile Computing, https://doi.org/10.1109/TMC.2015.2456895
3. S. Sankara Narayanan and Dr. S. Radhakrishnan, "Secure AODV to Combat BlackHole Attack in MANETs" 2013 International Conference on Recent Trends in Information Technology (ICRTIT) ISBN:978-1-4799-1024-3/13/$31.00 ©2013 IEEE
4. A. M. Kanthe, D. Simunic, R. Prasad, "A Mechanism for Detection of Gray Hole Attack in Mobile Ad Hoc Networks" International Journal of Computer Applications, Volume 53, Sep. 2012
5. Devu Manikantan Shila, Yu Cheng and Tricha Anjali, "Channel Aware Detection of Grayhole Attacks in wirless Mesh Networks", 978-1-4244-4148-8/09 IEEE "Globecom" 2009
6. Disha G. Kariya, Atul B. Kathole, Sapna R. Heda, "Detecting Blackhole and Gray hole Attacks in Mobile Adhoc Network using an Adaptive method", International Journal of Emerging Technology and Advanced Engineering, Volume 2, Issue 1, January 2012
7. Shuchita Upadhyaya and Karishma, "A Co-operative Mechanism to contrast Blackhole Attacks in Delay Tolerant Networks", IOSR Journal of Computer Engineering, E-ISSN: 2278-0661, P-ISSN: 2278-8727
8. Sangeetha. R and Krishnammal. N, "Detection of Routing Attacks in Disruption Tolerant Networks", The International Journal of Engineering and Science, Volume 2, 2013 ISSN: 2319-1813, ISBN: 2319-1805
9. Susan Basil, Panchami V, "Attack Identification in a Cryptographic system using Dynamic Taint Propagation", International Journal of Science, Engineering and Technology Research, Volume 4, Issue 4, April 2015

10. Haojin Zhu, Suguo Du, Zhaoyu Gao, Mianxiong Dong and Zhenfu Cao, "A Probabilistic Misbehavior Detection Scheme toward Efficient Trust Establishment in Delay-Tolerant Networks" IEEE Transcations on Parallel and Distributed System, Volume 25, Number 1, January 2014

11. Yinghui Guo, Sebastian Schildt and Lars Wolf, "Detecting Blackhole and greyhole Attacks in Vehicular Delay Tolerant Networks", IEEE, 978-1-4673-5494-3/2013

12. Surpriya Pustake, Dr. S. J. Wagh, D. C. Mehetre, "Grayhole Detection and Removal in MANETs by Pool Tile Method", International Journal of Science and Research, Volume 5, Issue 7, July 2016

Evaluation of Multi-label Classifiers in Various Domains Using Decision Tree

V. S. Tidake and S. S. Sane

Abstract One of the commonly used tasks in mining is classification, which can be performed using supervised learning approach. Because of digitization, lot of documents are available which need proper organization, termed as text categorization. But sometimes documents may reflect multiple semantic meanings, which represents multi-label learning. It is the method of associating a set of predefined classes to an unseen object depending on its properties. Different methods to do multi-label classification are divided into two groups, namely data transformation and algorithm adaptation. This paper focuses on the evaluation of eight algorithms of multi-label learning based on nine performance metrics using eight multi-label datasets, and evaluation is performed based on the results of experimentation. For all the multi-label classifiers used for experimentation, decision tree is used as a base classifier whenever required. Performance of different classifiers varies according to the size, label cardinality, and domain of the dataset.

Keywords Machine learning · Multi-label classification · Data transformation
Algorithm adaptation · Decision tree · Label cardinality

1 Introduction

One commonly used task in mining is *classification*. If a set of known instances, called train set, is used to train the model, then it is referred as *supervised learning*. Once the training and testing of the model are complete, it is useful for classification

V. S. Tidake (✉)
Department of Computer Engineering, MCERC, Savitribai Phule Pune
University, Nashik, India
e-mail: vaishalitidake@yahoo.co.in

S. S. Sane
Department of Computer Engineering, KKWIEER, Savitribai Phule Pune
University, Nashik, India
e-mail: sssane@kkwagh.edu.in

© Springer Nature Singapore Pte Ltd. 2018   117
S. Bhalla et al. (eds.), *Intelligent Computing and Information and Communication*,
Advances in Intelligent Systems and Computing 673,
https://doi.org/10.1007/978-981-10-7245-1_13

of unseen instances. Several distinct domains [1–8] like TC use supervised learning. Sometimes, a document may reflect multiple semantic meanings. Hence, unlike traditional classification, it may be associated with one or more than one class labels, which represents multi-label learning. It is the method of associating a set of predefined classes to an unseen document depending on its contents. Association of each input example with single-class label is termed as *SL (single-label) classification* or just classification. Depending on the total count of class labels involved, SL classification is either referred as a *binary single-label (BSL) classification* when the label space has only two class labels or *multi-class single-label (MSL) classification* if the label space includes more than two class labels. For example, a news document represented as a square in Fig. 1 may be related to either education (+) or health (−) category representing BSL (Fig. 1a) or one of education, health, and economy (^) categories representing MSL (Fig. 1b) [2, 6]. A news saying that "Yoga and meditation are crucial for the stress management of students" is related to education as well as health categories (+ −) representing *MLC (multi-label)* classification (Fig. 1c). Already many tools and algorithms are available to handle SL classification problems. Use of MLC in the recent past has been done for TC, prediction of gene function, tag recommendation, discovery of drug, [2–6], etc. So in the area of machine learning, it has gained the position of an upcoming research field.

This paper deals with a comparative study of MLC. Sections 2, 3, and 4 describe the metrics used for evaluation, two approaches used for MLC, about the experiments and results, respectively. Section 5 gives the concluding remarks.

2 Multi-label Classification (MLC)

A. Definition

Like SL, MLC uses supervised approach for learning. It is the task which relates an unseen instance considering its features to a set of predefined labels. Let C represents a set of disjoint labels. Let an instance be described by a vector f_j of

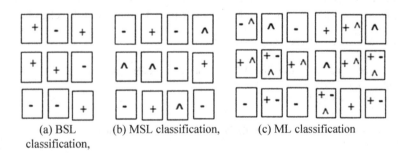

(a) BSL classification,　　(b) MSL classification,　　(c) ML classification

Fig. 1 Classification

features and belongs to a set y_j of labels. Let Q denotes a training set (x_j, y_j), and then obtain a function $f(x)$ for mapping each f_j vector to a set y_j of labels, where $y_j \subseteq C$ and $j = 1, 2... |Q|$.

B. Metrics used to measure performance

Let PL_i and AL_i denote set of predicted labels by a classifier and a set of actual labels for training instance x_i. Let T and C denote a test set and a set of disjoint labels, respectively. Let f denotes a classifier. ML learning uses metrics following metrics.

Hamming loss Most commonly used which is used to measure the number of times an instance and its associated label is not correctly classified. Expected value of hamming loss metric is small [2].

$$HL(f) = \frac{1}{|T|} \sum_{i=1}^{T} \frac{|B(PLi \ominus ALi)|}{|C|}, \tag{1}$$

where $B(.) = 0$ if AL_i and PL_i are same for all labels of instance i, else $B(.) = 1$. Here, $\ominus$ is used for symmetric difference.

Ranking loss This metric measures performance of ranking task which generates all labels in the order of relevance. It is used to measure the number of times an irrelevant label has been ranked above the relevant labels. Expected value of ranking loss metric is small [6].

$$RL(f) = \frac{1}{|T|} \sum_{i=1}^{|T|} \frac{1}{|ALi||\overline{ALi}|} |\{(y1, y2)|\mu(y1, xi) \geq \mu(y2, xi)\}|, \tag{2}$$

where $y1 \in ALi$ and $y2 \in \overline{ALi}$. Assume $\mu(q, r)$ denotes relevance of label q for an instance r and smaller value denotes more relevance.

One-error It counts the number of times a label generated by the classifier at the top rank does not appear in the correct labels associated with an input instance. The smaller the one-error, the better it is [6, 7]. Here, $B(.) = 1$ if (.) is true, else $B(.) = 0$.

$$OE(f) = \frac{1}{|T|} \sum_{i=1}^{T} B\left(\left(\arg \min_{y \in C} \mu(y, xi)\right) \notin ALi\right) \tag{3}$$

Coverage It measures how much down the list of labels generated by the classifier should be traversed to include all the labels relevant to an example assuming top-most labels appear at the start of the list. The less the value, the better is the result [6, 7].

$$CG(f) = \frac{1}{|T|} \sum_{i=1}^{|T|} \max_{y \in ALi} \mu(y, xi) - 1 \qquad (4)$$

Average precision It computes an average proportion of relevant labels which are ranked above a particular relevant label. The bigger the value, The better is the result [6, 7].

$$AP(f) = \frac{1}{|T|} \sum_{i=1}^{T} \frac{1}{|ALi|} \sum_{y \in ALi} \frac{|\{z \in ALi | \mu(z, xi) \le \mu(y, xi)\}|}{\mu(y, xi)} \qquad (5)$$

Subset Accuracy It is an average over all the instances which checks whether predicted label set of an instance is same as its actual label set [3, 5, 9].

$$SA(f) = \frac{1}{|T|} \sum_{i=1}^{T} B(PLi = ALi), \qquad (6)$$

where $B(.) = 1$ if AL_i and PL_i are same for all labels of instance i, else $B(.) = 0$.

Example-Based Recall, Precision, and F-Measure [2, 6, 7]:

$$ExRc(f) = \frac{1}{|T|} \sum_{i=1}^{|T|} \frac{|PLi \cap ALi|}{|ALi|},$$

$$ExPr(f) = \frac{1}{|T|} \sum_{i=1}^{T} \frac{|PLi \cap ALi|}{|PLi|}, \qquad (7)$$

$$ExF1(f) = \frac{1}{|T|} \sum_{i=1}^{|T|} \frac{2|PLi \cap ALi|}{|ALi| + |PLi|} \cdots$$

3   Various Methods

In the literature, various methods to perform multi-label learning have been developed and reported. Two broad categories used to perform MLC are the data transformation and the algorithm adaptation [4]. The data transformation approach involves transformation of an input instance into data which suits for many single-label traditional classifiers, whereas the algorithm adaptation approach involves transformation of SL classifier algorithm which suits multi-label data [2, 3].

So far various SL algorithms are developed by researchers. The data transformation approach utilizes these existing SL algorithms. The approach transforms data representation from multi-label to single-label which is acceptable by existing

SL classification algorithms. In different words, the data transformation operates on the fundamental concept of "*fit data to an algorithm*" [2]. Transformation does not change the algorithm, hence is said to be "*independent of an algorithm*" [3]. Algorithms like BR, LP, CC, ECC, RAkEL, and HOMER use this approach. The adaptation approach modifies the existing SL algorithms for managing multi-label data appropriately. In different words, the algorithm adaptation operates on the fundamental concept of "*fit an algorithm to data*" [2]. Since an algorithm, not data, is updated, this approach is said to be "*dependent on an algorithm*" [3]. Algorithms like BRkNN, MLkNN, ML-C4.5, and BP-MLL use this approach. Let C represents a set of labels.

Binary Relevance (BR)
It is the most widely used method for data transformation in which a multi-label problem is converted into $|C|$ binary SL classification problems. Each of the binary classifiers contributes its vote separately to do classification [4, 5]. BR has one disadvantage of not considering the association between labels (if any) as it treats every label individually [2, 6].

Label Powerset (LP)
Overcoming the drawback of BR for treating every label individually is removed in LP. It considers each different group of labels as a separate class and treats the entire problem as a multi-class single-label (MSL) problem [7].

Multi-label data is treated as multi-class data. For example, multi-label data having $|C|$ labels forms $2^{|C|}$ classes with different label combinations. Thus, LP considers multiple labels simultaneously and overcomes the drawback of BR [8]. However, the number of groups of classes formed increases with $|C|$. It results in distribution of the original data into different groups of classes. This distribution may result in scenario similar to class imbalance where few classes may belong to more number of instances, whereas some classes may belong to less number of instances. The situation may affect classifier accuracy. Also, higher value of C causes time complexity of LP to become worst.

Random k-Label sets (RAkEL) It is actually an ensemble of multiple LP classifiers having different combinations of all labels referred as label sets [1]. These k-size label sets help to remove the class imbalance drawback in LP. For multi-label data with $|C|$ labels, N label sets each of size k are formed randomly and separate LP classifier models are designed for them. Average of votes obtained from N models for each label is used for classification of an unseen instance. If it is more than a threshold, then prediction of that label is P; otherwise, it is A to represent the presence or absence of that label. However, classifier accuracy depends on the label sets which are selected randomly. Also, choosing N and k values may also affect the classifier performance.

Classifier Chain (CC) A weakness of BR not considering association between labels is removed in CC [7, 8]. Like BR, it transforms ML problem into $|C|$ SL problems and for each label C_j, a separate binary classifier B_j is designed. But the input for each classifier B_j is different. Each classifier B_j takes as input all feature vectors $f_{1...D}$ of all instances and predictions of all earlier classifiers also. In general, output O_{ij} of each classifier represents prediction of classifier B_i for $Class_i$ for instance j. O_{ij} takes values either P or A for class C_i of instance j. Accordingly, output of all classifiers is obtained. Thus, label information is passed from classifier B_i to B_j, and so on. Such organization takes into account associations among labels and thus overcomes the weakness of BR described earlier. But an important concern in CC is that sequence of considering labels may result in different classifier accuracies [8] affecting its performance to a great extent and guessing the best possible order is difficult.

Ensemble of Classifier Chains (ECC) Instead of depending on single chain of labels, ECC takes benefit of using multiple different order chains as well as ensemble. It obtains votes from a group of classifiers each using different chains and different set of instances, which improves the accuracy of prediction [8].

Calibrated Label Ranking (CLR) It is a modification of Ranking by Pairwise Comparison (RPC) [2, 6]. It augments the label set with a virtual label L_v. Then, it constructs $C(C-1)/2$ binary classifiers as in RPC. Each classifier B_{ij} outputs P for an instance if it contains label C_i and A if it contains label C_j, and does not consider instances having both or none of these labels in the pair (C_i, C_j) [5]. CLR also constructs C binary classifiers to represent relationship between each label C_i and a virtual label L_v. While classifying an unseen instance, votes are obtained from all these constructed classifiers to generate ranking of all labels having relevant and irrelevant labels separated by a virtual label.

Multi-Label k-Nearest Neighbors (MLkNN) It updates traditional kNN algorithm to process multi-label data. For classification of an unseen instance, it finds k-nearest neighbors. Then, statistical data like count of nearest neighbors for a training instance x associated with particular label and not associated with particular label is obtained for each label using computed k-nearest neighbors. Next, a rule based on Bayes theorem is applied to labels of an unseen instance [2, 10]. Further, it computes label information from obtained nearest neighbors with the help of posterior and prior probabilities. The MLkNN exhibits a limitation of not considering label relationship by processing each label separately.

Hierarchy Of Multi-labEl leaRners (HOMER) Each individual classifier works on smaller size distinct label set as compared to the original one, where each label set contains related labels together. Hierarchical distribution of these labels is an important feature of this classifier [2].

4 Experimentation and Results

A. Multi-label Datasets

MEKA is a WEKA-based project. An open-source library Mulan uses Java to perform multi-label data mining. Data sets from various domains are made available in MEKA, MULAN, and LibSVM [11–14]. Table 1 briefs some multi-label datasets used for experimentation along with their information [11]. Table 1 shows label cardinality of all datasets which denotes the average number of labels per example [2].

B. Parameter initialization

For BR, LP, CLR, CC, and ECC, C4.5 decision tree algorithm [15, 12] is used as base SL classifier. For HOMER and RAkEL, LP with C4.5 is used as a base classifier. HOMER is run with three clusters and random method. RAkEL [1] runs with ix models, 3 as size of subset and 0.5 as threshold. MLkNN is executed with 10 neighbors and 1 as smoothing factor. Cross-validation is used for evaluation with tenfolds [9, 11].

C. Results and Discussion

Experiments are carried out using Intel(R) Core(TM) i5-6200U CPU @2.30 GHz having 8 GB RAM and Windows10 and Java for programming with libraries from Mulan 1.5 [11] and WEKA 3.8.1 [12]. Figures 2, 3, 4, 5, 6, 7, 8, 9, and 10 show results obtained from execution of eight classifiers on eight datasets to measure nine performance metrics. Information in brackets shows criteria expected for that metric value. Legend for all charts is same as that shown in Fig. 10.

LP classifier produced memory error for mediamill dataset when running on the above-mentioned hardware. An attempt is done to compare results obtained with other work reported in the literature [7, 11, 16]. The variation in the results may be due to different base classifiers used or different parameter settings or different default parameters in different tools used for experimentation. For multimedia

Table 1 Multi-label datasets

Domain	Dataset	#attributes	#labels	#instances	Label cardinality	Label density
Biology	Yeast	103	14	2417	4.237	0.30
Biology	Genbase	1186	27	662	1.252	0.05
Text	Medical	1449	45	978	1.245	0.03
Text	Enron	1001	53	1702	3.378	0.06
Multimedia	Scene	294	6	2407	1.073	0.18
Multimedia	Corel5 k	499	374	5000	3.522	0.01
Multimedia	Emotions	72	6	593	1.868	0.31
Multimedia	Mediamill	120	101	43,907	4.375	0.04

Fig. 2 Hamming loss (small)

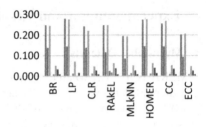

Fig. 3 Ranking loss (small)

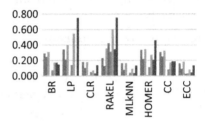

Fig. 4 Subset accuracy (large)

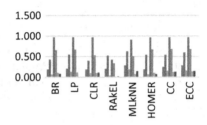

Fig. 5 One-error (small)

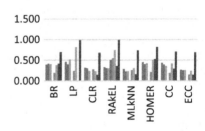

Fig. 6 Coverage (small)

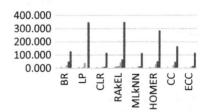

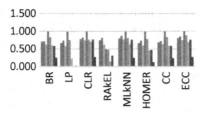

Fig. 7 Average precision (large)

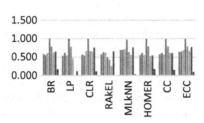

Fig. 8 Example-based precision (large)

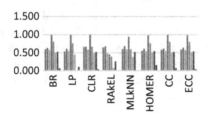

Fig. 9 Example-based recall (large)

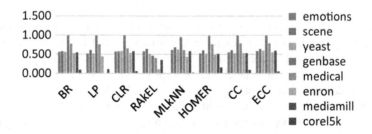

Fig. 10 Comparison of various classifiers for example-based F-measure (large)

domain, MLkNN achieved best results in terms of hamming loss followed by subset accuracy. Especially, mediamill and corel5 k datasets of multimedia domain having more label cardinality achieved better results on hamming loss and subset accuracy among all. Emotions dataset in multimedia domain showed good performance for all MLC methods using C4.5 [15] except LP and HOMER for all example-based metrics, namely hamming loss, example-based recall, precision, and F-measure as compared to ranking-based metrics, namely coverage, one-error, and average precision [7]. LP, RAkEL, and HOMER performed poor in all domains for all metrics. The reason can be that the decision tree classifier may not be able to utilize the label relationship to the required extent when used with LP. Drawback of LP showing poor performance if different classes are associated with different numbers of examples should not be ignored. Also, subset size and number of models used with RAkEL [1] may be too small for not considering label correlation properly. Already, LP with decision tree has shown poor results and same is taken as base classifiers for HOMER and RAkEL, which may be the reason behind their poor performance. ECC performed much better than CC, LP, CLR, RAkEL, and HOMER for all domains next to MLkNN for ranking loss, hamming loss, coverage, and one-error, but poor for the remaining metrics. It can be due to the power of ensemble which has been already proved to be better than single-label classifier in the literature [8]. For biology domain, BR, CC, and ECC showed best performance on all example-based measures followed by CLR for genbase dataset only. Yeast dataset has shown poor performance than genbase dataset. We can hypothesize that it may be due to higher label cardinality for yeast dataset as compared to genbase dataset. Similar may be the case for the text domain. Medical dataset in text domain has achieved better metric values especially in BR, CLR, CC, and ECC for the same reason as compared to enron dataset. MLkNN and CLR showed less misclassification of instance label pairs by giving smaller hamming loss than [7]. The cause may be different in number of neighbors and the base classifier selected in both classifiers, respectively.

5 Conclusion

There are two ways to design algorithms for ML classification. Data transformation methods transform data having multiple labels into single-label aiding traditional single-label methods. The adaptation approach updates the existing algorithms of learning for processing multi-label data. For all the multi-label classifiers used in this work, decision tree is used as a base SL classifier wherever necessary. Mediamill and corel5 k of multimedia domain having more label cardinality achieved good results on hamming loss and subset accuracy. For biology domain, BR, CC, and ECC showed best performance on all example-based measures on both datasets followed by CLR for genbase only. Medical in text domain has achieved better metric values especially in BR, CLR, CC, and ECC as compared to enron. ECC also showed effectiveness next to MLkNN but better than other methods, thus giving their votes to ensemble and adaptation, respectively. It will be interesting to see the effect of other base classifiers on different ML classifiers.

References

1. G. Tsoumakas, I. Katakis, and I. Vlahavas, "Random k-labelsets for multilabel classification," IEEE Trans. Knowl. Data Eng., vol. 23, no. 7, pp. 1079–1089, Jul. 2011.
2. M. L. Zhang and Z. H. Zhou, "A review on multi-label learning algorithms", IEEE Transactions On Knowledge And Data Engineering, Vol. 26, No. 8, August 2014.
3. A. de Carvalho and A. A. Freitas, "A tutorial on multi-label classification techniques," in Studies in Computational Intelligence 205, A. Abraham, A. E. Hassanien, and V. Snásel, Eds. Berlin, Germany: Springer, 2009, pp. 177–195.
4. G. Tsoumakas and I. Katakis, "Multi-label classification: An overview," Int. J. Data Warehousing Mining, vol. 3, no. 3, pp. 1–13, 2007.
5. G. Tsoumakas, I. Katakis, and I. Vlahavas, "Mining multilabel data", Data Mining and Knowledge Discovery Handbook, O. Maimon and L. Rokach, Eds. Berlin, Germany: Springer, 2010, pp. 667–686.
6. G. Tsoumakas, M. -L. Zhang, and Z. -H. Zhou, "Tutorial on learning from multi-label data," in ECML PKDD,Bled,Slovenia,2009[Online].Available: http://www.ecmlpkdd2009.net/wpcontent/uploads/2009/08/learning-from-multi-label-data.pdf.
7. G. Madjarov, D. Kocev, D. Gjorgjevikj, and S. Džeroski, "An extensive experimental comparison of methods for multi-label learning," Pattern Recognit., vol. 45, no. 9, pp. 3084–3104, 2012.
8. J. Read, B. Pfahringer, G. Holmes, E. Frank, "Classifier chains for multi-label classification", in: Proceedings of the 20th European Conference on Machine Learning, 2009, pp. 254–269.
9. Nasierding, Gulisong, and Abbas Z. Kouzani. "Comparative evaluation of multi-label classification methods." In Fuzzy Systems and Knowledge Discovery (FSKD), 2012 9th International Conference on, pp. 679–683. IEEE, 2012.
10. M. -L. Zhang and Z. -H. Zhou, "ML-KNN: A lazy learning approach to multi-label learning," Pattern Recognit., vol. 40, no. 7, pp. 2038–2048, 2007.
11. G. Tsoumakas, E. Spyromitros-Xioufis, J. Vilcek, and I. Vlahavas, "MULAN: A Java library for multi-label learning," J. Mach. Learn. Res., vol. 12, pp. 2411–2414, Jul. 2011.
12. M. Hall et al., "The WEKA data mining software: An update", SIGKDD Explor., vol. 11, no. 1, pp. 10–18, 2009.
13. C. -C. Chang and C. -J. Lin, "LIBSVM: A library for support vector machines," ACM Trans. Intell. Syst. Technol., vol. 2, no. 3, Article 27, 2011 [Online]. Available: http://www.csie.ntu.edu.tw/~cjlin/libsvm.
14. Tidake, Vaishali S., and Shirish S. Sane. "Multi-label Learning with MEKA", CSI Communications (2016).
15. Ross Quinlan (1993). C4.5: Programs for Machine Learning. Morgan Kaufmann Publishers, San Mateo, CA.
16. Cerri, Ricardo, Renato RO da Silva, and André CPLF de Carvalho. "Comparing methods for multilabel classification of proteins using machine learning techniques." In Brazilian Symposium on Bioinformatics, pp. 109–120. Springer Berlin Heidelberg, 2009.

An Effective Multilabel Classification Using Feature Selection

S. S. Sane, Prajakta Chaudhari and V. S. Tidake

Abstract Recently, multilabel classification has received significant attention during the past years. A multilabel classification approach called coupled k-nearest neighbors algorithm for multilabel classification (called here as CK-STC) reported in the literature exploits coupled label similarities between the labels and provides improved performance [Liu and Cao in A Coupled k-Nearest Neighbor Algorithm for Multi-label Classification, pp. 176–187, 2015]. A multilabel feature selection is presented in Li et al. [Multi-label Feature Selection via Information Gain, pp. 346–355, 2014] and called as FSVIG here. FSVIG uses information gain that shows better performance when used with ML-NB, ML-kNN, and RandSvm when compared with existing multilabel feature selection algorithms.This paper investigates the performance of FSVIG when used with CK-STC and compares its performance with other multilabel feature selection algorithms available in MULAN using standard multilabel datasets. Experimental results show that FSVIG when used with CK-STC provides better performance in terms of average precision and one-error.

Keywords Algorithm adaptation · Coupled label similarity · Feature selection
Information gain · Multilabel classification

S. S. Sane · P. Chaudhari (✉)
Department of Computer Engineering, K. K. Wagh Institute of Engineering
Education and Research, SPPU, Nashik, India
e-mail: p.prajakta42@gmail.com

S. S. Sane
e-mail: sssane@kkwagh.edu.in

V. S. Tidake
Computer Engineering, Matoshri College of Engineering and Research Center,
Savitribai Phule Pune University, Nashik, India
e-mail: vaishalitiake@yahoo.co.in

© Springer Nature Singapore Pte Ltd. 2018
S. Bhalla et al. (eds.), *Intelligent Computing and Information and Communication*,
Advances in Intelligent Systems and Computing 673,
https://doi.org/10.1007/978-981-10-7245-1_14

1 Introduction

Previously in supervised learning, each example is associated with only one instance and single label is assigned to this instance. As single label is assigned to the instance, this classification is known to be single-label classification. In some situations, where real-world object cannot be handled by single label, this problem can be solved by using multilabel classification. The purpose of multilabel classification is to discover the set of labels for unseen instances. In multilabel classification, input data instances are related to multiple class labels. For example, particular news article can be associated with economical article, political article, and business article.

Multilabel classification is used in several areas such as bioinformatics, text categorization, tag recommendation, image classification, direct marketing, medical diagnosis, query categorization, and protein function prediction [3, 4].

2 Related Work

Multilabel classification techniques are grouped into two categories that are problem transformation and algorithm adaptation method [5].

Problem transformation approach transforms the multilabel problem into a set of single-label problem. Problem transformation approach is algorithm independent. Several problem transformation methods such as binary relevance (BR) [6], label powerset (LP) [5], and classifier chain (CC) [7] are used in multilabel classification.

The algorithm adaptation method extends the available machine learning algorithms so as to handle multilabel data. The several algorithm adaptation approaches have been proposed that are multilabel k-nearest neighbor (ML-kNN) [8], binary relevance k-nearest neighbor (BR-kNN) [9], and IBLR [10].

ML-kNN and BR-kNN do not consider label correlation whereas IBLR considers label correlation but not good in terms of hamming loss.

Liu and Cao [1] proposed coupled k-nearest neighbor algorithm for multilabel classification (CK-STC). CK-STC is based on coupled label similarity [11] which updates ML-kNN algorithm which can handle label correlation. An advantage of CK-STC is that it considers label correlations and provides better performance than that of ML-kNN, IBLR, BSVM, and BR-kNN but it is more complex.

Same as traditional classification, multilabel classification also suffers from curse of dimensionality which can be handled by feature selection. Feature selection selects the relevant or more efficient features from original set of features. Traditional feature selection methods handle multilabel data by transforming multilabel problem into single-label problem and then apply feature selection.

Lee et al. [12] proposed multivariate mutual information-based feature selection. Due to transformation of multilabel problem into single-label problem, there is damage to original label structures and which reduces the performance of classifier.

So new method called information gain feature selection for multilabel data [2] can perform feature selection on multilabel data directly.

Multilabel classification tools MEKA [13] and MULAN [14] are available for handling multilabel data. MEKA is a multilabel extension to WEKA data mining tool. MULAN provides framework for implementations of many multilabel learning algorithms. In MULAN feature selection methods, multilabel data is transformed into single-label data and then traditional single-label dimensionality reduction technique is applied on these single-label data.

3 Implementation Details

3.1 FSVIG Multilabel Feature Selection Algorithm Details

FSVIG multilabel feature selection algorithm is implemented using information gain measure. Feature selection method removes the redundant and irrelevant features from database. Feature selection technique used in this work is reported in [2]. In this feature, selection filter approach is used. Information gain is used as information metric for measuring correlation degree between features and labels that are present in database. Information gain (IG) between features and entire label set is used to quantify the importance of features. IG is calculated with the help of label entropy H(L) and feature entropy H(fi). A bigger value of IG represents greater correlation between feature and labels. But in some situation, information gain of each label set may not be in same range of measurement. So for comparison, the normalized processing was performed on information gain (SU). Figure 1 shows pseudo-algorithm for feature selection FSVIG which is as follows.

3.2 CK-STC Multilabel Classification Algorithm Details

CK-STC algorithm for multilabel classification is reported in [1, 15]. In CK-STC algorithm, coupled label similarity is estimated using intra-coupling similarity between label value w_i^x and w_j^y and inter-coupling label similarity between label value w_i^x and w_j^y w.r.t feature value w_p^z. Inter-coupling label similarity is calculated with the help of co-occurrence frequency (CF) and intra-coupling similarity is calculated with the help of occurrence frequency (F). For every instance in training data, k-nearest neighbor is calculated. Coupled label similarity is considered while calculating prior probabilities and frequency arrays. Estimation of k-nearest neighbor for unseen instance is done. For each label, statistical information is calculated with the help of k-nearest neighbors of unseen instances. Unseen instance labels are evaluated via MAP (Maximum a posterior) rule. MAP rule is based on Bayes theorem.

```
Input:
F { f1,f2,f3,..........fn}
L {l1,l2,l3..............lq}
Output:
Relevant set of features
1.begin
2.Initialize IGS = null
3.for(i=1 to m )
4.IG(fi,L)= H(fi)+H(L)-H(fi,L)
5.SU( fi,L) = (2•IG(fi,L))/(H(fi)+H(L))
6.IGSi = SU( f i,L)
7.IGS = IGS ∪IGSi
8.End for
9.threshold = (1/m)∑ᵢ₌₁ᵐ IGSᵢ
10.Relevant feature = F
11.for(i=1 to m)Do
12.if  IGSi < threshold then
13.Relevant feature = Relevant feature - {fi}
14.end for
15.end
```

Fig. 1 Pseudo-algorithm for feature selection FSVIG

Figure 2 shows pseudo-algorithm for multilabel classification CK-STC which is stated as follows:

4 Experimental Setup

Table 1 shows benchmark multilabel datasets information including number of features, total number of labels, and number of instances. Experiments were performed on Genbase, Medical, and Enron dataset.

Experiments were performed on i5 processor and on Windows 7 operating system. Implementation of multilabel classification algorithm CK-STC and multilabel feature selection algorithm FSVIG was done in MS VISUAL STUDIO.NET and development tool is visual studio 10. In experiments, a proposed (CK-STC with feature selection) method was compared with existing CML-kNN method. Euclidean distance was used to calculate k-nearest neighbors. The experiments were performed on $k = 5, 7, 9$ and then the average is calculated. Tenfold cross-validation is performed on the above dataset.

Input:
Unlabeled instance (u_t), multilabel data.
Output:
Label set of unlabeled instance.
1. begin
2. Calculate intra coupling similarity $\delta^{ia}\left(w_i^x, w_j^y\right) = \frac{F(w_i^x).F(w_j^y)}{F(w_i^x)+F(w_j^y)+F(w_i^x).F(w_j^y)}$
3. Calculate inter coupling similarity $\delta^{ier}\left(w_i^x, w_j^y | w_p^z\right) = \frac{min(CF(w_p^{zx}), CF(w_p^{zy}))}{max(F(w_i^x), F(w_j^y))}$
4. Calculate coupled label similarity (CLS) $CLS\left(w_i^x, w_j^y\right) =$
$\delta^{ia}\left(w_i^x, w_j^y\right).\sum_{k=1}^m \delta^{ier}\left(w_i^x, w_j^y | w_k\right)$
5. for i=1 to m do
6. Estimate k nearest neighbor k(u_i) for each u_i
7. end for
8. for j=1 to q do
9. Estimate prior probability P(Hj) and P(-Hj)
10. Maintain frequency array
11. End for
12. Estimate k nearest neighbor k(u_t) for unseen instance u_t
13. for j = 1 to q
14 .Estimate unseen instance label statics
15. End for
16. Return label set of unlabeled instance u_t according to MAP rule
17. end

Fig. 2 Pseudo-algorithm for multilabel classification CK-STC

Table 1 Dataset Description

Datasets	Number of instances	Number of features	Number of total labels	Label cardinality	Label density	Feature type
Genbase	662	1185	27	3.378	0.064	Categorical
Medical	978	1449	45	1.252	0.046	Categorical
Enron	1702	1001	53	1.245	0.028	Categorical

5 Experimental Results

Performance of multilabel classification is measured in terms of hamming loss, one-error, and average precision. For each evaluation metric, "↓" indicates "smaller value has better results" and "↑" indicates "bigger value has better results". Bold value indicates winner of the classifier. Experiments were carried out in four sets as follows:

(1) **Multilabel classification using CK-STC without feature selection**
Table 2 shows results of CK-STC algorithm.

Table 2 Results of CK-STC

Dataset	One-error ↓	Hamming loss ↓	Average precision ↑
Enron	0.303	0.087	0.597
Genbase	0.008	0.003	0.991
Medical	0.158	0.013	0.874
Average	**0.156**	**0.034**	**0.821**

(2) **Multilabel feature selection using FSVIG algorithm**
Table 3 indicates results of feature selection FSVIG.

(3) **Multilabel classification using MULAN feature selection technique binary relevance attribute evaluator and its comparison with CK-STC with feature selection FSVIG**

The results of multilabel feature selection algorithm FSVIG and MULAN attribute selection algorithms such as binary relevance attribute evaluator algorithm represented in Tables 4, 5 and 6.

Table 4 indicates that CC algorithm with binary relevance attribute evaluator performs better than other algorithms w.r.t. hamming loss. Table 5 indicates that CK-STC algorithm with FSVIG feature selection technique performs better than other algorithms w.r.t. one-error. Table 6 shows that CK-STC algorithm with FSVIG feature selection technique performs better than other algorithms w.r.t average precision.

(4) **Multilabel classification using MULAN feature selection technique multiclass attribute evaluator and its comparison with CK-STC with feature selection FSVIG**

The results of multilabel feature selection algorithm FSVIG and MULAN attribute selection algorithms such as multiclass attribute evaluator algorithm represented in Tables 7, 8, and 9.

Table 7 indicates that CK-STC algorithm with FSVIG feature selection technique performs better than other algorithms w.r.t. hamming loss. Table 8 indicates that CK-STC algorithm with FSVIG feature selection technique performs better than other algorithms w.r.t. one-error. Table 9 indicates that CK-STC algorithm with FSVIG feature selection technique performs better than other algorithms w.r.t average precision.

Table 3 Results of feature selection as per FSVIG

Dataset	Number of features before feature selection	Number of features after feature selection
Enron	1001	596
Genbase	1185	316
Medical	1449	400

Table 4 Comparisons of CK-STC, feature selection FSVIG with MULAN binary relevance attribute evaluator w.r.t. hamming loss ↓

Dataset	Number of attribute used	CK-STC	FSVIG CK-STC	MULAN (binary relevance attribute evaluator)							
				CK-STC	BR	CC	RAKEL	ML-kNN	LP	BR-kNN	IBLR
Enron	596	0.087	0.054	0.052	0.049	0.052	0.047	0.052	0.069	0.056	0.055
Genbase	316	0.003	0.002	0.004	0.001	0.001	0.001	0.005	0.001	0.003	0.002
Medical	400	0.013	0.012	0.012	0.011	0.001	0.011	0.012	0.013	0.014	0.013
Average	–	0.034	0.022	0.023	0.020	**0.018**	0.020	0.023	0.027	0.025	0.024

Table 5 Comparisons of CK-STC, feature selection FSVIG with MULAN binary relevance attribute evaluator w.r.t. one-error ↓

Dataset	Number of attribute used	CK-STC	FSVIG CK-STC	MULAN (binary relevance attribute evaluator)							
				CK-STC	BR	CC	RAKEL	ML-kNN	LP	BR-kNN	IBLR
Enron	596	0.303	0.225	0.285	0.369	0.397	0.268	0.282	0.832	0.441	0.332
Genbase	316	0.008	0.007	0.013	0.003	0.003	0.010	0.006	0.006	0.016	0.013
Medical	400	0.158	0.146	0.192	0.190	0.197	0.238	0.200	0.262	0.283	0.231
Average	–	0.156	**0.126**	0.163	0.187	0.200	0.172	0.163	0.366	0.247	0.192

Table 6 Comparisons of CK-STC, feature selection FSVIG with MULAN binary relevance attribute evaluator w.r.t. average precision ↑

Dataset	Number of attribute used	CK-STC	FSVIG CK-STC	MULAN (binary relevance attribute evaluator)							
				CK-STC	BR	CC	RAKEL	ML-kNN	LP	BR-kNN	IBLR
Enron	596	0.597	0.617	0.643	0.613	0.595	0.622	0.643	0.211	0.578	0.626
Genbase	316	0.991	0.996	0.986	0.992	0.992	0.989	0.989	0.990	0.982	0.986
Medical	400	0.874	0.913	0.855	0.839	0.820	0.772	0.850	0.727	0.803	0.829
Average	–	0.821	**0.842**	0.828	0.815	0.802	0.794	0.827	0.642	0.788	0.813

Table 7 Comparisons of CK-STC, feature selection FSVIG with MULAN binary relevance attribute evaluator w.r.t. hamming loss ↓

Dataset	Hamming loss ↓										
	Number of attribute used	CK-STC	FSVIG	MULAN (multiclass attribute evaluator)							
			CK-STC	CK-STC	BR	CC	RAKEL	ML-kNN	LP	BR-kNN	IBLR
Enron	596	0.087	0.054	0.056	0.056	0.060	0.055	0.056	0.070	0.061	0.060
Genbase	316	0.003	0.002	0.004	0.001	0.001	0.004	0.004	0.001	0.038	0.002
Medical	400	0.013	0.012	0.012	0.010	0.010	0.011	0.011	0.011	0.013	0.013
Average	–	.034	**0.022**	0.024	0.022	0.023	0.024	0.024	0.027	0.037	0.025

Table 8 Comparisons of CK-STC, feature selection FSVIG with MULAN multiclass attribute evaluator w.r.t one-error ↓

Dataset	One-error ↓										
	Number of attribute used	CK-STC	FSVIG	MULAN (multiclass attribute evaluator)							
			CK-STC	CK-STC	BR	CC	RAKEL	ML-kNN	LP	BR-kNN	IBLR
Enron	596	0.303	0.225	0.366	0.442	0.473	0.383	0.367	0.954	0.565	0.431
Genbase	316	0.008	0.007	0.013	0.003	0.003	0.010	0.006	0.004	0.016	0.013
Medical	400	0.158	0.146	0.117	0.162	0.167	0.184	0.185	0.215	0.233	0.217
Average	–	0.156	**0.126**	0.185	0.202	0.214	0.192	0.186	0.391	0.271	0.220

Table 9 Comparisons of CK-STC, feature selection FSVIG with MULAN multiclass attribute evaluator w.r.t average precision ↑

Dataset	Number of attribute used	CK-STC	FSVIG CK-STC	MULAN (multiclass attribute evaluator)							
				CK-STC	BR	CC	RAKEL	ML-kNN	LP	BR-kNN	IBLR
Enron	596	0.597	0.617	0.587	0.568	0.549	0.515	0.590	0.108	0.555	0.569
Genbase	316	0.991	0.996	0.986	0.992	0.992	0.989	0.989	0.991	0.982	0.986
Medical	400	0.874	0.913	0.862	0.858	0.838	0.818	0.855	0.768	0.825	0.829
Average	–	0.821	**0.842**	0.812	0.806	0.793	0.774	0.812	0.622	0.787	0.795

Tables 6, 7, 8, and 9 indicate that feature selection technique FSVIG with CK-STC gives better performance in terms of one-error and average precision than feature selection algorithms reported in MULAN. Multilabel classification with FSVIG feature selection does not provide best performance in terms of hamming loss because of weak connection between labels in the datasets. Feature selection algorithms in MULAN library transform multilabel problem into single-label problem and on single-label dataset, feature selection methods are applied; so due to transformation there is damage to original label structures which reduces the performance of classifier.

For evaluating the statistical significance of the results, the Friedman test for paired data is used. The level of significance of the Friedman test was determined at $\alpha = 0.05$.

6 Conclusions

A lazy learning approach CK-STC has been reported in the literature for multilabel classification using coupled similarity between labels that improves the prediction performance, but irrelevant features present in database affect the prediction accuracy of classifier.

A multilabel feature selection FSVIG is incorporated in CK-STC algorithm. FSVIG uses information gain that shows better performance than existing multilabel feature selection algorithms when used with ML-NB, ML-kNN, and RandSvm classifier. This paper investigates a performance of FSVIG when used with CK-STC and compares its performance with other multilabel feature selection algorithms available in MULAN using standard multilabel datasets.

Experimental results show that FSVIG when used with CK-STC provides better performance in terms of average precision and one-error.

References

1. Chunming Liu and Longbing Cao, "A Coupled k-Nearest Neighbor Algorithm for Multi-label Classification", Springer PAKDD, Part I, LNAI 9077, pp. 176–187, (2015).
2. Ling Li, Huawen Liu, Zongjie Ma, Yuchang Mo, Zhengjie Duan, Jiaqing Zhou, Jianmin Zhao, "Multi-label Feature Selection 6a Information Gain", 345–355, (2014).
3. Tsoumakas, G., Katakis, I. "Multi-label classification: An over6ew" International Journal of Data Warehousing and Mining (IJDWM) 3(3), 1–13 (2007).
4. Gjorgji Madjarov, DragiKocev, Dejan Gjorgje6kj, Saso Dzeroski, "An extensive experimental comparison of methods for multi-label learning", Pattern Recognition 45 (2012) 3084–3104.
5. Grigorios Tsoumakas, Ioannis Katakis, and Ioannis Vlahavas, "Mining Multi-label Data", Data Mining and Knowledge Discovery Handbook, Springer, 2010, pp. 667–685.
6. Boutell, M.R., Luo, J., Shen, X., Brown, C.M. "Learning multi-label scene classification", Pattern recognition 37(9), 1757–1771 (2004).

7. Read, J., Pfahringer, B., Holmes, G., Frank, E.: Classifier chains for multi-label classification. Machine Learning 85(3), 333–359 (2011).

8. Zhang, M.L., Zhou, Z.H., "ML-kNN: A lazy learning approach to multi-label learning", Pattern recognition 40(7), 2038–2048 pp. 401–406. Springer (2007).

9. Spyromitros, E., Tsoumakas, G., Vlahavas, I., "An empirical study of lazy multilabel classification algorithms", Artificial Intelligence: Theories, Models and Applications, pp. 401–406. Springer (2008).

10. Cheng, W., HÂ´lullermeier, E., "Combining instance-based learning and logistic regression for multi-label classification", Machine Learning 76(2–3), 211–225 (2009).

11. Cao, L., "Coupling learning of complex interactions", Information Processing and Management (2014).

12. Lee, J., Kim, D.W. "Feature selection for multi-label classification using multivariate mutual information", Pattern Recognition Letters 34(3), 349 – 357 (2013).

13. Tidake Vaishali S. And Shirish S. Sane, "Multilabel Learning with MEKA", CSI communications (2016).

14. G. Tsoumakas, E. Spyromitros-Xioufis, J. 6lcek, and I. Vlahavas, "MULAN: A Java library for multi-label learning" Machine Learning. Res., vol. 12, pp. 2411–2414, Jul.(2011).

15. Prajakta Chaudhari and S. S. Sane, "Multilabel Classification Exploiting Coupled Label Similarity with Feature Selection", IJCA ETC (2016).

High-Performance Pipelined FFT Processor Based on Radix-2² for OFDM Applications

Manish Bansal and Sangeeta Nakhate

Abstract This paper introduces high-performance pipelined 256-point FFT processor based on Radix-2² for OFDM communication systems. This method uses Radix-2 butterfly structure and Radix-2² CFA algorithm. Radix-2 butterfly's complexity is very low and Radix-2² CFA algorithm reduces number of twiddle factors compared to Radix-4 and Radix-2. The proposed design is implemented in VHDL language, synthesized using XST of Xilinx ISE 14.1, and simulated using ModelSim PE Student Edition 10.4a successfully. Also, MATLAB code has been written and simulated with MATLAB R2012a tool. The computation speed of proposed design is observed to be 129.214 MHz after the synthesis process and SQNR is 50.95 dB.

Keywords FFT · CFA · Radix-2² · Complex multiplier · SDF
OFDM

1 Introduction

The FFT is the extensively used algorithms in digital signal communication systems. Nowadays, FFT is the prime building block for the communication systems especially for the orthogonal frequency division multiplexing receiver systems. OFDM is a multi-carrier technology for high-speed and high rate reliable data transmission in the wireless and wire communication systems such as LAN and wireless local area network (Fig. 1). WLANs are one of the most modern based on IEEE 802.11 standards [1]. Fast computational schemes for the implementation of FFT architectures have fascinated many researchers. Some of the approach to

M. Bansal (✉) · S. Nakhate
ECE Department, Maulana Azad National Institute of Technology,
Bhopal, Madhya Pradesh, India
e-mail: manishbansal2008@gmail.com

S. Nakhate
e-mail: sanmanit@gmail.com

© Springer Nature Singapore Pte Ltd. 2018
S. Bhalla et al. (eds.), *Intelligent Computing and Information and Communication*,
Advances in Intelligent Systems and Computing 673,
https://doi.org/10.1007/978-981-10-7245-1_15

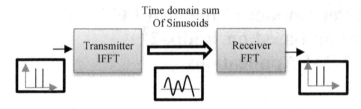

Fig. 1 OFDM system block diagram

design FFT is memory-based, pipeline-based, and general-purpose DSP. Memory based is most area efficient but it needs many computation cycles. Pipeline-based architecture possesses small chip area and high throughput rate with a lower frequency.

All fast Fourier transform designs can be isolated basically into three types of pipelined architectures—single-path delay commutator (SDC), multiple-path delay commutator (MDC), and single-path delay feedback (SDF). Among these architectures, the SDF architecture is more convenient as—(1) The SDF structure is more relevant for the fixed length FFT implementation and (2) SDF structure requires less number of delay elements as compared to SDC and MDC architectures [2].

This paper presents the design of pipelined single-path delay feedback FFT processor based on Radix-2^2 CFA for 256 point which can be used for OFDM applications. The Radix influences FFT architecture. A small Radix means simple butterfly structure and increases number of twiddle factors. A higher Radix means complex butterfly structure and reduces count of twiddle factors. In the proposed FFT Radix-2^2 architecture, Radix-2 butterfly is used which is simplest butterfly structure and Radix-2^2 uses less number of twiddle factors as Radix-4.

The controller of proposed design is simpler than other designs as overall architecture and all components are controlled by count signal only.

This paper is structured as follows—Sect. 2 represents the proposed FFT processor design based on Radix–2^2 Common Factor Algorithm (CFA). Section 3 represents data flow of proposed 256 point FFT processor. Section 4 represents the comparison. Finally, Sect. 5 represents the conclusion.

2 Proposed FFT Processor Design Based on Radix-2^2 CFA

Proposed processor is designed using Radix-2^2 common factor algorithm and Radix-2 butterfly structure which can be used for performing for 256-point FFT computation. Proposed processor is designed in SDF (Single delay feedback) pipeline architecture which is one the most efficient architectures and reduces complexity of FFT processor. Each stage of proposed design consists of Radix-2 butterfly structure (BU) and shift register (SRE). In this section, Radix-2^2 common factor algorithm, proposed FFT processor design, and required components are described for 256-point.

The DFT is defined for sequence of N as [2]

$$X(k) = \sum_{n=0}^{N-1} x(n) \cdot e^{\frac{-j2\pi nk}{N}}, \tag{1}$$

where $W_N^{nk} = e^{\frac{-j2\pi nk}{N}}$ twiddle factor which represents the N-th root. n and k are time and frequency indexes, respectively. The Radix-2^2 CFA for 256 point is formulated by using three-dimensional linear index mapping and CFA and is expressed as follows.

Divide and conquer 3-D linear index mapping is represented as

$$n = 128n_1 + 64n_2 + n_3 \tag{2}$$

$$k = k_1 + 2k_2 + 4k_3, \tag{3}$$

where

$$n_1, n_2 = 0, 1 \text{ and } n_3 = 0, 1, \ldots, 63$$

$$k_1, k_2 = 0, 1 \text{ and } k_3 = 0, 1, \ldots, 63$$

The common factor algorithm (CFA) forms [2]

$$X(k_1 + 2k_2 + 4k_3) = \sum_{n_3=0}^{63} \sum_{n_2=0}^{1} \sum_{n_1=0}^{1} x(128n_1 + 64n_2 + n_3) \cdot W_N^{nk}$$

$$= \sum_{n_3=0}^{63} \sum_{n_2=0}^{1} \sum_{n_1=0}^{1} x(128n_1 + 64n_2 + n_3) \cdot$$

$$W_N^{(128n_1 + 64n_2 + n_3)(k_1 + 2k_2 + 4k_3)} \tag{4}$$

The twiddle factor can be expressed as

$$W_N^{nk} = W_N^{(128n_1 + 64n_2 + n_3)(k_1 + 2k_2 + 4k_3)}$$

$$\underbrace{(-1)^{n_1 k_1}}_{BU} \underbrace{(-j)^{n_2 k_1}}_{PE} \underbrace{(-1)^{n_1 k_2}}_{BU} \underbrace{W_{256}^{n_3(k_1 + 2k_2)}}_{} \underbrace{W_{64}^{n_3 k_3}}_{PE} \tag{5}$$

where n_1, n_2, n_3 are the sampled terms of the input n and k_1, k_2, k_3 are the sampled terms of the output k. $(-1)^{n_1 k_1}$ and $(-1)^{n_1 k_2}$ are butterfly elements. $(-j)^{n_2 k_1}$ and $W_{256}^{n_3(k_1 + 2k_2)}$ are processing elements.

The detailed structure of proposed Radix-2^2 FFT processor for 256 point is shown in Fig. 2 which retains eight stages (S1, S2, S3, S4, S4, S5, S6, S7, and S8). SRE indicates shift register which stores values. In case of 256 point, first shift register stores 128 values, second shift register stores 64 values, third shift register stores 32 values, fourth shift register stores 16 values, fifth shift register stores 8 values, sixth shift register stores 4 values, seventh shift register stores 2 values, and last eighth shift register stores 1 value. Twiddle ROM has 189 number of twiddle factors values which are calculated based on W_{256}. Swap $(-j)^{n_2k_1}$ performs real sign inversion and swaps signed inverted real and imaginary values. Butterfly (BF) performs addition, subtraction, and by-pass operations. Complex multiplier multiplies input complex values with twiddle factor values.

Figure 3 shows the butterfly (BU) structure which performs operation between n-th value and $(n + N/2)$th value. On first 128 cycles, the signal T of Mux1, 2, 3 and 4 switch to position "0" in butterfly unit. The input data is moved to the shift registers until shift registers are filled. On next $N/2$ cycles, the signal T of Mux1, 2, 3 and 4 switch to position "1", and then butterfly structure starts addition/subtraction operation to compute 2-point DFT between the data stored in the SREs and incoming data [3]. The subtraction results are stored in same register; first half number of addition results are stored in next stage register and second half number of addition results start addition/subtraction with stored first half number of addition

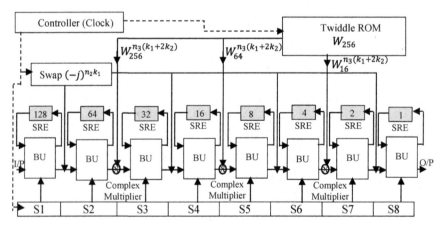

Fig. 2 Proposed FFT processor based on Radix-2^2 for 256-point

Fig. 3 Butterfly (BU) structure

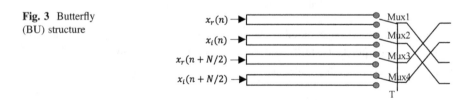

Fig. 4 Trivial (-j) multiplier (swap)

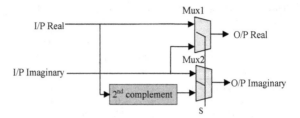

in the same stage. Thus, this process runs from current stage to next stage till last stage and butterfly element is connected in pipelined structure in order to compute a result in every clock cycle.

In Fig. 4, trivial $(-j)$ multiplier performs swapping of imaginary real value and sign inversion of real value. This swapping between imaginary and real values is controlled by multiplexer control signal "S" and sign inversion of real value is done by second compliment. Whenever $-j$ multiplication is required as per signal flow graph, Mux1 and Mux2 multiplexors switch to position "1" and then first, sign of real value is inverted and then inverted real value is swapped to imaginary and imaginary value is swapped to real. In this way, sign inversion and swapping are done instead of multiplication in case of $-j$ multiplication. This swap approach increases the performance and reduces the hardware logic.

3 Data Flow of Proposed 256-Point FFT Processor

Proposed FFT processor is designed for 256-point FFT. As shown in Fig. 2, all components—BFs, stages, SREs, Twiddle ROM, and swap—are configured by count control signal based on value of $n_1, n_2, n_3, k_1, k_2, k_3$ and all operations are controlled by count signal. Here, proposed FFT is designed which takes 33% less number of twiddle factors than conventional FFT processor [4].

Proposed FFT design uses eight stages, eight shift registers, and one twiddle ROM. First shift register stores 128 values, second shift register stores 64 values, third shift register stores for 32 values, fourth shift register stores 16 values, fifth shift register stores 8 values, sixth shift register stores 4 values, seventh shift register stores 2 values, and eighth shift register stores 1 value. Twiddle ROM has 189 number of twiddle factors value which are calculated based on W_{256} and generate $W_{256}^{n_3(k_1 + 2k_2)}$, $W_{64}^{n_3(k_1 + 2k_2)}$, and $W_{16}^{n_3(k_1 + 2k_2)}$ twiddle factors value. Swap $(-j)^{n_2 k_1}$ in first stage and twiddle factors $W_{256}^{n_3(k_1 + 2k_2)}$ operations in second stage are performed as per 256 point samples. Similarly, Swap $(-j)^{n_2 k_1}$ in third stage and twiddle factors $W_{64}^{n_3(k_1 + 2k_2)}$ operations in fourth stage are performed as per 64 points samples. Similarly, Swap $(-j)^{n_2 k_1}$ in fifth stage and twiddle factors $W_{16}^{n_3(k_1 + 2k_2)}$

Table 1 Processing elements value after stages at each position for FFT processor based on Radix-2^2 CFA

PE N	S1 stage $(-j)^{n_2 k_1}$	S2 stage $W_N^{n_3(k_1+2k_2)}$	S3 stage $(-j)^{n_2 k_1}$	S4 stage $W_N^{n_3(k_1+2k_2)}$	S5 stage $(-j)^{n_2 k_1}$	S6 stage $W_N^{n_3(k_1+2k_2)}$	S7 stage $(-j)^{n_2 k_1}$
256	$n_1, n_2, k_2, k_3 = 0, 1$ $n_3, k_3 = 0, 1,$ $2, \ldots 63 \& N = 256$		$n_1, n_2, k_2, k_3 = 0, 1$ $n_3, k_3 = 0,$ $1, 2, \ldots 15 \& N = 64$ and repeating for every next 64 points		$n_1, n_2, k_2, k_3 = 0.1$ $n_3, k_3 =$ $0, 1, 2, 3 \& N = 16$ and repeating for every next 16 points		$n_1, n_2, k_2, k_3 = 0, 1$ $n_3, k_3 = 0 \& N = 4$ and repeating for every next 4 points

operations in sixth stage are performed as per 16 points samples. Similarly, swap $(-j)^{n_2 k_1}$ in seventh stage is performed as per four points samples as mentioned in Table 1.

In the 256-point FFT design, first 128 point is stored in stage 1 register using butterfly structure in each clock and takes 128 clock. At 129th clock as x(128) point inputted into stage 1, butterfly start addition and subtraction between x(0) and x(128) point and this addition is stored in stage 2 register and subtraction is stored in stage 1 register. At 130th clock as x(129) point inputted into stage 1, butterfly again starts addition and subtraction between x(1) and x(129) point and this addition is stored in stage 2 register and subtraction is stored in stage 1 register. This process runs up to 192th clock. At 193th clock as x(192) point inputted into stage 1, stage 1's butterfly again starts addition and subtraction between x(64) and x (192) point and this addition is stored in stage 2 register and subtraction is stored in stage 1. At this clock, stage 2's butterfly also starts addition and subtraction and stores addition in stage 3 register and subtraction in stage 2 register. This process continuously runs up to 256th clock. At 256th clock, as x(255) point inputted stage 1's butterfly again starts addition and subtraction between x(127) and x(255) point and this addition is stored in stage 2 register and subtraction is stored in stage 1 and at this clock, buttery structure of stages 2, 3, 4, 5, 6, 7, and 8 also starts addition and subtraction and stores addition in stages 3, 4, 5, 6, 7, and 8 registers and subtraction in stages 2, 3, 4, 5, 6, and 7 registers. Stage 8 output is the final FFT design values.

The twiddle ROM in the proposed processor has 189 number of twiddle factors values which are used at even stages based on values of $n_1, n_2, n_3, k_1, k_2, k_3$ at each point and multiplied with incoming data signal as mentioned in Table 1. This approach is reduced extra hardware as same twiddle factors values of twiddle ROM are used at even stages by using symmetric $W_N^{k+N/2} = -W_N^k$ and periodicity $W_N^{k+N} = W_N^k$ property. In the proposed processor, twiddle factor W_N^0 value is bypassed instead of multiplying wherever is required to multiply as W_N^0 value is 1. These all approaches are not only reducing hardware but also increasing the speed of the proposed processor (Fig. 5).

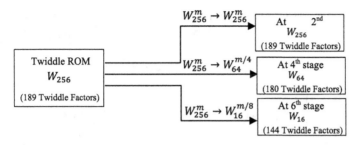

Fig. 5 Twiddle factors at even stages

4 Comparison

The architecture of proposed design is implemented in VHDL language, synthesized with xc7vh290t-2hcg1155 device and simulated in ModelSim and MATLAB to verify its results.

Table 2 shows proposed FFT design with conventional FFT design comparison. The results show that proposed architecture design used 513 twiddle factors and 18 real multipliers, while conventional design use 768 twiddle factors and 16 complex multiplier. 16 complex multiplier means 64(16*4) real multiplier. Thus, proposed design uses less twiddle factors compared to conventional design which causes less number of multiplier and less area.

Table 3 shows synthesized results of proposed design with conventional FFT design [4] between max clock frequency and other parameters. Max clock frequency of proposed design is 129.21 MHz which is higher speed than conventional design.

Figure 6a–c represents comparison of max frequency, number of twiddle factors, and number of multiplier, respectively, between proposed and conventional design. It represents 70% high frequency, 33% less number of twiddles factor, and 71% less number of multiplier of proposed design compared to conventional.

Figure 7 shows the register-transfer level (RTL) schematic view of proposed FFT processor which is generated while synthesizing VHDL code using xc7vh290t-2hcg1155 FPGA device by Xilinx 14.1 synthesis XST tool.

Table 2 FFT architectures parameters

Architecture	Number of point	Method	Radix	Number of stages	Number of twiddle factors	Number of real multiplier	Memory
FFT [4]	256	Pipeline	Radix-2^4	8	768	16*4	255
Proposed FFT	256	Pipeline	Radix-2^2	8	513	18	255

Table 3 Implementation parameters comparison

Parameters	FFT [4]	Proposed FFT
Number of point (N)	256	256
FPGA Kit	Xilinx Virtex2 1500	xc7vh290t-2hcg1155
Number of slices used	–	725
Number of slices Flip Flops used	–	538
Number of four input LUTs used	–	1345
Number of GCLK used	–	6
Max frequency (in MHz)	35.76	129.21
SQNR	–	50.95

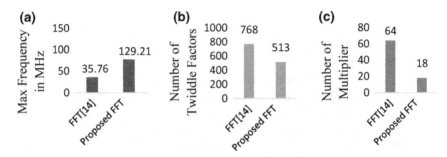

Fig. 6 Proposed and conventional FFT design comparison

Fig. 7 RTL schematic view of proposed FFT processor

5 Conclusions

The purpose of this research is to implement FFT processor for 256 point based on Radix-2^2 CFA for OFDM applications and use less number of twiddle factors than conventional FFT which reduces the memory and area and reduces number of multiplications. The less number of multiplications reduces processing time and increases maximum frequency. This maximum frequency of proposed design is higher which shows higher speed than conventional design.

References

1. Ahmed Saeed, M. Elbably, G. Abdelfadeel and M. I. Eladawy, "FPGA implementation of Radix-2^2 Pipelined FFT Processor," Proceedings WAV'09 Proceedings of the 3rd WSEAS international symposium on Wavelets theory and applications in applied mathematics, signal processing & modern science, 2009, pp. 109–114.
2. Jiang Wang and Leif Arne Ronningen, "An Implementation of Pipelined Radix-4 FFT Architecture on FPGAs," Journal of Clean Energy Technologies, Vol. 2, No. 1, January 2014, pp. 101–103.
3. Wei-Hsin Chang and Truong Nguyen, "An OFDM-specified lossless FFT architecture," IEEE Transactions on Circuits and Systems I, 2006, pp. 1235–1243.
4. Chih-Peng Fan1, Mau-Shih Lee and Guo-An Su, "Efficient low multiplier cost 256-point FFT design with Radix-2^4 SDF architecture," Journal of Engineering, National Chung Hsing university, Vol. 19, No.2, 2008, pp. 61–74.

An Image Processing Approach to Blood Spatter Source Reconstruction

Abhijit Shinde, Ashish Shinde and Deepali Sale

Abstract Blood spatter analysis is a part of Criminal Justice system and subpart of the forensic science. Traditional blood spatter analysis has a problem with the crime scene contamination. The crime scene contamination can led to the unacceptance of the evidences. The blood spatter analysis is the process which heavily relies on the expertise of the forensic scientist. The human intervention also creates the problem of errors and misjudgments. Use of image processing to the whole will deal with automation of removing human factor. The proposed method takes the image from blood spatter using image processing and reconstructs the source of the blood. The proposed methodology uses the Otsu's method for thresholding and Hough transform for edge detection.

Keywords Blood spatter trajectory · Image processing · Trajectory analysis
Blood spatter analysis · Forensic image processing

1 Introduction

Forensic science is an integral part of the justice system because it helps in collecting the evidences on a scientific basis. The blood pattern analysis or Blood spatter analysis is one of the tools used for reconstruction of the crime scene.

A. Shinde (✉)
Bhima Institute of Management and Technology, Kagal, Kolhapur 416202, India
e-mail: abhijitvshinde@gmail.com

A. Shinde
Sinhgad College of Engineering, Pune-41, Maharashtra, India
e-mail: shindeashishv@gmail.com

D. Sale
Dr. D. Y. Patil Institute of Engineering Technology, Pimrpi, Pune- 18, India
e-mail: deepalisale@gmail.com

© Springer Nature Singapore Pte Ltd. 2018
S. Bhalla et al. (eds.), *Intelligent Computing and Information and Communication*,
Advances in Intelligent Systems and Computing 673,
https://doi.org/10.1007/978-981-10-7245-1_16

The BPA traditionally has predominant skill of the Forensic Scientist. The National Academy of Sciences, USA has emphasized the need of more standardization in the field and the better training for the analysts, so that the reliability of BPA cannot be questionable in the courts [1, 2].

The blood spatter originates in the violent crime scenes. The source of blood is an exit wound of blood. The blood which impacts on the wall leaves the stain pattern which is blood spatter. The blood droplets will be circular if the impact angle is 90°, otherwise they will be elliptical. The shape of the blood droplets depends upon the gravity, the speed, blood viscosity, air drag, and surface tension of the blood drop [1, 2].

The blood spatter analysis has been neglected by the researchers of the image processing and only few research is available which was highlighted by Abhijit and Sale [3].

Abhijit and Sale [4] has explained all steps needed to be carried out in blood spatter analysis but did not used any dataset as a standard dataset that is absent. So the dataset was generated to test the output of the proposed methodology. Giovanni Acampora et al. stressed the need for the automation of blood spatter analysis process. Authors had developed the cognitive robot for the purpose of the blood spatter analysis but the research work fails to propose the methodology blood spatter analysis [5].

The basis blood spatter scenario is explained in Fig. 1; projections l_1 and l_2 are extensions of major axes of blood droplet, respectively. Projections l_1 and l_2 meet at point P_1 which is on y-z plane. Point P_1 is the projection of the blood source on yz plane. Angles θ_1 and θ_2 are the angles between sources l_1 and l_2, respectively. The traditional blood spatter analysis calculates these angles by taking tan inverse of major and minor axes of blood drop [1].

Fig. 1 3D projection representation of blood spatter analysis

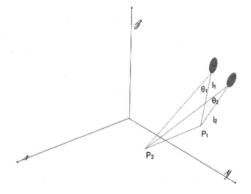

2   Methodology

2.1   *Orthorectification and Interpolation*

The plane of the camera and the plane of the surface of the blood spatter may not be parallel to each other which results in the error in determining the dimensions of the ellipses. To bring the images in the same plane, the orthorectification is performed, using reference point [6]. The blue dots are placed in four corners; these corners form the square. As the planes of the camera and surface are not parallel, the dots on corner form the trapezoid which is rectified, as shown in Fig. 2.

2.2   *Blood Droplet Segmentation*

The next step in the process will be blood droplet segmentation. This step finds out the belonging of the pixel to its corresponding blood droplet. This step performs not only the segmentation but also the background removal.

The method which will be suitable for the above-mentioned condition is adaptive Otsu's Method. This method will not only segment the blood droplets but also make the background black. The one more advantage of using Otsu's method is that it eliminates the small droplet and the large drops in the process of segmentation.

The large drops do not take the shape of the ellipses but takes the shape of the circle, whereas smaller droplets cannot be used for the mathematical processing of finding major and minor axes. So if we considered that the spatter creates the Gaussian distribution, then after applying area criteria we use blood droplets which are segmented and their distribution will be in very narrow range centered around the mean of all segmented areas [7]. This has been illustrated in Fig. 3.

Fig. 2 The blood spatter image before and after the orthorectification

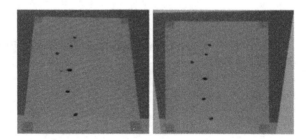

Fig. 3 Blood droplet
segmentation

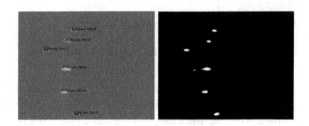

2.3 Ellipse Detection

The Hough transform is used to detect the ellipse. The ellipse detection and fitting
help us to find only the point which satisfies the equation of the ellipse, whereas we
need to develop the algorithm to find out the needed parameters, which are lengths
of major and minor axes, and the orientation of the ellipse [8].

We need two pints to find out the centroid of the ellipse and the length of the
major axis. It can be given by Eq. 1

$$a = \sqrt{(x_2 - x_1)^2 + (y_2 - y_1)^2} \tag{1}$$

$$\alpha = \tan\left(\frac{y_2 - y_1}{x_2 - x_1}\right) \tag{2}$$

Here, a is the length of the major axis and the α is the orientation of the ellipse
(McLaughlin [8]). We need the second point to find out the length of the minor axis.
It can be given by equations

$$b = 2\sqrt{\frac{a^2 d^2 \sin^2 \tau}{a^2 - d^2 \cos^2 \tau}} \tag{3}$$

$$\cos\tau = \frac{a^2 + d^2 - f^2}{2ad} \tag{4}$$

Here, b is the length of the minor axis (Yuen et al. [9]). Other parameters can be
observed in Fig. 4.

Fig. 4 Ellipse geometry, f_1
and f_2 are the ellipses foci [10]

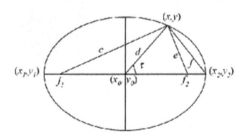

2.4 Major and Minor Axes Analysis of Individual Ellipses

The major axis and minor axis will help us to draw the projection in 3D. Here, we have to find out the projection for the y-z plane and x-z plane separately. The y-z plane is the plane of the surface which has blood spatter; the x dimension is the dimension coming out of the y-z plane. For the y-z plane, we need the orientation of each of the major axes, whereas for the x dimension projection, we need the lengths of major and minor axes. Figure 5 shows the major and minor axes detection along with ellipses.

2.4.1 The y-z Plane Projection

As we discussed in the literature survey, the major axis is useful for the determining the direction of the travel of the blood droplet. If we extend the major axis of each of the blood droplets, then they will meet each other. Theoretically, projections are supposed to meet at single point but practically they do not. So we need to make some approximation. So we put forward the concept of the profile line. We draw the mean line with mean orientations of all projections, and draw a perpendicular line to it, known as profile line. We will move the profile line in x-y direction. The profile line will measure the cumulative distances of each line with other. Wherever the cumulative distance is minimum, we will take that point as the common intersection point. It can be expressed in Eq. 5 and can be represented as shown in Fig. 6a, b [2].

$$\min\left(\frac{1}{N-1}\left[\sum_{n=1}^{N-1}(y_0 - y_n)\right]\right) \geq 0 \tag{5}$$

The projection then made with a single point as the intersecting point. This is shown in Fig. 6c. This is the projection of the source on the x-z plane, i.e., on the surface of the blood spatter.

Fig. 5 Major and minor axes tracking

(a)   **(b)**   **(c)**

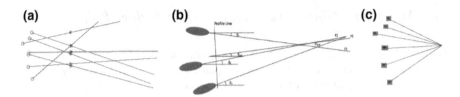

Fig. 6 a Plotting of profile line. **b** Schematic showing profile plane. **c** The common point of conversions

2.4.2 Projection in the *x* Dimension

Once we found out the 2D projection, we need to convert it to 3D. To get the position in 3D, we need to draw projection from the common intersecting point and a projection making an angle with major axis. The shape of the blood droplet changes with impact angle as shown in Fig. 7.

We can see at 90° blood drop is perfectly 90°, whereas with an angle other than 90° it forms ellipse and the major axis length increases with increase in angle.

The traditional method calculates the angle by Eq. 6:

$$\theta = \sin^{-1}\frac{b}{a} \tag{6}$$

We have considered that the angle will be proportional to the ratio of the major and minor axes:

$$\theta \propto \frac{b}{a} \tag{7}$$

We found out the ratio of a/b for the angles from 10° to 70° with interval of 5° and by training the program we found out the ratios for intermediate angles.

2.4.3 Plotting of the Projections

Figure 8 shows the plotting of the paths in *y-z plane* and *x-z plane*. The coordinates got from this are used to generate the 3D projection as shown in figure.

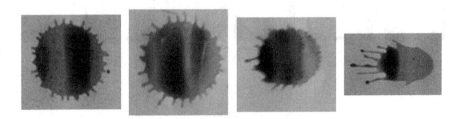

Fig. 7 Angle of impact affection of the shape of the blood drops

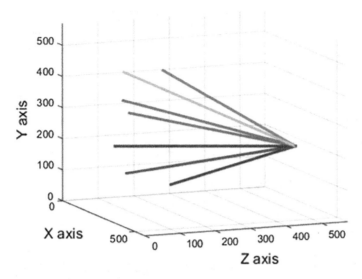

Fig. 8 3D projection

2.5 Gravity Correction

The gravity affecting the blood droplet makes its path parabolic. The above assumed steps have considered the flight of the blood as the straight line which is needed to be corrected. This flight under the influence of gravity make blood drop to strike the wall at a position lower than it would had been under no gravity condition. This height adjustment has to be done. This height is added in z direction as the z direction represents the height. The processes have been illustrated using Fig. 9, and the output after correction is shown in Fig. 10.

The calculation for the gravity correction can be given by Eq. 7 [4]:

$$Z_A = Z_D + \sqrt{\left\{ z - \frac{1}{2} \times 9.8 \times \left(\frac{Y}{3.2} \right)^2 \right\} - Y^2} \tag{7}$$

3 The Dataset

The dataset was generated using the method described in research apparatus (Acampora et al. [5]). The dataset consists of 35 samples of blood spatter describing various scenarios reconstructed. Figure 11a–c represents groups which contain perfectly shaped ellipses; Fig. 11d, e shows very few uneven and non-elliptical shapes, while 11f very high degree of inconsistency in the formation of elliptical spatter.

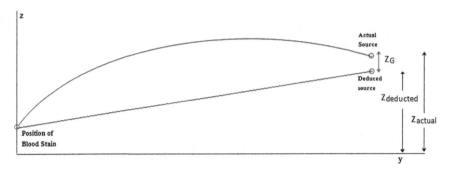

Fig. 9 The schematic of gravity correction

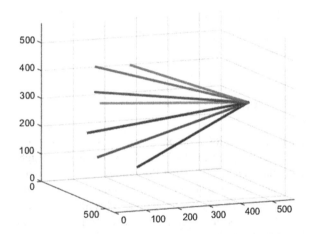

Fig. 10 3D projection with gravity correction

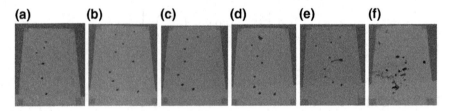

Fig. 11 a, b and **c** Blood spatter with perfectly shaped ellipses. **d** and **e** Blood spatter with one blood drop having uneven shape. **f** Messy blood spatter resembling actual crime scene blood spatter

Table 1 Error estimation

Sr no.	Image	Actual distance (mm)	Estimated distance (mm)	Error in distance estimation (%)	Actual angle	Estimated angle	Error (%)
1	Figure 11a	1300	1343.7814	3.367	30°	36.2349°	20.783
2	Figure 11b	1300	1235.7948	−4.9389	40°	35.473°	−11.31
3	Figure 11c	1300	1317.6864	1.36049	30°	29.5632°	−1.45
4	Figure 11d	1300	1374.6556	5.74274	35°	40.6501°	16.14
5	Figure 11e	1300	1241.3125	−4.5144	35°	34.6944°	−0.87
6	Figure 11f	1300	1436.5138	10.5011	50°	55.0622°	10.12
		Mean error in distance estimation (%)		5.2	Mean error in angle estimation (%)		10.2

4 Result

Table 1 shows estimations of errors for the mean distance from the source and the blood droplet and angles of projection. We have considered the constant distance of 1300 mm, varying the values of angle for the estimation of error.

Average estimation error for the angle of projection for various samples approximates at 10.2%. While in a case it has reached to 20%, the error is a certain degree of prediction which is used to compensate mathematical errors. The average estimation error in distance calculation is 5.02% only and has highest error percentage of 10.5%.

5 Conclusion and Future Scope

Results of proposed methodology have replicated the manual process of blood spatter analysis. The result also reflects the less error margin, whereas the biggest success of the proposed methodology is that it has eliminated the human factor from the process which will help in avoiding crime scene contamination and human error. This will help in making strong evident in criminal justice. The 3D modeling proposed will also help in increase in reliability.

The absence of standard database has limited result analysis. In a further step, the development of the methodology will be in the generation of extensive dataset, checking results, an extension of a model for spatter stains on multiple planes, and generation of parabolic flight trajectory.

Disclaimers The fake blood used for experimentation was made up of non-animal food products such as corn syrup, sugar, coco-powder, and artificial food coloring. The resulting product is usually used for forensic experiments. No human being or animal was harmed for the experiment. The procedure for generating blood spatter is similar to one followed by many other authors [1, 2].

References

1. Anita Wonder BS, "Blood Dynamics", Anita Y. Wonder, Academic Press, 2001 ISBN-10: 0127624570, 1st edition August 2001, pp. 36–51.
2. P. Joris, W. Develter, E. Jenar, D. Vandermeulen, W. Coudyzer, J. Wuestenbergs, B. De Dobbelaer, W. Van de Voorde, E. Geusens, P. Claes, "A novel approach to automated bloodstain pattern analysis using an active bloodstain shape model", Journal of Forensic radiology and image, Volume 2, Issue 2, April 2014 pp. 95.
3. Abhijit Shinde, Dr. Deepali Sale, "Blood Spatter Trajectory Analysis for Spatter Source Recognition Using Image Processing", 2016 IEEE Conference on Advances in Signal Processing (CASP), Cummins College of Engineering for Women, Pune, pg. 375–380, Jun 9–11, 2016, DOI: 10.1109/CASP.2016.774619934-39.
4. Abhijit Shinde, Dr. Deepali Sale, "Blood Spatter Source Reconstruction Using Image Processing", IETE National Journal of Innovation and Research (NJIR), ISSN 2320–8961, Volume III, Issue II, December 2015.
5. Giovanni Acampora, Autilia Vitiello, Ciro Di Nunzio, Maurizio Saliva, Luciano Garofano, "Towards Automatic Bloodstain Pattern Analysis through Cognitive Robots", conference—IEEE International Conference on Systems, Man, and Cybernetics, 2015.
6. Wen-Yuan Chen, Yan-Chen Kuo, Chen-Chung Liu and Yi-Tsai Chou, "Region-Based Segmentation Algorithm for Color Image Using Generated Four-Connectivity and Neighbor Pixels Comparing Criterion", Proceedings of the 2005 Workshop on Consumer Electronics and Signal Processing, 2005.
7. G. Gordon, T. Darrell, M. Harville, J. Woodfill, "Background estimation and removal based on range and color", Proceedings of the IEEE Computer Society Conference on Computer Vision and Pattern Recognition, (Fort Collins, CO), June 1999.
8. Robert A. McLaughlin, Technical Report, "Randomized Hough Transform: Improved Ellipse Detection with Comparison", University of Western Australia, 1997.
9. H. K. Yuen, J. Illingworth, J. Kittler, "Detecting partially occluded ellipses using the Hough transform", Image and Vision Computing—4th Alvey Vision Meeting archive, Volume 7 Issue 1, February 1989, pp. 31–37.
10. Yonghong Xie; Dept. of Comput. Sci., Nevada Univ., Reno, NV, USA; Qiang Ji, "A new efficient ellipse detection method", Pattern Recognition, 2002. Proceedings 16th International Conference on (Volume: 2), 2002, pp. 957–960.

Overlapping Character Recognition for Handwritten Text Using Discriminant Hidden Semi-Markov Model

Ashish Shinde and Abhijit Shinde

Abstract The field of handwritten character recognition has always attracted a large number of researchers. The proposed methodology uses Discriminant Hidden Semi-Markov Model for tackling the problem of recognition of handwritten characters. Preprocessing on the input image such as denoising and adaptive thresholding is done for input conditioning, followed by segmentation for finding the area which contains text. The text image is then passed through the second stage of segmentation, which separates overlapping characters. Then, these segmented characters are digitized using feature extraction. For feature extraction, Discriminant Hidden Semi-Markov Model is used. For feature matching and character extraction, the proposed methodology uses KNN Classifier. The training feature library, consisting 180 samples of each character in the capital and small, processed using training algorithm of Discriminant HsMM. Paragraphs of 80–120 characters are processed in recognition module. The 86% average accuracy rate is achieved for a large set of characters.

Keywords Overlapping character segmentation · Word recognition
Preprocessing techniques · Handwritten text

1 Introduction

The pathway to higher end digital handheld devices like mobile, PDAs, etc. Hassle free is one of the most researched areas of interest in digital devices nowadays. One of such areas is input through handwritten characters or phrases. The features of

A. Shinde (✉)
Sinhgad College of Engineering, Pune-41, Maharashtra, India
e-mail: shindeashishv@gmail.com

A. Shinde
Bhima Institute of Management and Technology, Kagal,
Kolhapur, Maharashtra 416216, India
e-mail: abhijitvshinde@gmail.com

© Springer Nature Singapore Pte Ltd. 2018
S. Bhalla et al. (eds.), *Intelligent Computing and Information and Communication*,
Advances in Intelligent Systems and Computing 673,
https://doi.org/10.1007/978-981-10-7245-1_17

handwritten character make these tasks comparatively difficult due to high variance in any person's handwriting. The level of difficulty further increases when the phrases or sentences are written without giving much importance to the separateness of characters which causes them to overlap. A solution is to separate the characters and recognize them separately. Hence, this research proposes a person-dependent handwritten character recognition using Discriminant Hidden Semi-Markov Model.

HMM was proposed by Stratonovich [1] in the field of mathematical modeling of speech recognition. Later it was applied in DNA sequencing, character recognition, and many other applications. The use of HMM in character recognition has been really popular and successful hence. The switch to Discriminant Hidden Semi-Markov model from HMM happened very recently due to the shortcomings of HMM. The most prominent of them is hidden states of model. The important feature of HsMM over HMM is the visibility of intermediate states. Figure 1 shows general structure and observation scenario in HsMM. HMM is called hidden as the processing states of the model for each input or event are hidden and hence are difficult to understand and improve for a better result. HsMM overcomes this obstacle by generating intermediate outputs which are observable and can also be processed on. These observations are complex to understand directly and usually need processing before the presentation. This implies that, unlike in HMM where a single state produces one observation, a state in HsMM produces a sequence of characters.

In most cases of handwritten text, there is unreliability of recognition models due to the incomplete, ambiguous, or imprecise contents of the handwritten text images. So, performing handwritten text recognition aims toward capturing the organization of the complexity and it has to deal with a high degree of tolerance toward uncertainty. To summarize it is quite a challenging task to design a simplified algorithm which may be optimally cost effective.

The appropriate feature extraction algorithm selection is the most important step, because it decides the degree of recognition performance [2].

HMM is utilized for feature recognition of the words and the optimal paths algorithm for classification. When tested on a group of 98 words, this system achieved 72.3% success [3]. Another application of HMM in OCR was presented by Huang et al. [4] with character segmentation-based word recognition approach

Fig. 1 General structure and observation scenario in HsMM

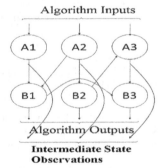

(2001), who used CDVDHMM to build character models. This technique suffers from slow speed to train and in operation.

Kamble et al. (2014) for the purpose of extraction of features have used the "Rectangle Histogram Oriented Gradient representation" [5]. The recognition rate for printed characters for this method nears at 95%.

The work by Deepjoy Das et al. has used SVM and HMM for the "Online character recognition system for Assamese language". The authors further performed a recognition performance analysis for both the models [6]. After training, 181 different Assamese Stokes Recognition models were generated.

In a study, Yarman-Vural and Arica [7] have overviewed the present status of CR research. The recognition results of different algorithms appear successful. Most of the recognition accuracy rates reported are over 85%.

Most of the above algorithms work specifically for printed characters with less variations. Consequently, the problem of touching and overlapping characters are not handled by many researchers and therein lies their shortcomings. We propose an innovative perspective and solutions for this problem. The preprocessing addresses the problem of touching characters, while the HsMM algorithm's efficient training module solves the problem of variation in handwriting pattern of a person and provides a higher recognition rate.

2 Methodology

The proposed approach to recognize handwritten characters adopts a Discriminant Hidden Semi-Markov field model for overlapping character segmentation and recognition [6]. In the proposed problem, segmentations are modelled as the duration of output or intermediates output state. The proposed methodology finds segmentation and corresponding labeling for each of initial segments of image most likely to be successful in separating characters.

2.1 Preprocessing

2.1.1 Thresholding

The proposed adaptive binarization methodology relies on the calculation of local mean of block segments of the image. The technique is based calculation of local threshold using formula

$$T_{\text{sub},P} = \mu_{\text{sub},P}\left(1 + \alpha * \log N_{\beta,P}\right) \tag{1}$$

2.1.2 Skew Correction

The algorithm here uses entropy scoring measurement for the image. A simple way to do it is, first decide constraints for the maximum skew angle. Measure least entropy score for each angle change in previously decided step-size using formula. A sum of entropies is calculated for each step-size. The smallest of entropy sum is for least skewed text image.

2.1.3 Character Segmentation

Projection profiles and reverse water reservoir have been used for finding overlapping area of characters. Projection profiles are one-dimensional arrays used to store the count of pixel each with value "1" in binary images which belong to each character's area when normal X–Y axis profiles are created. Each value in the projection array is generated with the number of pixels above a predefined threshold. The reverse water reservoirs are created by considering characters upside down and imagining raining over them. The sections which look like lakes will most probably hold water in them and hence these areas are called water reservoirs.

If the characters are isolated then the process is aborted, otherwise, for connected characters the overlapping space is determined. The cut-set are selected based on both the algorithms, where a bottom of the reverse reservoir and initial cut-sets are used in the following algorithm.

Step 1.   *A segmented line sub-image is input*

Step 2.   *A vertical Projection of Image is calculated*

$$Vproj(j) = \sum_{i=1}^{no\ of\ columns} I(i,j)\ For\ j = 1\ to\ no.\ of\ rows\ \text{--------- (2)}$$

Step 3.   *Binarization of vertical projections*

Step 4.   *A profile of lowermost edge of whole image is generated*

$$Lprofile(1,j) = (max(i_{TRUE}),j)\ For\ all\ j\ \qquad \text{--------- (3)}$$

Step 5.   *Using a reverse gravity water reservoir to find overlap area and its local reverse bottom is chosen as area of segmentation*

Step 6.   *The point at which binarized '1' of vertical projection and reverse bottom is chosen as cut-point*

Step 7.   *Search nearest −k and +k reverse bottom which is higher than selected local reverse bottom as cut-point*

    a.   *If local reverse bottom with higher than selected bottom is not found, then cut-point remains same*

    b.   *If local reverse bottom with higher than selected bottom is found, then cut-point is made larger to cover initial and final local bottoms and all pixels in between.*

Step 8.   *Pixel value of column near and at cut-point and between later found bottom are set to background color*

2.2   Character Feature Recognition Using HsMM

For recognition of each character, a given segmentation and corresponding labeling is generated using similarity functions. The similarity function captures the appearance of a character by means of gradient features. Additional functions describing character properties such as bounding boxes overlap between characters and gaps between characters are also used (Fig. 2).

A segmentation in this model will induce a set of unknowns y and a corresponding set of similarity functions $\{CCR\}_{C \epsilon c}$. Each unknown $y_i \epsilon Y(T)$ generates a function $Y \left(CG, CS^A_{r,t}, CS^B, CG^G_{n,r} \right)$. The sample in Fig. 3 shows one word of a sentence. In that word, there exist five separated regions with the probability of characters present which are given labels. Notice that character segment regions, $y4$, which show the "blank space" character. Considering "blank spaces" between words, a special character which will be recognized which will enable us to differentiate between two sequential words and arrange the words in a defined index. As shown in the figure, there are multiple overlaps (as for $y1$ and $y2$) and multiple gaps (as for $y2$ and $y3$).

Fig. 2 Water reservoirs forming in a word with boundaries. Initial cut-points and final cut-points of segmentation

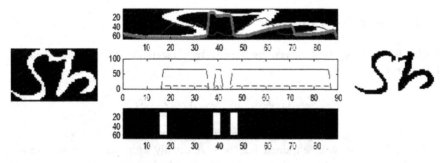

Fig. 3 Example of input, reservoir formation and output of proposed projection profiles and reverse water reservoir technique for character segmentation

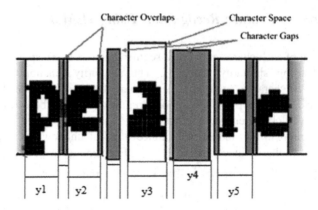

Fig. 4 Segments of a word with 5 characters into 5 character segment regions (annotated by $y1$ to $y5$), which can be segmented for recognition

$$CG(y|x, CC)\alpha \exp\left\{\sum_{C \in c} CS_C(y_c, x; CC)\right\} \tag{4}$$

where x is a segment of the image being conditioned upon and C a feature function for character learned with training. The notation y_C will indicate sub-function of character function CS_C (Fig. 4).

2.3 Feature Function Extraction

Here, it is proposed to use five sub-functions for calculating the discriminant feature of the character. Which are Appearance of Character, Segment Block, the overlap between character segments, the gap between character segments, and character gradients.

2.3.1 Appearance of Character

Each character segment, given by CS, is recognized using pretrained Feature Function CC,

$$CS_{r,t}^A(y, x; CC) = CC_{r,t}^A(y) \cdot F_{r,t}(x) \tag{5}$$

for a character segment given by 'y' and the segment width from r to t.

Thus, the model will be trained that "w"s are usually wide while "i"s are narrow. The features $F_{r,t}(x)$ of each character image are calculated from r to t, which are overlapping consecutive character spaces but the flexible design of the recognition model enables researchers to use overlapping segments of the sub-image even if they are extending the character segment.

2.3.2 Character Gradients

Feature function energies can be easily trained in the form of gradients. Each neighboring series of character gives us a bigram score in form of overlapping or nonoverlapping segment energy, $CS^B(y', x; CC) = CC^B(y', y)$.

2.3.3 Overlap in Characters

A pair of neighboring segments may have an overlap or character gap between them. In such case of neighboring character having overlap, an segment flexibility is added, $U_{n,r}^o(CC^O) = CC_{n,r}^O$, of which's value depends upon number of pixels in overlapping region annotated by characters by n & $n + 1$. Thus, the resulting bounding boxes may have an overlap to allow a soft boundary selection.

2.3.4 Gap Between Characters

As declared above, neighboring characters can also have a small gap (desirable for seamless segmentation) between them. The gap is scored by a learning function, $CG_{n,r}^G(x; \theta^G) = \sum_{i=n}^r CC^G \cdot F_i(x)$ which a count of number of horizontal pixel between consecutive characters.

2.3.5 Sparse-Based Parsing

After creating a sub-image from the segmentation to have a best coverage for a character, the optimal output of comparison function $SLS(n, t, n + 1, t + 1)$ should tend to infinity or be as large valued as possible, where the character $t + 1$ is fixed and the segmentation boundaries n and $n + 1$ are varied. However, the function only creates a competitive test between closely matching characters in possible recognition scenarios. For example a given segment, the term $SLS(n, t, n + 1, k)$ may represent character "I" (capital I) which matches to feature functions of "i" (small i), "t", "l" (small L) or "1" (one).

$$CS(n+1, t+1 | n, t, x, CC, I)\alpha \exp\{C(n, t, n+1, t+1)/n - t + 1. \qquad (6)$$

KNN used here as classifier makes the decision of selection of class based on the entire training dataset or all subsets of them. The output class in our case is a character. Training has been done with 180 samples of each character which makes 9360 sample dataset with 52 labels repeating 180 times each.

3 Experiment and Analysis

180 samples have been used for each character for the training of both small and capital sets. The recognition was done on 15 different samples of each containing 80–120 samples of characters more than 50% of which were overlapped with one other character and 20% overlapped with two other characters (both left and right neighbor).

The segmentation of lines overlapping characters gives 96% average accuracy in perfect segmentation while 3% shows undersegmentation and 1% shows overseg-mentation which leaves very small segments which do not affect recognition rate as they are not used during recognition (Fig. 5).

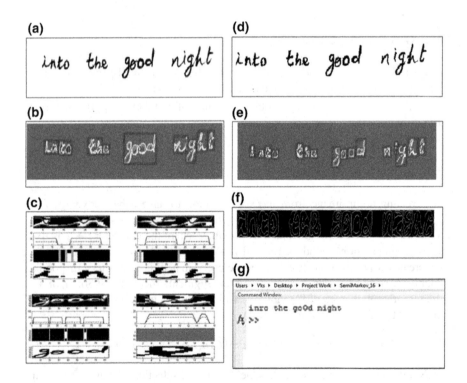

Fig. 5 Stage-wise results of methodology. **a** Initial overlapping characters. **b** Initial segmentation. **c** Application of modified water reservoir technique for separation of overlapping characters. **d** Segments of separated characters. **e** Skeletalisation of segmentation result. **f** Recognized text

Table 1 Change in recognition rate with increase in the number of training samples

Number of training sample/character	Recognition rate for a sample of 84 characters	
	True positive (%)	False positive (%)
5	48	48
12	51	45
48	63	33
60	71	28
120	92	6
180	96	1

4 Results and Discussion

The recognition rate for a sample containing 84 characters is used to extract values described in Table 1. When classified with training dataset created using 5 training samples for each character is recognition rate of 48% was achieved. Similarly, when classified with training dataset created using 12 training samples for each character is recognition rate achieved was 51%. After using samples of 180 different slant varying from $-5°$ to $+15°$ for each character for training the recognition rate reached to 98% true positive and 1% false positive for the same input sample.

The attained average recognition rate is 86% aggregated for all samples. All of the samples contain 50% or more characters, which are overlapping and are separated using the proposed algorithm with almost perfect segmentation with very small chance of over-segmentation.

5 Conclusion

The HsMM can be implemented efficiently for purpose of handwritten and overlapping or touching character recognition. The recognition rate depends upon multiple things such as user's handwriting consistency especially in text slants and area of overlapping characters. It also depends upon no of training samples and efficiency of preprocesses. The model can now be implemented in anybody's handwriting if samples of their handwriting are available in abundance for learningthe pattern. This model hence has a high applicability for handwritten CR in any language but also has a requirement such as preprocessing and high training time.

References

1. Stratonovich, R. L. (1960). "Conditional Markov Processes". Theory of Probability and its Applications. 5 (2): 156–178.

2. Dasu, Nagendra Abhinav, "Implementation of hidden semi-Markov models" UNLV Professional Papers. Paper 997, 2011.
3. Chen, X., & Yuille, A. L., "Detecting and reading text in natural scenes", Proc. IEEE Conf. on Comp. Vis. and Pat. Recog. 2004, pp. 366–373.
4. Huang, X., Hon, H. W., & Acero, A., "Spoken language processing: A guide to theory, algorithm, and system development", Prentice Hall PTR., 2001.
5. Ravinda S. Hegadib, Parshuram M. Kamblea, "Handwritten Marathi character recognition using R-HOG Feature", Procedia Computer Science 45, ICACTA, 266–274, 2015,.
6. Deepjoy Das, SRM Prasanna, Rituparna Devi, Subhankar Ghosh, Krishna Naik, "Performance Comparison Of Online Handwriting Recognition System For Assamese Language Based On HMM And SVM Modelling", CSCP 2014, pp. 87–95, 2014.
7. Fatos T. Yarman-Vural, Nafiz Arica, "An Overview of Character Recognition Focused on Off-Line Handwriting", IEEE Transactions on Systems, Man and Cybernetics—Part C: Applications and Reviews, Vol. 31, No. 2, 2001.

A Literature Survey on Authentication Using Behavioural Biometric Techniques

Rajvardhan Oak

Abstract With technological advancements and the increasing use of computers and internet in our day to day lives, the issue of security has become paramount. The rate of cybercrime has increased tremendously in the internet era. Out of the numerous crimes, identity theft is perhaps the one that poses the most dangers to an individual. More and more voices strongly declare that the password is no longer a reliable IT security measure and must be replaced by more efficient systems for protecting the computer contents. Behavioural biometrics is an emerging technology that resolves some of the major flaws of the previous scheme. This paper is the first stage of a project which aims to develop a novel authentication system using behavioural biometrics. It presents a comprehensive survey of various techniques and recent works in the respective fields.

Keywords Information security · Biometrics · Authentication mechanism

1 Introduction

Biometrics can be characterized basically as the estimation of human qualities. Biometric identifiers are then unmistakable, quantifiable qualities used to name and depict people [1]. Famous cases of biometric confirmation are retina examinations, fingerprint tests and DNA tests.

Biometrics verification is a strategy used by coordinating an individual's hereditary attributes or behavioural qualities with information that have already been learned, enlisted into a layout and organized in a framework database or on a token [2]. It can likewise be characterized as the idea of recognizing oneself by something that you know, something that you have, or something that you are [3]. That is, it identifies the innate qualities.

R. Oak (✉)
Department of Computer Engineering, Pune Institute of Computer Technology, Pune, India
e-mail: rvoak@acm.org

© Springer Nature Singapore Pte Ltd. 2018
S. Bhalla et al. (eds.), *Intelligent Computing and Information and Communication*,
Advances in Intelligent Systems and Computing 673,
https://doi.org/10.1007/978-981-10-7245-1_18

For a parameter to be called a biometric identifier, it must satisfy some properties [1, 4]:

(1) Universality: Every person possesses that particular characteristic. For example, DNA can be called a biometric parameter, whereas birthmarks cannot.
(2) Uniqueness: The characteristic is different for every person. For example, while a fingerprint is a biometric, eye colour and blood group are not.
(3) Permanence: The characteristic does not disappear or change with time. For example, DNA is biometric but hormone levels are not.
(4) Collectability: It is possible to obtain readings of the characteristic using sensors in a feasible, fast and highly accurate manner. For example, voice can be called a biometric characteristic, but factors such as confidence and self-esteem cannot.
(5) Circumvention: The parameter is forgery-proof and it is nearly impossible to replicate it.

Biometric parameters may be classified into two types: Physiological and behavioural biometrics [1]. Physiological characteristics are those that are anatomic and biological properties of an individual. They include fingerprints, facial recognition, iris scan, voice recognition, palm veins, DNA, etc. [5]. These are the traditional means by which an individual's identity is verified.

Behavioural biometrics, on the other hand, refers to factors such as gait, GUI interaction, Haptics [6], programming style, registry access, system call logs, mouse dynamics, etc. [7, 8]. It depends on an individual's inward qualities and attributes [9].

The advantages of this technique over other customary biometric approaches are as follows [3, 5, 7]:

(1) It provides persistent security. The authentication process is not complete after the login, but there is continuous monitoring.
(2) Behaviours can be collected surreptitiously without alerting the user.
(3) No special hardware is necessary to identify the behaviours.
(4) It is difficult to replicate the behaviour, hence making identity theft less likely.

2 Classification of Behavioural Biometrics

Depending on the nature of the parameters collected and evaluated, behavioural biometric mechanisms can be subdivided into five types (Fig. 1).

(1) Authorship Based: It is based on the analysis of a work produced by the user. The system identifies styles and characteristics particular to a user as he writes/draws and verification is done based on the matches of these characteristics.
(2) HCI Based: It is based on the traits and mannerisms exhibited by the user while interacting with the system such as mouse movements or touchscreen strokes.

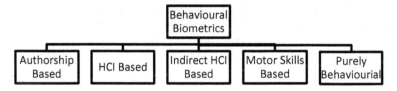

Fig. 1 Classification of behavioural biometrics

Every person applies different strategies, shortcuts and conventions while using a computer. A collection of such interaction particularities serves as a base for constructing a user identity and for authentication.

(3) Indirect HCI: It is very similar to HCI-based interaction. In indirect HCI-based systems, the system monitors the effects of the normal HCI actions. All user actions leave certain low-level digital evidence in the system in the form of system call traces, audit logs, execution traces, call stack analysis, etc.

(4) Motor-Skills based: It refers to the way in which the user utilizes the muscles. It is presumably the best inquired about of all behavioural biometric techniques. Human movements are a combination of muscle action, bone action and impulses travelling in the nervous system.

(5) Purely Behavioural: It is not based on the body part metrics or intrinsic behaviour. Rather, it is based on the fact that human beings utilize different strategies, innovative ideas, critical thinking and creative thinking. It attempts to quantify such traits and use them for authentication.

3 Evaluation Metrics

The following parameters are used as a measure of the effectiveness of the biometric systems:

(1) *FAR*: It is the false acceptance rate [5]. It is the ratio of the amount of attack instances incorrectly labelled as authentic to the total number of attack instances [5, 10]. Value of FAR must be as small as possible. It gives a measure of the percentage of attacks that could not be identified by the biometric system.

(2) *FRR*: It is the false rejection rate. It is the ratio of the amount of authentic interactions that were incorrectly classified as attacks and the total number of authentic, valid instances [10]. It gives a measure of the percentage of authentic interactions that were incorrectly classified as intrusive.

(3) *ROC*: It is the Receiver/Relative operating characteristic. It is a plot which represents a compromise between FAR and FRR. The matching algorithm of a biometrics mechanism has an established threshold which evaluates how close to the learned format an instance must be so that it qualifies as authentic [5]. Increasing the threshold reduces the FAR but increases the FRR. Decreasing

the threshold will reduce FRR but lead to higher FAR. The ROC helps to identify the optimum threshold so that both FAR and FRR are minimized.

(4) *EER/CER*: At this rate, both acceptance and rejection rates are numerically the same [5, 10]. ROC gives the value of EER. Ideally, the value of EER should be small [5].

4 Survey of Techniques

4.1 Keystroke Dynamics

It is a type of HCI-based behavioural biometric parameter. The keyboard is the primary input device used by humans to interact with a system. Different individuals have varied characteristics with respect to the speed of typing, error rate, use of certain key combinations, use of the touch typing method, etc. [10]. As a result of these differences, it is possible to verify the identity of the user. For example, the touch typing method is used by veteran typists, whereas novices use a hunt-and-peck technique which utilizes just two fingers. [7, 10]. Keystroke dynamics are based on the two important parameters: Flight Time and Dwell Time [8, 10]. Flight time is the time gap between releasing a key and pressing the next one, and the latter is the time for which a key is pressed. A large amount of research regarding keystroke dynamics for verification has been carried out. Bartolacci in 2005 and Curtin in 2006 have studied keystroke dynamics for long text analysis [4, 7, 10]. A study for email authorship identification was carried out by Gupta et al. in 2005. In [11], the author constructs digraphs (consisting of two adjacent characters) which he classifies into seven different categories and calculates mean latency for each category. This gives a measure of the programming experience of the user [11]. The authors in [12] have deduced that keystroke on integrating it with accelerometer biometrics, has a false acceptance rate of just 7%.

4.2 Mouse Dynamics

The mouse is probably the most important device after the keyboard which aids humans in interacting with the Graphical User Interface (GUI). In general, mouse dynamics refer to the characteristics of different individuals to use different pointing devices like mouse and light pen in different manners [13]. A number of different mouse gestures can be analyzed, such as single clicks, double clicks, scrolling, drag and drop and stillness [7]. To acquire the features, a software program intercepts the low-level events occurring because of mouse dynamics, along with associated attributes such as timestamps and cursor coordinates [10, 13]. At the high level, the gathered information would incorporate abnormal state itemized data about the

GUI-related activities of the client, for example, left the tap on the begin menu, double tap on explorer.exe, close notepad.exe window, and so forth [7]. Other factors such as velocities in a horizontal, vertical and tangential direction along with the angular direction, tangential acceleration and jerk also form a part of the captured data [10]. From this event log, various statistical and kinematic features are extracted which are used to build a user profile. Pusara and Brodley in 2004 have proposed an approach in which split the mouse event data is classified into movements of the mouse wheel and clicks on different entities on the screen [7]. Gamboa and Fred in 2003 have described an approach in which identification and authentication of humans are carried out by analyzing the human–mouse interaction in online gaming. In [14], mouse movements were captured as functions of timestamps and graphical coordinate values and analysis was done using support vector machines with an error rate of 1.3%.

4.3 Haptics

It identifies with the feeling of touch. Nowadays, intelligent cell phones are equipped with various sensing elements such as an accelerometer, gyroscope, computerized compass and high-resolution camera [6, 9]. As a result, it is possible to measure several physical quantities during use. The parameters measured are direction, pressure, force, angle and position of the user's interactions [7]. In [15], a biometric authentication system based on the haptics was built by integrating it with fuzzy logic. A combination of three factors such as hold-time, inter-key behaviour and finger pressure was proposed in [16]. As much as 30 behavioural features were proposed by Frank et al. (2012) [17]. Xu et al. (2014) have proposed a continuous and passive mechanism which achieved an error of less than 1% [15]. Sitova et al. (2015) introduced Hand Movement Orientation and Grasp (HMOG)-based system. Furthermore, Buriro et al. (2016) have developed a system which analyses micro-movements of a phone, and the exact points on the screen which are pressed [15]. In [18], a characteristic and consistent metric is known as Index of Individuality has been proposed which uses Gaussian Process Models to capture data.

4.4 Gait

It refers to a person's way of walking about. It is a muscle controlled biometric parameter. In gait-based biometrics, parameters such as kinematic patterns, knee ankle movements, moments, angles, hunch, etc. [19, 20]. It is a complex spatio-temporal activity which permits biometric identification of individuals at a distance generally via recorded video [7, 19]. The capture of data may be carried out by floor sensors, machine vision systems, or wearable sensors [20]. Gait is a factor which is subject to several variations from person to person depending on age, gender, bone density, waddling, muscle strength, fat percentage and energy level [7]. As a result, this is one of the biometric parameters which is nearly impossible to replicate. Typical features may include arm swing, walking speed, stooping of the back, step size, head-foot distance and head-pelvis distance [7, 19]. By using three different approaches of signal correlation, frequency domain and data distribution statistics [15, 21], it was found by Mantyjarvi et al. (2005) that the lowest error of 7% was achieved with signal correlation method [21]. Gafurov et al. (2006) developed a method to identify the person using an accelerometer attached to the leg at an error rate of 10%. Derawi et al. (2010) attached the accelerometer to the hip and an error of 20% was seen. In [22], a gait-based WiFi signature system was proposed using a simple Naïve–Bayes classifier with a correct identification rate of 87%. Cola et al. [23] a device is worn on the wrist and identification was done with an error rate of 2.9%.

4.5 Log Files

Operating systems generally maintain exhaustive log files which contain records of every small activity initiated by the user. Such log file entries contain fields such as the identity of the user who fired a command, the timestamp, CPU usage and other associated parameters [7]. In a system based on the audit files, there is a high chance of false positives due to routine, legal activities such as adding new users, changing network settings or change in permissions [7]. Hence, in these systems, there is an overhead of informing the authentication program of such possibilities. Network level logs which maintain traffic and various attributes such as protocol, sequence number, length, correction checksum, etc. serve to identify intruders in the system. In the training phase, a profile is built which identifies certain behaviours as normal. A field known as 'alert flag' is set if any abnormal activity is detected. In [24], the authors describe a five-step process: (i) Formatting data, (ii) Compare degree of similarity, (iii) Clustering, (iv) Retranslation and (v) Detection, in order to identify insider threats using log entries. [25] suggests the creation of a distributed Control Flow Graph (CFG) by extracting template sequences from log files This is the normal, expected behaviour. All activities are analyzed by comparison with this CFG.

5 Comparative Study

(Table 1)

Table 1 A comparison of major behavioural biometric approaches

Behavioural approach	Typical parameters analysed	Results obtained
Keystroke dynamics	• Flight Time • Dwell Time • Error rate • Typing technique	FAR of 7% [11]
Mouse dynamics	• Mouse wheel velocity • Types of clicks	FAR of 1.3% [14]
Haptics	• Direction • Force • Pressure of touch	FER of 1% [7]
Gait	• Arm swing • Head-foot distance • Knee angle • Head-pelvis distance	FER of 2.9% [23]
Log Files	• Network traffic • CPU usage • Dump files	No quantified results yet

6 Conclusion

Behavioural biometrics analyses various parameters of a person's behaviour such as gait, stride, typing patterns, mouse patterns, etc. As it is very difficult to replicate, these systems have a high-security. systems with very low errors have been developed. Clearly, behavioural biometrics promises to usher in a new era in the domains of computer and information security.

References

1. L. Wang, X. Geng, L. Wang, and X. Geng. Behavioral Biometrics For Human Identification: Intelligent Applications. Information Science Reference—Imprint of: IGI Publishing, Hershey, PA, 2009
2. Michelle Boatwright, Xin Luo, "What Do We Know About Biometrics Authentication?", InfoSecCD '07: Proceedings of the 4th annual conference on Information security curriculum development 2007
3. James L. Wayman, Anil K. Jain, Davide Maltoni, and Dario Maio," Biometric Systems: Technology, Design and Performance Evaluation", Springer, ISBN: 1852335963
4. Tarik Mustafi´c, Arik Messerman, Seyit Ahmet Camtepe, Aubrey-Derrick Schmidt, Sahin Albayrak, "Behavioral Biometrics for Persistent Single Sign-On", Proceedings of the 7th ACM workshop on Digital identity management, 2011
5. K P Tripathi, "A Comparative Study of Biometric Technologies with Reference to Human Interface", International Journal of Computer Applications (0975 – 8887) Volume 14– No.5, January 2011
6. Wolff, Matt. Behavioral Biometric Identification on Mobile Devices. Foundations of Augmented Cognition. Springer Berlin Heidelberg, 783–791, 2013

7. Yampolskiy, R.V. and Govindaraju, V. (2008) 'Behavioural biometrics: a survey and classification', Int. J. Biometrics, Vol. 1, No. 1, pp. 81–113

8. Ioannis C. Stylios, Olga Thanou, Iosif Androulidakis, Elena Zaitseva, "A Review of Continuous Authentication Using Behavioral Biometrics", Proceedings of the SouthEast European Design Automation, Computer Engineering, Computer Networks and Social Media Conference 2016

9. Esther Vasiete, Vishal Patel, Yan Chen, Larry Davis, Ian Char, Rama Chellappa, Tom Yeh, "Toward a Non-Intrusive, PhysioBehavioral Biometric for Smartphones", MobileHCI 2014, Sept. 23–26, 2014, Toronto, ON, CA

10. Monika Bhatnagar, Raina K Jain, Nilam S Khairnar, "A Survey on Behavioral Biometric Techniques: Mouse vs Keyboard Dynamics", Proceedings of the International Conference on Recent Trends in engineering & Technology - 2013(ICRTET'2013)

11. Juho Leinonen, Krista Longi, Arto Klami, Arto Vihavainen, "Automatic Inference of Programming Performance and Experience from Typing Patterns", Proceedings of the 47th ACM Technical Symposium on Computing Science Education, 2016

12. Kyle R. Corpus, Ralph Joseph DL. Gonzales, Alvin Scott Morada, Larry A. Vea, "Mobile User Identification through Authentication using Keystroke Dynamics and Accelerometer Biometrics", Proceedings of the IEEE International Conference on Mobile Software Engineering and Systems, 2016

13. Zach Jorgensen, Ting Yu, "On mouse dynamics as a behavioral biometric for authentication", Proceedings of the 6th ACM Symposium on Information, Computer and Communications Security, 2011

14. Nan Zheng, Aaron Paloski, Haining Wang, "An Efficient User Verification System Using Angle-Based Mouse Movement Biometrics", ACM Transactions on Information and System Security, Vol 18, Issue 3

15. Andrea Kanneh, Ziad Sakr, "Biometric user verification using haptics and fuzzy logic", Proceedings of the 16th ACM international conference on Multimedia, 2008

16. Saevanee H., Bhatarakosol, P., (2008). User Authentication Using Combination of Behavioral Biometrics over the Touchpad Acting Like Touch Screen of Mobile Device. International Conference on Computer and Electrical Engineering, 2008. Page(s): 82–86

17. Frank, M., Biedert, R., Ma, E., Martinovic, I., Song, D., (2012). Touchalytics: On the Applicability of Touchscreen Input as a Behavioral Biometric for Continuous Authentication. IEEE Transactions on Information Forensics and Security. 2012. (Volume:8, Issue: 1). Page (s): 136–148

18. Daniel Buschek, Alexander De Luca, Florian Alt, "Evaluating the influence of Targets and Hand Postures on Touch-based behavioural biometrics", Proceedings of the ACM Conference on Human Factors in Computing Systems, 2016

19. Benabdelkader, C., Cutler, R., and Davis, L. 'Person identification using automatic height and stride estimation', IEEE International Conference on Pattern Recognition 2002

20. S Laxmi, Tata A S K Ishwarya, S Sreeja, "Exploring Behavioural Type Biometrics: Typing Rhythm, Gait, Voice", International Journal of Innovative Research in Computer and Communication Engineering, Vol.4, Issue 10, October 2016

21. Mantyjarvi, J., Lindholm, M., Vildjiounaite E., Makela, S.M., Ailisto, H. A. (2005). Identifying users of portable devices from gait pattern with accelerometers. IEEE International Conference on Acoustics, Speech, and Signal Processing, 2005. (Volume:2). Page(s): ii/973–ii/976 Vol. 2

22. Yan Li, Ting Zhu, "Gait-Based WiFi signatures for Privacy-Preserving", Proceedings of the 11th ACM on Asia Conference on Computer and Communications Security, 2016

23. Gugliemlo Cola, Marco Awenuti, Fabio Musso, Alessio Vecchio, "Gait-based Authentication using a wrist-worn device", Proceedings of the 13th International Conference on Mobile and Ubiquitous Systems: Computing, Networking and Services, 2016

24. Markus Wurzenberger, Florian Skopik, Roman Fiedler, Wolfgang Kastner, "Discovering Insider Threats from Log Data with High-Performance Bioinformatics Tools", Proceedings of the 8th ACM CCS International Workshop on Managing Insider Security Threats, 2016
25. Animesh Nandi, Atri Mandal, Shubham Atreja, Gargi B. Dasgupta, Subhrajit Bhattacharya, "Anomaly Detection Using Program Control Flow Graph Mining from Execution Logs", Proceedings of the 22nd ACM SIGKDD International Conference on Knowledge Discovery and Data Mining, 2016

Number System Oriented Text Steganography in English Language for Short Messages: A Decimal Approach

Kunal Kumar Mandal, Santanu Koley and Saptarshi Mondal

Abstract Information or data transfer between nodes are the necessary part of computing but the biggest need is to secure the data from evil sources. There are many technique but they are insufficient in case of information larceny. Our approach with number system oriented text steganography for short messages is the outcome for preventing such purpose. The combination of mathematical, computation rules, and knack of us hides the data from view. The approach used to hide data is innovative among data transfer protocols as word to word or rather alphabet to alphabet including numbers and special characters is derived. This approach makes data invisible when moving in any one of the ways such as SMS, WhatsApp, Email, or Facebook messenger. Good mixtures of any number, characters form a pair of sets that can be used to hide information. Security agencies like navy, army or air force can use such kind of techniques transferring data from one node to another for sake of setting aside their native soil.

Keywords Text steganography · Pair of number · Information hiding
Secret message

K. K. Mandal
Mankar College, Mankar, West Bengal, India
e-mail: kunal@turiyan.com

S. Koley (✉)
Budge Budge Institute of Technology, Bardhaman, West Bengal, India
e-mail: santanukoley@yahoo.com

S. Mondal
Indian Railways, Asansol, West Bengal, India
e-mail: saptarshi.crj@gmail.com

© Springer Nature Singapore Pte Ltd. 2018   183
S. Bhalla et al. (eds.), *Intelligent Computing and Information and Communication*,
Advances in Intelligent Systems and Computing 673,
https://doi.org/10.1007/978-981-10-7245-1_19

1   Introduction

Today, the data/information transfer is done using e-media right through the globe. Short data/information transfer between different nodes (especially text data) has grown up exponentially in the last decade, which is nearly about 561 billion per month [1]. WhatsApp and Facebook messenger are responsible for transferring text data size of 60 billion practically [2]. Therefore, this enormous dimension of data makes us to do something for transmitting data safely. Now, it is a trial on English language but not limited to it. In case of producing language other than English, the alphabets of that language will be processed as an image. Steganography is the best approach so far for securing a message from external obstruction. Here, the existence of the message will be acquainted with the addressee barely.

This technique provides a way that anticipated recipient recognizes the subsistence of the message(s). This is just a obscuring a file, message, image, or video within another. In other way, steganography is a way of covering a message surrounded by an extra message, so that nobody can have a notion of being its presence [3]. Messages can be perceived by the predetermined addressee only. Polygraphia and Steganographia are a grand inscription by Trithemius. Trithemius is the person to start off the expression "Steganography". It is originated from two Greek words steganos, meaning "covered", and graphein, meaning "to write". We may find the black and white facts on the subjects of steganography by uncovering legendary Herodotus story of sending messages on shaved heads by their slaves.

This world of internetworking transfers data and information through worldwide network, besides the security threats growing exponentially. This is need of invention for securing data as much as possible. The protection of data or information on network plays a significant role as data transfer increases on Internet. Privacy, secrecy, and integrity of data are mandatory for defending in opposition to legitimate access by intruders. We say need is the father of inventions; this results in a massive development on the ground of data/information hiding. They followed numerous techniques resembling cryptography, steganography, and coding for safety measures. Conversely, it can be said without any doubt that steganography portrays further concentration than others. Steganography is a combination of fine art of putting sensitive data out of sight and science of covering information so that its existence cannot be identified [4]. Encoding is performed on furtive messages which is a technique where the subsistence of the information is out of sight for unintended users. In comparison to different existing data transmission schemes, steganography is the finest approach on secreted data communication through internet. Developing a protected messaging in an entirely untraceable mode is the main purpose of steganography. In addition to set up and to stay away from portrayal of uncertainty to the diffusion of a secret data is through digital communication channels [4]. This technique is not to ensure others to be acquainted with the hidden information, but to make certain further thoughts of the existence of information. Suspicion of surreptitious message in a shipper standard for a steganographic means fall short the entire procedure at once. Categorical of submission like digital image can be purposeful [5] together with copyright protection, feature tagging, and undisclosed communication. Copyright notice or watermark can be entrenched and delimited by an image to be acquainted with it as intellectual possessions [6]. Several illicit efforts to use this image can be acknowledged by digging out the watermark.

2 Steganography

Steganography is the procedure of hiding messages by which the dispatcher of that memorandum and his beneficial counterpart recognize the subsistence of the message. The beauty of this process is that it does not create a center of consideration for the hackers. Steganography is not a new process, it was devised even in primeval era and these olden techniques are identified physical steganography [7]. There are a number of illustrations for such processes which are messages concealed in messages in various supplementary modes, messages carved in secret inks, messages shaped on envelopes in areas covered by stamps, etc. This elderly process is termed as digital steganography. These contemporary schemes comprise put out of sight of messages contained by piercing pictures, implanting messages surrounded by arbitrary data, inserting images among the message contained by video files, etc. [8]. In addition, network steganography is drawn on the telecommunication networks. These include the procedures resembling steganophony (covering a message in Voice-over-IP exchanges) and WLAN steganography (techniques for conveying steganograms in wireless local area networks) [9, 10].

Steganography can be classified according to its importance and goals. Various types of steganography are shown in the diagram (Fig. 1).

The study of "hide from view" of any information is known as cryptography. On the other hand, steganography deals with arranging hidden messages so that only the sender and the receiver are acquainted with the presence of the message. In steganography, only the sender and the receiver know the reality of the message, while in cryptography the survival of the encrypted message is able to be seen to the world [11]. For the above reason, steganography gets rid of the unnecessary concentration to the hidden message. It can be clearly understood that steganography exercises on hiding both the message as well as the content whereas cryptographic methods try to protect the content of a message only [12]. By coalescing cryptography and steganography, one can accomplish improved safekeeping.

3 Previous Work

Several efforts are associated here to illustrate in this segment on information hiding by the side of their benefits and drawbacks.

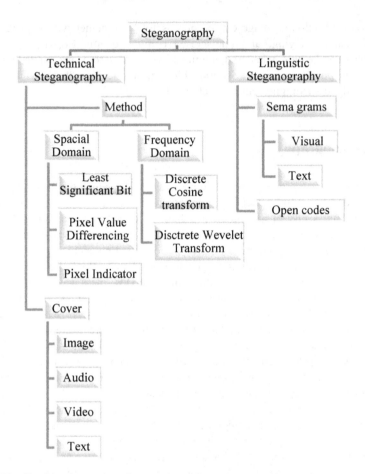

Fig. 1 Classification of steganography

3.1 Use of Markup Languages in Text Steganography

To hide any data or information using text steganography, the markup languages are used as a part of its existing services it provides [12]. The case inattentiveness of tags of any HTML document may be an example of the characteristics of it. Now as an illustration, the <BR> tag ought to be used as <br> or it may be as <Br>.

3.2 Text Steganography in Explicit Characters in Words

Here in this mode, one or a few of the word or a group of characters (words) are chosen from all or some paragraphs [13]. For example, a hidden or secret message shapes by the manner as the initial words of every paragraph are chosen. This is a process by placing the first characters of a particular word side by side. An unwanted receiver may find discrepancy in the text. Similarly, a series of text after some intervals can be decided for hiding with steganography.

3.3 Line Shifting Technique

In this system, the lines of the text are vertically shifted to some extent of degrees [14, 15]. For example, some selected appearances are being altered. They can be 1/695 inch up or down in the text and information are concealed by generating a secreted distinctive nature of the text. This could be unknown to others. Bare eyes cannot find this altered text. It can be assured that printed texts are sufficient for this technique.

3.4 The Altering of Words

The techniques of horizontally shifting words and by changing distance among words, information are hidden in the text [14, 16]. The distance between words keeps changing in a technique that is acceptable for texts only. Filling of a line is quite common due to change of distance, as a cause of it can be identified less. Security is constrained in this method.

3.5 Syntactic Scheme

This technique can be used to hide information by changing the punctuations likely full stop (.) and comma (,) in appropriate positions. There is an example: "When I sing well, ladies feel sick", now altering the punctuation and we could see "When I sing, well ladies feel sick". One can hide information in a text file [17, 18]. These methods are inconsequential and also have need of recognizing appropriate positions of setting punctuation symbols.

3.6 Semantic Schemes

The semantic scheme hides information within the text using the synonym of words for definite words [19]. The foremost improvement and significance of this scheme is the protection of information. It cannot be violated in case of retyping or using OCR programs (contrary to methods listed under Sects. 3.3 and 3.4). However, the significance of the transcript may change.

4 Recommended Method

Steganography for the text communication can have quite a lot of approach in Bengali text, Hindi text, or several languages. The text written with English alphabets (may be in any language) is converted to ASCII number system. A model number system is being defined that can cover the given text from external view. Our representation is different than others. This decimal representation of ASCII can be applied on any combination of text A to Z, a to z, 0 to 9 and other special characters but we have restricted numbers and special characters. In this demonstration, any number is represented by two digits ordered pair. Now these two digits can add 1 individually and thus send these pairs. At the moment of extraction using these two added digits are subtracted by 1 and apply the formula to get the final value. This concept will be clearer by looking at the following example.

Let us assume a secret message—"BOMB IN MARKET"

The counting is done by taking any quantity of the commodity as (Table 1).

Such type of text message is very common mixed with several languages like Bengali, Hindi, and English. They are written using English alphabets. Now, the first word "BOMB" represented in ASCII is 66 (B), 79 (O) 77 (M) and 66 (B). At this point, the number 66 will be written as (10, 11). As per proposed algorithm, we add 1 to each number in the pair and finally this will be (11, 12), which can be sent to other node. Here, the pair of number is again subtracted as it was added and becomes (10, 11). The computation is done by the formula [{M * (M + 1)}/2] + N for an ordered pair (M, N) [20]. Now the value of (10, 11) is given by

$$[\{10 \times (10+1)\}/2] + 11$$
$$= [\{10 \times 11\}/2] + 11$$
$$= [110/2] + 11$$
$$= 55 + 11$$
$$= 66$$

This concept pursues a different way of counting, which means in ancient times some counted commodities are exchanged by the same and equal number of some different commodity (Figs. 2 and 3).

The numbers are divided the group in such a manner that the first group can contain only one digit/number, i.e., 1 (as a single number). In the same way, second group can contain two digits/numbers, i.e., 2, 3 and the nth group can contain "n" numbers. Thus, if we want to represent the number given by the second position in the 11th group then it is represented as (10, 2). The formula gives us the value as (10 * 11)/2 + 2 = 110/2 + 2 = 57. This can be verified by counting the second position from left side of the 11th group.

Table 1 Grouping applicable in proposed number system

Group number	Group elements
1	(1)
2	(2, 3)
3	(4, 5, 6)
4	(7, 8, 9, 10)
5	(11, 12, 13, 14, 15)
6	(16, 17, 18, 19, 20, 21)
7	(22, 23, 24, 25, 26, 27, 28)
8	(29, 30, 31, 32, 33, 34, 35, 36)
9	(37, 38, 39, 40, 41, 42, 43, 44, 45)
10	(46, 47, 48, 49, 50, 51, 52, 53, 54, 55)
11	(56, 57, 58, 59, 60, 61, 62, 63, 64, 65, 66)
12	(67, 68, 69, 70, 71, 72, 73, 74, 75, 76, 77, 78)
...	...

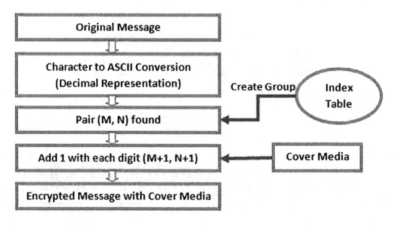

Fig. 2 Proposed encryption technique

5 Technical Background

In this technique, the cover media is a pair of number with text. Here, it is hiding one message inside the couple. Sender only knows what the message is, how to hide the message, how to extract and all. It also knows what encoding technique is applied to hide the message inside the pair of number. Receivers end must know what the decoding technique is and how to extract the message within the pair. An equation used in this paper is $[\{M\,(M+1)\}/2] + N$. There some exceptions as in the case of spaces (Hex value 32), no such pair would be created. Obtain an illustration as the secret message (Figs. 4 and 5).

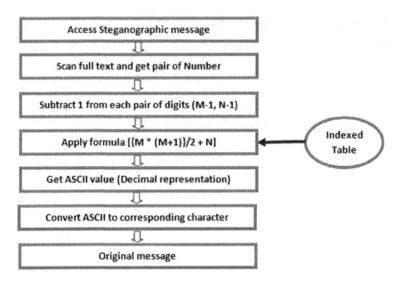

Fig. 3 Proposed decryption technique

"BOMB IN MARKET"

Fig. 4 Original text

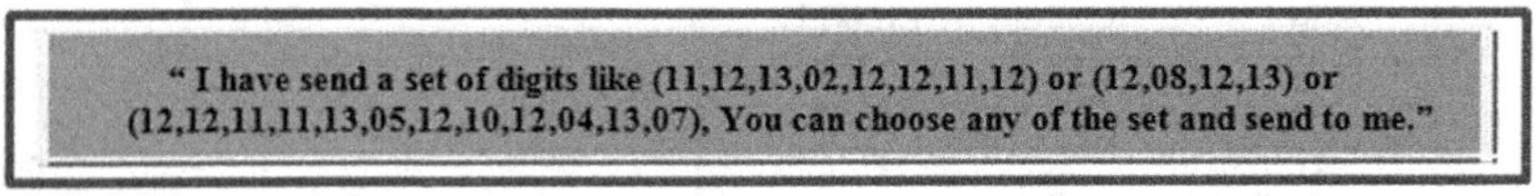

" I have send a set of digits like (11,12,13,02,12,12,11,12) or (12,08,12,13) or (12,12,11,11,13,05,12,10,12,04,13,07), You can choose any of the set and send to me."

Fig. 5 Steganographic message with cover text

6 Proposed Algorithm

We have presented a newly developed algorithm for encryption as well as decryption technique. These algorithms will help us to determine the techniques for hiding data in different ways.

6.1   Encoding Algorithm

Step 1: Start.

Step 2: Initiation of the short message.

Step 3: Characters from the message are taken from the beginning and proceed for the next character.

Step 4: If the character is a blank go to Step 8. If it is a special character, it remains the same and concatenated with the set produced in Step 7. Else it is converted it into corresponding ASCII value.

Step 5: ASCII value is converted into a pair of number (M, N) where M = Group number of a fully completed cluster. N = Position from the left most digit of the next group. They can be found by comparing with the given table for the proposed number system.

Step 6: This pair of number (M, N) is added by 1 individually as (M + 1, N + 1).

Step 7: Added pairs are kept one by one as a set.

Step 8: The set concludes when a blank space (ASCII code—32) is accomplished.

Step 9: If the last character is not reached go to Step 3.

Step 10: Before sending, the data will be prepared as a number of sets depending upon the words within the message.

Step 11: End.

6.2   Decoding Algorithm

Step 1: Start.

Step 2: Initiation of the sets as cipher text.

Step 3: Sets from the encoded message are taken from the beginning and proceed for the next. If there are no sets left, go to Step 10.

Step 4: Numbers within sets are taken as pair from the beginning one by one. If there is no pair within the set, go to Step 3.

Step 5: The pair of number (M, N) is subtracted by 1 individually as (M − 1, N − 1).

Step 6: This pair of number (M − 1, N − 1) will be applied with a proposed formula $(((M * (M + 1))/2) + N)$.

Step 7: Output of the formula is an ASCII number.

Step 8: ASCII will be converted into its corresponding character.

Step 9: When last pair of the set will reach go to Step 3, else go to Step 4.

Step 10: Before sending, the data will be prepared as a number of sets depending upon the words within the message.

Step 11: End.

7 Conclusion

Exploring an ancient art is steganography for hiding information from past to future. On a large scale, it is used in military, diplomatic, personal, and intellectual property applications. A technically different way of decimal number system is introduced in this paper where we are using steganography and a pair of numbers with text as a cover media. Here, we have tried to hide a secret message to develop a number system and the proposed formula is itself an invention. Our encryption and decryption of the original message from cover are also a problematic task for intruders as another up-to-the-minute algorithm we are using. We will try to explore steganography from its most primitive instances through probable future application.

Future Scope
The work presented in this paper is optimistically within the defined scope, but research work in no way ends, therefore, future research work is expected to survey horizons beyond the scope of this paper. It is tricky to acquire a steganography that satisfies both security and robustness. Therefore, we will make an effort to enhance the security and robustness for steganography.

References

1. HOW MANY TEXTS DO PEOPLE SEND EVERY DAY? Source:https:// www.textrequest. com/blog/many-texts-people-send-per-day/, Last Access: May (2017).
2. Messenger and WhatsApp process 60 billion messages a day, three times more than SMS, Source: http://www.theverge.com/2016/4/12/11415198/facebook-messenger-whatsapp-number-messages-vs-sms-f8-2016, Last Access: May (2017).
3. K. Tutuncu, A.A. Hassan, New approach in E-mail based text steganography, Int. J. Intell. Syst. Appl. Eng., 3 (2015), pp. 54–56.
4. A.A. Mohamed, —An Improved Algorithm for Information Hiding Based on Features of Arabic Text: A Unicode Approach‖ A Egyptian Informatics Journal 15, 79–87, 2014.
5. Staff, CACM. 2014. Know your Steganographic Enemy. Communications of the ACM, Vol. 57 No. 5, Page 8.
6. Kaur, S., Bansal, S., Bansal, R.K. Steganography and classification of image steganography techniques. International Conference on Computing for Sustainable Global Development (INDIACom), 2014. pp. 870–875.
7. Zielińska, E., Mazurczyk, W. and Szczypiorski, K. 2014. Trends in Steganography. Communications of the ACM, Vol. 57 No. 3, Pages 86–95.
8. Abdelmgeid Amin Ali and Al - Hussien Seddik Saad, "New Text Steganography Technique by using Mixed-Case Font", International Journal of Computer Applications (0975 – 8887) Volume 62– No.3, January 2013.
9. Chen, T. S. Chen, M. W. Cheng, A New Data Hiding Scheme in Binary Image, in Proc. Fifth Int. Symp. on Multimedia Software Engineering. Proceedings, pp. 88–93 (2003).
10. B. Feng, W. Lu and W. Sun, "Secure Binary Image Steganography Based on Minimizing the Distortion on the Texture", Information Forensics and Security, IEEE Transactions, vol. 10, (2015), pp. 243–255. (2001).

11. Z. Pan, X. Ma and X. Deng, "New reversible full-embeddable information hiding method for vector quantisation indices based on locally adaptive complete coding list", Image Processing, IET, vol. 9, no. 1, (2015), pp. 22–30.
12. G. Doërr and J.L. Dugelay, A Guide Tour of Video Watermarking, Signal Processing: Im-age Communication, vol. 18, Issue 4, pp. 263–282, (2003).
13. G. Doërr and J.L. Dugelay, Security Pitfalls of Frame by Frame Approaches to Video Wa-termarking, IEEE Transactions on Signal Processing, Supplement on Secure Media, vol. 52, Issue 10, pp. 2955–2964, (2004).
14. K. Gopalan, Audio Steganography using bit modification, Proceedings of the IEEE Inter-national Conference on Acoustics, Speech, and Signal Processing, (ICASSP '03), vol. 2, 6–10 April 2003, pp. 421–424, (2003).
15. M. Nosrati, A. Hanani and R. Karimi, "Steganography in Image Segments Using Genetic Algorithm", Advanced Computing & Communication Technologies (ACCT), (2015), pp. 102–107.
16. R. Aqsa, M. Muhammad, M. Saas and S. Nadwwm, "Analysis of Steganography Technique using Least Significant Bit in Graysclae Images and their ectension to Colour Images", Journal of Scientific Research and Report, vol. 9, no. 3, (2016), pp. 1–14.
17. N. Provos and P. Honeyman, Hide and Seek- An introduction to Steganography, IEEE Security and Privacy, pp 32–44, May/June (2003).
18. T. Moerland, Steganography and Steganalysis, www.liacs.nl/home/tmoerlan/privtech.pdf, May15 (2003).
19. K. Bennett, Linguistic Steganography: Survey, Analysis, and Robustness Concerns for Hiding Information in Text, Purdue University, CERIAS Tech. Report (2004–13).
20. K. K. Mandal, A. Jana, V. Agarwal, A new approach of text Steganography based on ma-thematical model of number system, Circuit, Power and Computing Technologies (ICCPCT), IEEE, pp. 1737–1741, 20–21 March (2014).

Novel Robust Design for Reversible Code Converters and Binary Incrementer with Quantum-Dot Cellular Automata

Bandan Kumar Bhoi, Neeraj Kumar Misra
and Manoranjan Pradhan

Abstract This work, we employ computing around quantum-dot automata to construct the architecture of the reversible code converters and binary incrementer. The code converter and binary incrementer are made up of Feynman gate and Peres gate, respectively. We have presented the robust design of Ex-OR in QCA, which is used for the construction of code converters and binary incrementer. The layouts of proposed circuits were made using the primary elements such as majority gate, inverter, and binary wire. A novel binary-to-gray converter design offers 59% cell count reduction and 36% area reduction in primitives improvement from the benchmark designs. Being pipeline of PG gate to construct the 1-bit, 2-bit, and 3-bit binary incrementer, we can use this robust layout in the QCA implementation of binary incrementer. By the comparative result, it is visualized that the binary incrementer such as 1-bit, 2-bit, and 3-bit achieved 60.82, 60.72, and 64.79% improvement regarding cell count from the counterpart.

Keywords Reversible logic · Quantum-dot cellular automata · Nanotechnology
Code converter, binary incrementer

1 Introduction

The current microelectronics has grown for the past years, but it faces problems in miniature, power, and design cost. Nanoelectronics have an alternative to tackle these problems [1]. The challenges task today is the construction of robust design.

B. K. Bhoi (✉) · M. Pradhan
Veer Surendra Sai University of Technology, Burla, India
e-mail: bkbhoi_etc@vssut.ac.in

M. Pradhan
e-mail: manoranjan66@rediffmail.com

N. K. Misra
Institute of Engineering and Technology, Lucknow, India
e-mail: neeraj.mishra@ietlucknow.ac.in

© Springer Nature Singapore Pte Ltd. 2018   195
S. Bhalla et al. (eds.), *Intelligent Computing and Information and Communication*,
Advances in Intelligent Systems and Computing 673,
https://doi.org/10.1007/978-981-10-7245-1_20

QCA computing is suitable for the efficient computing around nanometer scale [2]. This paper introduces the robust design with emerging QCA technology which has a potential to the difficulty of conventional CMOS technology [3].

The essence of reversible circuits is accessible to the new generation designer regarding no loss of information. Landauer [4] proved that conventional irreversible logic gates have information loss, i.e., energy dissipate. Bennett [5] proved that for a digital circuit construct from reversible logic gates no information loss also negligible energy dissipation. Moreover, the quantum circuit is to be developed by reversible gates [6].

This paper proposes a robust design of Ex-OR which uses the QCA technology are presented to be used in the construction of reversible code converter and binary incrementer. The design was driven toward less latency, less cell count, minimum clock zones, and no crossover. QCADesigner tool was adapted toward achieving these QCA primitives' results. However, more cell complexity means more power dissipation similarly low complexity increase the computation speed [7]. The complete quantitative analysis of converters and binary incrementer are explored to less complexity means faster computational speed. The results show reduction of QCA primitives when compared with previous work reported herewith. The reversible binary incrementer introduced in this paper will be useful in various fields, including digital signal processing and ALU. Whereas, reversible code converter will be applicable in a change of information from one format to another. Moreover, reversible code converter and binary incrementer will be useful in digital electronics computing where dedicated code converter and binary incrementer modules are demanded. The proposed idea of this article is described as follows.

- A novel two-input Exclusive-OR (Ex-OR) gate is proposed which is used to design an efficient Feynman gate (FG) in QCA.
- The efficient Feynman gates are used to construct the basic designs of a code converter and incrementer.
- Using proposed two-input Ex-OR design of QCA, the 1-bit, 2-bit, and 3-bit binary incrementers have been constructed and reported better QCA primitives as compared to previous designs.
- Half adder and n-bit binary incremental are designed using the proposed Peres gate.

The rest of the article is organized as follows: Sect. 2 discusses previous work. Section 3 presents novel design architecture and the results are discussed and in Sect. 4 comparative statistics is presented. In Sect. 5, conclusion is presented.

2   Previous Work

The architecture of reversible code converter and binary incremental in QCA framework are drawing attention in the previous works. Many works of these designs have been proposed. In 2016, Das et al. [8] proposed binary incrementer in

QCA. The 1-bit design consists of 97 cells, 0.075 μm^2 area, seven majority voter, four inverter, and four clocking zones. The 2-bit incrementer design consists of 14 majority voter, 8 inverters, 196 cell count, 0.272 μm^2 area, and 4 clocking zones. Several reversible code converters in QCA have been proposed in [9, 10]. Das et al. [9] present the code converters in QCA. In these designs, multilayer design is used for layout, with suffers from high latency and complexity. Misra et al. [10] presents the 3-bit reversible binary-to-gray and gray-to-binary code converter in QCA. The design consists of 118 and 112 cell count in binary-to-gray and gray-to-binary code converter, respectively. Several efforts have been made in optimizing the code converter and binary incrementer in QCA technology for efficient reversible logic computing. In this work, the robust QCA design of the proposed reversible circuits is introduced first by using Ex-OR. Such a proposed QCA design shows the robust in the reliability as a whole.

3 Building Blocks of Code Converter and Binary Incremental

This section introduces robust architecture of reversible code converter and binary incrementer. Toward the QCA design of code converter and binary incrementer, one robust design of Ex-OR is presented which incurs zero latency.

3.1 Ex-OR Gate in QCA

Ex-OR operation is commonly used in many digital logic applications such as designing arithmetic circuits, pseudorandom number generation, correlation, and sequence detection. In this work, we are presenting a new two-input Ex-OR gate. Figure 1a, b illustrates the gate symbol and QCA layout of this gate respectively. Figure 2c illustrates the simulation result which indicates that there is a zero latency.

Our proposed QCA layout of two-input Ex-OR gate requires 13 cell count with an overall area of 0.02 μm^2. This gate will be highly cost efficient in designing complex digital circuits. The comparison result is presented in Table 1. This is the first Ex-OR gate that utilizes less cell count and zero latency as compared to previous.

3.2 Reversible Code Converter Implementation in QCA

To design a code converter, we used the two-input Ex-OR gate in logic output construction. In the first type, code converter is binary-to-gray. In this design, three

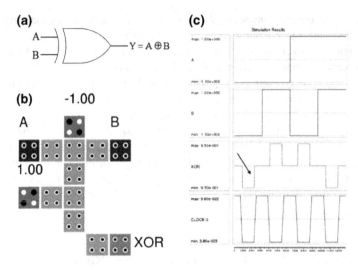

Fig. 1 Ex-OR gate **a** functional diagram **b** quantum-dot cell layout **c** timing waveform

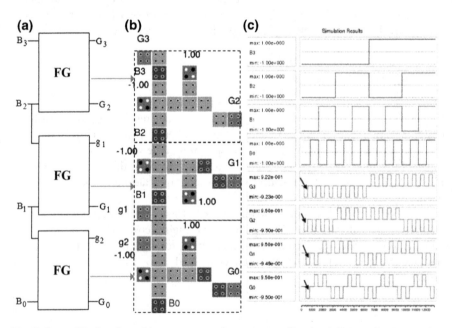

Fig. 2 Reversible four-input binary-to-gray code converter **a** functional diagram **b** quantum-dot cell layout **c** timing waveform

Table 1 Comparison of two-input Ex-OR gate QCA layouts

Design	Cell count	Latency	Area (μm^2)
[11]	67	1.25	0.06
[12]	29	0.75	–
[13]	18	0.75	0.01
[14]	14	0.5	0.02
New	13	0.5	0.02

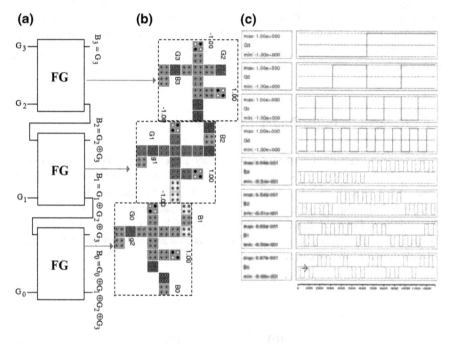

Fig. 3 Reversible four-input gray-to-binary code converter **a** Functional diagram **b** quantum-dot cell layout **c** timing waveform

Ex-OR gates were settled in the specified logic manner. The function diagram and cell layout of binary-to-gray are exhibited in Fig. 2a, b. Figure 2c presents the result of binary-to-gray converter. In this result, settled outputs are correctly with zero latency. The second code converter is gray-to-binary, which was drawn in Fig. 3a. The gray-to-binary design in QCA is presented in Fig. 3b, which was utilized by the less cell count, latency, and two clock zones. Figure 3c presents simulation result of proposed gray-to-binary. In this result attained correctly logic signal, zero latency is achieved in output B_3, B_2, and B_1, whereas B_0 after one clock cycle delay.

3.3 Reversible Binary Incrementer in QCA

In the proposed scheme, we have shown pipeline design of binary incrementer in size of 1-bit, 2-bit, and 3-bit in QCA. With a pipeline design in QCA of binary incrementer, we utilize Peres gate as a building block. Implementation of Peres gate in QCA has been presented in Fig. 4b. It is noted that the latency of the design in large size order of binary incrementer has been diminished. Peres gate design in QCA has been achieved by use of Ex-OR. Peres gate is a reversible gate having three inputs and three outputs as shown in Fig. 4a. Here, the input to output mapping is $P = A$, $Q = A \oplus B$, $R = AB \oplus C$ and the quantum cost of this gate is four. The simulation result of the Peres gate is shown in Fig. 4c. In simulation result, it is shown that P and Q outputs have zero latency, whereas R output has one latency (clock delay). This quantum-dot cell of PG is achieved by taking three majority gates and two inverters.

3.4 Reversible 1-Bit Binary Incrementer

In the quantum-cell design of reversible binary incrementer only by normal cell, and fix polarization cell are utilized. Most of the existing binary incrementers are based on coplanar or multilayer [8]. In this work, we are targeting the normal and fix cells based design. Figure 5a shows the binary incrementer block for one bit. The proposed Peres gate implemented reversible binary incrementer circuit is shown in Fig. 5b. Figure 5c, d shows the cell layout and simulation results related to binary incrementer, respectively.

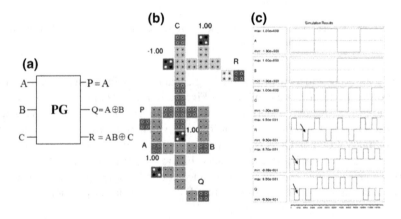

Fig. 4 Reversible PG **a** functional diagram **b** quantum-dot cell layout **c** timing waveform

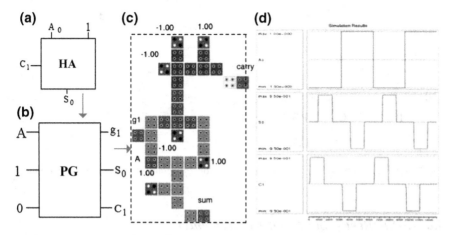

Fig. 5 Binary incrementer **a** Half adder as Binary incremental **b** Peres gate as an incremental **c** quantum-dot cell of PG as half adder **d** timing waveform

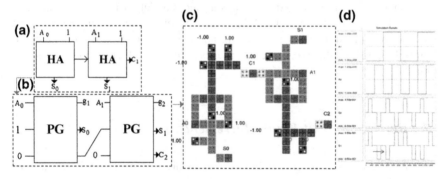

Fig. 6 2-bit reversible binary incrementer **a** functional diagram **b** quantum-dot cell **c** timing waveform

3.5   2-Bit Reversible Binary Incrementer

A 2-bit binary incrementer is implemented with cascading two half adders as shown in Fig. 6a. Here A_1, A_0 is 2-bit input binary number and S_1, S_0 represents incremented output. A structure of 2-bit binary incrementer using Peres gates is shown in Fig. 6b. QCA schematic and simulation results of the proposed 2-bit binary incrementer are drawn in Fig. 6c and Fig. 6d respectively. In timing waveform result, arrow sign indicates the output signal that has arrived at that point after two latency.

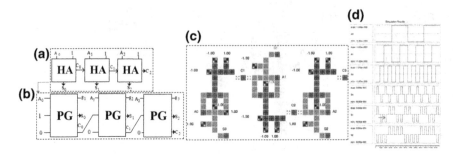

Fig. 7 Reversible 3-bit binary incrementer **a** functional diagram **b** quantum-dot cell **c** timing waveform

3.6 3-Bit Reversible Binary Incrementer

The 3-bit binary incrementer is implemented using three number of cascaded HA as shown in Fig. 7a. Here, $A_2A_1A_0$ represents 3-bit input number and $S_2S_1S_0$ represents 3-bit output number. The 3-bit reversible binary incrementer with the PG's is shown in Fig. 7b. Its proposed QCA layout and simulation results are drawn in Fig. 7c, d, respectively. The QCADesigner has two type of engines that are available for simulation. Coherence vector and the bistable approximation are two engines [15, 16]. In this work, we use a bistable engine which is a default set of parameters.

3.7 Reversible n-Bit Binary Incrementer

The structure of 1-bit reversible binary incrementer requires one Peres gate as shown in Fig. 5 and it has one garbage output. Two numbers of Peres gates are used in designing 2-bit binary incrementer having two garbage outputs as shown in Fig. 6. Figure 7 shows that three Peres gates are used in the design of 3-bit reversible binary incrementer, and it has three garbage outputs. Similarly to design n-bit reversible binary incrementer n-number of Peres gates are used with n-number of garbage outputs. The structure of n-bit binary incrementer using PG's is shown in Fig. 8.

4 Comparative Statistics

Table 1 explores the comparison of the quantum-dot cell circuits of reversible binary-to-gray and gray-to-binary converters. Table 2 shows that the new binary-to-gray converter has an improvement of 59% in cell count and 36%

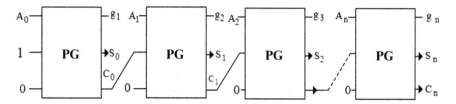

Fig. 8 Functional diagram of n-bit reversible binary incrementer

Table 2 Comparison of QCA layouts of reversible binary-to-gray code converters

	Design	Capability	Cell count	Area	Crossover	Latency
Binary-to-gray	[10]	3-bit	90	0.95	0	2
	[17]	4-bit	192	0.34	0	2
	[9]	3-bit	118	0.38	1	0.75
	[13]	4-bit	108	0.11	1	0.75
	New	4-bit	44	0.07	0	0.5
Gray-to-binary	[10]	3-bit	194	0.28	0	4
	[17]	4-bit	269	0.69	0	6
	[9]	3-bit	99	–	0	0.75
	[13]	4-bit	69	0.10	0	0.75
	New	4-bit	50	0.09	0	1

improvement in the total area over the previous design [9]. This circuit also has no crossover and 33% improvement in latency [9]. The proposed reversible gray-to-binary code converter has 27% improvement in cell count and 10% improvement in the area over the previous work [9].

The proposed quantum-dot cell of PG and 1-bit binary incrementer have 60 and 6% improvement in terms of cell count and area over [8], respectively. Our proposed reversible 2-bit binary incrementer has 60 and 52% improvement in cell count and area over existing QCA layout [8], respectively. Similarly, the proposed reversible 3-bit binary incrementer has 64 and 34% improvement in cell count and area compared to existing designs [8], respectively. The overall evaluation is drawn in Table 3. It is realized that there is a significant enhancement of results in the proposed QCA layout designs.

Table 3 Comparison of Peres gate and reversible binary incrementer QCA layouts

		MV No	INV No.	CC	Area (μm^2)	Crossover	Latency
Peres gate	[8]	7	4	97	0.075	1	1
	New	3	2	38	0.070	0	1
1-bit reversible incrementer	[8]	7	4	97	0.075	1	1
	New	3	2	38	0.070	0	1
2-bit reversible incrementer	[8]	14	8	196	0.272	2	2
	New	6	4	77	0.130	0	2
3-bit reversible incrementer	[8]	21	12	321	0.274	3	3
	New	9	6	113	0.18	0	2

5 Conclusion

This paper has reported the novel designs of reversible code converters and binary incrementer using QCA technology. All layouts are implemented in QCADesigner tool. The simulation results of all the QCA layouts have confirmed the functionality of the designs. All these designs have achieved a higher efficiency compared to previous works available in the literature. We also presented optimized designs of Feynman gate and the Peres gate in QCA. These two are basic gates having a low quantum cost which can be used for designing reversible circuits in QCA technology. Therefore, all our proposed QCA layouts can be used as an effective architecture for designing efficient digital circuits with the power of reversibility and QCA technology.

References

1. Lent, C.S., Taugaw, P.D.: A device architecture for computing with quantum dots. Proc. IEEE 85, 541–557 (1997).
2. Misra, N.K., Sen, B. and Wairya, S.: Designing conservative reversible n-bit binary comparator for emerging quantum-dot cellular automata nano circuits. Journal of Nanoengineering and Nanomanufacturing, 6(3), 201–216 (2016).
3. Allan, A., Edenfeld, D., Joyner, W.H., Kahng, A.B., Rodgers, M. and Zorian, Y.: 2001 technology roadmap for semiconductors. Computer, 35(1), 42–53 (2002).
4. Landauer, R.: Irreversibility and heat generation in the computating process. IBM J. Res. Dev. 5(3), 183–191 (1961).
5. Bennett, C.H.: Logical reversibility of computation. IBM J. Res. Dev. 17, 525–532 (1973).
6. Misra, N.K., Sen, B. and Wairya, S.: Towards designing efficient reversible binary code converters and a dual-rail checker for emerging nanocircuits. Journal of Computational Electronics, 16(2), 1–17 (2017).
7. Misra, N.K., Sen, B., Wairya, S. and Bhoi, B.: Testable Novel Parity-Preserving Reversible Gate and Low-Cost Quantum Decoder Design in 1D Molecular-QCA. Journal of Circuits, Systems and Computers, 26(9), 1750145 (2017).

8. Das, J. C., & De, D.: Novel low power reversible binary incrementer design using quantum-dot cellular automata. Microprocessors and Microsystems. 42, 10–23 (2016).
9. Das, J.C., De, D.: Reversible Binary to Grey and Grey to Binary Code Converter using QCA. IETE Journal of Research. 61(3), 223–229 (2015).
10. Misra, N.K., Wairya, S. and Singh, V.K.: Optimized Approach for Reversible Code Converters Using Quantum Dot Cellular Automata. In Proceedings of the 4th International Conference on Frontiers in Intelligent Computing: Theory and Applications (FICTA) Springer India, 367–378 (2016).
11. Angizi, S., Alkaldy, E., Bagherzadeh, N., Navi, K.: Novel robust single layer wire crossing approach for exclusive or sum of products logic design with quantum-dot cellular automata. Journal of Low Power Electronics. 10, 259–271 (2014).
12. Singh, G., Sarin, R.K., Raj, B.: A novel robust exclusive-OR function implementation in QCA nanotechnology with energy dissipation analysis. Journal of Computational Electronics. 15(2), 455–465 (2016).
13. Karkaj, E. T., Heikalabad, S.R.: Binary to gray and gray to binary converter in quantum-dot cellular automata. Optik-International Journal for Light and Electron Optics. 130, 981–989 (2017).
14. Sasamal, T.N., Singh, A.K., Ghanekar, U.: Design of non-restoring binary array divider in majority logic-based QCA. Electronics Letters. 52(24), (2016).
15. Sen, B., Dutta, M., Goswami, M., Sikdar, B.K.: Modular Design of testable reversible ALU by QCA multiplexer with increase in programmability. Microelectronics Journal. 45(11), 1522–32 (2014).
16. Sen, B., Dutta, M., Some, S., Sikdar, B.K.: Realizing reversible computing in QCA framework resulting in efficient design of testable ALU. ACM Journal on Emerging Technologies in Computing Systems (JETC). 11(3), 30 (2014) https://doi.org/10.1145/2629538.
17. Rafiq, B.M., Mustafa, M.: Design and Simulation of Efficient Code Converter Circuits for Quantum-Dot Cellular Automata. Journal of Computational and Theoretical Nanoscience. 11(12), 2564–2569 (2014).

Routing with Secure Alternate Path Selection for Limiting the Sink Relocation and Enhanced Network Lifetime

Renuka Suryawanshi, Kajal Kapoor and Aboli Patil

Abstract The main challenge of wireless sensor network is its lifetime. In this type of network, single static sink node is present; a sensor device node requires more energy for estimating information packet specifically those that are available in the area of the sink node. Such nodes separate the energy so fast due to the numerous tone traffic patterns and at the end they die. This uneven event is named as hotspot issue which gets more real as the numbers of sensor nodes increase. Generally, replacement of such energy sources is not a feasible and cost-effective solution. For this problem, there is one solution regarding to distance. If the distance among sensor and sink node is minimized, the energy consumption will be effectively reduced. This paper presents the solution for enhancing network lifetime with energy saving of sensor nodes. Here, we also discuss on the limitations and advantages of previous methods. The sensors nodes consume more battery power which is at minimum distance from sink node. Therefore, energy of sensor nodes in network will quickly consume their energy. So that, the lifetime of a sensor nodes will be produced. To overcome this drawback of this system, we propose alternate shortest path technique. To enhance the efficiency of energy along with network lifespan, this approach is used. Furthermore, we developed a novel technique known as Energy Aware Sink Relocation (EASR) for remote base station in WSN if the energy of alternate path is going to die. This system exploits information recognized with the remaining energy of sensor nodes battery for increasing the range of transmission of sensor node and relocation technique for the sink node in network. Some calculated numerical and theoretical calculations are given to demonstrate that the EASR strategy is used to increase the network energy of the remote system essentially. Our system proposes secure data sending using ECC algorithm and increases more network lifetime.

R. Suryawanshi (✉) · K. Kapoor · A. Patil
Department of Computer Engineering, MITCOE, Pune, India
e-mail: renukasuryawanshi743@yahoo.com

K. Kapoor
e-mail: kajal.kapoor@mitcoe.edu.in

A. Patil
e-mail: aboli.patil@mitcoe.edu.in

© Springer Nature Singapore Pte Ltd. 2018
S. Bhalla et al. (eds.), *Intelligent Computing and Information and Communication*,
Advances in Intelligent Systems and Computing 673,
https://doi.org/10.1007/978-981-10-7245-1_21

207

Keywords Wireless sensor networks · Cluster head · Base station
Cache-based system · Sensor nodes

1 Introduction

In the wireless network of sensor, nodes are outlined by heavy deployment of the huge types of sensor nodes in a particular geographical area. The information captured from sensor nodes is transmitted to monitor the station. These monitoring stations worked as a sink or base station. This base station is placed far from actual sensing area. To transfer this sensed information from source to base station, concept of multi-hop routing and flooding is used. With the multiple numbers of base stations, the total number of hops can get minimized. This will result in minimized energy consumption by sensor node. The minimum energy consumption of sensor nodes will improve lifetime of sensor network along with high rate of packet transmission to base station. So the communication nodes deployment and the different sink nodes are treated as most important components in the lifetime in wireless sensor network.

The WSN have various applications such as climate observing, battlefield investigation and inventory, manufacturing progressions. For the maximum amount of time, the sensor environment needs to be intolerant. In the remote network system, the sensor devices are not present to replace when their batteries get drained. The battery exhausted from nodes can be brought several issues, for example, take coverage hollow space moreover, communication hollow space issues. Consequently, some WSN systems are busy in planning proficient approach to keep the energy of sensor nodes, as an instance, drawing schedule of cycle for sensor nodes, which is used to permit some of nodes and enter into the die state to moderate power of energy. Now sensor node does not damage the running sensing process of the wireless network. The efficient design of energy algorithms aims at balancing the depletion of the battery exploitation strength of every sensor node or consuming a limited data aggregation technique for mixture of sensory information into a unit to decrease the number of message transmitted to prolong the wireless network lifetime. The enormous majority of such system scans coincide in the network system work. The another methodology is used for the purpose of storing energy as well as utilizing remote sensors to maintain the locations of region with aggregating lifetime network energy of nodes.

For enhancing the lifespan of network as well as efficiency of energy, we propose a shortest alternate path mechanism in this paper. For transmitting data from sender node to sink node, safely use alternative route and ECC algorithm. When energy level is less than given threshold for alternate path, we trigger the relocation of sink. Scheme for sink relocation is explored here, which decides when and where to relocate the position of sink. The mathematical performance evaluations are calculated to determine the proposed sink relocating scheme which is beneficial for enhancing the network lifetime of a system. The simulation of technique is used in project to check out the precaution of the EASR technique. This type of approach can work to improve the lifetime of a WSN system. The sink node relocation will be prolonging the battery usage of nodes.

Section 2 illustrates the related work studied for our new topic. Section 3 demonstrates the details of project implementation, definitions of terms and in addition the documentation can express the proposed system undertakings in this paper. Section 4 includes conclusions and represents future work of project.

2 Literature Survey

In this part, we illustrate the previous techniques proposed by the authors for WSN system.

Sara and Sridharan [2] represent a survey of routing schemes in remote sensor nodes networks. In WSN's author review the challenges for routing protocol designs were discussed. The comprehensive research of individual routing technique is classified into three stages depending upon the structure of network such as flat, hierarchical, location-based routing, etc.

Somasundara et al. [3] investigate a network system which depends on the utilization of mobile components and reduces the utilization of the energy constrained nodes at the time of communication and enhances useful network. Similarly, their approach gains the advantages in sensor networks and inadequately deployed sensors in network. They demonstrate how their procedure supports to reduce energy utilization at energy controlled nodes. After that, for enhancing the performance of energy, author illustrates their framework model which uses their proposed way.

Sensor network deployment is highly challenging because of the aggressive as well as volatile nature of consumption environments. Mousavi et al. [4] implemented two routines for the self-deployment of mobile sensors. Basically, author developed a randomized way that offers both simplicity and applicability to different environments.

Akyildiz et al. [5] describe idea of network formed by sensors. These sensors have combined microelectromechanical technology, wireless communication and digital physics. First, the sensing tasks and applications of sensor networks are examined, and a comparative analysis of things influencing the look of sensor networks is given.

The main benefit of sensor node networks is their self-organizing nature as well as autonomous process and possible architectural alternatives suitable for a different types of data-centric driven applications. During this article, Jain and Agrawal [6] deal with the presenting an outline of this state of the art inside the field of wireless sensor networks.

Tian and Georganas [7] have introduced inclinations to a node scheduling scheme, which can reduce the energy utilization of complete system, therefore growing network time period, by characteristic redundant nodes with respect to sensing coverage of network, moreover distributing them an offline operation mode. This offline mode has minimum energy consumption than the online mode.

Hong et al. [8] implemented a capable route setup for distributed sensing element network utilizing the similar process between the wireless and multi-hop communications network concerning instruments and rovers and therefore the packet radio network is utilized as a typical ad hoc networking surroundings.

Huang and Jan [9], for enlarging the lifetime of network's, presented an Energy Aware Cluster Based Routing Algorithm (ECRA) in WSN's. This algorithm chooses some nodes as cluster heads to construct Voronoi outlines and move the cluster head load balancing in every cluster of nodes. To improve the execution of the ECRA, a two-tier architecture (ECRA-2T) is designed. The reproductions demonstrate that both the ECRA-2T as well as ECRA algorithm perform better than other routing schemes such as direct communication, static clustering, and LEACH.

Shah and Rabaey [10] developed an energy aware routing mechanism that depends upon sub-optimal paths which give substantial gain. Additionally, the experimental results show the increment in lifetime of network over practically similar plans like directed diffusion routing. The more elegant degradation of service with time in a fairer manner to overcome the burning energy of nodes were showed.

In planning sensor networks, sensor deployment is a primary problem. Wang et al. [11], they review and make use of disseminated self-deployment protocols for mobile sensors. The protocols are proposed to estimate the target positions of the sensors after finding coverage of hole in network where the sink is ready to move.

3 Implementation Details

In this field, we illustrate the overview of system, algorithmic steps of system, and mathematical formulation of the proposed system.

3.1 System Overview

Figure 1 represents the architecture of the proposed system. System works as described in the following:

- Network Generation
  In this phase, user can generate vertices or nodes. These nodes are connected by edges.
- Path Generation:
  It creates all possible routes from source to sink node after creating source as well as sink.
- Shortest Path Selection:

Fig. 1 Proposed system
architecture

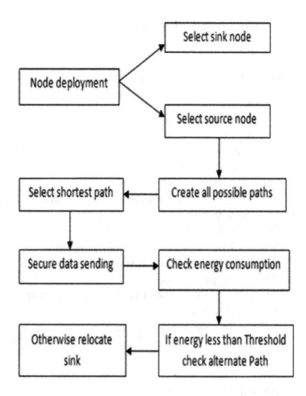

After generation of all possible paths from source to destination node, shortest path is selected on the basis of minimum weight of edge.

- Generation and distribution of Keys:
  At key generation center, keys are generated and distributed to all nodes belongs to shortest path.
- Encryption of data:
  At every node, collected data is encrypted by using ECC algorithm for security purpose.
- Estimation of Energy Consumption:
  After collection of data or sending of data or any type of action, consumed energy is calculated at each node.
- Data Authentication:
  Sink node checks the authenticated data after determining the hash value at source node.
- Data Decryption:
  After receiving the data from source node, sink node decrypts the data for further processing. For decryption, sink node has the decryption key.

3.2 Mathematical Model

The proposed system (PS) is represented as
  PS= {DN, SN, SiD, PA, SpS, AK, DS}

1. Deployment of nodes
   $DN = \{dn_1, dn_2,...,dn_n\}$
   where
   DN is the set of number of deployed nodes.
2. Source node selection
   $SN = \{sn_1, sn_2,...,sn_n\}$
   where SN is a set of sources selected at each runtime.
3. Sink node selection
   $SiD = \{Sid_1, Sid_2,....,Sid_n\}$
   Where, SiD is a set of sink nodes selected at each runtime.
4. All paths from source to destination
   $PA = \{pa_1, pa_2,....,pa_n\}$
   where PA is a set of all n number of paths from source to destination.
5. Shortest path selection
   $SpS = \{sps_1, sps_2, sps_3,....,sps_n\}$
   where SpS is the set of all possible shortest path from source to destination at each runtime.
6. Authentication with keys
   $AK = \{ak_1, ak_2,....,ak_n\}$
   where AK is a set of n number of keys generated and distributed to each node for authentication.
7. Data sending from source to destination.
   $DS = \{ds_1, ds_2, ds_3,....,ds_n\}$
   where DS is the set of all data packets securely routing through shortest path.

3.3 Algorithm

The proposed scheme works as follows:

- Algorithm 1: Proposed algorithm description
    1. The network graph is generated such as $G_i(V_i, E_i)$
       where V_i is the set of vertices and E_i is the set of all connecting edges to vertices.
    2. Select sender and destination node from all sensor nodes.
    3. Produce all possible paths from selected source to destination node.
    4. Among all generated possible path, select the one shortest path based on weight factor.

5. Generate and distribute public–private key pair for source and destination node.
6. Perform data sending at source node through selected shortest path.
7. Encrypt the data with the private key before actually sending.
8. Estimate energy consumed by each node belongs to shortest path.
9. Decrypt the private key and authenticate received data at destination.
10. If energy node in path is going to die then select alternate path among shortest path.
11. Resend the data from source to destination node through alternate path and also calculate energy consumed by path.
12. Again energy may expire of alternate path.
13. When energy is minimized, use Energy Aware Sink Relocation (EASR) technique to relocate sink node at other place.

Explanation: Algorithm 1 describes the levels of proposed system. Primarily, with sensor nodes, source and sink node network is created. Then generate all routes from source to sink node and for data sending purpose choose the shortest path. Sensor nodes are not working properly if energy utilization is greater. Therefore, systems must choose optional communication path between source and destination node and also estimate energy consumed by each node in network. By using the ECC algorithm, encrypt the data with the secret key. With the help of its hash value, data is validated. Only verified data is accepted by sink node. Decrypt the received data with the appropriate public key. If again energy is evacuated and path is expired, then repeat the procedure of sink relocation.

- Algorithm 2: ECC Algorithm

    1. Sender and receiver node calculate edB = S = (s1, s2).
    2. Sender node sends a message M E to receiver node as follows:
    3. Compute L such that, (s1 * s2) mod N = L.
    4. Compute L * M = C.
    5. Send C to sender node.
    6. Receiver node receives C and decrypts as follows:
    7. Compute (s1 * s2)modN = L.
    8. Compute (L-1)mod N, Where N = E.
    9. L-1*C = L-1*L*M = M.

3.4 Experimental Setup

Basically, the system is constructed with the help of java framework with version JDK 1.8 on window platform. IDE is used for developing the system, Netbeans with version 1.8 is used as a development tool. For generating the network, Jung simulation tool is used, by using this tool, network is created with the number of sensor node.

4   Results Discussion

Figure 2 describes the evaluation graph among ratio of energy consumption for existing as well as proposed system. In the designed system, energy utilization is less than the energy utilization in the existing one (Table 1).

Figure 2 represents the evaluation of designed system and existing system on the basis of accuracy. The accuracy of designed system is higher than the existing system.

Table 2 illustrates the outcomes for time consumption for developed site classifier and train classifier for designed and existing system.

Figure 3 exhibits the comparison between designed system and existing system. The time required for implementing the proposed system is less than the existing system.

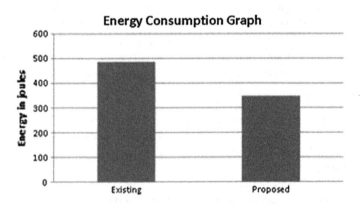

Fig. 2 Energy consumption graph comparison

Table 1 Energy consumption

	Existing system	Proposed system
Energy consumption	484	345

Table 2 Network lifetime

	Existing system	Proposed system
Total energy	3645	5068

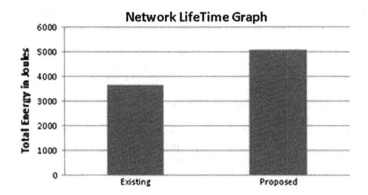

Fig. 3 Network lifetime comparison

5 Conclusion

This paper implemented the different methods to improve the lifetime of network. A relocatable sink is one approach to enhance the lifetime of network but still it has its own limitations as sink relocation involves more energy so we have proposed alternate shortest path technique which optimizes all nodes in the network system and also enhances lifetime of network by limiting the number of sink relocating actions. In addition, we also proposed secure data sending and node authentication for communication purpose. In future, we can increase lifetime of a network and secure the network by providing security.

References

1. Chu-Fu Wang, Jau-Der Shih, Bo-Han Pan, and Tin-Yu Wu, "A Network Lifetime Enhancement Method for Sink Relocation and Its Analysis in Wireless Sensor Networks, IEEE sensors journal, vol. 14, no. 6, June 2014.
2. G. S. Sara and D. Sridharan, Routing in mobile wireless sensor network: A survey, Telecommun. Syst., Aug. 2013.
3. A.A. Somasundara, A. Kansal, D. D. Jea, D. Estrin, and M. B. Srivastavam, Controllably mobile infrastructure for low energy embedded networks, IEEE Trans. Mobile Comput., vol. 5, no. 8, pp. 958973, Aug. 2006.
4. H. Mousavi, A. Nayyeri, N. Yazani, and C. Lucas, Energy conserving movement-assisted deployment of ad hoc sensor networks, IEEE Commun. Lett., vol. 10, no. 4, pp. 269271, Apr. 2006.
5. I. F. Akyildiz, W. Su, Y. Sankarasubramaniam, and E. Cayiric, Wireless sensor networks: A survey, Comput. Netw., vol. 38, no. 4, pp. 393422, Mar. 2002.
6. N. Jain and D. P. Agrawal, Current trends in wireless sensor network design, Int. J. Distrib. Sensor Netw., vol. 1, no. 1, pp. 101122, 2005.

7. D. Tian and N. D. Georganas, A node scheduling scheme for energy conservation in large wireless sensor networks, Wireless Commun. Mobile Comput., vol. 3, no. 2, pp. 271290, Mar. 2003.

8. X. Hong, M. Gerla, W. Hanbiao, and L. Clare, Load balanced energyaware communications for Mars sensor networks, in Proc. IEEE Aerosp. Conf., vol. 3. May 2002, pp. 11091115.

9. S. C. Huang and R. H. Jan, Energy-aware, load balanced routing schemes for sensor networks, in Proc. 10th Int. Conf. Parallel Distrib. Syst., Jul. 2004, pp. 419425.

10. R. C. Shah and J. Rabaey, Energy aware routing for low energy ad hoc sensor networks, in Proc. IEEE Wireless Commun. Netw. Conf., vol. 1. Mar. 2002, pp. 350355.

11. G. L. Wang, G. H. Cao, and T. L. Porta, Movement-assisted sensor deployment, in Proc. IEEE Inf. Commun. Conf., Aug. 2004, pp. 24692479.

12. Chu-Fu Wang, Jau-Der Shih, Bo-Han Pan, and Tin-Yu Wu, A Network Lifetime Enhancement Method for Sink Relocation and Its Analysis in Wireless Sensor Networks, ieee sensors journal, vol. 14, no. 6, June 2014.

Gene Presence and Absence in Genomic Big Data for Precision Medicine

Mohamood Adhil, Mahima Agarwal, Krittika Ghosh, Manas Sule
and Asoke K. Talukder

Abstract The twenty–first-century precision medicine aims at using a systems-oriented approach to find the root cause of disease specific to an individual by including molecular pathology tests. The challenges of genomic data analysis for precision medicine are multifold, they are a combination of big data, high dimensionality, and with often multimodal distributions. Advanced investigations use techniques such as Next Generation Sequencing (NGS) which rely on complex statistical methods for gaining useful insights. Analysis of the exome and transcriptome data allow for in-depth study of the 22 thousand genes in the human body, many of which relate to phenotype and disease state. Not all genes are expressed in all tissues. In disease state, some genes are even deleted in the genome. Therefore, as part of knowledge discovery, exome and transcriptome big data needs to be analyzed to determine whether a gene is actually absent (deleted/not expressed) or present. In this paper, we present a statistical technique to identify the genes that are present or absent in exome or transcriptome data (big data) to improve the accuracy for precision medicine.

Keywords Big data · Algorithms · Genomics · Multimodal distribution
Exome analysis · Transcriptomics analysis · Gaussian mixture model
Precision medicine

M. Adhil · M. Agarwal · K. Ghosh · M. Sule · A. K. Talukder (✉)
Interpretomics, 5th Floor, Shezan Lavelle, 15 Walton Road, Bangalore 560001, India
e-mail: asoke.talukder@interpretomics.co

M. Adhil
e-mail: mohamood.adhil@interpretomics.co

M. Agarwal
e-mail: mahima.agarwal@interpretomics.co

K. Ghosh
e-mail: krittika.sasmal@interpretomics.co

M. Sule
e-mail: manas.sule@interpretomics.co

© Springer Nature Singapore Pte Ltd. 2018
S. Bhalla et al. (eds.), *Intelligent Computing and Information and Communication*,
Advances in Intelligent Systems and Computing 673,
https://doi.org/10.1007/978-981-10-7245-1_22

1 Introduction

Biology and medicine used to be a data-poor science. With the completion of the Human Genome Project (HGP), biology has entered the quantitative data-rich era. This has led the implosion in the amount of data being generated. In precision, medicine of twenty-first-century genetic tests of body fluid (blood) or tissues (for cancer or various neoplastic tissues) are being included in *evidence-based medicine* (EBM) as "molecular pathology" tests. The analysis of molecular pathology tests needs to overcome complex big data algorithmic challenges. The whole genome data for a human ranges upward of 300 GB (Giga Bytes) based on the sequencing depth. The volume of data generated by the protein coding region (exome) is of the order of 18–20 GB while the transcriptome (RNA-Seq) data measuring the protein expression of the coding region is of the same order as well [1]. To discover the *actionable insight* or a *biomarker* from this data, it is necessary to perform extensive statistical and exploratory data analysis [2], Although this "Big data" in form of DNA, gene expression, or the pathway data has been collected and recorded in a wide variety of databases, the intricate understanding as a holistic process to address the an individual at a system level still remains elusive. This is due to lack of explicit correlations amongst these different data forms. Key questions that still confront biologists include (a) the functional roles of different genes and the cellular processes in which they participate; (b) which genes are functionally present or absent within a sample and how they interact to trigger a phenotype; (c) how gene expression levels differ in various cell types and states, as well as how gene expression is changed by various disease pathologies and their treatments.

A phenotype is any observable characteristic of an organism related to its external appearance or behavior, both physical and biochemical; whereas, the genotype is the state of the DNA inside the body, which is transcribed and manifests in RNA and subsequently in proteins; and finally how the nutrients and environmental conditions make the cells in the body function in harmony. The linking of genotype or mutations in the gene and the gene expression to phenotype is the fundamental aim of modern genetics [3]. In other words, what is the genetic root of the abnormal manifestation (disease state) of an individual? The focus currently lies on study of interactions amongst genes supported through the expression data and phenotype data through integrative analysis. The inherent complexity of phenotypes makes high-throughput phenotype profiling a very difficult and laborious process. Phenotype–genotype association analysis problems are complicated by small sample size of data coupled with high dimensionality. Many complex phenotypes, such as cancer, are the product of not only gene expression, but also gene interactions [4].

The *central dogma of molecular biology* explains how the genetic information flows within the biological system starting from DNA to protein expression and functional implications of the same. Changes in this flow of information at the gene level, transcript level, or protein level can lead to the genetic causation of diseases.

The monitoring of gene functions and their interactions can provide an overall picture of the genes being studied, including their expression level and the activities of the corresponding protein under certain conditions. To accomplish this, functional genomics provides techniques characterized by high-throughput or large-scale experimental methodologies combined with statistical and computational analysis of the results [5]. Current methods in evidence-based medicine require advanced technologies for evidence search and discovery. Next generation sequencing provides convenient and high-throughput means for accomplishing systemic investigation and functional genomics compared to more traditional methods as q-PCR and Sanger based methods for mutation detection. Genes expressed in cells in different organs (tissues) are different. This implies that though all cells were born from one mother cell (with genetic information carried down half from father and half from mother), their functions differ vividly due to a very sophisticated process of cellular differentiation and gene expression. Apart from this, environmental cues, epigenetics play an important role in gene regulation. A core challenge in molecular biology therefore is to identify the genes that are expressed in a cell within a tissue (or organ) and to measure their expression levels [6]. The major challenge is to determine, if a gene is not expressed in a tissue—is it due to normal phenotype or a disease state phenotype. Therefore, the analytics software must be able to determine which genes are expressed (present) and not expressed (absent) in the tissue of interest from the experiment data. In order to do this, it must be able to identify and eliminate the noise in the expression values. In microarray experiments, special negative control probes are designed based on the knowledge of the nucleotide sequence of a gene. During the analysis of microarray data, these negative control probe measures are used in the normalization process to eliminate the noise. Gene level presence/absence calls in microarrays are made using these negative probe sets (no hybridization) and cutoffs generated using p-values to eliminate the noise [7].

The RNA-Seq procedures in NGS have been widely used to estimate the expression of genes in specific tissues and under particular condition. In this technique, the RNA is extracted from the human tissue of interest, libraries of sequences are prepared and then sequenced using NGS. The data generated in such experiments are in the order of 18–30 GB, generally consisting of around 30 million reads. In principle, the NGS derived reads are statistically mapped across a genome to estimate the regions mapped. However, the mapping of reads is statistical and not devoid of system bias. Since, genomic regions differ in terms of sequence complexity, the regions with lower complexity present lower potential for mapping than others. Moreover, there are repetitive regions which are considered "unmappable" owing to the absence of unique reads originating from those regions which can map them back. Such regions extend bias in alignment by mapping algorithms leading to noise in mapping across the genome [8]. This leads to some regions having over represented coverage whereas other regions may be less covered.

Transcriptome data presents an added challenge due to the phenomenon of splicing. The alignment requires considering "splice aware" alignment, wherein reads from transcripts can be fragmented and aligned to the reference genome from where it had originated [9]. These exert a selective bias and sometimes give rise to nonspecific alignments in regions and transcripts which are not expressed under a particular condition thus leading to incorrect assessment. Similar, features can be expected in any enriched library being sequenced out from the human genome. However, accuracy is a must when application space is health care.

In case of RNA-Seq and NGS gene expression data, the absence of negative control probe-set as in microarray studies, there is a need to discover which genes are expressed. Therefore, the need to algorithmically eliminate the noise generated holds prime importance. In this paper, we present a novel statistical algorithm that does this job accurately. Here, we improve the accuracy of this calling of genes by introducing the algorithm that acts as an additional filter to the initial results provided by transcriptomic experiments. The algorithm works to identify mapping artifacts thereby refining the genomic coverage for RNA sequencing analysis by estimating the presence or absence of alignment of reads and hence gene expression. The presence/absence call algorithm presented here is part of the reference based RNA-Seq analysis of the iOMICS software [10] and uses statistical techniques to discover the genes that are not explicitly expressed in the sample under investigation.

2 Methods

The following key points have been considered while developing the algorithm for identification of presence/absence of genes in RNA-Seq data:

1. Parallel computing technique has been used to calculate the Reads Per Nucleotide (RPN) to reduce the time taken by many folds.
2. The algorithm works both on single sample ($n = 1$) and cohort data ($n = N$) of assembled RNA-Seq data to identify the sample level presence/absence.
3. Genes are also optionally annotated with tissue level baseline gene expression values from external databases.

Because the sequence reads generated by sequencing machine are few hundreds of nucleotides (generally 300 nucleotides from two 150 nucleotide reads), these sequence reads often align to nonspecific regions. Also, reads that are erroneous but not filtered by the cleaning up procedure will align to regions as noise. Due to fundamental properties of cells, a gene may be expressed low, generating lesser number of reads. Therefore, the challenge is to detect whether less number of read is due to lower expression of the gene or due to miss-alignment. Here, our challenge is to devise a mathematical model that will be able to detect the various expression profiles within the experimental data. Before calculating RPN value for each gene, the raw sequence quality and alignment statistics are verified for data integrity.

Gaussian mixture model (GMM) which is a parametric model is used to divide the whole population into subpopulation. Here, the whole population refers to all genes in the sample with logarithmic reads per nucleotide and subpopulation refers to presence, absence, and marginal expression [11]. GMM is preferred mainly due to its nature of distinguishing clusters with definite boundaries using p-value and with no preassignment of data point into a cluster. GMM is an unsupervised model similar to k-means, but unlike k-means we do not need to assign a data point (gene) to a single cluster. GMM parameters are estimated using the iterative Expectation–Maximization (EM) algorithm [12]. GMM is a weighted sum of M components' Gaussian densities as given in Eq. 1. The components are predefined before the model fitting.

$$p(\mathbf{x}|\lambda) = \sum_{i=1}^{M} w_i g\left(\mathbf{x}|\boldsymbol{\mu}_i, \sum_i\right), \tag{1}$$

where

1. x is a D-dimensional vector of continuous value
2. w_i, $i = 1,..., M$, are the mixture weights,
3. $g(x|\mu_i, \Sigma_i)$, $i = 1,..., M$, are the component Gaussian densities. Each component density is a D-variate Gaussian function given in Eq. (2)
4. μ is the mean vector, Σ_i covariance matrix and mixture weight satisfy the constraint $\Sigma_{i=1 \text{ to } M} w_i = 1$

$$g(\mathbf{x}|\boldsymbol{\mu}_i, \boldsymbol{\Sigma}_i) = \frac{1}{(2\pi)^{D/2}|\boldsymbol{\Sigma}_i|^{1/2}} \exp\left\{-\frac{1}{2}(\mathbf{x} - \boldsymbol{\mu}_i)'\boldsymbol{\Sigma}_i^{-1}(\mathbf{x} - \boldsymbol{\mu}_i)\right\}, \tag{2}$$

Here, the number of components is two (presence and absence). μ (mean) initially will be zero and x will be the vector of RPN values. Model fit is performed with two components; mean and covariance are calculated for those two components. After model fitting, RPN values are iterated and assigned into two distributions. The GMM is parameterized by the mean vectors, covariance matrices, and mixture weights from all component densities.

The noise distribution is identified using the RPN value calculated from intergenic reads aligned to reference genome due to by chance or error. The RPN for noise distribution is calculated using the coordinates of intergenic regions throughout the genome. Once we have the noise distribution, it is used as a threshold or boundary for labeling the gene as present, absent, and having marginal expression. This is done by expectation–maximization to find the maximum likelihood, where the mean and covariance of noise distribution is compared with the mean and covariance of the genes calculated from the GMM model. Values of genes where a presence/absence label cannot be assigned with high confidence using the given p-value cutoff are labeled as marginally expressed. The p-value and confidence level is calculated for three labels (Presence, Absence and Marginal). The significance level (*p*-value) depends on the user, where it can be less than 0.05.

3   Results and Discussion

The RNA-Seq reads are aligned to the exonic regions (coding regions) of genes. The number of reads aligning to the exonic regions of a gene is proportional to the expression level of the gene and the length of the exonic region. However, due to sequencing errors and the statistical nature of reference assembly, errors are incorporated into the read alignment process. This results in leakage or spillover in the data. There will be a few reads which will align to genes as alignment artifacts (noise). Similarly, there will also be some reads that will align to the intergenic regions also. In order to identify the noise, the algorithm uses the feature Reads Per Nucleotide (RPN) statistic (Eq. 3). RPN is defined as

$$\text{RPN} = R/L \text{ where } R \text{ is the number of reads in a region of length } L \text{ nucleotides} \quad (3)$$

It may be noted that while the read alignment at a whole genome level is modeled using Poisson distribution, the RPN distribution is log-normal.

The RPN statistic is calculated for the genic as well as intergenic regions to identify the noise. The algorithm models the RPN for both exonic (collapsed to genic level) and noncoding/intergenic regions, and uses it to generate *p*-values for each gene and obtain an RPN cutoff (Fig. 1). Using this RPN cutoff, the sample level presence/absence of a gene expression is estimated.

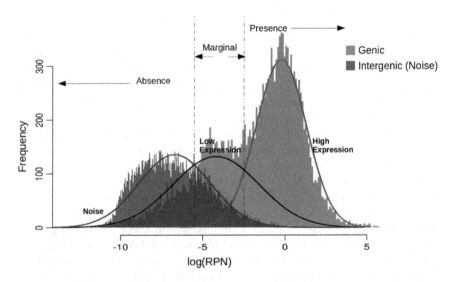

Fig. 1 RPN distribution across the genome for genic and intergenic (noise) along with regions of absence, marginal, and presence

3.1 Case Study (RNA)

In order to test and validate our algorithm, we have used RNA-Seq data from 20 samples of the lymphoblastoid cell lines generated by Lappalainen et al. [13] available from ArrayExpress (E-GEUV-1). The RPN statistical model for each sample results in two distributions, viz., genic (red) and intergenic (blue). The gene level presence/absence expression for each sample can be estimated using the genic distribution showing two expression profiles through p-value cutoffs. Figure 1 shows the distribution of log RPN for the genic and intergenic regions for the sample HG00096. Table 1 shows a few records from the final PA-call output for this example dataset. This table gives log (RPN), presence/absence call, and the call confidence. Baseline tissue gene expression values (Fragments Per Kilobase per Million mapped reads (FPKM) shown in last column of Table 1) from Expression Atlas [14] have been used to cross validate the results.

In the above table, "-Inf" in log (RPN) column represents that no reads were aligned to the particular gene. Using the log (RPN) cutoff value obtained from the genic and intergenic distributions, Presence (P)/Absence (A)/Marginal (M) expression calls are identified as shown in Sample PA Call (column 3). The confidence levels (ranging from 0 to 1) for these calls are given in column 4 (Call Confidence). The results can be compared with the tissue baseline expression values (Column 5) based on the tissue of origin.

3.2 Estimating Presence/Absence of Genes Using Exome Data

Variations in the DNA are responsible for various phenotypes. When these variations become pathogenic, they manifest into cancer and many other diseases. These

Table 1 The gene level presence/absence of expression for 10 genes as estimated by the algorithm

Gene	log (RPN)	Sample PA call	Call confidence $(1 - p)$	Tissue baseline (FPKM)
ACTL9	-Inf	A	1	–
C140R64	−10.2146419812	A	1	–
ACTB	5.7318105034	P	0.9999999902	3439
RPS29	3.7241632036	P	0.9999984735	63
MBD2	1.9461681249	P	0.9999332766	31
PARN	0.7630505519	P	0.9994225476	14
CARD11	0.1472238208	P	0.9984111854	23
OR2T5	−3.7027823593	M	0.8505041566	0.1
OR2T3	−3.7802292327	M	0.8738975872	0.3
HIST1H3I	−3.8649313979	M	0.8996086247	0.2

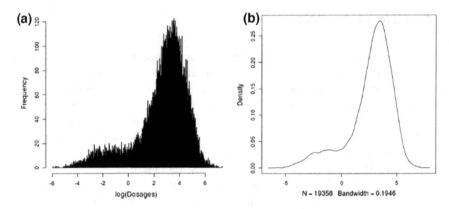

Fig. 2 Mixture of two populations in the DNA data. **a** Distribution plot of localized CNV in exome data. **b** Density plot of localized CNV in exome data

variations may include single base substitutions (single nucleotide polymorphism—SNP), small insertions, and deletions (indels), or rearrangements and copy number variations (CNV) together with epigenetic changes [15]. All these variations are heterogeneous in nature and represented through categorical variables. For quantitative analysis of CNV, however, we used quantitative matrix of the read abundance at each exonic location. We use exonic limits as breakpoints to quantify these CNV. The measure of DNA variation is presented as SPKMG (Sequence Per Kilobase of exon, per Megabase of the mappable Genome) and already reported in our earlier publication [16]. We showed that this quantitative measure of DNA demonstrates a normative pattern with higher level of statistical evidence with phenotypic association. We also showed that localized CNV works very well to discover DNA biomarkers [16] for both single patient and a population. The deletion of a gene in exome can be assumed similar to low expression as in RNA. The amplification of genes is similar to high expression. Therefore, in principle the localized CNV will show similar characteristics like the RNA. The frequency distribution and the density plot of localized CNV are shown in Figs. 2a, b, respectively. The localized CNV for this graph are taken from the familial breast cancer data generated by Gracia-Aznarez et al. [17]. This also indicates that the same model described earlier for RNA-Seq will be applicable to identify the genes that are amplified or deleted in a DNA-Seq data.

4 Conclusion

Genomic data processing faces combinations of complex challenges related to big data, high dimensionality, and multimodal distribution. Knowledge discovery in big data genomics is often top-down. Therefore, for genomics-based molecular pathology tests for clinical that can lead towards precision medicine, we must address all these challenges algorithmically. Here in this paper, we have presented the presence–absence algorithm for genomic big data analysis. We have shown that RNA data from RNA-Seq experiments and the localized CNV data from exome are multimodal. The limited data samples and high dimensionality problems can be addressed efficiently using statistical methods. We have presented here an analytic algorithm for processing such genomic data and identified accurately the presence or absence of genes in both RNA and DNA data from NGS experiments, where the volume of a single patient data is upward of 10 GB. We also showed that the multimodal challenges in genomic data can be addressed using the Gaussian mixture model. We compared the result of our model with the experimental results from other experiments and bibliomic data. Results from our model match with the experiments performed by other independent studies. This shows that our algorithms we designed are able to solve all these challenges of precision medicine accurately and can be used in clinical setups.

References

1. Eisenstein, Michael. "Big data: the power of petabytes." *Nature* 527.7576 (2015): S2–S4.
2. Bock, Hans-Hermann, and Edwin Diday, eds. *Analysis of symbolic data: exploratory methods for extracting statistical information from complex data*. Springer Science & Business Media, 2012.
3. Morley, Michael, et al. "Genetic analysis of genome-wide variation in human gene expression." *Nature* 430.7001 (2004): 743–747.
4. Ried, Thomas, et al. "Genomic changes defining the genesis, progression, and malignancy potential in solid human tumors: a phenotype/genotype correlation." *Genes, Chromosomes and Cancer* 25.3 (1999): 195–204.
5. Kitano, Hiroaki. "Computational systems biology." *Nature* 420.6912 (2002): 206–210.
6. Maniatis, Tom, Stephen Goodbourn, and Janice A. Fischer. "Regulation of inducible and tissue-specific gene expression." *Science* 236 (1987): 1237–1246.
7. Komura, Daisuke, et al. "Noise reduction from genotyping microarrays using probe level information." *In silico biology* 6.1, 2 (2006): 79–92.
8. Schwartz, Schraga, Ram Oren, and Gil Ast. "Detection and removal of biases in the analysis of next-generation sequencing reads." *PloS one* 6.1 (2011): e16685.
9. Trapnell, Cole, et al. "Differential gene and transcript expression analysis of RNA-seq experiments with TopHat and Cufflinks." *Nature protocols* 7.3 (2012): 562–578.
10. iOMICS-Research Version 4.0.
11. Reynolds, Douglas. "Gaussian mixture models." *Encyclopedia of biometrics* (2015): 827–832.
12. Moon, Todd K. "The expectation-maximization algorithm." *IEEE Signal processing magazine* 13.6 (1996): 47–60.
13. Lappalainen, Tuuli, et al. "Transcriptome and genome sequencing uncovers functional variation in humans." Nature 501.7468 (2013): 506–511.
14. Petryszak, Robert, et al. "Expression Atlas update—an integrated database of gene and protein expression in humans, animals and plants." *Nucleic acids research* (2015): gkv1045.
15. Pleasance, Erin D., et al. "A comprehensive catalogue of somatic mutations from a human cancer genome." Nature 463.7278 (2010): 191–196.

16. Talukder, Asoke K., et al. "Tracking Cancer Genetic Evolution using OncoTrack." *Scientific Reports* 6 (2016).
17. Gracia-Aznarez, Francisco Javier, et al. "Whole exome sequencing suggests much of non-BRCA1/BRCA2 familial breast cancer is due to moderate and low penetrance susceptibility alleles." *PloS one* 8.2 (2013): e55681.

A Survey on Service Discovery Mechanism

Gitanjali Shinde and Henning Olesen

Abstract A wide range of web services are available, with a deployment of a huge number of the Internet of Things (IoT) applications, e.g., smart homes, smart cities, intelligent transport, e-health and many more. In the IoT paradigm, almost every device around us will be on the Internet to provide specific services to end users. A large number of services will be available to help end users, however, providing and selecting appropriate service from a massive number of available services is a challenge of IoT. Service discovery plays a vital role in web service architecture. Traditionally, keyword search has been used for service discovery. The traditional approaches of service discovery cannot be applied as it is for web services in the IoT environment as devices providing, and accessing services are resource constrained in nature, e.g., limited size, limited energy source, limited computational capability, and limited memory. Hence, there is need of a lightweight and semantic service discovery approach for the IoT. This work analyzes different service discovery approaches. This analysis will focus on the key requirements of service discovery mechanism needed for the IoT environment.

Keywords Internet of Things (IoT) · Service description · Service discovery
Semantic service discovery · User preference

1 Introduction

In the IoT environment, devices around us are connected to the Internet to provide services to the end users. IoT devices can be anything around us, i.e., sensors, RFID tag, RFID reader, actuators, etc. The different IoT devices may use different

G. Shinde (✉) · H. Olesen
Center for Communication, Media and Information Technologies (CMI),
Aalborg University Copenhagen, 2450 Copenhagen SV, Denmark
e-mail: gis@es.aau.dk

H. Olesen
e-mail: olesen@cmi.aau.dk

© Springer Nature Singapore Pte Ltd. 2018
S. Bhalla et al. (eds.), *Intelligent Computing and Information and Communication*,
Advances in Intelligent Systems and Computing 673,
https://doi.org/10.1007/978-981-10-7245-1_23

wireless/wired technology to transmit data to cluster head, i.e., WLAN, ZigBee, NFC, Zwave, WiFi [1]. These devices may be different in sizes, computational capability; hence heterogeneity will be the major challenge to design WS. IoT will be deployed in large number and almost in all application domains, i.e., e-health, smart home, the smart city [2], intelligent transport, logistics and many more. Hence, a large number of devices will be working with each other over the Internet to provide services to the end users. End users access a web service (WS) depending on their requirements, hence selecting appropriate service from a large number of available services is an essential design parameter for WS.

The remainder of this paper is structured as follows: the Motivation is discussed in Sect. 2, the Detailed Literature of Service Discovery Mechanisms is discussed in Sect. 3, Comparative Analysis of literature work is presented in Sect. 4, and the Discussion is summarized in Sect. 5.

2 Motivation

Predicting and providing appropriate service depending on context and user preference is the necessity of IoT environment. The user may be mobile and will require accessing services via his/her smartphone; therefore the lightweight WS discovery process is an important design parameter of WS architecture for IoT. In the IoT environment, devices around us sense information and depending on this information services are built. Every IoT device is not capable of building and providing WS as IoT devices are resource constrained in nature. Hence, few devices are grouped together as a cluster, and one computationally strong device is selected as cluster head to perform complex task [3], i.e., building and providing WS. The information senses by the IoT devices are raw and unstructured in nature, so there is need of describing services depending on the raw and unstructured information by converting it into a structured format. The need of clustering IoT devices and lightweight WS discovery is shown in Fig. 1.

The traditional WS discovery methods are not applicable to IoT environment due to its different requirements, hence it is important to review different WS discovery mechanisms. WS are the description of operations that perform a certain task. WS are language and platform independent. WS can perform a simple task or complex task. A WS has three components, which include service provider, service registry, and service requestor. Service provider describes services using a standard Extensible Markup Language (XML), i.e., Web Service Description Language (WSDL). A service is advertised by publishing it to a registry. Universal Description Discovery and Integration (UDDI) is the registry used to publish web services. End users are the service requestors, requesting for web service using find operation. WS has a four-phase life cycle, which includes Build, Deploy, Run, and Manage.

The build phase of life cycle plays very important role in the WS life cycle to provide appropriate service to end user when the description of WS is matched with

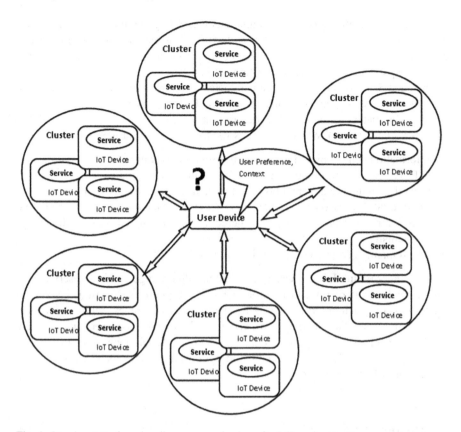

Fig. 1 Requirement of service discovery mechanisms for IoT environment

requirements of the end user. WS uses the Simple Object Access Protocol (SOAP) to access WS, and it is not suitable for the WS architectures in the IoT environment as a SOAP-based approach are heavyweight in nature. Traditionally, syntactic service discovery mechanism is used to discover the WS. A syntactic service discovery mechanism results in a list of services among that user has to select appropriate service depending on his/her requirement, and this is not feasible approach for IoT environment. To provide appropriate service to end user, the service description should be done in a semantic way, so that in the service discovery phase, a service can be searched very efficiently.

The WSDL, SOAP, and UDDI standards are used to build service architecture. However, the WSDL is not designed for semantic description. WSDL describes WS in the form of a port, port type, and bindings. Hence, it is very difficult to specify semantic details about WS, i.e., nonfunctional requirements, QoS requirements, availability, reliability, throughput, response time, and mean time before failure. This invokes the need for a semantic description of services. To provide the semantic descriptions, different approaches are invoked, i.e., OWL-S, WSDL-S.

However, these methods have few limitations, i.e., a huge number of services exist are described using WSDL format and it is not feasible to add semantic in the existing services, every new service cannot have a semantic description with it and service requestor may be unaware of terms related to the service. UDDI is used to publish a WS, however, the runtime parameters cannot be stored in the UDDI and there is no universal registry for storing WS. Hence, it is very difficult to store runtime parameters to improve WS description [4].

A few challenges of WS architectures for IoT environment are as follows:

- Information sensed by IoT device is heterogeneous, raw, and unstructured
- IoT devices are resource constrained in nature, i.e., limited energy source, limited computational capability, limited memory, and limited size
- Traditional WS discovery mechanism is heavyweight and keyword search based
- User-centric approach is not considered in the traditional service discovery mechanisms
- The semantic service description is required for WS.

In the near future, IoT devices around us will offer a huge number of services to end users for betterment and comfortable life. The end user must get services depending on their requirement. However, the IoT comes with challenges mentioned above. This invokes the need for a lightweight and semantic service architecture for IoT. In the next section, review of different approaches of service discovery is done. This analysis helps find out design parameters of service architecture for IoT environment.

3   Related Work

The service description is the first and important step of service architecture as the efficiency of service discovery is dependent on the service description. In the literature, service description is done by three different approaches, such as functional, nonfunctional, and data structural [5]. To provide appropriate service depending on a user's requirement, the need of betterment in a service description is invoked. Semantic service description is the step to fulfill the need mentioned above. The semantic service description can be done using the following two approaches: annotation based and semantic language based. The annotation-based approach improves the expressivity of service description by providing an annotative description for data types, inputs, and output of WS. In the IoT paradigm, devices around us will provide services to end users. A large number of services will be available as IoT applications will be deployed in all domains. In order to provide appropriate and useful service to the end user, efficient service discovery mechanisms are needed. Traditionally, services are registered in the UDDI registry; the services are searched using the keyword search mechanism. The UDDI-based service architecture faces the limitation due to a centralized approach, hence different ways of deploying service architectures are discussed by Demian et al. [6], i.e., centralized and distributed. Centralized approach is categorized into the portal and UDDI-based architecture. UDDI-based architecture is further categorized into hybrid and portal-based architecture. Distributed architecture approach is categorized into three different ways, such as agent-based, Internet-based, and point to point (P2P) based architecture. Among these different ways of WS architecture, appropriate architecture is used depending on the requirement of the application.

The ontology-based services discovery is proposed by Paliwal et al. [7]. Service discovery is achieved by functional level service categorization and semantic enhancement of service request. The service categorization is done at UDDI by performing the following two steps: parameter based service refinement and semantic similarity based matching. Service refinement is done by combining semantic characteristics with syntactic characteristics in WSDL description. In this approach, each service is represented as vector, i.e., Service Description Vector (SVD). Service refinement is done by improving semantic contents of the SVD. Services having similar functionality may be grouped in different categories at UDDI, such services from different categories are grouped together in the cluster. These clusters are labeled depending on their functionality for efficient service discovery. To label the cluster, an ontology concept is used. Services from the cluster are mined to find the pattern to satisfy the user's request. Hyperclique patterns are used for this purpose. A user's request is enhanced using Latent Semantic Indexing (LSI) technique. In this approach, the service refinement and discovery is dependent on the WSDL file. It is impractical that every existing service will have a semantic description with WSDL file.

Logic-based and knowledge-based service search and discovery are proposed by Cassar et al. [8]. The latent factor based approach is used to describe the service. Latent factors are recovered from service description by probabilistic machine learning approach. A latent factor is related to the semantic description of web service and can be expressed as probability distribution function. In this, Ontology Web Language for Services (OWL-S) is used for service description. To retrieve latent factor from the description, it is converted into Service Transaction Matrix (STM). The STM provides a training set, and it is served as input to probabilistic machine learning technique to extract latent factors. Clustering of relevant services is done on the basis of latent factors using Latent Dirichlet Allocation (LDA). Service discovery process works efficiently on clustered services.

Meditskos et al. [9] proposed service profile ontology-based service description. Object-based structural matching technique and logic reasoning methods are used for match making process of service discovery. Ontology roles are combined with description logic reasoning for improved service discovery process. The comparison between the Complex Concept (CC) and Service Profile (SP) description of WS description is proposed in this work. The four ontology levels are used to describe WS, such as Service, Service Profile, Service Process, and Service Grounding. In the profile-based discovery mechanisms, the service request and service advertisement are both represented as the profile. The profile-based discovery gives discovery results very efficiently.

Bootstrapping ontology mechanism for service discovery is proposed by Segev et al. [10]. The Term Frequency and Inverse Document Frequency (TF/IDF) methods are used to evaluate the WSDL description. These methods review the service description from internal and external perspective. Lastly, free text description method is used to resolve the inconsistencies in the service description ontology. It works in four steps: token extraction, token evaluation, concept evocation, and ontology evolution. Initial filtering and extraction of the labels from WSDL document are done in the token extraction step. In the second step, the tokens are evaluated by TF/IDF and Web Context Extraction (WCE) method. The TF/IDF method searches for the token that presents most frequently in its WSDL document, but least present in WSDL documents of other services, i.e., token that can represent service in a unique way. In a WCE method, group of tokens are extracted from WSDL file of service that can represent the context of service. In the concept evocation, step tokens extracted from both methods are evaluated, and tokens present in both methods are selected to describe the service. In the last step, current ontology is compared with newly extracted descriptors. In this work, the description of WS is improved for efficient discovery.

Chakraborty et al. [11] proposed service discovery for pervasive computing. In the pervasive environment, usually services are advertised in a broadcast way. However, the global broadcasting has limitations, such as broadcasting of services cannot be applied to the large-scale network as it works poorly with increasing network size and it uses the more resources and power. In the proposed method, solution to overcome the drawbacks of global broadcasting is discussed. It works in two steps: service caching and group-based advertisement of the service request. Services are cached at every peer and service requests are forwarded in the group-based approach. Every group of nodes advertises the services provided by the group in an abstract way. Hence, service request can be easily forwarded to the appropriate group. The group-based approach minimizes the efforts to search appropriate node by providing the service demanded by the user.

WS description, advertisement, execution, and discovery using DARPA Agent Markup Language (DAML) are discussed by McIlraith et al. [12]. Descriptive API is used to provide service specification, service capabilities, interfaces, and pre-requisites for service execution. DMAL Agent-based technology is used in this work to discover WS depending on the user preferences and constraints. Zisman et al. [13] proposed the need and solution of runtime service discovery. Service replacement is needed due to various reasons, such as unavailability or malfunctioning of service, changes in the structure of interface or functionality of service, and better service is available to complete the associated task. The pull and push model of runtime service discovery is used to replace the service that is no longer needed. The requirements of Quality of Service (QoS) parameters for WS description and discovery are discussed by Kritikos et al. [14]. QoS parameters are categorized into two types, such as domain dependent and domain independent. QoS broker service architecture is proposed in this work. The broker plays the middle layer between user and WS. Analysis of different QoS parameters helps to describe WS semantically with nonfunctional qualities.

Guinard et al. [15] proposed the RESTful service architecture. Proposed approach is targeted for the IoT environment. The IoT devices are resource constraints in nature, hence RESTful approach is used for WS. Device profile (DP) is generated for each IoT device that provides services. DP provides services to the users with minimal user interaction. Minimal user interaction is the need of the IoT environment to achieve lightweight service architecture. The context awareness is used as a key parameter for service discovery as in future network large number of services will be available. Among such large number of services, only appropriate service should be provided to the user, hence user preference and context are the parameters for service discovery. In the proposed work, these requirements of the future network are focused, e.g., RESTful services, context, and user preferences. However, the access methodology for WS is not addressed in this work. Universal Presence Server (UPS) based lightweight RESTful WS architecture is discussed by Chunyan et al. [16]. The proposed architecture includes the following three entities: Request Manager (RM), Publication Manager (PM), and Subscription Manager (SM). The PM publishes the service description in the publication database. The SM retrieves the list of services present from the UPS database. The SM dynamically decides the interface agent to interact with the end user. The dynamic decision of message interface makes this approach lightweight for accessing and interacting with WS.

A vast literature work is done in the service architecture field to achieve the dynamic, on-demand, lightweight, and context-aware service discovery. Thirteen different approaches of service discovery are analyzed by Debajyoti et al. [17], such as context-aware WS discovery, publish-subscribe model, keyword clustering, service request expansion method, graph ranking, layer-based semantic discovery, WS discovery techniques for heterogeneous networks, WS indexing, structural case based reasoning, agent and QoS based discovery, collaborative tagging, using finite automaton, and hybrid approach for WS discovery.

When the service discovery method is not capable of discovering service based on a user's requirement, then approximate WS is suggested. For such type of process, the behavioral matching of WS is required. The user's requirements are described using the behavioral graph and the WS discovery process is converted into graph matching process. From the analysis of different techniques of WS discovery in this work, the importance of QoS parameters is focused. Using different semantic and syntactic discovery approach, a large number of services can be discovered for the same type of function. Selecting one WS for a user's requirement will be more dependent on the nonfunctional parameters, such as QoS parameters.

A survey of QoS-based WS discovery approaches is done by Phalnikar et al. [18]. The detailed categorization of QoS parameters is discussed in this work. Ontology and syntactic approaches of WS Discovery with QoS parameters are compared, and the requirements of ontology-based WS discovery are proposed in the work, which include the automatic WS Discovery, WS invocation, and WS composition. WS description is the base of WS discovery process, the efficiency of discovery process increases with the semantic description of WS. Bitar et al. [19] reviewed different ways of WS description methods. The need of lightweight semantic description is proposed in this work. WS described using semantic method then that can be efficiently discovered through semantic discovery. In the near future, huge number of WS will be present around us, hence a need of semantic WS discovery is invoked; tremendous research work is going in the direction to discover efficient protocol for WS discovery.

Review of different semantic WS discovery approaches is discussed in the Ibrahim et al. [20] categorized the service matching processes into three types depending on the heterogeneity level in the matching process, i.e., algebraic, deductive, and hybrid approach. In an algebraic approach, the degree of similarity between WS descriptions is taken as a parameter to discover appropriate service for user's requirement. Structured graphs are used to find out the degree of similarity between WS descriptions. In a deductive approach, the WS discovery is based on logic, such as description logic and first-order logic. In this approach instead of semantic different, other elements of WS description are considered for service matching process, which include Preconditions and Effect matching (PE matching), Input and Output matching (IO matching), and Input, Output, Precondition and Effect matching (IOPE).

4 Comparative Analysis

In the above section, different prominent techniques of WS discovery are discussed. In view of realizing IoT application, there is need of dynamic WS architecture, hence this literature survey will help to figure out the designing parameters of WS architecture for IoT environment. In the IoT environment, end users may be mobile and may not be aware of services provided at that location. In each location, there may be a large number of services present. Among these services, end users must be made aware of only a few services that are required for them. Therefore, the context-aware and user preference aware WS framework is required in the IoT environment. Thinking ahead, the IoT devices that will provide and access WS are resource constraint in nature. Thus, there is a requirement of lightweight WS architecture. That is why user preference, context awareness, and lightweight framework parameters are taken for comparative analysis.

Table 1 shows whether approach support these three parameters and summarizes the applicability of WS discovery approaches for IoT environment.

UPerf—User Preferences, CAW—Context Awareness and LW—Lightweight.

Requirements of WS architecture for IoT environment are as follows:

- Semantic description of WS
- Automatic on-demand WS discovery
- User-centric approach
- Context awareness approach
- Lightweight registration process for WS
- RESTful approach.

Table 1 Analysis of WS discovery approaches

Approaches	Discovery technique	UPref	CAW	LW
[7]	Semantic with Latent Semantic Indexing	No	Yes	No
[8]	Semantic with Service Transaction Matrix and Latent Dirichlet Allocation	No	Yes	No
[9]	Service profile ontology-based	No	Yes	No
[10]	Term Frequency and Inverse Document Frequency and web context generation	No	Yes	No
[11]	Group-based service request forward approach	No	No	Yes
[12]	DARPA agent Markup Language	Yes	Yes	No
[13]	pull and push model	No	Yes	No
[14]	Quality of Service	Yes	Yes	No
[15]	RESTful Approach	Yes	Yes	Yes
[16]	Universal Presence Server	No	No	Yes

5 Conclusions and Future Work

In order to provide WS framework for IoT environment, it is important to determine the requirements of WS architecture for IoT network. Hence, we have reviewed the literature for service description and service discovery methods. In the literature, tremendous work is done for service architecture, however, these approaches are not appropriate to apply for IoT settings as devices that will provide and access services in IoT paradigm are resource constrained in nature. This survey work will help to determine the designing parameters for WS architecture for IoT network. The user-centric approach is required in the near future as a large number of services will be available, among such huge number of services user must get appropriate service depending on a user's requirements. The future plan is to explore a semantic service description method and lightweight, user-centric, and context-aware service discovery methods for IoT network.

References

1. J. Gubbi, R. Buyya, S. Marusic, and M. Palaniswami, "Internet of Things (IoT): A vision, architectural elements, and future directions," *Futur. Gener. Comput. Syst.*, vol. 29, no. 7, pp. 1645–1660, Sep. 2013.
2. A. Zanella, N. Bui, A. Castellani, L. Vangelista, and M. Zorzi, "Internet of Things for Smart Cities," *IEEE Internet Things J.*, vol. 1, no. 1, pp. 22–32, Feb. 2014.
3. G. Shinde and H. Olesen, "Interaction between users and IoT clusters : Moving towards an Internet of People, Things and Services (IoPTS)," in *World wireless Research Forum Meeting 34*, 2015.
4. M. K. Nair and V. Gopalakrishna, "Look Before You Leap: A Survey of Web Service Discovery," *Int. J. Comput. Appl.*, vol. 7, no. 5, pp. 975–8887, 2010.

5. H. Omrana, I. El Bitar, F. -Z. Belouadha, and O. Roudies, "A Comparative Evaluation of Web Services Description Approaches," in *2013 10th International Conference on Information Technology: New Generations*, 2013, pp. 60–64.

6. D. A. D'Mello and V. S. Ananthanarayana, "A Review of Dynamic Web Service Description and Discovery Techniques," in *2010 First International Conference on Integrated Intelligent Computing*, 2010, pp. 246–251.

7. A. V. Paliwal, B. Shafiq, J. Vaidya, Hui Xiong, and N. Adam, "Semantics-Based Automated Service Discovery," *IEEE Trans. Serv. Comput.*, vol. 5, no. 2, pp. 260–275, Apr. 2012.

8. G. Cassar, P. Barnaghi, and K. Moessner, "Probabilistic Matchmaking Methods for Automated Service Discovery," *IEEE Trans. Serv. Comput.*, vol. 7, no. 4, pp. 654–666, Oct. 2014.

9. G. Meditskos and N. Bassiliades, "Structural and Role-Oriented Web Service Discovery with Taxonomies in OWL-S," *IEEE Trans. Knowl. Data Eng.*, vol. 22, no. 2, pp. 278–290, Feb. 2010.

10. A. Segev and Q. Z. Sheng, "Bootstrapping Ontologies for Web Services," *IEEE Trans. Serv. Comput.*, vol. 5, no. 1, pp. 33–44, Jan. 2012.

11. D. Chakraborty, A. Joshi, Y. Yesha, and T. Finin, "Toward Distributed service discovery in pervasive computing environments," *IEEE Trans. Mob. Comput.*, vol. 5, no. 2, pp. 97–112, Feb. 2006.

12. S. A. McIlraith, T. C. Son, and Honglei Zeng, "Semantic Web services," *IEEE Intell. Syst.*, vol. 16, no. 2, pp. 46–53, Mar. 2001.

13. A. Zisman, G. Spanoudakis, J. Dooley, and I. Siveroni, "Proactive and Reactive Runtime Service Discovery: A Framework and Its Evaluation," *IEEE Trans. Softw. Eng.*, vol. 39, no. 7, pp. 954–974, Jul. 2013.

14. K. Kritikos and D. Plexousakis, "Requirements for QoS-Based Web Service Description and Discovery," *IEEE Trans. Serv. Comput.*, vol. 2, no. 4, pp. 320–337, Oct. 2009.

15. D. Guinard, V. Trifa, S. Karnouskos, P. Spiess, and D. Savio, "Interacting with the SOA-Based Internet of Things: Discovery, Query, Selection, and On-Demand Provisioning of Web Services," *IEEE Trans. Serv. Comput.*, vol. 3, no. 3, pp. 223–235, Jul. 2010.

16. C. Fu, F. Belqasmi, and R. Glitho, "RESTful web services for bridging presence service across technologies and domains: an early feasibility prototype," *IEEE Commun. Mag.*, vol. 48, no. 12, pp. 92–100, Dec. 2010.

17. D. Mukhopadhyay and A. Chougule, "A survey on web service discovery approaches," in *Advances in Intelligent and Soft Computing*, 2012, vol. 166 AISC, no. VOL. 1, pp. 1001–1012.

18. R. Phalnikar and P. A. Khutade, "Survey of QoS based web service discovery," in *2012 World Congress on Information and Communication Technologies*, 2012, pp. 657–661.

19. I. El Bitar, F. -Z. Belouadha, and O. Roudies, "Review of Web services description approaches," in *2013 8th International Conference on Intelligent Systems: Theories and Applications (SITA)*, 2013, pp. 1–5.

20. I. El Bitar, F. -Z. Belouadha, and O. Roudies, "Semantic web service discovery approaches: overview and limitations," Sep. 2014.

Analysis of Multiple Features and Classifier Techniques Combination for Image Pattern Recognition

Ashish Shinde and Abhijit Shinde

Abstract Automatic visual pattern recognition is complex and highly researched area of image processing. This research aims to study various pattern recognition algorithms, cloth pattern recognition is presented as research problem and to find out best combination suited for the cloth pattern recognition problem. The dataset is collected from CCNY clothing pattern dataset and contains 150 samples of each category (Patternless, Striped, Plaid, and Irregular). The presented study compares all combinations of three different feature extraction techniques and three classifier techniques. Feature extraction techniques used here are Radon Feature Extraction, projection of rotated gradient, and quantized histogram of gradients. The classifiers used are KNN, neural network, and SVM classifier. The highest recognition rate is achieved using Radon Signature feature and KNN classifier combination which reaches to 93.7% of accuracy.

Keywords Clothing pattern recognition · Texture analysis · Image feature extraction · Classifier

1 Introduction

Image pattern recognition is a tool for machine or a computer system to understand the pattern which makes it possible humans to recognize objects. Pattern recognition is also a simplest way of distinguishing between multiple objects depending on a single property which is unique to that object. Patterns are repetitive visual

A. Shinde (✉)
Sinhgad College of Engineering, Pune 41, Maharashtra, India
e-mail: shindeashishv@gmail.com

A. Shinde
Bhima Institute of Management and Technology, Kagal,
Kolhapur 416216, Maharashtra, India
e-mail: abhijitvshinde@gmail.com

© Springer Nature Singapore Pte Ltd. 2018
S. Bhalla et al. (eds.), *Intelligent Computing and Information and Communication*,
Advances in Intelligent Systems and Computing 673,
https://doi.org/10.1007/978-981-10-7245-1_24

markers present on an object's surface. Here, a comparative study is presented to show compatibility between a feature recognition system and classifier.

Feature extraction techniques are basis of smart machine vision algorithms. The understanding of data by machine usually comes from a property of data type by which it can be distinguished from other image. For image pattern recognition, this is the most important step too. Features can be of various types. Features such as color, edges, size, etc., which are primary visual features while histogram, projections, spatial texture, etc., are much complex types of feature [1].

Multiple combinations of feature extraction and classifier are compared in this research to ensure a broader understanding of machine vision algorithms.

A lot of efforts have already been taken in terms of pattern recognition for cloth classification. Hidayati et al. [2] proposed a methodology for classification of clothes based on genres. Davis et al. [3] proposed a new method for "texture analysis based on the spatial distribution of local features" for unsegmented image textures. Generalized co-occurrence matrices (GCM) are used for texture description and feature extraction.

Eisa et al. [4] proposed the use of the LBP feature for analyzing texture feature. This feature extraction technique presents an effective way for both structural and statistical texture analysis. In approach proposed by Kalantidis et al. [5], they adopt the locality sensitive hashing (LSH), a very popular and effective high-dimensional indexing technique. Here, it is applied specifically for cloth pattern classification.

2 Feature Extraction for Pattern Recognition

In this study, three different feature extraction methods have been considered. Radon Signature characterizes the directionality feature of clothing patterns. Projection of Rotated Gradient takes x-projection of images which are rotated stepwise at a given angle to create a multidirectional projection array. Histogram of Oriented Gradient takes histogram of gradient taken one by one from multiple direction of original image.

2.1 Radon Signature Features

Radon Signature is feature extracted using the Radon transform [6] which is commonly used to extract information of the principle orientation of an image. The image is then rotated to find dominant direction represented by projection of the image intensity along a radial line oriented at a specific angle. A two-dimensional Radon transform of function is given as [6]:

$$R(r, \emptyset) = \int\limits_{-\infty}^{\infty} \int\limits_{-\infty}^{\infty} I(x, y)\delta(r - x\cos\emptyset - y\sin\emptyset)dxdy \tag{1}$$

where 'Φ' is the projection line angle and 'r' is the perpendicular distance of a projection line to the origin.

To ensure the consistency in variance of Radon transform for multiple projection orientations, Radon transform has been applied on the image selected by maximum radius area instead of the complete image. The large intra-pattern variations of clothing are also reflected as shown in Fig. 1.

The probable orientation of image is represented by variance in radon signature calculated by equation [6]

$$\text{var}(r, \emptyset_i) = \frac{1}{N}\sum_{j=1}^{N-1}\left(R(r_j, \emptyset_i) - \hat{R}(r, \emptyset_i)\right)^2 \tag{2}$$

where

$$\hat{R}(r, \emptyset_i) = \frac{1}{N}\sum_{j=1}^{N-1}R(r_j, \emptyset_i), \tag{3}$$

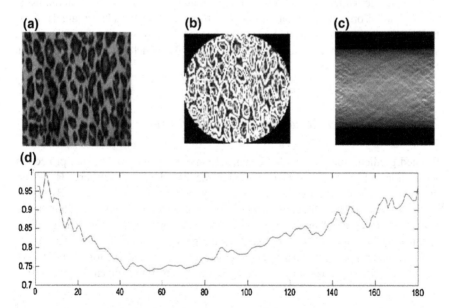

Fig. 1 Radon signature feature extraction. **a** Input image. **b** Disk area selected for calculating radon signature. **c** Radon signature of image calculated at 0–180°. **d** Radon signature variance used as extracted image feature

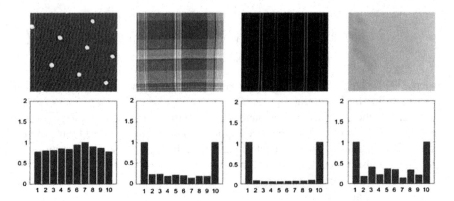

Fig. 2 Different cloth pattern images (Irregular, Plaid, Striped and Patternless) and their corresponding radon signature variance

where $R(r, \Phi)$ is value of perpendicular projection at distance r and angle Φ and $\hat{R}(r, \Phi)$ is its expected value.

The Radon Signature feature array is extracted by concating all variances signatures acquired at all projection directions

$$[\text{var}(r, \emptyset_1), \text{var}(r, \emptyset_2), \text{var}(r, \emptyset_3), \text{var}(r, \emptyset_4), \ldots \text{var}(r, \emptyset_{N-1})]$$

To enhance the large variations in image intensity, image has been filtered using the X/Y-Sobel operator to create the gradient image as in Eq. (1). Figure 1b illustrates the area selected for Radon transform and Fig. 1c demonstrates the radon signature. The selected feature is shown in Fig. 1d which is variance in Radon Signature (Fig. 2).

2.2 Projection of Rotated Gradient Features

Rotated gradient images are edge enhanced versions of multidirectional projection of an image [7]. An image can be enhanced by taking its gradient. However, gradient taken from a single direction will only enhance its limited edges. By taking gradient from multiple directions ensures image is collectively enhanced and hence gives higher degree of information about the features of that image. The gradient enhancement of an image is done using Sobel operator given by [1, 2, 1; 0, 0, 0; −1, −2, −1] which is x-directional Sobel operator and its rotated version [1, 0–1; 2, 0, −2; 1, 0, −1] is a y-directional Sobel operator. The resulting image is given by

$$G(x,y) = \{f(x,y) \odot s(x,y)\} \odot \hat{S}(x,y), \tag{4}$$

where $s(x, y)$ and $\hat{s}(x, y)$ are x-direction and y-direction sobel operator.

The rotation of image here is by 90° each time hence there will four concated gradient images in total. The projections of gradient are

$$P(x)_i = \sum_{j=1}^{\text{no.ofcolumns}} G(x, y_j)_i \tag{5}$$

2.3 Quantized Histogram of Orientated Gradient Features

Quantized oriented gradient is an enhanced version of HOG features. HOG has robustness to change of illumination and attains high computational accuracy in detection of variously textured objects [8]. HOG features are calculated by taking histogram of image created in by generating gradient images with different orientation. Here too gradients were calculated using Sobel operator given by [1, 2, 1; 0, 0, 0; −1, −2, −1] which is x-directional Sobel operator and its rotated version [1, 0−1; 2, 0, −2; 1, 0, −1] which is a y-directional Sobel operator (Figs. 3, 4 and 5).

The histograms were however quantized at bin of 8, i.e., each histogram array consisted of eight values. From this histogram, a single gradient descriptor was calculated by calculating highest value of each bin. This value describes the average orientation of that sub-image. Each of such histograms was calculated for whole image by creating a block of 8 × 8, which gives 64 values for an angle theta. The theta was varied seven times further to calculate the whole feature vector (Fig. 6).

The calculated HOG features from each block are as follows (where length of each vector is 8 and $n = 64$):

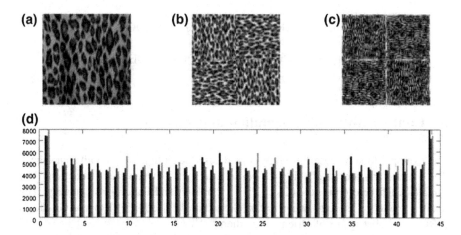

Fig. 3 Sample pattern image and its projection of gradient feature. **a** Input image. **b** Rotated version of input image. **c** Gradient image of each rotated image. **d** Projection of gradient images

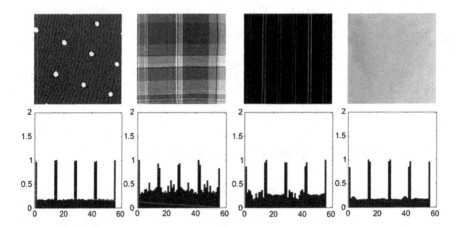

Fig. 4 Different cloth pattern images (Irregular, Plaid, Striped and Patternless) and their corresponding projection of gradient feature

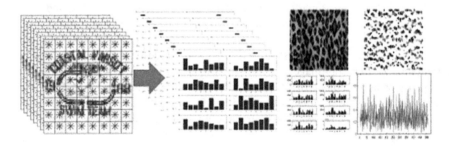

Fig. 5 Flow of calculation of Q-HOG feature

$$\mathrm{HOG}\{8, n, m\} = [\mathrm{HOG}_1, \mathrm{HOG}_2, \mathrm{HOG}_3, \ldots, \mathrm{HOG}_n]_{m=1,2\ldots8} \tag{6}$$

$$\mathrm{QHOG}\{n, m\} = [\mathrm{QHOG}_1, \mathrm{QHOG}_2, \mathrm{QHOG}_3, \ldots, \mathrm{QHOG}_n]_{m=1,2\ldots8} \tag{7}$$

3 Comparative Implementation and Analysis

The example to showcase this study uses dataset of cloth patterns with four labels. The evaluations are done using CCNY dataset of cloth patterns. It contains 150 samples each belonging to Striped, Plaid, Random, and Patternless category. The feature extraction is done using Radon Signature, Projection of Rotated Gradients (PRG), and Histogram of Oriented Gradients (HOG).

The research aims toward comparison of compatibility between feature extraction and feature classifier for given example of pattern dataset of cloths. CCNY Clothing

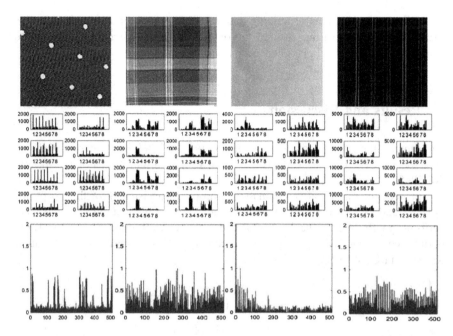

Fig. 6 Example image of each pattern in first row. Second row is histograms of all 8 directed gradient of each image in row one. Third row is final feature vector for each image in row one

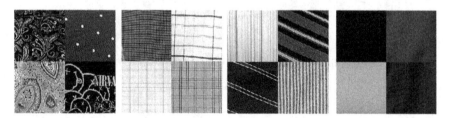

Fig. 7 Example images for each pattern. Irregular, Plaid, Striped and patternless in order

Pattern Dataset includes four different typical clothing pattern designs: Plaid, Striped, Patternless, and Irregular with 150 images in each category. The resolution of each image is 140 × 140. Figure 7 shows example images in each category.

Each of the dataset is split in training set and recognition set with varying percentage of split. The training dataset images are processed using all pattern recognition techniques described in Sect. 2. Here, three different classifiers are used namely:

1. Feedforward neural network with 10 hidden layers and Levenberg–Marquardt optimization,
2. KNN classifier with neighborhood-1 and
3. Multilabel SVM classifier.

Different combinations of training data set splitting as well as multiple feature extraction-classifier pairing ensured thorough analysis and study of both feature extraction technique and classifier techniques.

4 Results and Discussion

150 samples have been considered belonging to each label, i.e., pattern types. The data was split for Training Data-Test Data in various combinations of 10–90, 30–70, 50–50, and 70–30% of dataset.

The recognition for all combinations shows a consistent growth except for Q-HOG and NN combination which do not show consistent growth. NN also performs poorly in combination with PRG. SVM however less accurate than KNN is quite consistent with varying size of training data when used in combination RadSig and Q-HOG (Fig. 8; Table 1).

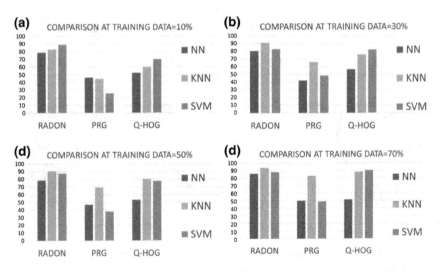

Fig. 8 Comparison of recognition accuracy of all feature extraction and classifier combinations at **a** 70% Training data **b** 50% Training data **c** 30% Training data **d** 10% Training data

Table 1 Change in recognition rate in percent with increase in number of training samples

Training sample used	70%			50%			30%			10%		
	NN	KNN	SVM	NN	KNN	SVM	NN	KNN	SVM	NN	KNN	SVM
RADON	85.8	93.7	88.1	78.6	90.5	87.4	80.2	90.5	82.6	78.6	82.6	88.9
PRG	50	82.6	49.3	46.9	69.1	38.1	42.1	65.9	48.5	46.1	44.5	25.4
Q-HOG	51.6	88.1	90.5	53.2	80.2	77.8	56.4	75.4	81.8	52.4	59.6	69.9

5 Conclusion

The most consistent results were achieved with Radon Signature Feature-SVM classifier. Highest results however were obtained at 70% training data using Radon Signature Feature-KNN features. Also KNN performed comparatively better than all other classifiers in all combinations consistently.

In terms of operation speed training of neural network took longest during training. KNN do not need special training while its recognition time was shortest making it the fastest classifier technique. Q-HOG took longest for feature training while Radon Signature feature took shortest. In terms of the algorithm complexity represented mathematical calculations, the PRG proved simplest while Q-HOG hardest.

All considerations resulted in conclusion that RadSig-KNN combination is best feature extraction-classifier combination for presented scenario.

The aim of research is to find out which combination of feature extraction-classifier produces best results for a given problem. The same scenario can be recreated using a different dataset of patterns of other types, which may include wood pattern, wall color pattern, skin color pattern, etc. The presented research statement is flexible and can be introduced as a methodology of comparative analysis of a number of image processing problems.

References

1. Van Gool L, Dewaele P, Osterlinck A (1983). Texture analysis anno 1983. Computer Vision Graphics Image Processing 29(12):336–57.
2. S. Hidayati. W. Cheng, and K. Hua, "Clothing Genre Classification by Exploiting the Style Elements," In Proc. ACM Multimedia, 2012.
3. Larry S. Davis, Steven A. Johns, J. K. Aggarwal, "Texture Analysis Using Generalized Co-Occurrence Matrices", IEEE Transactions on Pattern Analysis and Machine Intelligence (Volume: PAMI-1, Issue: 3, July 1979), pp. 251–259.
4. M. Eisa, A. ElGamal, R. Ghoneim and A. Bahey, "Local Binary Patterns as Texture Descriptors for User Attitude Recognition," International Journal of Computer Science and Network Security, vol. 10 No. 6, June 2010.
5. Y. Kalantidis, L. Kennedy, L. Li, "Getting the Look: Clothing Recognition and Segmentation for Automatic Product Suggestions in Everyday Photos," ICMR, ACM 978-1-4503-2033-7/13/04, April, 2013.
6. K. Khouzani and H. Zaden, "Radon Transform Orientation Estimation for Rotation Invariant Texture Analysis," IEEE Trans. on Pattern Analysis and Machine Intelligence, vol. 27, no. 6, pp. 1004–1008, 2005.
7. Lin-Lin Huanga, Akinobu Shimizua, Yoshihoro Hagiharab, Hidefumi Kobatakea, "Gradient feature extraction for classification-based face detection", Pattern Recognition Volume 36, Issue 11, November 2003, Pages 2501–2511.
8. S.K. Uma, Srujana B.J., "Feature Extraction for Human Detection using HOG and CS-LBP methods", International Journal of Computer Applications (0975–8887), 2015, pp. 11–14.

Smart and Precision Polyhouse Farming Using Visible Light Communication and Internet of Things

Krishna Kadam, G. T. Chavan, Umesh Chavan, Rohan Shah
and Pawan Kumar

Abstract Recently, Polyhouse farming has taken the place of traditional farming. This technique reduces dependency on rainfall and dramatically changing environmental conditions and makes the optimum use of land and water resources in a controlled environment. Manually controlled system requires a lot of attention and care. To overcome this limitation, novel technologies like Internet of Things (IoT) and sensor network (SN) can be used. Case studies show that use of electromagnetic or radio spectrum for communication degrades the growth and quality of crops. The visible light communication (VLC) provides an effective solution to this problem using visible light spectrum for communication. VLC outperforms when compared with traditional communication techniques in terms of available frequency spectrum, reliability, security, and energy efficiency. LED source used in VLC acts as communication element as well as is used for illumination. This paper proposes use of sensor network to sense the data like temperature, humidity, soil moisture, and luminosity along with VLC as a medium of communication from node to the network gateway. IoT performs its part in automating polyhouse countermeasures and fertilizer suggestion based on the data processed in cloud for Business Intelligence and Analytics. Thus, the proposed system will ensure automated and sustainable polyhouse farming using VLC and IoT.

Keywords Internet of Things · Sensor network · Fertilizer suggestion
Cloud · Visible light communication · Business intelligence · Analytics

K. Kadam (✉) · G. T. Chavan · U. Chavan · R. Shah · P. Kumar
Sinhgad College of Engineering, Pune, India
e-mail: kjk12895@gmail.com

G. T. Chavan
e-mail: gtchavan.scoe@sinhgad.edu

© Springer Nature Singapore Pte Ltd. 2018
S. Bhalla et al. (eds.), *Intelligent Computing and Information and Communication*,
Advances in Intelligent Systems and Computing 673,
https://doi.org/10.1007/978-981-10-7245-1_25

1   Introduction

Globally, traditional agriculture is one of the most practiced occupations. It heavily depends on human labor and is surviving on the mercy of environment that is deteriorating day by day. Technological advances in agriculture are providing much easier ways to farm efficiently. Automation of various farming devices in open farms like a sprinkler and drip irrigation has gained a lot of popularity in recent times. Despite its popularity, it fails to address the issue of unpredictable environmental conditions. Unpredictable environmental conditions make it difficult to take up farming as primary occupation resulting in less agricultural yield forcing farmers to commit suicide. Hence, farmers are moving to a novel environment for farming, viz., Polyhouse farming (a.k.a. Greenhouse Farming). Polyhouse provides a healthy environment with controlled environmental conditions to grow high quality crops. It provides suitable temperature, humidity, soil moisture, and luminosity for the entitled crop. Polyhouse, seemingly a perfect solution falls short to resolve all the limitations. Major limitations include the harmful effect of chemical fertilizers and pesticides on living beings and in turn animal husbandry, effect of radio frequency on the health of crops when radio communication is used for sensor data transfer and no accumulation of sensor and statistical data. These limitations propose a threat when devising a green technology for agricultural industry.

The chemical fertilizers are used to gain extra yield, but malignant effects of them are now surfacing out. Chemical fertilizers reduce the fertility of soil and causes harmful effect on human beings who apply it in the field. Fertilizer suggestion ability is descended down the generations and requires expertise.

PLC based parameter regulation was devised to automate activities based on sensor data. Wireless communication technologies like Wi-Fi and ZigBee are used to transfer data from sensor nodes to sink node. This nearly crunched and unsafe RF spectrum proves to be an obstacle as it affects the health of crops when exposed for a longer duration [1].

The sensor network deployed gathers very important data which is used to regulate environmental conditions inside polyhouse. This data is not accumulated to a nonvolatile storage and thus loses its significance. Thus, it can only be used for local automation and no knowledge is extracted out of it.

Dan et al. [2] uses the sensor network that is designed for collecting greenhouse parameters like temperature, pressure, light, humidity, and CO_2 and a control system that contains fan control, curtain control, fertilizer control, and sprinkler. Nakutis et al. [3] propose a methodology for separation of data collection and automated control system through IoT gateway using OPC UA server for controlling remote agricultural processes. Chaudhary et al. [4] propose IoT-based case study that uses GPRS for communication and GPS for remote monitoring. FuBing [5] provides agriculture Intelligent System based on IoT. Pavithra et al. [6] show the other applications of IoT such as home automation. Chavan et al. [7] propose a system with PLC and WSN for monitoring environmental parameters of polyhouse. Chaudhary et al. [8] use Programmable System on Chip for precision agriculture.

Sachdeva et al. [9] design a farmer friendly software for polyhouse automation using LabVIEW. Khandekar et al. [10] use mobile telephony for monitoring and control of polyhouse through Internet. Purnima et al. [11] provide a system for automatic irrigation using GSM Bluetooth. Borade et al. [12] provide a system with star network applied for polyhouse automation.

Nowadays, Internet of Things (IoT) and cloud computing are gaining a lot of popularity. IoT is one of the very few novel technologies that enables things to communicate via the Internet and gives a new dimension to automation. The data can now be processed and coupled with cloud platforms. Cloud allows to store and process big data. The knowledge, when extracted out of this data, can be used to analyze it and make predictions on future strategies using business intelligence (BI).

So, we propose a system that takes the benefit out of these novel technologies to automate polyhouse activities using IoT, data processing and analytics using cloud, and green communication using Visible Light Communication (VLC). The proposed system provides precise organic fertilizer suggestion based on NPK parameters of analyzed soil.

The rest of the paper is organized as follows: Sect. 2 describes the methodology for the implementation of the system. Section 3 describes experimental results and analysis. Section 4 concludes the paper and Sect. 5 describes the scope for future work.

2 Methodology

When combining novel technologies for the betterment of agriculture, various factors such as cost-effectiveness, scalability, and analytics capability should be considered. Embedded systems allow the use of compact computation units to be deployed on field making the system fairly cost-effective. Use of cloud technology delivers modularity and scalability. Cloud also allows to store big data and to operate on that data and get knowledge extracted. The environmental data gathered by using sensors needs to be communicated with the central gateway and here VLC comes into the picture. It enables green communication between sensor nodes to the network gateway. Figure 1 shows system architecture of the proposed system. The overall system can thus be divided into four modules, viz., sensor network (SN), VLC, cloud computing and analytics, and fertilizer suggestion.

2.1 Sensor Network

Polyhouse farming is focused on environmental parameter control. The major parameters which favor the healthy growth of crops are temperature, humidity, luminosity, and soil moisture. The appropriate and durable sensors for the above parameters are available in the market. The sensors used are LM35 (Temperature

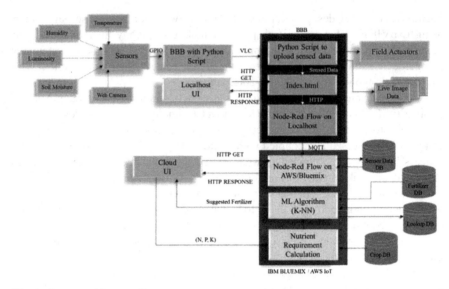

Fig. 1 System architecture diagram

Sensor), SY-HS 220 (Humidity sensor), bh1750fvi-e (Light Intensity Sensor) and Soil moisture sensor.

To control the environment inside the polyhouse, the corresponding actuators need to be controlled based on sensed data. These sensor–actuator pairs are ventilation fans and curtains for temperature control, Fogging System for humidity control, Shade Nets for Light intensity control, and Drip Irrigation System for soil moisture control.

Beaglebone Black (BBB) is used to attach sensors. Figure 2 shows the circuit diagram to attach sensors and actuators to the BBB. The LM35 temperature sensor is used since it has a 0.5 °C ensured accuracy and has low cost due to wafer-level trimming. While, SY-HS 220 humidity sensor is used because it has an accuracy of 5%RH and it operates at 30–90% RH. Bh1750fvi is used since it has an inbuilt amplifier and an ADC that can be interfaced with BBB's I^{2}C interface. It gives wide range and high resolution for sensing light intensity. Soil moisture sensor used has to be interfaced with the analog input of BBB and this reading can be mapped to the soil moisture. For controlling actuators thresholds are used. These are crop specific and can be found on NIC website. MCT2E Optocoupler is used to separate the relay driver circuit from BBB. ULN2803A is used as a relay driver IC. It allows to drive eight relays at a time. To control any actuators, 12 V DC relay is used. For a polyhouse of size 60 m × 40 m, soil moisture sensors are placed at every 15 m along the length and 10 m along the width. Temperature sensors are placed at an interval of 30 m along the length. A humidity sensor and a light intensity sensor are placed at the center of the polyhouse. User interface provided by the web application allows the user to control any on-site actuators remotely. This gives the control in hands of the farmer, when system misbehaves due to any reason.

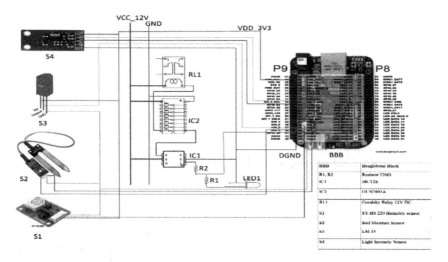

Fig. 2 Sensor attachment circuit diagram

2.2 Visible Light Communication (VLC)

Communicating sensed data from the sensor node to the network gateway is quintessential. Various communication technologies can be used and all of them have their own advantages and limitations. The wired network helps with higher data rate while lacks when concerned with scalability. Wireless networks such as Wi-Fi and ZigBee, on the other hand, are suitable for scaling the system as required. Dan et al. [2] provide an IoT-based automation mechanism which works using ZigBee for communication. These wireless technologies are based on radio frequency and can have harmful effects on growth of crops after prolonged exposure [13]. To enable green communication that uses natural visible light frequency spectrum, VLC can be used.

VLC works on the principle of transmitting data using LEDs and receiving that data using photodetectors. Various techniques such as On-Off Keying (OOK), Pulse Width Modulation (PWM), and NRZ-OOK are suitable for the transmission [14]. On the receiver side, photodiodes are used to sense the light intensity. These intensity values are then demodulated. The proposed system uses NRZ-OOK. Figure 3 shows the block diagram of VLC module. BBB contains the python script for transmitter. For a zero in binary, MOSFET 2 and MOSFET 3 are turned off while for a one in binary, MOSFET 2 and MOSFET 3 are turned on. On the receiver side, light intensity sensor with a photodiode detects the data sent from LED. The receiver python script running on receiver BBB demodulates the intensity values by mapping them in binary. For this, reference intensity value is sensed before the start of each packet.

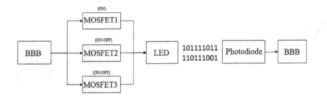

Fig. 3 VLC block diagram

2.3 Cloud Computing and Analytics

The cloud platform is the backbone of IoT as it enables remote connectivity and monitoring through the Internet. Even latest polyhouse automation technologies lack in providing a mechanism to store big data generated by IoT and process that data for business intelligence. The traditional system just allows to control and regulate a single polyhouse. As agriculture is an industry based on experience and knowledge of farmers, why not to use data analytics and business intelligence to share and plan agricultural activities based on the success of a polyhouse and data gathered from farming cycle. Among various available cloud platforms, Amazon Web Services (AWS) was chosen since it provides special flavor for AWS IoT and a cloud instance to host the web application. A web application in python is developed to provide user interface where the live environmental parameters along with the status of actuators are displayed. Data from network gateway is communicated to a thing on AWS IoT using MQTT protocol with the help of Node-RED. This data is then made available on Amazon EC2 instance where the web application is running. This data then can be provided to a BI system to analyze and devise future business plans and to share the recipe of success to other polyhouses.

2.4 Fertilizer Suggestion

The essential part of any farming cycle is providing a proper fertilizer in a precise amount. To provide an appropriate fertilizer, farmers depend on inherited knowledge and experience. A novice farmer when lacks the both, gets into a paralyzed situation. The government has come up with lots of soil testing laboratories where farmers can visit with their soil samples to get them tested. The basic report of any soil test shows the nutrient values that are present and a fertilization amount of predefined fertilizers. But the traditional fertilizer suggestion methods are not precise and can sometimes have an adverse effect on the produce. Often, suggested fertilizers are chemical or inorganic fertilizers. These fertilizers have proved to have an adverse effect on fertility of soil and on health of living beings that come in contact with them.

Proposed system provides precise organic fertilizer suggestion. Organic fertilizer has very less amount of fillers and is not harmful to any form of living beings. The algorithm for fertilizer suggestion is as follows:

```
Program Fertilizer_Suggestion (N, P, K, crop, age)
  Begin
     [N₀, P₀, K₀] := get_optimum_npk_values (crop, age)
     [Nᵣ, Pᵣ, Kᵣ] := [N₀ - N, P₀ - P, K₀ - K]
     If ([Nᵣ, Pᵣ, Kᵣ] in lookup_table)
        Return lookup_table.suggested_fertilizer
     Else
        Var min := 999
        Repeat for each f in fertilizer_DB
           Euclidean_distance:= sqrt ((f.N - Nᵣ) ² + (f.P -
           Pᵣ) ² + (f.K - Kᵣ) ²)
           If (Euclidean_distance < min)
           min := Euclidean_distance
           Suggested_fertilizer:= f
           Add_to_lookup_table (f)
        Return Suggested_fertilizer
  End.
```

The above algorithm takes nitrogen (N), phosphorus (P), potassium (K), selected crop and age of the crop as input. Farmer inputs these values obtained from soil analysis on the web page hosted on the cloud. First, the optimum values required for the crop at the particular age are taken from database present for all of the crops. According to this, required macro-nutrient values are calculated. If these are already present in lookup table then suggested fertilizer is returned from lookup table. If it is not in the lookup table, Euclidean distance is calculated from the database of nutrient values of organic fertilizers present in the market. The fertilizer with the minimum Euclidean distance is hence returned and displayed on web page.

3 Experimental Results and Analysis

The proposed system is fully implemented and is ready to be deployed on field. The prototype of the system is shown in Fig. 4.

The UI for the system is shown in Fig. 5. It shows the appropriate status of actuators and live temperature parameters. The web application is hosted on Amazon EC-2 instance. Figure 6 shows the fertilizer suggestion module for the crop tomato. The module gives 98% accuracy in suggesting fertilizer. It is ensured that the suggested fertilizer is the nearest one that fulfills the nutrient requirement.

Fig. 4 System prototype (A: Sensor network with VLC Transmitter, B: Display of Transceiver python script running on BBB, C: Cloud Server running remotely, D: Web Application UI)

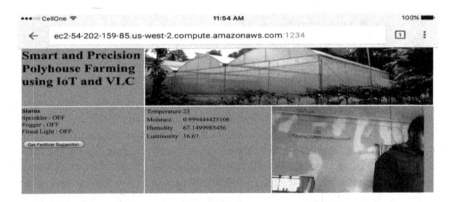

Fig. 5 Screenshot of working UI of the web application

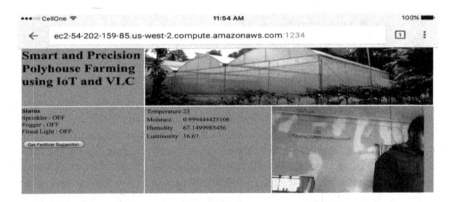

Enter N-P-K values and age of crop(days)

Nitrogen [0.2]
Phosphurus [0.1]
Potassium [0.1]
Age [33]

(Submit)

Suggested Fertilizer:
['Seaweed (Ascophyllum nodosum)', 1.9, 0.25, 3.68]

Fig. 6 Screenshot of working fertilizer suggestion module

Table 1 Sensor–actuator pairs and operating thresholds for tomato crop

Sensor	Actuator	Threshold for control
Temperature sensor	Ventilation fans and curtains	15–29 °C
Humidity sensors	Fogging system	60–90% RH
Light intensity sensor	Shade nets	10,000–140,000 lx
Soil moisture sensor	Drip irrigation system	50–90% SM

Table 1 shows sensor–actuator pairs along with their operating thresholds for sample crop Tomato.

4 Conclusion

The system proposed in this paper automates activities inside the polyhouse and reduces the human requirement for monitoring environmental parameters. The system gathers real-time data for analytics and for decision-making using Business Intelligence. It also provides organic fertilizer suggestions based on N-P-K values of analyzed soil, thus reduces harmful effects of chemical fertilizers on the health of land, nature, and living beings. Use of VLC eliminates the effect of radio frequency on crop health.

5 Future Scope

The proposed system will be implemented on site for a couple of years thus providing a bigger picture considering BI and Analytics. The data gathered from longer time span allows to map success rate to analyzed data and ensure a well-defined success recipe. The further cost reduction can be achieved by developing a custom system on chip (SOC) with optimum processing power. Higher data rates will be achieved in VLC module by using a high quality LED and a fast sensing photodiode.

References

1. Special article: Effect of ultra-short radio waves on plant growth: Science 17 Nov 1939: Vol. 90, Issue 2342, pp. 470–471 (1939).
2. LIU Dan, Cao Xin, Huang Chongwei, JI Liangliang: Intelligent Agriculture Greenhouse Environment Monitoring System Based on IOT Technology: 2015 International Conference on Intelligent Transportation, Big Data & Smart City, IEEE (2015).
3. Ž. Nakutis, V. Deksnys, I. Jauruševiius, E. Marcinkeviius, A. Ronkainen, P. Suomi, J. Nikander, T. Blaszczyk, B. Andersen: Remote Agriculture Automation using Wireless Link

and IoT Gateway Infrastructure: 26th International Workshop on Database and Expert Systems Applications, IEEE (2015).

4. Rashi Chaudhary, Jalaj Ranjan Pandey, Prakhar Pandey, Prakhar Pandey: Case study of Internet of Things in area of Agriculture, 'AGCO's Fuse Technology's 'Connected Farm Services, IEEE (2015).

5. FuBing: Research on the Agriculture Intelligent System Based on IoT: 2012 International Conference on Image Analysis and Signal Processing (IASP), IEEE (2012).

6. Pavithra. D, Ranjith Balakrishnan: IoT based Monitoring and Control System for Home Automation: 2015 Global Conference on Communication Technologies, IEEE (2015).

7. Sachin. V. Chavan, Sumayaa. C. Pathan, Suhas. N. Patil and Bhimrao. P. Ladgaonkar: design of LM4F120H5QR based node for wireless sensor network to monitor environmental parameters of polyhouse: International Journal of Advances in Engineering & Technology (2015).

8. D.D. Chaudhary, S.P. Nayse, L.M. Waghmare: application of wireless sensor networks for greenhouse parameter control in precision agriculture: International Journal of Wireless & Mobile Networks (IJWMN) Vol. 3, No. 1 (2011).

9. Kanwal Sachdeva, Neeraj Kumar, Rajesh Kumar, Rohit Verma: Monitoring and Controlling of Environmental Parameters of Polyhouse Based On Lab VIEW: International Journal of Innovative Research in Science, Engineering and Technology Vol. 4, Issue 9 (2015).

10. Yogesh R. Sonawane, Sameer Khandekar, Bipin Kumar Mishra, K. K. Soundra Pandian: Environment Monitoring and Control of a Polyhouse Farm through Internet: IIT Kanpur (2016).

11. Purnima, S.R.N. Reddy: Design of Remote Monitoring and Control System with Automatic Irrigation System using GSM-Bluetooth: International Journal of Computer Applications Volume 47– No. 12 (2012).

12. Kiran E. Borade, Prof. C.S. Patil, Prof. R.R. Karhe: Polyhouse Automation System: International Journal of Advanced Research in Computer Science and Software Engineering Volume 3, Issue 8 (2013).

13. Natural News: http://www.naturalnews.com/043238_Wi-Fi_routers_radiation_plant_growth.html.

14. Krishna Kadam, Prof. Manisha R. Dhage: Visible Light Communication for IoT: iCATccT, pp. 275–278, IEEE (2016).

Acceleration of CNN-Based Facial Emotion Detection Using NVIDIA GPU

Bhakti Sonawane and Priyanka Sharma

Abstract Emotions often mediate and facilitate interactions among human beings and are conveyed by speech, gesture, face, and physiological signal. Facial expression is a form of nonverbal communication. Failure of correct interpretation of emotion may cause for interpersonal and social conflict. Automatic FER is an active research area and has extensive scope in medical field, crime investigation, marketing, etc. Performance of classical machine learning techniques used for emotion detection is not well when applied directly to images, as they do not consider the structure and composition of the image. In order to address the gaps in traditional machine learning techniques, convolutional neural networks (CNNs) which are a deep Learning algorithm are used. This paper comprises of results and analysis of facial expression for seven basic emotion detection using multiscale feature extractors which are CNNs. Maximum accuracy got using one CNN as 96.5% on JAFFE database. Implementation exploited Graphics Processing Unit (GPU) computation in order to expedite the training process of CNN using GeForce 920 M. In future scope, detection of nonbasic expression can be done using CNN and GPU processing.

Keywords CBIR · CNN · GPU processing · Emotion detection

1 Introduction

Facial expression which is a form of nonverbal communication plays an important role in interpersonal relations. Facial expressions represent the changes of facial appearance in reaction to persons inside emotional states, social communications, or intentions. Recently, active research is going on in the area of Automatic Facial

B. Sonawane (✉) · P. Sharma
Nirma University, Ahmedabad, India
e-mail: bhakti.sonawane@sakec.ac.in

P. Sharma
e-mail: priyanka.sharma@nirmauni.ac.in

© Springer Nature Singapore Pte Ltd. 2018
S. Bhalla et al. (eds.), *Intelligent Computing and Information and Communication*,
Advances in Intelligent Systems and Computing 673,
https://doi.org/10.1007/978-981-10-7245-1_26

Expression Recognition (FER). It is useful in the field of medicine, human emotion analysis, etc., and will be one of the best steps for improving Human Machine Interaction (HMI) systems. Many factors make emotion detection as challenging problem. Among them is the problem of an unavailability of the standardized database for FER. Benchmark database that can fulfill the various requirements of the problem domain in order to become standard database for future research is a tough and challenging exercise [1]. Most of existing databases expressions is posed and not spontaneous. Biggest challenge is to capture spontaneous expressions on images and video. Like different subjects express the same emotions at different intensities and sometimes laboratory conditions become hurdle for the subject to display spontaneous expressions. Another major challenge is labeling of the data which is a time-consuming process and possibly error prone also. Challenges involved in capturing and recognizing spontaneous nonbasic expression are more than basic expressions. Most of the FER has lack of rotational movement freedom [1]. Here, our aim is to present an approach based on CNNs for FER and a systematic comparison of five different 12-layer CNNs. The input to CNN is an image to predict the facial expression label which should be one of these labels: anger, happiness, fear, sadness, disgust, and neutral. Among various database, JAFFE database is chosen for implementation as this is the most commonly used in other automatic FER systems.

2 Background

Generally, face detection, facial feature extraction, and facial expression classification are the parts of an automatic FER system using traditional machine learning techniques. In the face detection step, given an input image system performs some image processing techniques on it in order to locate the face region. In feature extraction step, from located face, geometric features and appearance features are the two types of features that are generally extracted to represent facial expression. Geometric features describe shape of face and its components like lips, nose or mouth corners, etc. Whereas appearance features depict the changes in texture of face when expression is performed. Classification is the last part of the FER system which based on machine learning theory. Output of previous stage which is a set of features retrieved from face region is given as an input to the classifier like Support Vector Machines, K-Nearest Neighbors, Hidden Markov Models, Artificial Neural Networks, Bayesian Networks, or Boosting Techniques [2]. Some expression recognition systems classify the face into a set of standard emotions. Other system aims to find out movements of the individual muscle that the face can produce. In [1], author provided an extensive list of researches between 2001 and 2008 on FER and analysis. Since past few years, there were several advances to perform FER using traditional machine learning methods which involve different techniques of face tracking and detection, feature extraction, training classifier, and classification. In [3], to provide a solution for low resolution images, framework for expression recognition based on appearance features of selected few prominent facial patches which are active when emotion are expressed is proposed. In order to get discriminative features for classification, salient patches are obtained after processing selected patches further. One-against-one classification task is performed using these features and recognition of the expression is done based on majority vote. Experimentation of the proposed method is carried on CK and JAFFE facial expression databases. In [4], feature extraction is done using PCA along with LBP and SVM classifier used to obtain results. Database used are JAFFE database and MUFE database and obtained results show that both PCA and LBP gave high performance together. In [5], live video stream frames are extracted containing face using Gabor feature extraction method and neural network and modified k means with PCA is used for classification of emotion. JAFFE database is used for simulation of framework. In [6] Active Appearance Models (AAM) were used to identify the face and extract its graphic features. For expression prediction, HMM is used and to identify the person in the image, K-NN is used. In [7], improved Directional Ternary Patterns (DTP) feature extraction and SVM classifier are used for real-time purpose emotion detection by facial expressions on JAFFE database.

In FER using traditional machine learning techniques, programmer has to be very specific about what he is interested which involves laborious process of feature extraction. Domain knowledge is expected for feature extraction. Thus, success rate of system depends on programmer's ability to accurately define a feature set. In addition, whenever the problem domain changes the whole system needs to change requiring a redesign of the algorithm from the start [8]. In [9], authors proposed a novel FER system based on features resulting from principal component analysis (PCA) which are fine-tuned by applying particle swarm. The best classification result achieved was 97% for CK database.

Most of researcher also used neural networks as its ability to extract undefined features from the training database. Most of the time it is observed that if neural networks that are trained on large amounts of data are able to extract generalized features well to scenarios that the network has not been trained on. In [10], constructive training algorithm for MLP neural networks has been proposed as classification step for the FER system. Experiments carried on three well-known databases show that the best recognition rate has been obtained using the constructive training algorithm as compared to the fixed MLP architecture. In [11], authors proposed Neural Network and K-NN based model for facial expression classification. For extraction of facial features on JAFEE database, ICA is used. Recent approaches include increased use of deep neural networks (neural networks with many numbers of hidden layers) for automatic FER problem. With growing computing power, for finding complex patterns in images, sound, and text, deep neural network architectures provide learning architecture similar to the development of brain-like structures which can learn multiple levels of representation and abstraction. Extreme variability patterns with robustness to distortions and simple geometric transformations are recognized by CNNs which are deep neural networks. It has been proven by a wide range of applications that are using CNN such as face detection, face recognition, gender recognition, and so forth that minimal domain knowledge of the problem at hand is sufficient to perform efficient pattern recognition tasks [8, 12–16]. CNNs have become the traditional approach for researchers examining vision and deep learning. Starting with LeNet-5 [17], variations of this basic design are prevalent in the image

classification literature with the best results. The recent trend is to increase the number of layers and layer size for larger datasets such as ImageNet and use of dropout in order to deal with the problem of overfitting [18, 19].

In [20], authors proposed network consists of two convolutional layers each followed by max pooling with next four inception layers and conducted experiments on seven publicly available facial expression databases. In [21], two different deep network models are proposed, for extraction temporal appearance features from image sequences and for extraction temporal geometry features from temporal facial landmark points. A new integration method for combining these two models is required in order to boost the performance of the FER. For Emotions in the Wild [22] contest for static images in [23], multiple deep convolutional neural networks are trained as committee members and combine their decisions, generating up to 62% test accuracy.

3   JAFFE Database and Proposed CNN Architecture

Proposed CNN architectures are tested on JAFFE database set of facial expression images for posed emotions (six different emotions and neutral face displays) of 10 Japanese female subjects. These six expressed emotions are the basic emotions given by Ekman and Friesen [24]. Figure 2 shows sample images from JAFFE database [25]. Expressed emotion seems to be universal across human ethnicities and cultures which are happiness, sadness, fear, disgust, surprise, and anger. The grayscale images are 256×256 pixels size. The images were labeled into $6 + 1 = 7$ emotion classes. Some head pose variations can be featured by these images [26].

Five different 12-Layer CNNs architectures are proposed for facial expression classification up to seven different basic emotions. In all five CNNs, input layer is followed by convolutional layer with different filter size and number of filters. This layer is followed by relu layer and max pooling layer with pool filter size as 2 which outputs maximum among the four values. Max pooling layers are trailed by convolutional layer with different filter sizes and number of filters for different CNNs again followed with relu and max pooling layer. Next layer is a fully connected layer with a number of output neurons which varies in different CNNs and followed by relu layer. And last fully connected layer is with seven output neurons and output of this layer is given to final softmax and classification layer. Table 1 shows detail regarding proposed 5 different CNNs. For CNN_1, CNN_2, CNN_3, and CNN_4, number of training images are 164 and testing images are 26. In CNN_1 and CNN_2, training images are repeated to increase the total number of training image set. For CNN_5, number of training images are 178 and testing images are 35.

Table 1 Details for proposed CNNs for emotion detection

		Convolution	Convolution	Fully connected	Fully connected
Sr. No.	CNNS	Layer 1	Layer 2	Layer 1	Layer 2
01	CNN 1	3(3 × 3)	9(3 × 3 × 3)	512	7
02	CNN 2	3(12 × 12)	6(9 × 9 × 3)	16	7
03	CNN 3	3(12 × 12)	6(9 × 9 × 3)	16	7
04	CNN 4	3(15 × 15)	6(9 × 9 × 3)	16	7
05	CNN 5	3(15 × 15)	6(9 × 9 × 3)	16	7

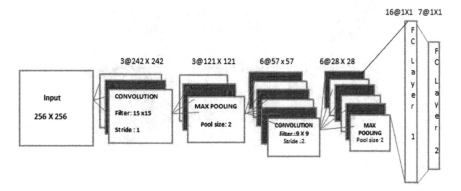

Fig. 1 Proposed CNN_5 architecture

In Table 1, numbers in convolutional layer column give number of feature maps generated with filter size and numbers in fully connected layer column gives number of output neurons. Thus in CNN_1, number of training images is 164 and first convolutional layer generates 3 feature map using 3 × 3 filter size. Second convolutional layer generates 9 feature map using 3 × 3 filter size. Fully connected layer 1 has output neurons 512 and last fully connected layer has number of output neurons as 7 representing basic emotion.

Proposed 12 layers architecture are shown in Fig. 1. Similar architecture is (referring Table 1) for CNN_1 CNN_2, CNN_3, and CNN_4 only with different filter size, feature map, and number of output neurons in fully connected layer.

4   Results and Discussions

The proposed design is tested on a 2.40 GHz Intel i7-5500U quad core processor; 8 GB RAM, with windows 10, 64 bit system. As CNN is computationally intensive which requires GPU processing for faster computation, thus system used was CUDA-enabled NVIDIA GPU with compute capability higher than 3.0, DirectX Runtime Version 12 (graphics card GeForce 920 M). MATLAB used for

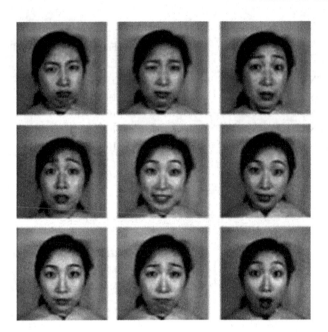

Fig. 2 JAFFE sample images [25]

Table 2 Comparison for CNNs

Sr. No.	CNNS	Accuracy (%)
1	CNN 1	80.76
2	CNN 2	88.46
3	CNN 3	73.07
4	CNN 4	88.46
5	CNN 5	96.15

implementation and system tested on JAFFE database From Tables 1 and 2 it is observed that more image detail gets captured using 12×12 or 15×15 filters. In CNN_1 and CNN_2, number of training set is increased by repeating that set but its effect is similar to increasing the epoch during training. If first layer captures good detail from the input image of 256×256 size and number of epoch is more this leads final accuracy. Maximum accuracy achieved in CNN_5 in which total number of images to be trained is more as compared to other CNNs.

5 Conclusion and Future Scope

Minimal preprocessing involved as CNNs are designed to recognize visual patterns directly from pixels of images. This is completely in contrast with the conventional pattern recognition tasks in which prior knowledge of the problem at hand is needed in order to apply a suitable algorithm to extract the right features. In this paper, five new CNN architectures have been proposed for automatic facial expression recognition. Among them, CNN_5 resulted in the highest accuracy with a classification accuracy of 96.15% achieved on test samples of JAFFE database using 2.5 GHz i7-5500U quad core processor, 8 GB RAM with GeForce 920 M. Proposed architecture can be extended for detection of nonbasic expression in future.

References

1. Face Expression Recognition and Analysis: The State of the Art Vinay Kumar Bettadapura Columbia University.
2. Nazia Perveen, Nazir Ahmad, M. Abdul Qadoos Bilal Khan, Rizwan Khalid, Salman Qadri, Facial expression recognition Through Machine Learning, International Journal of Scientific & Technology Research, Volume 5, Issue 03, March 2016, ISSN 2277-8616.
3. S L Happy, Aurobinda Routray Automatic Facial expression recognition Using Features of Salient Facial Patches, IEEE transactions on Affective Computing, VOL. 6, NO. 1, January–March 2015.
4. Muzammil Abdulrahman, Alaa Eleyan, Facial expression recognition Using Support Vector Machines, in Proceeding of 23nd Signal Processing and Communications Applications Conference, PP. 276–279, May 2015.
5. Debishree Dagar, Abir Hudait, H. K. Tripathy, M. N. Das Automatic Emotion Detection Model from Facial Expression, International Conference on Advanced Communication Control and Computing Technologies (ICACCCT), ISBN No. 978-1-4673-9545-8,2016.
6. G. Ramkumar E. Logashanmugam, An Effectual Facial expression recognition using HMM, International Conference on Advanced Communication Control and Computing Technologies (ICACCCT), ISBN No. 978-1-4673-9545-8,2016.
7. S. Tivatansakul, S. Puangpontip, T. Achalakul, and M. Ohkura, "Emotional healthcare system: Emotion detection by facial expressions using Japanese database," in Proceeding of 6th Computer Science and Electronic Engineering Conference (CEEC), PP. 41–46, Colchester, UK, Sept. 2014.
8. A.R. Syafeeza, M. Khalil-Hani, S.S. Liew, and R. Bakhteri, Convolutional neural network for face recognition with pose and illumination variation, Int. J. of Eng. and Technology, vol. 6 (1), pp. 44–57, 2014.
9. Vedantham Ramachandran, E Srinivasa Reddy, Facial Expression Recognition with enhanced feature extraction using PSO & EBPNN. International Journal of Applied Engineering Research, 11(10):69116915, 2016.
10. Boughrara, Hayet, et al. "Facial expression recognition based on a mlp neural network using constructive training algorithm." Multimedia Tools and Applications 75.2 (2016): 709–731.
11. Hai, Tran Son, and Nguyen Thanh Thuy. "Facial expression classification using artificial neural network and k-nearest neighbor." International Journal of Information Technology and Computer Science (IJITCS) 7.3 (2015): 27.
12. Shih, Frank Y., Chao-Fa Chuang, and Patrick SP Wang. "Performance comparisons of facial expression recognition in JAFFE database". International Journal of Pattern Recognition and Artificial Intelligence 22.03 (2008): 445–459.
13. C. Garcia and M. Delakis, "Convolutional face finder: a neural architecture for fast and robust face detection," IEEE Transactions on Pattern Analysis and Machine Intelligence, vol. 26, pp. 1408–1423, 2004.

14. S. Chopra, R. Hadsell, and Y. LeCun, "Learning a Similarity Metric Discriminatively, with Application to Face Verification," in In Proceedings of CVPR (1) 2005, 2005, pp. 539–546.
15. T. Fok Hing Chi and A. Bouzerdoum, "A Gender Recognition System using Shunting Inhibitory Convolutional Neural Networks," in International Joint Conference on Neural Networks, 2006, pp. 5336–5341.
16. Caifeng Shan, Shaogang Gong, Peter W. McOwanb, Facial expression recognition based on Local Binary Patterns: A comprehensive study, Image and Vision Computing, V. 27 n. 6, pp. 803–816, May 2009. https://doi.org/10.1016/j.imavis.2008.08.005].
17. Y. LeCun, B. Boser, J. S. Denker, D. Henderson, R. E. Howard, W. Hubbard, and L. D. Jackel. Backpropagation applied to handwritten zip code recognition. Neural Comput., 1 (4):541551, Dec. 1989.
18. O. Russakovsky, J. Deng, H. Su, J. Krause, S. Satheesh, S. Ma, Z. Huang, A. Karpathy, A. Khosla, M. Bernstein, et al. Imagenet large scale visual recognition challenge. arXiv preprint arXiv:1409.0575, 2014.
19. C. Szegedy, W. Liu, Y. Jia, P. Sermanet, S. Reed, D. Anguelov, D. Erhan, V. Vanhoucke, and A. Rabinovich. Going deeper with convolutions. arXiv preprint arXiv:1409.4842, 2014.
20. Mollahosseini, Ali, David Chan, and Mohammad H. Mahoor. "Going deeper in Facial expression recognition using deep neural networks." Applications of Computer Vision (WACV), 2016 IEEE Winter Conference on. IEEE, 2016.
21. Jung, Heechul, et al. "Joint fine-tuning in deep neural networks for facial expression recognition." Proceedings of the IEEE International Conference on Computer Vision. 2015.
22. Dhall A, Murthy OVR, Goecke R, Joshi J, Gedeon T (2015) Video and image based emotion recognition challenges in the wild: Emotiw 2015. In: Proceedings of the 2015 ACM on International Conference on Multimodal Interaction, ACM, pp 423–426.
23. B. Kim, J. Roh, S. Dong, and S. Lee, Hierarchical committee of deep convolutional neural networks for robust facial expression recognition, Journal on Multimodal User Interfaces, pp. 117, 2016.
24. P. Ekman and W. Friesen. Constants Across Cultures in the Face and Emotion. Journal of Personality and Social Psychology, 17(2):124129, 1971.
25. Michael J. Lyons, Shigeru Akemastu, Miyuki Kamachi, Jiro Gyoba. Coding Facial Expressions with Gabor Wavelets, 3rd IEEE International Conference on Automatic Face and Gesture Recognition, pp. 200–205 (1998).
26. Lyons M., Akamatsu S., Kamachi M., and Gyoba J. Coding Facial Expressions with Gabor Wavelets. In Third IEEE International Conference on Automatic Face and Gesture Recognition, pages 200205, April 1998.

Research Issues for Energy-Efficient Cloud Computing

Nitin S. More and Rajesh B. Ingle

Abstract Everyone is using highly computing devices nowadays and indirectly getting part of cloud computing, IOT, and virtual machines being either as a client or server. All this was possible because of virtualization techniques and upgradation of technologies from networking to ubiquitous computing. However, today, there is a need of considering the cost of computing devices versus the cost of power consumed by computing devices. Everybody required mobile of more computing capacity and along with that more battery backup. This also is needed to be thought about cloud computing, as energy consumption by data centers was increased year to year which also causing footprints of CO_2 behind. According to Gartner, worlds 2% CO_2 emission was only due to IT industry. This paper deals with investigating such research issues for energy-efficient cloud computing.

Keywords Energy efficient cloud computing · Research issues
Virtualization · Data centers · Green computing · CO_2

1 Introduction

Cloud computing has changed the information technology industry by many ways and thus also contributed in the environment hazards indirectly. In cloud computing, client request resource may be infrastructure, platform or software resource and servers situated across the datacentres will serve the requested resource as per the software level agreement (SLA) existing between client and server. On the other hand, cloud data centers end up with consuming lots of power required for cooling, operating, and maintaining data centers and emitting ample amount of CO_2. In

N. S. More (✉) · R. B. Ingle
Pune Institute of Computer Technology, Savitribai Phule Pune University, Pune, India
e-mail: nsmore27@gmail.com

R. B. Ingle
e-mail: rbingle@pict.edu

© Springer Nature Singapore Pte Ltd. 2018
S. Bhalla et al. (eds.), *Intelligent Computing and Information and Communication*,
Advances in Intelligent Systems and Computing 673,
https://doi.org/10.1007/978-981-10-7245-1_27

2020, energy consumption by data centers worldwide was predicted to be 4% of the worldwide electricity use and is likely to cultivate in the upcoming years [1].

This paper tries to deal with presenting novel research issues for energy-efficient cloud computing. As new researchers are coming out to work on cloud computing they indeed go to find the issues to be work on, this paper will suggest some issues for them. Energy is the parameter of concern mostly for wireless or battery operated device but now the time has come to consider it for cloud computing as well. Energy utilization in cloud data centers can be optimized by reducing the need for physical servers present in data centers and there are many ways for it, but along with better quality of service (QOS) and without violating software level agreement (SLA). The planned methodology should be enough scalable, distributed, and effectual for managing the power utilization trade-off. Figure 1 shows how IT industries paying on Servers, Power, and Cooling Management. Industries are acquiring cloud computing for their companies instead of going for actual investments in servers, and other computing devices. It is clearly predictable that cost on server spending is decreasing due to virtualization techniques but power and cooling cost is increasing day by day which is concerned a lot and cannot be underestimated as it will create disaster for all human beings with issues like the greenhouse effect, global warming, etc. This motivates for further vital research issues

- pointing energy efficient virtual machine (VM) topologies
- identifying energy efficient attributes and finding relations between them
- investigate energy efficient software design for cloud
- optimized VM migration algorithms
- verifying energy efficient techniques against QOS and SLA.

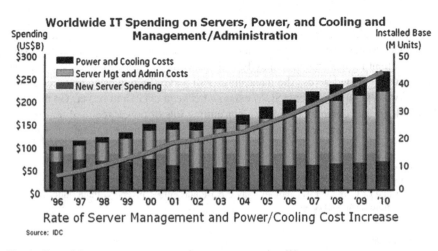

Fig. 1 Cost of data center versus cost of power consumption [3]

2 Survey of Work Done

Further, Table 1 shows the gap analysis of energy efficient cloud computing techniques referred during survey, which includes 123 Springer journal/conference papers, 78 IEEE, 5 ACM, and 12 other papers published in various journals or conferences. The survey also shows that the regular power load consumption of the companies surveyed increased from 2.1 MW in 2011 to 2.9 MW in 2013 [1]. In the age of weather change and global warming, it is unacceptable to avoid power consumption and hence, it is essential to review existing body of knowledge. Table 1 shows comparison of various power saving strategies in the cloud.

As tabulated in Table 1, Beloglazov et al. [3] studied various hardware and software techniques and proposed VM consolidation algorithms which do not affect SLA. Bergmann et al. [4] used swarm intelligence and dynamic voltage scaling (DVS), scheduling jobs technique, and achieving energy efficiency without disturbing QOS. Nguyen et al. [5] proposed a genetic algorithm for VM placement. Kulseitova et al. [6] observed many techniques with server and network as a scope and strategies like virtualization, dynamic voltage frequency scaling (DVFS), dynamic power management, and sleep mode for the servers which are not active, etc., but without any secondary objective.

Priya et al. [7] used a technique sleep mode for network as a scope and by monitoring link load as a metric they do not consider any another second objective other than energy efficiency.

Strunk et al. [8] proposed fault tolerant switching off strategy based on CPU load metric with vm migration, they also consider QOS. Rahman et al. [9] proposed virtual network embedding with vm migration they consider bandwidth metric. Cavdar et al. [10] used green routing strategy for switch and link load metrics along with QOS, SLA and CO_2 emission parameters are considered.

Mahadevan et al. [11] used server load consolidation with network sleep mode by considering CPU and link load metrics for server load migration with QOS as a secondary objective. No one has considered vm topologies effect on power consumption. The above-discussed strategies are mainly focusing on saving power and making datacenters more efficient by considering server and network as its scope, however in future due to the requirement of high bandwidth and network connectivity of datacenters, the power requirement of datacenter will go beyond our imagination.

To overcome the above-mentioned concern, Koutitas et al. [2] mention how actually power consumption happens in datacenters, most of the part of power are consumed by computing servers 70%, CPU 43% and switches 25%, i.e., network devices. There still are lots of research need to be done to design such modern algorithm and techniques which achieve better energy efficiency along with consideration of QOS, SLA, and VM consolidation by defining VM topologies which are fault tolerant and also helps nature, mankind by minimizing CO_2 emission.

Table 1 Gap analysis of energy efficient techniques in cloud computing

Reference	Scope	Strategy	Metrics	Migration	Secondary objective	Fault tolerant	VM topologies	CO_2 Emission
Beloglazov et al. [3]	Server	Virtualization, VM migration	Physical machine utilization	VM	SLA	–	–	–
Bergmann et al. [4]	Server	DVFS, Swarm intelligence	Task execution time	–	QOS	–	–	–
Nguyen et al. [5]	Server	VM Migration	Server load	VM	–	–	–	–
Kulseitova et al. [6]	Network	Sleep mode	Queue length at switch	–	–	–	–	–
Priya et al. [7]	Network	Sleep mode	Link load	–	–	–	–	–
Strunk et al. [8]	Server	Switched off	CPU load	VM	QOS	C	–	–
Rahman et al. [9]	Network	Virtual network embedding	Bandwidth	VM	–	–	–	–
Cavdar et al. [10]	Network	Green routing	Switch and link load	–	QOS & SLA	–	–	C
Mahadevan et al. [11]	Server and Network	Server load consolidation with network sleep mode	CPU and link load	Server load	QOS	–	–	–

3 Mathematical Model for Energy Efficient Cloud

For achieving mathematical model for energy consumption in cloud, there are ample of parameters for concentrating on power consumption of cloud, let us consider energy of virtual machine (vm) while running, migrating and when vm idle, type of task submitted to vm, energy utilization of vm at time of reading data as well writing data, energy consumption of resources, all these technical issues and their power usage are important to consider. The above-mentioned content can be placed in mathematical equation as follows. Let us consider total energy consumed by cloud is denoted by E_c then

$$E_c = E_{vmrunning} + E_{vmmigrating} + E_{vmidle}, \tag{1}$$

where
Nomenclature

$$E_{vmrunning} = N_t.e_t + R_t.e_{rt} \tag{2}$$

$$E_{vmmigrating} = f(nu, pm) \tag{3}$$

$$E_{vmidle} = E_{vm} \tag{4}$$

E_c	The energy consumption of cloud
$E_{vmrunning}$	The energy consumption of vm when vm is running
$E_{vmmigrating}$	The energy consumption of vm while migrating
E_{vmidle}	The energy consumption of vm when vm is idle
N_t	The number of task t executing at vm
e_t	The energy consumed by task t
R_t	The number of resources used during task t
e_{rt}	The energy consumed by resources during task t
nu	The network utilization
pm	The physical machine
E_{vm}	The energy of vm when task t = 0.

Equation 1 says that energy of the cloud is equal to the sum of energy consumed by vm in three states, i.e., vm running, vm migrating and when vm is idle. Equation 2 says energy consumed by vm during running is the sum of product of two things, one is the product of number of task and energy required by that task and second is the product of number of resources required for task and energy required by resources like disk, memory, etc., which can be more expressed with different equations. Equation 3 says the energy of vm which is migrating is a function of energy used by network utilization and physical machine on which vm running. Lastly, Eq. 4 stating the energy of vm when the task is zero and vm is idle.

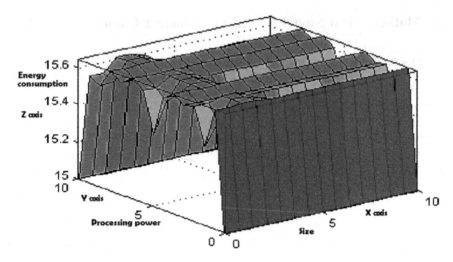

Fig. 2 Cloud entities relationship with energy consumption

4 Simulation Result with MATLAB

For identifying energy-efficient attributes in the cloud and finding relationship of them classification of cloud entities like a physical machine, vm, hypervisor, memory, hard disk, motherboard, switches, etc., done in three sets as

1. Size = [large, medium, small]
2. Processing power = [good, average, minimum]
3. Energy consumption = [high, medium, low]

Based on these three sets, 27 rules are derived in MATLAB fuzzy toolbox by considering size and processing power of cloud entities as input and energy consumption as output. Then membership functions are created by considering appropriate values for all three sets. By processing all 27 rules in rule viewer an output is obtained in surface viewer as given in Fig. 2, showing on x-axis size of cloud entities, y-axis processing power of cloud entities by considering zero to ten numbering for small, medium, large for size and minimum, average, good for processing power as input and output on z-axis, i.e., energy consumption of cloud. The result shows that as the size of cloud entities increasing energy consumption also increased and it will vary with processing power.

5 Research Issues

Shortening all of the above studies, stating some of the research issues which can be considered for further enhancements in the green cloud computing. They are stated as follows:

- To invest more for renewable energy use in cloud data centers.
- To design algorithms and techniques for energy efficient cloud without affecting QOS, SLA.
- To investigate and propose fault tolerant heuristics approaches for live virtual machine migration.
- To design energy efficient software for cloud applications without hurting security and privacy of the user.
- To inspect each and every entity in the cloud for energy efficiency.

Methodologies and techniques to be adapted in order to address the research issues

1. Study of existing energy efficient techniques.
2. Gap Analysis.
3. Propose innovative Algorithms and Techniques.
4. Experimenting and Evaluating proposed Algorithm and Techniques.
5. Come up with new dynamic, scalable, and fault-tolerant energy-efficient computing techniques.

6   Conclusion

This paper briefly analyzes and reviews the different strategies for energy-efficient cloud computing along with the motivation of it. Finding research issues and doing a gap analysis of existing energy efficiency techniques. A mathematical model for energy efficient cloud which gives the basic equation of calculating energy consumed at cloud by considering different activities of the virtual machine was discussed. Experiment with MATLAB simulator for investigating interrelationships between cloud entities and energy consumption using fuzzy logic was carried out. Lastly, different research issues along with methodologies and techniques to be adapted for achieving them were discussed. There are still other relevant issues that need to be comprehensively inspected for energy-efficient cloud computing.

References

1. Infoworld, http://www.infoworld.comtgreen-itkoomeys-law-computing-efficiency keeps-pace-moores-law-172681 access date 12 December 13.
2. G. D. Koutitas, P. Demestichas, Challenges for Energy Efficiency in Local and Regional Data Centers, Journal of Green Engineering, pp. 1– 32, 2010.
3. Anton Beloglazov, Energy-efficient management of virtual machines in data centers for cloud computing, Ph.D. thesis submitted to Melbourne university, February 2013.
4. Neil Bergmann, Yuk Ying Chung, Xiangrui Yang, Zhe Chen, Wei Chang Yeh, Xi- angjian He, and Raja Jurdak. 2013. "Using swarm intelligence to optimize the energy consumption

for distributed systems". Modern Applied Science 7, 6 (2013), 5966. DOI:http://dx.doi.org/10.5539/mas.v7n6p59.

5. Quang-Hung Nguyen, Pham Dac Nien, Nguyen Hoai Nam, Nguyen Huynh Tuong, and Nam Thoai. 2013. "A genetic algorithm for power-aware virtual machine allocation in private cloud. In Information and Communicatiaon Technology", Lecture Notes in Computer Science, Vol. 7804. Springer, Berlin, 183191.

6. Kulseitova, A., Ang Tan Fong, "A survey of energy-efficient techniques in cloud data centers," ICT for Smart Society (ICISS), 2013 International Conference on, vol., no., pp. 1, 5, 13–14 June 2013.

7. Priya, B., Pilli, E.S., Joshi, R.C., "A survey on energy and power consumption models for Greener Cloud," Advance Computing Conference (IACC), 2013 IEEE 3rd International, vol., no., pp. 76, 82, 22–23 Feb. 2013.

8. Strunk A., "Costs of Virtual Machine Live Migration: A Survey," Services (SER- VICES), 2012 IEEE Eighth World Congress on, vol., no., pp. 323, 329, 24–29 June 2012.

9. Rahman, A., Xue Liu, Fanxin Kong, "A Survey on Geographic Load Balancing Based Data Center Power Management in the Smart Grid Environment," Communications Surveys & Tutorials, IEEE, vol. 16, no. 1, pp. 214, 233, First Quarter 2014.

10. Cavdar, D., Alagoz, F., "A survey of research on greening data centers," Global Communications Conference (GLOBECOM), 2012 IEEE, vol., no., pp. 3242, 3–7 Dec. 2012.

11. P. Mahadevan, P. Sharma, S. Banerjee and P. Ranganathan, Energy aware network operations, in Proceedings of the 28th IEEE international conference on Computer Communications Workshops, Rio de Janeiro, Brazil, 2009, pp. 25–30.

12. T. Kaur and I. Chana, "Energy Efficiency Techniques in Cloud Computing: A Survey and Taxonomy" ACM Computing Surveys, Vol. 48, No. 2, Article 22, Publication date: October 2015.

13. Ching-Hsien Hsu, Kenn Slagter, Shih-Chang Chen, and Yeh-Ching Chung. 2014. Optimizing energy consumption with task consolidation in clouds. Information Sciences 258 (Feb. 2014) 452462.

14. L. Wang, S. U. Khan, D. Chen, J. Koodziej, R. Ranjan, C.-z. Xu, and A. Zomaya, "Energy-aware parallel task scheduling in a cluster," Future Generation Computer Systems, 2013.

15. White paper by T systems intel, "DataCenter 2020: first results for energy-optimization at existing data centers", july 2010.

16. RichaSinha et al, Int. J. Comp. Tech. Appl., Vol 2 (6), ISSN: 2229-6093, 2012, pp. 2041-2046.

17. Dzmitry Kliazovich, Pascal Bouvry, and Samee U. Khan. 2013. DENS: Data center energy-efficient networkaware scheduling. Cluster Computing 16, 1 (2013), 6575, 2013.

Implementation of REST API Automation for Interaction Center

Prajakta S. Marale and Anjali A. Chandavale

Abstract AVAYA Interaction Center is developed for communication between customer and the company. The Interaction Center provides business-class control of contact center communications across several channels like voice, email, web chat. Chat is a progressively popular communication channel due to growth in the era of Internet. Existing chat generator tool supports only IE browser. So, there is a need that the tool should be browser independent. The paper proposes the tool based on REST API. In the proposed system, the REST API of Chat media of CSPortal is captured through SOAPUI tool. The request–response of REST API is captured for different resources like /csportal/cometd/of CSPortal Chat Media. The result shows the proposed approach is more efficient than existing tool and chat is live for more than 2 min.

Keywords AVAYA interaction center · SoapUI · REST API, etc.

1 Introduction

A REST (Representational State Transfer) API (Application Programming Interface) explicitly captures supremacy of HTTP (Hypertext Transfer Protocol) methodologies. A RESTful API [1] is application program interface which uses HTTP requests to GET, POST, PUT, and DELETE data. Mostly browser uses REST technology can be thought of as the language of the Internet. REST allows users to connect and interact with cloud services. Social media or E-commerce websites nowadays mostly uses RESTful APIs. Transactions break down for creating sequences of small modules by using REST APIs. Each module directs a particular part of the transaction. This module provides developers a lot of flexibility, but it can be challenging

P. S. Marale (✉) · A. A. Chandavale
MIT College of Engineering Pune, Pune, India
e-mail: prajaktamarale@gmail.com

A. A. Chandavale
e-mail: Anjali.chandavale@mitcoe.edu.in

© Springer Nature Singapore Pte Ltd. 2018
S. Bhalla et al. (eds.), *Intelligent Computing and Information and Communication*,
Advances in Intelligent Systems and Computing 673,
https://doi.org/10.1007/978-981-10-7245-1_28

for developers. For communication in between client and agent of Interaction center of AVAYA, CSPortal webpage is available. Here, our main focus is on the CSPortal API. We are going to capture different responses from providing request and resource parameters of API through SOAPUI tool. Rest of the paper is divided into following sections. Section 2 presents the literature survey. Section 3 describes the proposed system. In Sect. 4, actual result is outlined, Sect. 5 describes conclusion, and finally acknowledgment.

2 Related Work

Hamad et al. [2] evaluated a RESTful web service for mobile devices. Author developed string concatenation and float number addition web services by using RESTful and conventional SOAP web service. The performance evaluation results show the advantages of using RESTful web services over SOAP web services for mobile devices are higher. RESTful web service offers a perfectly good solution for the majority of implementations, with higher flexibility and lower overhead. Belqasmi et al. [3] present a case study which is based on a comparison in between SOAP-based web services and RESTful web services. This case study is related to the end-to-end delays for executing the different conferencing operations like creating a conference, add participant, etc. The delays via the REST interface are three to five times less than when using SOAP-based web service.

Chou et al. [4] presented algorithms and methods to materialize the extensibility of REST API under hypertext-driven navigation. First, the author described a Petri-Net based REST Chart framework. Second, the author described and characterized a set of important hypertext-driven design patterns. Third, the author presented a novel differential cache mechanism that can significantly reduce the overhead of clients in hypertext-driven navigation.

We studied that SOAP-based services are much harder to scale than RESTful services. SOAP protocol uses XML for all messages, which make the message size much larger and less efficient than REST protocol. RESTful web service provides efficient, flexible, better performance than SOAP-based web services.

3 Proposed Work

CSPortal Webpage is for customer for interaction between customer and agent of Interaction Center. The CSPortalWebAPI is a client-side JavaScript API included in the CSPortalWebAPI SDK. The high-level architecture for the custom client and various other components involved. Client and agent can communicate through REST API Façade. For communication purposes, REST Protocol is used. SoapUI tool can test this REST protocol communication effectively. The API enables web-based clients to integrate various collaboration options available in Interaction

Center, such as Email, LiveChat, and Schedule Call Back. The Live Chat API is required for initiating the chat between client and agent. The actual end-to-end flow of CSPortal chat media is depicted in Fig. 1 which is described below.

1. Customer escalates the chat by calling escalatemedia API through SOAPUI tool. The chat lands on the Apache server.
2. CSPortal Client tomcat redirector escalates the incoming chat to CSPortal Server via HTTP protocol.
3. CSPortal Server redirects the chat to remote ICM server. ICM sends makecall request to ICMBridge and sends the chat request to the attribute server.
4. Attribute server creates eDU making a request to EDU Server. It sends the incoming event to WebACD. WebACD sends onhold to ICMBridge.
5. ICMBridge forward onhold messaging to ICM. ICM sends onhold messages to HTMLClient Servlet. HTMLClient receives the onhold messages. The messages are displayed in the transcript.
6. WebACD contacts workflow server to run the Qualify Chat flow. WebACD sends task assignment message to agent via paging server.
7. Agent accepts the task assignment. Agent enters room. ICM notifies the CSPortal Server on Tomcat Server which in turn notifies the CSPortal client and connected message is sent to Customer. CommetD polling takes place to publish messages on the server to the CSPortal Javascript running on SoapUI as events from the server.

SoapUI is the most useful open-source, cross-platform API Testing tool. SoapUI captures different API's required for the chat media of CSPortal. Some API's resources are listed below.

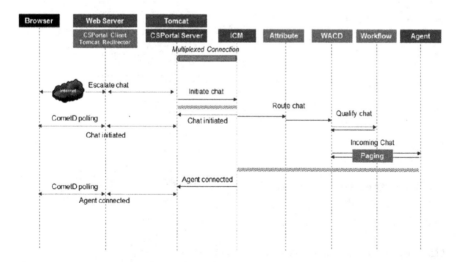

Fig. 1 End-to-end flow of CSPortal API [1]

(1) Csportal/cometd/handshake—CometD handshake request to establish connection. (2) Csportal/cometd—CometD request to send commands to AIC chat server. (3) Csportal/cometd/connect—CometD request to periodically poll and fetch the Chat Events from AIC Chat Server.

For implementation of automation of REST API, we provide URI, resources and request parameter as input to SOAPUI tool and get response. All the requests to servers go through SoapUI Tool instead of browsers. So for establishing chat, we are not dependent on the browsers.

4   Result

A transaction is breaking down into sequences of small modules. Single chat scenario considers different modules which considered as a different API's. To capture chat scenario through SOAPUI, we need to call some set of API's. For Chat purpose, we need to call listed API's in Sect. 3. Request and response of some API's is depicted in the Fig. 2.

The response we have captured through SOAPUI tool has been compared with logs at server side. The request goes to the server side, we can check through logs. Through SOAPUI, we are getting higher chat duration as compared to existing chat generator tool. Existing chat generator tool works on only IE instance. Here by using SoapUI tool, we eliminate the browser dependency (Fig. 3).

Fig. 2   Request–response of csportal/cometd/(Command: open) API

Fri Apr 28 19:06:39 IST 2017:DEBUG:> > '"clientId":"s1w0rbzqp60tl51d5pz26119wzp",[\n]"
Fri Apr 28 19:06:39 IST 2017:DEBUG:> > '"connectionType":"long-polling",[\n]"
Fri Apr 28 19:06:39 IST 2017:DEBUG:> > '"id":"8"[\n]"
Fri Apr 28 19:06:39 IST 2017:DEBUG:> > "]"
Fri Apr 28 19:06:39 IST 2017:DEBUG:< < "HTTP/1.1 200 OK[\r][\n]"
Fri Apr 28 19:06:39 IST 2017:DEBUG:< < "Date: Fri, 28 Apr 2017 13:28:03 GMT[\r][\n]"
Fri Apr 28 19:06:39 IST 2017:DEBUG:< < "Server: Apache[\r][\n]"
Fri Apr 28 19:06:39 IST 2017:DEBUG:< < "X-FRAME-OPTIONS: SAMEORIGIN[\r][\n]"
Fri Apr 28 19:06:39 IST 2017:DEBUG:< < "Pragma: no-cache[\r][\n]"
Fri Apr 28 19:06:39 IST 2017:DEBUG:< < "Cache-Control: no-cache, no-store, must-revalidate[\r][\n]"
Fri Apr 28 19:06:39 IST 2017:DEBUG:< < "Expires: Thu, 01 Jan 1970 00:00:00 GMT[\r][\n]"
Fri Apr 28 19:06:39 IST 2017:DEBUG:< < "Content-Type: application/json;charset=UTF-8[\r][\n]"
Fri Apr 28 19:06:39 IST 2017:DEBUG:< < "Content-Length: 367[\r][\n]"
Fri Apr 28 19:06:39 IST 2017:DEBUG:< < "[\r][\n]"
Fri Apr 28 19:06:39 IST 2017:DEBUG:< < "[{"data":[{\"chatmesg\":[{\"userType\":\"system\",\"userName\":\"System\",\"message\":\"Initiating Call, Please Wait.\",\"url\":\"\",\"datetime\":\

Fig. 3 Logs that shows correct response for request through SOAPUI tool

Table 1 Chat duration

Channel	Existing chat duration	Proposed chat duration
Live chat	1–2 min	3–4 min

Sometimes, our request might be misplaced or it may not go to appropriate server so through SoapUI we can get accurate response and we get clear understanding of the request (Table 1).

5 Conclusion

The paper has concentrated on REST API's. Different responses are captured for different resources like /csportal/cometd/connect of CSPortal REST API. We have not attempted to compare the effectiveness of tools related to the REST API. We believe that the automating REST API through SOAPUI tool improved the accuracy and chat is alive for more than 2 min. We have included a list of references enough to deliver a more detailed understanding of the approaches described. We ask for pardon to researchers whose important contributions may have been neglected.

Acknowledgements The author would like to thank Mr. Harshad Apshankar for his guidance. The work presented in this paper is supported by AVAYA PVT LTD.

References

1. http://searchcloudstorage.techtarget.com/definition/RESTful-API.
2. Hatem Hamad, Motaz Saad, and Ramzi Abed "Performance Evaluation of RESTful Web Services for Mobile Devices" in International Arab Journal of e-Technology, Vol. 1, No. 3, January 2010.
3. Fatna Belqasmi, Jagdeep Singh, SuhibBanimelhem, Roch H. Glitho "SOAP-Based Web Services vs. RESTful Web Services for Multimedia Conferencing Applications: A Case Study" in IEEE Internet Computing, Volume 16, Issue 4, July–Aug. 2012.
4. Li Li, Wu Chou, Wei Zhou and Min Luo "Design Patterns and Extensibility of REST API for Networking Applications", in IEEE Transactions on Network and Service Management, 2016.

Predictive Analysis of E-Commerce Products

Jagatjyoti G. Tuladhar, Ashish Gupta, Sachit Shrestha,
Ujjen Man Bania and K. Bhargavi

Abstract For the past few years, there has been increasing trend for people to buy products online through e-commerce sites. With the user-friendly platform, there is loop hole which does not guarantee satisfaction of the customers. The customers have the habit of reading the reviews given by other customers in order to choose the right product. Due to high number of reviews with mixture of good and bad reviews, it is confusing and time-consuming to determine the quality of the product. Through these reviews, the vendors would also want to know the future trend of the product. In this paper, a predictive analysis scheme is implemented to detect the hidden sentiments in customer reviews of the particular product from e-commerce site in real-time basis. This serves as a feedback to draw inferences about the quality of the product with the help of various graphs and charts generated by the scheme. Later, an opinion will be drawn about the product on the basis of the polarity exhibited by the reviews. Finally, prediction over the success or failure of the product in the regular interval of the timestamp is done using time series forecasting method. A case study for iPhone 5s is also presented in this paper highlighting the results of rating generation, sentiment classification, and rating prediction.

Keywords Predictive analysis · Product rating · Amazon · Review
Polarity · Preprocessor · Sentiment identifier

J. G. Tuladhar (✉) · A. Gupta · S. Shrestha · U. M. Bania · K. Bhargavi
Department of Computer Science and Engineering, Siddaganga Institute of Technology,
Tumkur, Karnataka 572103, India
e-mail: jagat.tula@gmail.com

A. Gupta
e-mail: 1foraashish@gmail.com

S. Shrestha
e-mail: ssachit3761@gmail.com

U. M. Bania
e-mail: ujjenms.bania@gmail.com

K. Bhargavi
e-mail: bhargavi.tumkur@gmail.com

© Springer Nature Singapore Pte Ltd. 2018   279
S. Bhalla et al. (eds.), *Intelligent Computing and Information and Communication*,
Advances in Intelligent Systems and Computing 673,
https://doi.org/10.1007/978-981-10-7245-1_29

1 Introduction

Predictive analysis is the advanced analysis technique that is used to predict unspecific future events. It uses techniques like machine learning, statistical algorithm, artificial intelligence, and data mining in order to analyze present-day data so that the predictions on future outcome are based on historical data. The goal of the predictive analysis is to do an in-depth analysis and to go further to know what would happen in order to provide the best assessment of what may happen in the future [1–3].

Predictive analysis has become a greater tool which is used to establish customer responses and purchases and promote multiple opportunities. This model helps to attract business, retain it, and increase their most valuable customers. A number of the companies use this model to predict and manage their resources. These analytics help organizations to function in a more efficient manner [4, 5].

In this paper, the reviews given by the customers over a particular product are extracted. Sentiment analysis is performed on the reviews to generate a product rating on a weekly basis. Predictive analysis is used to predict the rating for the nth week using the records of the previous weeks. This results in the prediction of the success and failure of the product.

The rest of the paper is organized as follows: Sect. 2 gives some of the related works, Sect. 3 provides proposed scheme to be used in this paper, Sect. 4 discusses the proposed algorithm in detail, Sect. 5 deals with the case study and finally Sect. 6 draws the conclusions.

2 Related Works

Mining and summarizing customer reviews were discussed in [6–8]. This sums up an array of positive words and negative words focuses on user reviews. The negative array of words has 4783 and the positive array of words has 2006. Both arrays also contain various words which are spelled incorrectly that are frequently used in social media posts and tweets. Sentiment taxonomy deals with categorization problem, and features containing sentiment, views or opinions information need to be identified before the categorization. Here, they considered only the polarization of the reviews, i.e., whether it is positive or negative. However, the rating of the product was not generated by extracting the sentiments in the preprocessed reviews [9].

Sentimental analysis using subjectivity summarization was discussed in [10]. In this, a unique machine learning methodology was applied that applies sentence categorization techniques to just the subjective part of the sentence. The sentences in the document can be either objective or subjective and that apply a novel machine learning classification tool to extract results. This can avert the polarity classifier from taking into consideration trivial or ambiguous text. Subjectivity contents can be extracted from the document that are provided to the users as sum up of the sentiment content of that document. The result represent the subjectivity extracted and created sentiment information of that documents in a very compact way with discussed on sentence-level categorization [11].

Sentiment analysis and opinion mining were discussed in [12]. Here in the text or the sentence, the problem was to classify the text or sentence into a particular sentimental polarity, negative, or positive. It mainly focuses on text or sentence which has three levels of sentimental polarity classification, and they are the entity and aspect level, the sentence level, and document level. The entity and aspect level focuses on what exactly users dislike or like from their reviews and opinions. Likewise, the sentence level focuses on each and every sentence with sentiment classification. While, the document level focuses on the content of the whole document and expresses the positive or negative sentiment. Here, the written text categorization and their polarity are discussed however the rating of the product was not generated and predictive analysis was not taken into consideration [13].

3 Proposed Scheme

The scheme for predictive analysis of e-commerce products is given in Fig. 1. The customer reviews on electronic gadgets are scrapped from various e-commerce websites like Amazon, Flipkart, E-bay, and Snapdeal using Python-based Web Scrapper and stored in review log file. The scheme mainly consists of four functional modules. They are Preprocessor, Sentiment Identifier, Rating Generator, and Rating Predictor.

The Preprocessor filters the reviews based on the part of speech exhibited by each and every token in the review. The Sentiment Identifier classifies the review as positives, negatives, or neutrals based on their polarity. The Rating Generator rates the review based on the weekly sentiment provided by the customer over the product. Finally, the Rating Predictor aggregates the previous rating on the monthly scale to predict the success and failure of the product.

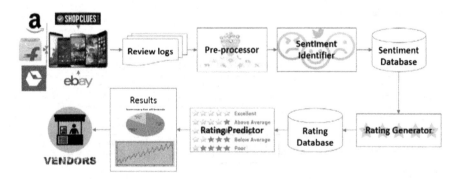

Fig. 1 Scheme for predictive analysis of e-commerce products

4 Algorithm

In this section, the detailed discussion on every functional module of the proposed scheme along with their algorithm is given.

Preprocessor

In the Preprocessor module, **Tokenization** is a process which breaks down the review into tokens of words separated by spaces between them. Then appropriate part of speech like nouns, pronouns, verbs, adjectives, adverbs, and prepositions will be tagged to each of the tokens in the review using NLTK (Natural Language Tool Kit) **Tagging** process. From the tagged tokens, the based words are converted using NLTK **Stemming** process is carried out on the review log via Suffix-stripping algorithm shown in Fig. 5. The reviews are then **Filtered** based on the parts of speech. The stop words in the reviews are then removed using stop word remover process. The working of the Preprocessor is given in Algorithm 1 and functioning of the Preprocessor is given in Fig. 2. A sample example depicting the process of tokenization and tagging is given in Figs. 3 and 4.

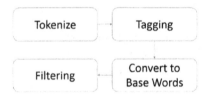

Fig. 2 Preprocessor

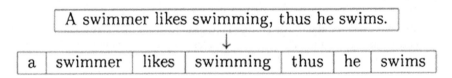

Fig. 3 Example of tokenization

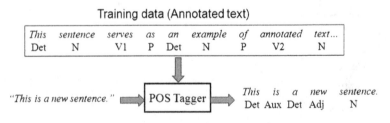

Fig. 4 Example of tagging

Fig. 5 Example of stemming

if the word ends in 'ed', remove the 'ed'

if the word ends in 'ing', remove the 'ing'

if the word ends in 'ly', remove the 'ly'

Fig. 6 Sentiment identifier

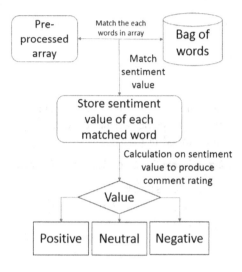

Algorithm 1: Preprocessor

```
Step 1: Start
Step 2: for each review Ri from the review log RL
            use word_tokenize to generate the list of
                                            tokens T
            tag Part-Of-Speech to each ti ∈ T {(t_i,
                POS_i), (t_j, POS_j), (t_k, POS_k), …}
            convert to base words using stemming
                                            process
            filter the words based on Part-Of-Speech
            remove stop words in the review
        end for
Step 3: Output the pre-processed review PR
Step 4: End
```

Sentiment Identifier

The Sentiment Identifier takes the preprocessed list as input and compares each element in the list with the Bag of Words with the reference from Naïve Bayes classifier. If the words in the preprocessed list match with the Bag of Words, the sentiment value of those bag of words is extracted. The reviews are classified into positive, negative or neutral based on their sentiment value using **Naïve Bayes**

Fig. 7 A sample of
sentiment identifier output

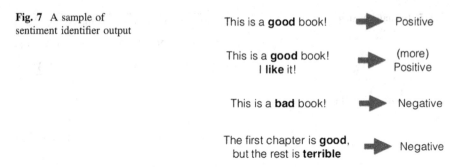

classifier. The working of Sentiment Identifier is given in Algorithm 2, the functioning of the Sentiment Identifier is given in Fig. 6 and a sample example is demonstrated in Fig. 7.

Algorithm 2: Sentiment Identifier

```
Step 1: Start
Step 2: Intialize a bag of words in sentiment
        database
Step 3: for each PRi ∈ PR
            match PR[i] in PR with the bag of words
            if match is not found
                discard the item
            else
                sentiment_value[] = Extract the value
                                    from the bag of words
            end if
            calculate ∑sentiment_value[]
        end for
Step 4: if ∑sentiment_value > 0
            PR is "positive"
        else if ∑sentiment_value < 0
            PR is "negative"
        else
            PR is "neutral"
        end if
Step 5: Store PR in Sentiment Database
Step 6: End
```

Rating Generator

The Rating Generator assigns a rank to each of the preprocessed sentiment database based on the mean of the sentiment values exhibited by every review over a week. The working of Rating Generator is given in Algorithm 3, the functioning of the Rating Generator is given in Fig. 8 and a sample example is demonstrated in Fig. 9.

Fig. 8 Rating generator

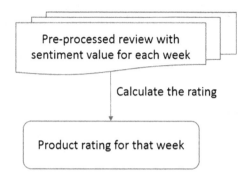

Calculate the rating

Product rating for that week

Fig. 9 A sample of rating
generator output

Good

7.4

from <u>548 reviews</u>

Algorithm 3: Rating Generator

```
Step 1: Start
Step 2: Accept pre-processed PR review along with the
        sentiment value.
Step 3: for each PR_i ∈ PR
            sum= ∑sentiment value {PR_1, PR_2, PR_3, …}
            average = sum / total no. of reviews with
                            sentiment value in that
                            time period
        end for
Step 4: Rate the product according the average
Step 5: Output product rating of every review
Step 6: End
```

Rating Predictor

The Rating Predictor forecasts the rating of the product in the upcoming time using time series forecasting technique. The calculation of the rating for the nth week uses the records of the previous weeks. This results in the prediction of the success and failure of the product. The working of Rating Predictor is given in Algorithm 4, the functioning of the Rating Predictor is given in Fig. 10.

Fig. 10 Rating predictor

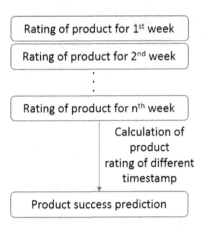

Algorithm 4: Rating Generator

```
Step 1: Start
Step 2: Accept the product rating
Step 3: for each product rating at time Δt
            if product rating Δt-1<=product rating Δt
            Product will be a success
        else
            Product will fail
        end if
    end for
Step 4: End
```

5 Result

In this section, the implementation details of the proposed work along with the results obtained are discussed. The tools used are Python-based Web Scrapper, used to extraction of reviews from the Amazon Website real time, NLTK for preprocessing, text processing API for Bag of Words comparison, and finally ARIMA model for Trend prediction modeling.

1000 reviews on Iphone 5s have been taken into consideration during implementation of the proposed work, the most common keywords used while expressing the reviews were battery life, touch screen and price of the Iphone 5s. Negative sentiment was expressed in reviews by using adjectives like bad, poor, worst, and horrible whereas positive sentiments were expressed by using sentiments like nice, good, best, excellent, and amazing.

The Rating Generator output over Iphone 5s is shown in Fig. 11. It was found out that majority of the reviews were above average (5). The Sentiment Identifier output over those reviews is depicted in Fig. 12 where around 450 reviews exhibited positive sentiment, 250 reviews exhibited negative sentiments and remaining were found to be neutrals. The Rating Predictor output of Iphone 5s over next 2 months is shown in Fig. 13. The ARIMA modeling output is shown in Fig. 14. It was found that Iphone 5s receives high rating in starting month and later the rating will remain saturated.

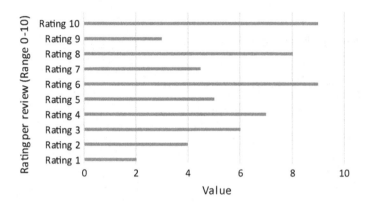

Fig. 11 Rating generator output

Fig. 12 Sentiment identifier output

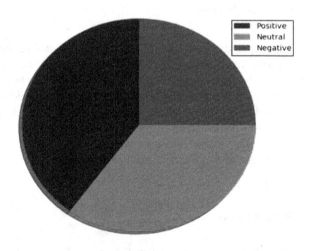

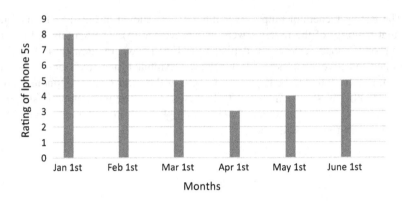

Fig. 13 Rating prediction output

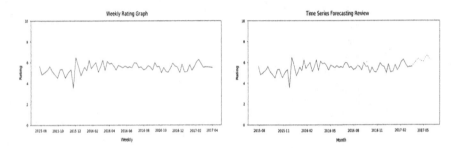

Fig. 14 Prediction output

6   Conclusion

This paper presents a novel scheme for predicting the rating of the product in E-commerce websites. The aim of this paper is to provide the vendor a framework to quickly gain insight about what customer in their field think about the product. It brings out the fact that there are sentiment values hidden in the reviews which can be used to generate a rating for the product. With the help of time series algorithm, the future market trend of the electronic products is predicted. This is a practical way to use unstructured data in the form of review to inform a vendor on how customers think about and react to product.

References

1. Mohammad Salehan and Dan J. Kim, "Predicting the Performance of Online Consumer Reviews: A Sentiment Mining Approach", Elsevier Journal on Information Processing and Management, 2016.

2. ShahriarAkter and Samuel FossoWamba, "Big data analytics in e-commerce: A systematic review and agenda for future research", Electronic Markets 26 173–194, 2016.
3. Wikipedia, https://en.wikipedia.org/wiki/Predictive_analytics.
4. Pavithra B, Dr. Niranjanmurthy M, Kamal Shaker J and Martien Sylvester Mani F, "The Study of Big Data Analytics in E-Commerce", International Journal of Advanced Research in Computer and Communication Engineering, 2016, Vol. 5, Special Issue 2.
5. Galit Shmueli and O. Koppius, "Predictive Analytics in Information Systems Research", MIS Quarterly, 2011, Vol 35 No. 3 pp. 553–571.
6. Hu M and Liu B, "Mining and summarizing customer reviews", In Proceedings of the tenth ACM SIGKDD international conference on Knowledge discovery and data mining, 2004, pp 168–177.
7. Rui Xia, FengXu Jianfei Yu, YongQi and ErikCambria, "Information Processing and Management Polarity shift detection, elimination and ensemble: A three-stage model for document-level sentiment analysis", Elsevier Journal on Information Processing and Management, 2016, vol 52, Issue 1, pp 36–45.
8. Hu M and Liu B, "Mining and summarizing customer reviews", In Proceedings of the tenth ACM SIGKDD international conference on Knowledge discovery and data mining, 2004, pp 168–177.
9. Yorick Wilks and Mark Stevenson. "The grammar of sense: Using part-of-speech tags as a first step in semantic disambiguation". Journal of Natural Language Engineering, 1998, pp 135–144.
10. Pang B and Lee L, "A sentimental education: Sentiment analysis using subjectivity summarization based on minimum cuts". In: Proceedings of the 42nd Annual Meeting on Association for Computational Linguistics, Stroudsburg, PA, USA, 2004, pp 1–8.
11. Jeonghee Yi, Tetsuya Nasukawa, Razvan Bunescu, and Wayne Niblack, "Sentiment analyzer: Extracting sentiments about a given topic using natural language processing techniques", In Proceedings of the IEEE International Conference on Data Mining (ICDM), 2003.
12. Liu B, "Sentiment Analysis, and Opinion Mining. Synthesis Lectures on Human Language Technologies", Morgan & Claypool Publishers, 2012, vol. 5, no. 1, pp 1–67.
13. Ellen Riloff and Janyce Wiebe, "Learning extraction patterns for subjective expressions", Proceedings of conference on Empirical methods in natural language processing, 2003, pp. 105–112.

User Privacy and Empowerment: Trends, Challenges, and Opportunities

Prashant S. Dhotre, Henning Olesen and Samant Khajuria

Abstract Today, the service providers are capable of assembling a huge measure of user information using big data techniques. For service providers, user information has become a vital asset. The present business models are attentive to collect extensive users' information to extract useful knowledge to the service providers. Considering business models that are slanted towards service providers, privacy has become a crucial issue in today's fast-growing digital world. Hence, this paper elaborates personal information flow between users, service providers, and data brokers. We also discussed the significant privacy issues like present business models, user awareness about privacy, and user control over personal data. To address such issues, this paper also identified challenges that comprise unavailability of effective privacy awareness or protection tools and the effortless way to study and see the flow of personal information and its management. Thus, empowering users and enhancing awareness are essential to comprehending the value of secrecy. This paper also introduced latest advances in the domain of privacy issues like User-Managed Access (UMA) that can state suitable requirements for user empowerment and will cater to redefine the trustworthy relationship between service providers and users. Subsequently, this paper concludes with suggestions for providing empowerment to the user and developing user-centric, transparent business models.

Keywords Big data · Personal information · Privacy · Privacy by design
Awareness · Trust

P. S. Dhotre (✉) · H. Olesen · S. Khajuria
CMI, Aalborg University Copenhagen, Copenhagen, Denmark
e-mail: psd@es.aau.dk

H. Olesen
e-mail: olesen@cmi.aau.dk

S. Khajuria
e-mail: skh@cmi.aau.dk

© Springer Nature Singapore Pte Ltd. 2018
S. Bhalla et al. (eds.), *Intelligent Computing and Information and Communication*,
Advances in Intelligent Systems and Computing 673,
https://doi.org/10.1007/978-981-10-7245-1_30

1 Introduction

Every day, Internet users send ten billion text messages and make one billion posts. This sharing of messages and post is said as "mass-self-communication" by Manuel Castells [1]. This communication contains personal information. Hence, along with protection of personal information, there is a strong need to notify users about data collection, its use and what happens next to their data.

The developed privacy protection principles like W3C P3P [2], Privacy by Design (PbD) [3], "Laws of Identity" from Kim Cameron [4] emphasize on notification of data collection, user control over data, minimum data disclosure, and user-defined consent for data use. However, today user has no control over their personal information. Service providers have already assigned a value to personal data. They apply it to build accurate customer profiles, do target advertising, sell and share collected information or analyze information to other entities without user notification and consent [5–7].

Personal data is becoming an integral voice of service use on the network. The present business models focused less on user privacy. The current framework for the protection of user privacy includes ENISA [8], OpenID [9], UMA [10] focused on minimum data disclosure to have fine-grained control.

Data brokers make their revenue by selling user information to publishers and advertisers [11]. Hence, big data is now the big business. Practices of data brokers' raise privacy issues. The end users are unaware of the involvement of data brokers in the bargain. There is no notification to end user about the user information collection, manipulation, and sharing. Besides, a recent study taken in India highlighted the fact that the user privacy concerns could be noteworthy [7]. 65% of respondents are unaware of privacy issues, rights, and knowledge. Moreover, 80% of respondents are concerned about the solicitation of personal data. This indicates that they want to experience more about the personal information aggregation.

Therefore, there is a strong need to empower users in understanding the value of privacy and flow of personal information between service providers and user. This will offer an opportunity to both users and service provider to come together on a common platform and to focus on accessing, sharing, and managing personal information in a manipulated way. Without user awareness and empowerment, it is exceedingly hard to protect users' privacy. From the user's perspective, the new business model should consider the minimum information collection, users' rights, consent, and choice. Likewise, from the service providers' point of opinion, it is as significant to identify valuable consumers/users and building a long relationship that establishes trust between them.

In this report, we introduced some of the principal challenges for user empowerment like how to visualize information flow, controlled bargain, and user consent gain. Also, we presented research guidelines that address challenges like the formation of a visualization tool, semiautomated negotiation process where both (users and service providers) will receive an equal return in information bargain.

2 Users, Data Brokers, and Privacy Concerns

2.1 Internet User

A user has different attributes and is widely diffused. As shown in Fig. 1, the user interacts with social media, online subscribers, etc., by disclosing a set of user attributes. Depending on context, the set of attributes forms the partial identity of the user. Hence, an identity comprised multiple identities [12]. The user must share their identity with the separate service provider. At the other end, data brokers collect, store, manage, manipulate, and share extensive information about end users without direct interaction with them.

In many instances, it is shared and sold to data brokers [13]. Still, worries about privacy have arisen with the increasing flow of personal information in many directions to different companies. Thus, it gets more and more hard to protect the user's data privacy.

2.2 How Do Data Brokers Work?

Data brokers collect the user data from the discrete sources like governments (central and state government), openly available media (social media, the Internet, blogs), and commercial sources (phone companies, banks, car companies, and so on). After analysis of gathered data, they separate the users as sports loving, interested in cancer solutions, trekking enthusiasts, blended with political posts, food choices, and hence along. Besides, they organize collected information into

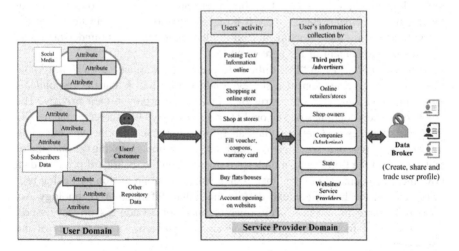

Fig. 1 Information flow between user, service provider, and data broker

different segments, which are individually identifiable dossiers that containing, e.g., marital status, location, income, shopping, hobbies, job, and travel plan. Then, segmented information is ready to sell to other data brokers, advertisers, or government sectors without the user's direct awareness [6].

2.3 Who Are the Data Brokers?

Acxiom, the largest data broker, has gathered information relating to 700 million customers worldwide and 3000 data segments for nearly every U.S. user [14]. In the Indian setting, "Zifzi" company provides IT and marketing solutions by providing 12 categories of Indian databases (electronic mail addresses, mobile numbers, business seekers, credit card holders, statewide and citywide databases, and so on) [15]. People can buy information like profiles, company, religion, usernames, salary and income sources, medical history, drug abuses, and sexual orientation from many information brokers. Internet users do not recognize that the greatest menace to personal data privacy comes from them.

3 State-of-the-Art Technologies and Frameworks

In the field of privacy awareness and protection, there have been a long practice and activities. Some of the research includes:

The users do not really know to what extent their data are collected, used, and dealt. Consent and privacy conditions (terms and conditions, agreements) are not effective for reading because of the length and complexity of language. According to the Futuresight [16], 50% of mobile users are giving their consent without reading the agreement. Hence, there is a requirement to raise the awareness of the users by analyzing these documents and implementing an easy-to-understand way to visualize the content of the agreement.

Lorrie Cranor said that "It is difficult to protect your privacy even if you know how" [17]. The conclusion of a survey conducted by her is that the respondents were unfamiliar with the tools, struggled, and failed to protect their privacy. The article on "The Economics of Personal Data and Privacy: 30 years after the OECD Privacy Guidelines" [11] revisits the Fair Information Practice principles including transparency, purpose specification, and use limitation. Looking at the major reclamation of the European Union legal framework on the security of personal data [18], there is a demand to strengthen individual rights of protection by raising user awareness. In the Indian context, despite the Indian Information Technology Act 2000 [19], there is a lack of consumer organization to address privacy issues. The DSCI [20] focuses on a privacy awareness program for employees of the governing body.

3.1 Tools to Support Privacy

"Disconnect me" is a software program which can offer you with a list of third parties like ad networks, analytics, and marketing companies, who is tracking your clicks, your interests, page views, etc. This information goes to data brokers [6]. Online Permission Technologies have provided an online privacy guard called MyPermission for Android and Apple devices [21]. MyPermission provides a menu for users to handle all service permissions. So, the depth of these companies and darkest corners has only been investigated to a very limited extent. There is a need to exploit the concepts of such tools and provide new enhancement for them that will allow the user to decide what to do with their personal data.

3.2 Privacy Enhancing Technologies

The European Network and Information Security Agency (ENISA) carried out a study and survey on security, privacy, and usability of online seals [8]. Their findings reaffirmed: "At the moment, it is too difficult for consumers to compare different information practices of online service providers" and also identified five main challenges of the seals: awareness, standards, validity checks, usability, and presence.

A framework of User-Managed Access (UMA) [10] helps the users to find out the value of their personal data and its flow control. Different scenarios were discussed to show fine-grained control of user attributes [12]. The OAuth protocol for delegation of authorization is becoming the standard mode to gain access to protected resources for websites and apps using tokens [22]. Also, ongoing research addresses Big Privacy and the Personal Data Ecosystem (PDE) based on PbD [3]. Seven architectural elements are a personal cloud, semantic data exchange, trust framework, identity and data portability, data by reference or subscription, pseudonyms, and contractual data anonymization.

4 Privacy Issues from Users and Service Providers View

The initial research of the granted project from the National Science Foundation's Secure and Trustworthy Computing initiative program [SaTC] represents online privacy issues filed in FTC enforcement actions [23]. The complaints are examined and separated as established in Table 1. Most of the descriptive claims relate to data disclosure without user consent and partial implementation measures for the user privacy protection.

Table 1 Summary of privacy issues filed in FTC enforcement actions

Classification grouping (based on privacy violation claims)	# of claims	Examples of complaint	Remark/Issues
Unauthorized disclosure of personal information	51	Use of personal information for financial gain (identity theft)	Use, disclosure, processing of personal information
		Use of personal information for unspecified purposes	
		Unfair personal information collection practices	
		Unauthorized use and sale of information	
		Unauthorized third party access without protection	
		Disclosed, offered/sold personal information	
Surreptitious collection of data	111	No notification of privacy changes	Notice, information collection, opt-in and opt-out
		Sharing of online shopping carts to third parties	
		Failed to reveal types of information collection	
		Unnecessary information collection	
		Failed to disclose the information collection sources.	
		Obtain customers financial information	
		Without user consent, gathering of personal information through false way	
		Failed to notify the software installation that pass on personal information	
		Continued to collect and display information even after opt-out	

(continued)

Table 1 (continued)

Classification grouping (based on privacy violation claims)	# of claims	Examples of complaint	Remark/Issues
Inadequate	275	Failed to develop, implement, and sustain a complete information security program for protection of information	Threat detection and prevention, employees and training, access credentials and authorization, Breach response, Information disposal
		Failed to work on security in the design of software delivered to consumer	
		Failed to implement acceptable security policies for sensitive information	
		Failed to implement satisfactory measures to detect, monitor and prevent unauthorized access to information	
		Failed to train/inform employees regarding consumer information privacy	
		Failed to monitor outbound network traffic to recognize and block unauthorized transfer sensitive information	
		Failed to implement complete policies regarding appropriate collection, handling, and removal of personal information	
		Failed to evaluate information security program in light on the possible risks	
Wrongful retention of personal information	6	Storage of personal information for a long time unnecessary	Information store

4.1 User Understanding and Awareness About Privacy

There is a simple ignorance and user's apathy toward their privacy because privacy is something whose value is completely understood only after it has been taken away. Users are unaware of the repercussions of exposing personal information [24, 25]. The users are unaware of online practices and do not recognize what happens to their data once it is picked up by the service suppliers. It is found that many consumers claim that they do not like anyone knowing anything about themselves or their habits unless they decide themselves to share that information [23, 24].

4.2 Privacy Policy Understanding

Privacy policies are convoluted and are characterized by a lot of jargons which are difficult for the naive users to understand. Privacy policies are complicated, take too long to understand and read [24]. This is a blindly accepted item on websites despite the user express increasing concern about personal data collection practices [26]. This document is important and acts as a single source of information to know how service providers collect, use, and share their data. The challenge is to be capable of studying and visualizing privacy policy in an easygoing manner.

4.3 User Control

Most web service providers do not have selective data submission, i.e., the user does not have complete control over the amount of personal information they want to enter. It is often mandatory to enter, birthday, phone number, organization, and so on while signing up for any service [12].

4.4 Service Provider's Issues

The current approach of the service provider relies on the formation of the business model that requires end users to provide maximum information from the initial moment of dealing. Thus, the challenge for service providers is to get valuable customers from a large pool of available customers by asking least/minimum information from them, providing privacy assurance for maintaining a long-term relationship with them. Therefore, privacy issues from the users and service provider's view are shown in Table 2.

Table 2 Summary of privacy issues

Entity	Issues
User side	Users disclose more information than needed
	Users are unaware of online practices and privacy and privacy policy
	Users are unaware of privacy protection tools
	Users are not aware of what happens to their data
	Users are not notified for collection and sharing personal information in a visible way
	Users are unable to control and manage their data
	Users do not give consent or no option for opt-out
Service provider side	Maintaining the privacy of customers' information
	Binding a customer connection (trust establishment)

5 New Challenges and Recommendations

To empower the users and protect user's privacy, the new research challenges are summarized in Table 3.

5.1 User Awareness and Personal Flow Visualization

To improve end user privacy decisions, a new visualization tool for measuring privacy can be developed using visual cues schemes. Visual cues about the privacy

Table 3 New challenges for privacy awareness and protection

Type	New challenges
User awareness and visualization of personal data	How do we create awareness among the end users about privacy and online practices?
	How to enhance end users' understanding and control over their personal data
	How do we visualize the flow of personal data?
	How do we create awareness among end users about data collection sources, methods, purposes of service providers, and their relations with other entities?
Assurance and truth	How do we ensure that service provider will not use information for any other purposes than what user have given consent for?
Privacy	How do we protect user's privacy?
User consent gaining	How to increase the general acceptability of the user–provider liaison?
User-centric businesses with transparency	How do we provide user-centric business cases to the service provider?
	How do we allow transparency of the personal data management? How will the user be enabled to see where their data are flowing and for what purposes?

intrusiveness of services and mobile applications must communicate using a visualization tool. This should focus on the completeness of privacy policy and development of a dashboard to assist users in making consent judgments. Machine learning-based technologies will facilitate the users to figure out the terms and privacy policies.

A bargaining mechanism must be implemented to make an informed decision on how much users accept to disclose for what in return. It should include easy-to-use and easy-to-understand semiautomated negotiation process, where the user can determine the benefits from service providers and be given whether they care to furnish this information.

5.2 Assurance and Truth

The "accept or reject" strategy of service providers should change in future. The service provider must notify the purpose of data information to end user and should have a panel to see the data collected from the user, its purpose, and a list of the third party involved. They should not share the information without notification and proper consent of Internet user.

5.3 Privacy Protection

PbD [3] is a major initiative from Canada, providing a set of principles for incorporating privacy protection in applications and services from the beginning, without sacrificing essential functionality. One may think of the analogy with a personal privacy firewall, where all ports are held in private by default, and you only open the ports that are needed.

Hence, on that point is a substantial demand to consider privacy-by-design principles in the fundamental concept of the privacy protection solution. The solution should be implemented that must comply with the existing regulations as per data protection directives and the framework proposed by the county commission.

5.4 User Consent Gaining

End user consent is gaining the desired level of permission from them while providing services. User consent is received through long, complicated text mentioned in privacy policies presented to end users in an ineffective way. Thus, a new method must furnish an effective and efficient means of requesting consent from users. This is a very challenging task because knowledge, reading time, and location of the user change from person to person. Hence, cognitive science and human–computer interface principles will assist in this research.

5.5 User-Centric Businesses with Transparency

The marketing strategy of the service provider should focus on specific and needy users' instead of looking at every user. This will facilitate the service providers to know their valuable consumers. User choices and interests must be protected from other parties. A novel mechanism must allow for transparency of the personal information management process through only transactions. It must assist end users and service providers in their trustworthiness to build better relations. History-based reputation will help users in building trust between user and service supplier.

6 Privacy Protection Parties and Responsibilities

Figure 2 represents interested and concerned entities for the protection of personal information privacy: Individual users, service providers/website developers, data brokers/advertisers/third party app developers, and government and governing bodies.

6.1 Individual Users

The user needs to be more mindful of their personal information disclosure when they are online. The minimum information should be shared with the other communicating parties. Before availing any services, the user must read privacy policies set by the service suppliers. They must be cognizant of the online activity tracking entities. A user must use privacy protection tools on a veritable base. Besides, they should render a thought to the consequences of information sharing when they are online.

Fig. 2 Privacy protection parties

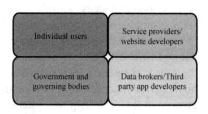

6.2 Service Providers

Service providers are responsible for providing a panel where an end user can see what information has been collected, the purpose of collection and if shared, the entity names. A new mechanism must include grant, revoke and change their consent to have control over their information before availing a service. A new user-centric business model must be established to provide services and hide information from additional and unnecessary entities. Likewise, a new platform may be provided where two partied can strongly negate on information interchange to have a win-win position.

6.3 Government and Governing Bodies

The government and governing bodies must play a leading role and enforce standardization and legitimate efforts. Data storage and transfer standardization must include secure communication, fine-grained access control, selected attribute-based authentication and authorization, an agreeable user consent mechanism to reduce barriers for illegal entries. Also, government bodies establish a common access point where an end user can file complaints of privacy breaches.

6.4 Data Brokers/Advertisers/Third Party App Developers

To gain explicit user consent, application developers are responsible for asking the required information from end users in a clean way. The list of services/features must be presented along with the authorization they wish to activate each service. Also, they must be able to convey the data collection methods, its purpose, and list the entities to which the data is shared.

7 Conclusion

Aggregation and analysis of user data through big data techniques are a gift to both service providers and private users. Nonetheless, present techniques are not satisfactory enough to support fine-grained access control and privacy-guaranteed data management. In today's information society, there is a compelling need to empower the user and enhance user awareness about privacy by building up and deploying novel privacy mechanism based on digital technologies. In this paper, we provided guidelines that will help to implement privacy awareness mechanisms to enhance the end users' understanding and control of their personal data during online transactions by the use of visualization tools. Also, innovative research method should enable service providers and app developers to engage in explaining their requests for, treatment and security of end users' personal data, ultimately increasing the general acceptability of the user–provider liaison. Introducing the possibility of assigning a value for users to pay and service providers to charge for privacy at desired levels will benefit both the parties.

We also identified a need to redefine the relationship between users' personal data and service providers' business models on data analytics by focusing on economic values of end users' data. This relation should be ethically corrected, comply with regulations and profitable to all. The responsibilities of privacy protection parties must lead to a solution that must consider user's awareness about privacy, empowering users, effective access control and authentication mechanisms to the users' personal data, understanding of the terms of services and build a comprehensive trust.

In the future work, the authors will do an analysis of privacy policy for different service providers and develop an improved privacy awareness tool to assist users in managing their personal information and trust toward service providers.

References

1. M. Kuneva, "European Commission - PRESS RELEASES - Press release - Meglena Kuneva European Consumer Commissioner KeynoteSpeech Roundtable on Online Data Collection, Targeting and Profiling Brussels, 31 March 2009." [Online]. Available: http://europa.eu/rapid/press-release_SPEECH-09-156_en.htm. [Accessed: 17-Dec-2016].
2. "P3P: The Platform for Privacy Preferences." [Online]. Available: http://www.w3.org/P3P/. [Accessed: 17-Oct-2016].
3. A. Cavokian and D. Reed, "Big Privacy:Bridging Big Data andthe Personal Data Ecosystem Through Privacy by Design," 2013. [Online]. Available: https://www.ipc.on.ca/images/Resources/pbd-big_privacy.pdf. [Accessed: 28-Jan-2017].
4. Kim Cameron, "The Laws of Identity -." [Online]. Available: http://www.identityblog.com/stories/2005/05/13/TheLawsOfIdentity.pdf. [Accessed: 21-Sept-2016].
5. "Sharing Information: A Day in Your Life| Consumer Information." [Online]. Available: http://www.consumer.ftc.gov/media/video-0022-sharing-information-day-your-life. [Accessed: 18-Dec-2016].
6. "The Data Brokers: Selling your personal information." [Online]. Available: http://www.cbsnews.com/news/data-brokers-selling-personal-information-60-minutes/.
7. "Getting to know you| The Economist." [Online]. Available: http://www.economist.com/news/special-report/21615871-everything-people-do-online-avidly-followed-advertisers-and-third-party.
8. "Deliverables — ENISA." [Online]. Available: https://www.enisa.europa.eu/activities/risk-management/emerging-and-future-risk/deliverables. [Accessed: 12-Jun-2015].
9. "OpenID Connect | OpenID." [Online]. Available: http://openid.net/connect/. [Accessed: 29-Dec-2016].
10. "Home - WG - User Managed Access - Kantara Initiative." [Online]. Available: https://kantarainitiative.org/confluence/display/uma/Home. [Accessed: 29-Jan-2017].
11. A. Acquisti, "The economics of Personal Data and the Economics of Privacy," *Econ. Pers. Data Priv. 30 years after OECD Priv. Guidel.*

12. H. Olesen and S. Khajuria, "Accessing and Disclosing Protected Resources: A User-Centric View," in *2015 IEEE 81st Vehicular Technology Conference (VTC Spring)*, 2015, pp. 1–5.
13. "Sharing Information: A Day in Your Life, Fed. Trade Comm'n,." [Online]. Available: http://www.consumer.ftc.gov/media/video-0022-sharing-information-day-your-life. [Accessed: 19-May-2016].
14. "Acxiom, Annual Report 2013." [Online]. Available: http://d3u9yejw7h244g.cloudfront.net/wp-content/uploads/2013/09/2013-Annual-Report.pdf.
15. "Indian Database." [Online]. Available: http://www.zifzi.com/index.php?route=product/category&path=20. [Accessed: 18-May-2016].
16. Futuresight, "Futuresight:, 'User perspectives on mobile privacy, Summary of research findings.'" [Online]. Available: http://www.gsma.com/publicpolicy/wp-content/uploads/2012/03/futuresightuserperspectivesonuserprivacy.pdf.
17. "Why Privacy Is So Hard to Get | MIT Technology Review." [Online]. Available: http://www.technologyreview.com/view/526421/self-defense/. [Accessed: 12-Jun-2016].
18. "Commission proposes a comprehensive reform of the data protection rules." [Online]. Available: http://ec.europa.eu/justice/newsroom/data-protection/news/120125_en.htm. [Accessed: 12-Nov-2016].
19. "Ministry of Law, Justice and Company Affairs (Legislative Department) New Delhi, the 11," pp. 1–13, 1994.
20. "Frameworks | Data Security Council of India." [Online]. Available: https://www.dsci.in/dsci-framework. [Accessed: 02-May-2016].
21. "MyPermissions - About." [Online]. Available: http://mypermissions.com/whoweare. [Accessed: 17-Aug-2016].
22. "OAuth - The Big Picture by Greg Brail & Sam Ramji." [Online]. Available: https://itunes.apple.com/us/book/oauth-the-big-picture/id515116202?mt=11. [Accessed: 29-Apr-2016].
23. "CLIP- Center on Law and Information Policy," 2014. [Online]. Available: www.fordham.edu/download/.../id/.../privacy_enforcement_actions.pdf.
24. P. S. Dhotre and H. Olesen, "A Survey of Privacy Awareness and Current Online Practices of Indian Users," *WWRF34, St. Clara, California, US*, Apr. 2015.
25. X. Liang, K. Zhang, X. Shen, and X. Lin, "Security and privacy in mobile social networks: challenges and solutions," *IEEE Wirel. Commun.*, vol. 21, no. 1, pp. 33–41, Feb. 2014.
26. J. R. Reidenberg, T. Breaux, L. F. Cranor, B. French, A. Grannis, J. T. Graves, F. Liu, A. M. McDonald, T. B. Norton, R. Ramanath, N. C. Russell, N. Sadeh, and F. Schaub, "Disagreeable Privacy Policies: Mismatches between Meaning and Users' Understanding," Aug. 2014.

Evolution Metrics for a BPEL Process

N. Parimala and Rachna Kohar

Abstract A Business Process Execution Language (BPEL) process in Service-Oriented Architecture (SOA) evolves over time. The study of evolution helps in analyzing the development and enhances maintenance of a BPEL process. In this paper, we study the evolution using metrics which measure the changes and the quantity in which the changes have occurred. A process could evolve on its own or because of the evolution in the partner services. In both the cases, changes could be to the internal logic of the process or involve changes in the interaction with the partner services. The evolution metrics proposed in this paper are—BPEL Internal Evolution Metric (BEM_I) and BPEL External Evolution Metric (BEM_E). The time complexity for computing the metrics is linear. The metrics are theoretically validated using Zuse framework and are found to be above the ordinal scale. In our previous work, metrics were proposed for a single service under evolution. The cohesiveness, between the changes of an evolving service and an evolving BPEL process which uses this service, is demonstrated using metrics.

Keywords BPEL process · Service · Evolution · Metrics

Electronic supplementary material The online version of this chapter (doi:10.1007/978-981-10-7245-1_31) contains supplementary material, which is available to authorized users.

N. Parimala (✉) · R. Kohar
School of Computer and Systems Sciences, Jawaharlal Nehru University, New Delhi, India
e-mail: dr.parimala.n@gmail.com

R. Kohar
e-mail: rachnajnu20@gmail.com

© Springer Nature Singapore Pte Ltd. 2018
S. Bhalla et al. (eds.), *Intelligent Computing and Information and Communication*,
Advances in Intelligent Systems and Computing 673,
https://doi.org/10.1007/978-981-10-7245-1_31

1   Introduction

In SOA, different services are composed together to realize a business process. It is used to achieve specific consumer business requirements. BPEL is considered as a standard language to represent these processes. In the current fast growing business market, a BPEL process evolves quite frequently. This evolution is driven by different factors. One of the factors is the changing demands of today's market. Other factor is improving the process in a way so that its quality or performance may be increased.

Evolution in software has been studied to understand the nature and amount of changes in its source code [1, 2]. Similarly, in SOA, the evolution of a BPEL process can be studied to analyze the nature of evolution. In this paper, we propose to study the evolution along two axes. These are "what" and "by how much" in its code. Here, "what" refers to nature of changes made in BPEL process and "by how much" refers to the amount of these changes. The answers to these questions help to have an idea of changes in the offered functionalities of BPEL process under evolution.

The changes that can occur in a BPEL process could be addition, deletion, modification, split and merge [3–6] in its different activities. These changes may be concerning the interactions with external services or with the internal logic. Therefore, there are two categories of changes in a BPEL process—Internal and External changes.

As an example, consider a Travel Booking BPEL process, TB version1 which coordinates with the consumer, Employee and the Airline web service and provides the functionality of booking the flight for an Employee. Activities in this version are 'receive TB request', 'invoke Employee service' (to retrieve Employee Travel details), 'receive Employee service' (to receive travel status), 'invoke Airline service (for travel booking), 'receive Airline service' (receive booking details), 'invoke consumer' (to reply travel booking confirmation)'. Let us say that the Airline service has added a Privilege functionality which provides the Employee an opportunity to book a Hotel, rent a Car or subscribe for a Magazine along with the discount given by the Airline. The BPEL Process accommodates the newly offered functionality by the Airline service for the Employee. Therefore, in TB version2, 'invoke Airline service' (for availing Privileges) activity and 'invoke consumer' (for returning Privilege confirmation details) activity are added. These changes involve interactions with the external partner services. Let us say that a new version TB version3 is made in which a wait activity is added in the process to wait for some duration to perform functionalities of TB version2. This is an internal change in the process. Clearly, there are two types of BPEL evolution, one is external in nature and other is internal in nature.

The changes need to be measured to give us a measure of evolution of BPEL process. Metrics provide a quantitative measure of quality of software [7, 8]. They are computed by considering its successive releases [9]. The evolution of a BPEL process is measured by taking into account its different versions. When a BPEL process evolves, the evolution can be measured form the consumer as well as the producer point of view. In [10], we have proposed metrics from the consumer point of view. In this paper, the metrics show the evolution for the producer of the BPEL process. As brought out above, two categories of changes—Internal and External changes are proposed. Metrics for these categories, namely, BPEL Internal Evolution Metric (BEM_I) and BPEL External Evolution Metric (BEM_E) are proposed.

Changes in a service may demand changes in a BPEL process. We measure changes using metrics. The cohesiveness of these changes is shown using metrics proposed for a service [11] and those proposed in this paper for a BPEL process.

The paper is organized as follows. In Sect. 2, related work is discussed. BPEL process is described in Sect. 3. Evolution Metrics are defined in Sect. 4. In Sect. 5, the formal validation of metrics using Zuse [12] framework is given. In Sect. 6, metrics computation algorithm is presented. In Sect. 7, the cohesiveness of changes between a BPEL process and its partner service is demonstrated. Finally, the paper is concluded in Sect. 8.

2 Related Work

Metrics are a standard of measurement of a process or a product [7, 8]. In SOA, metrics have been proposed for a composite service to measure its performance, reusability, granularity, coupling etc. [13–18].

In [13], a set of metrics is proposed for the composite service providers taking into account quality features such as availability in terms of time, reliability in terms of process requests and performance to measure throughput, discoverability etc. Metrics are presented in [14] for a composite service to measure its resource quality in terms of performance behavior, utilization of network and management of its versions. The metrics proposed in [15] measures its performance and evaluate its efficiency and effectiveness based on runtime data. Metrics proposed in [16] considers Service level agreement between consumer and provider of the composite service to measure how much performance, reliability and availability are actually met. Four metrics are defined in [17] to quantitatively measure the granularity appropriateness of a BPEL process considering attributes such as business value, reusability, context independency etc. The work in [18] measures decoupling considering black-box parameters of service stateness, service interface required, service interface provided invocation modes, self-containment, implicit invocation and binding modes.

Whereas the above are concerned with measuring QoS attributes, the work in this paper measures structural changes of a BPEL process when it evolves. To the best of our knowledge, metrics have not been proposed for measuring evolution of a process.

3 BPEL Process

In SOA, different services may be composed together through orchestration using BPEL. This process defines the activities and order of execution of these activities to attain a specific business objective. In this paper, BPEL 2.0 standard is used to categorize changes as well as to compute metrics.

4 Evolution Metrics

The evolution of a BPEL process is analyzed along internal and external changes. These changes are defined below.

1. Internal Changes: Changes in BPEL activities such as **If, while, assign** etc. may occur. These changes are internal to the BPEL process itself i.e. they are not concerned with the interaction with the services. These changes are categorized as internal changes. For example, addition of a wait activity is an internal change.
2. External Changes: A BPEL process uses services to accomplish the required business functionalities. It interacts with the services via **invoke, receive, reply** activities. Any change in this interaction is classified as external change. For example, addition of an **invoke** activity is an external change.

Table 1 provides detailed list of changes in the activities under both categories.

Now specification of metrics for measuring the evolution of a BPEL process is considered. The metrics are defined for both the categories of changes. When a process evolves, its new version is created. Metrics are computed for changes across different versions of a BPEL process. Now, to define the metrics, let there be two different versions of a BPEL process i.e. x and $x + 1$. Let, the table, which contains all the evolution data between these changes for an activity i stored in $T_{x,x+1}$. As an example, number of modification, deletion may occur in a flow activity. Let there be 3 wait activities added in $T_{x,x+1}$. This will make value for C_i as 3. When the context of the changes is unambiguous $T_{x,x+1}$ is not mentioned while defining the metrics.

Table 1 Category of changes

Category	Type of change: Activities in the BPEL process
External	Add/Delete/Modify/Split/Merge: invoke, receive, reply
Internal	Add/Delete/Modify/Split/Merge: rethrow, wait, if, pick, forEach, while, repeatUntil, flow, assign, throw, exit, sequence, empty

1. Internal Evolution Metric (BEM_I):

$$BEM_I = \frac{\sum\limits_{i}^{n} w_i * c_i}{n} \qquad (1)$$

where n is the count of types of changes which occur in a process under the category "Internal", C_i denotes the total number of changes for ith type of change within this category and w_i is the weight of ith type of change. The weight is computed as a proportion of the number of changes for each type of change within the category to the total number of changes in the category. For example, if in a BPEL process, two **wait** activities are added (C_1) and one **wait** activity is deleted (C_2) then count of types of changes is 2. C_1 is 2 and w_1 is 2/3; C_2 is 1 and w_2 is 1/3. So, $BEM_I = (2/3) * 2 + (1/3) * 1 = 1.67$

2. External Evolution Metric (BEM_E):

$$BEM_E = \frac{\sum\limits_{j}^{m} w_j * c_j}{m} \qquad (2)$$

where m is the count of types of changes which occur in a process under the category "External", C_j denotes the total number of changes for jth type of change within this category and w_j is the weight of jth type of change. For example, if **invoke** activities are added and modified then there are two types of changes i.e. $n = 2$. If 3 **invoke** activities are added and 2 are modified then C_1 is 3 & w_1 is 3/5 and C_2 is 2 & w_2 is 2/5. So, $BEM_E = (3/5) * 3 + (2/5) * 2 = 2.2$

5   Metrics Formal Validation

Theoretical validation of the proposed metrics is presented in this section using a software measurement i.e. Zuse framework [12]. Axiomatic approach is used in this framework to determine the scale of a metric. Measurement scale of a metric is used to analyze the metric values as well as its empirical properties. The four scales are nominal, ordinal, ratio and absolute [12]. Zuse framework is summarized in Table 2.

5.1   *BEM$_I$ Metric Formal Validation*

Let, there be a process P having n versions. Between any two process versions, evolution data is computed and stored in a file. This file is referred to as a change

Table 2 Summarized Zuse framework [12]

Modified extensive structure
Axiom1: $(A, \cdot >=)$ (weak order)
Axiom2: $A1 \text{ o } A2 \cdot > = A1$ (positivity)
Axiom3: $A1 \text{ o } (A2 \text{ o } A3) \approx (A1 \text{ o } A2) \text{ o } A3$ (weak associativity)
Axiom4: $A1 \text{ o } A2 \approx A2 \text{ o } A1$ (weak commutativity)
Axiom5: $A1 \cdot > = A2 \Rightarrow A1 \text{ o } A \cdot > = A2 \text{ o } A$ (weak monotonicity)
Axiom6: If $A3 \cdot > A4$ then for any $A1, A2$, then there exists a natural number n, such that $A1 \text{ o } nA3 \cdot > A2 \text{ o } nA4$ (Archimedean axiom)
Independence conditions
C1: $A1 \approx A2 \Rightarrow A1 \text{ o } A \approx A2 \text{ o } A$ and $A1 \approx A2 \Rightarrow A \text{ o } A1 \approx A \text{ o } A2$
C2: $A1 \approx A2 \Leftrightarrow A1 \text{ o } A \approx A2 \text{ o } A$ and $A1 \approx A2 \Leftrightarrow A \text{ o } A1 \approx A \text{ o } A2$
C3: $A1 \cdot > = A2 \Rightarrow A1 \text{ o } A \cdot > = A2 \text{ o } A$, and $A1 \cdot > = A2 \Rightarrow A \text{ o } A1 \cdot > = A \text{ o } A2$
C4: $A1 \cdot > = A2 \Leftrightarrow A1 \text{ o } A \cdot > = A2 \text{ o } A$, and $A1 \cdot > = A2 \Leftrightarrow A \text{ o } A1 \cdot > = A \text{ o } A2$
Modified relation of belief
MRB1: $\forall A, B \ \varepsilon \ \ddot{3} \colon A \cdot > = B$ or $B \cdot > = A$ (completeness)
MRB2: $\forall A, B, C \ \varepsilon \ \ddot{3} \colon A \cdot > = B$ and $B \cdot > = C \Rightarrow A \cdot > = C$ (transitivity)
MRB3: $\forall A \subseteq B \Rightarrow A = <\cdot B$ (dominance axiom)
MRB4: $\forall (A \supset B,$ $A \cap C = \varnothing) \Rightarrow (A \cdot > = B \Rightarrow A \cup C \cdot > B \cup C)$
MRB5: $\forall A \ \varepsilon \ \ddot{3} \ \acute{A} \colon A \cdot > = 0$ (positivity)

file. Let, there be two process versions i.e. x and $x + 1$. Let, their change file is denoted by $F_{x,x+1}$. Let F be the set of all change files.

Consider BEM_l. The measure BEM_l is a mapping: $BEM_l \colon F \to R$ such that the following holds for all change files $F_{x,x+1}$, $F_{x,x+1} \ \varepsilon \ F \colon F_{x,x+1} \bullet >= F_{x,x+1} \Leftrightarrow BEM_l(F_{x,x+1}) >= BEM_l(F_{x,x+1})$.

Concatenation operation is used to combine the values of two metrics. A combination rule is used to determine the behavior of metric. This behavior is required to validate the metric for the different axioms. In the proposed metrics, concatenation operation is denoted as follows for the combination rule.

$$BEM_l(F_{x,x+1} \text{ o } F_{y,y+1}) = BEM_l(F_{x,x+1} \cup F_{y,y+1})$$

where $F_{x,x+1} \cup F_{y,y+1}$ is a file which contains changes (distinct) in $F_{x,x+1}$ and $F_{y,y+1}$. This means that a common change in two different files appears once in their concatenated file.

5.2 BEM_I and the Modified Extensive Structure

Axiom 1: The binary relation • >= is known to be of weak order when it is transitive and complete. Let $F_{1,2}$, $F_{3,4}$ and $F_{5,6}$ be the three change files where $F_{1,2}$, $F_{3,4}$, $F_{5,6}$ ε F. It must be true that either $BEM_I(F_{1,2})$ >= $BEM_I(F_{3,4})$ or $BEM_I(F_{3,4})$ >= $BEM_I(F_{1,2})$. Therefore, completeness property is fulfilled. Next, is the transitivity property. If $BEM_I(F_{1,2})$ >= $BEM_I(F_{3,4})$ and $BEM_I(F_{3,4})$ >= $BEM_I(F_{5,6})$ then it is obvious that $BEM_I(F_{1,2})$ >= $BEM_I(F_{5,6})$. Thus, transitive property is also accomplished. Therefore, BEM_I fulfills Axiom 1.

Axiom 2: $BEM_I(F_{1,2}$ o $F_{3,4})$ >= $BEM_I(F_{1,2})$. It can be seen that when two files are combined then the value of the metric BEM_I is larger than the value of the metric for each of those files. This proves Axiom 2 for BEM_I.

Axiom 3: When weak associativity rule is applied to metric BEM_I, the formulation of the rule becomes, $BEM_I(T_{1,2}$ o $(T_{3,4}$ o $T_{5,6})) = BEM_I((T_{1,2}$ o $T_{3,4})$ o $T_{5,6})$. In this paper, the concatenation operation for the metric is Union operation. It is known that the union operation is associative, therefore, $BEM_I(T_{1,2} \cup (T_{3,4} \cup T_{5,6})) = BEM_I((T_{1,2} \cup T_{3,4}) \cup T_{5,6})$. Axiom 3 is satisfied.

Axiom 4: The weak commutative axiom for the metric BEM_I is stated as $BEM_I(F_{1,2}$ o $F_{3,4}) = BEM_I((F_{3,4}$ o $F_{1,2})$. It is known that the union operation is commutative. Hence, BEM_I fulfills Axiom 4.

Axiom 5: The property of weak monotonicity is stated as $BEM_I(F_{1,2})$ >= $BEM_I(F_{3,4})$ ⇒ $BEM_I(F_{1,2}$ o $F_{5,6})$ >= $BEM_I(F_{3,4}$ o $F_{5,6})$. To prove $BEM_I(F_{1,2} \cup F_{5,6})$ >= $BEM_I(F_{3,4} \cup F_{5,6})$ (given $BEM_I(F_{1,2})$ >= $BEM_I(F_{3,4})$), let the count of common changes between $F_{1,2}$ and $F_{5,6}$ be more than the count of common changes between $F_{3,4}$ and $F_{5,6}$. Since these common changes appear once after applying concatenated operation, then the resultant metric computed based on their concatenated files be $BEM_I(F_{3,4} \cup F_{5,6})$ >= $BEM_I(F_{1,2} \cup F_{5,6})$. Therefore, BEM_I does not fulfill this axiom.

Axiom 6: Idempotent property is considered here to prove this axiom. A metric is idempotent going by definition of concatenation operation i.e. $BEM_I(F_{1,2}$ o $F_{1,2}) = BEM_I(F_{1,2})$. Therefore, BEM_I does not fulfill this axiom.

It is concluded that modified extensive structure is not fulfilled by BEM_I.

5.3 BEM_I and the Independence Conditions

C1: To prove this condition, it has to be shown that $BEM_I(F_{1,2}$ o $F_{5,6}) = BEM_I(F_{3,4}$ o $F_{5,6})$ and $BEM_I(F_{5,6}$ o $F_{1,2}) = BEM_I(F_{5,6}$ o $F_{3,4})$ given $BEM_I(F_{1,2}) = BEM_I(F_{3,4})$. Now, $BEM_I(F_{1,2} \cup F_{5,6})$ may or may not be equal to

$BEM_I(F_{3,4} \cup F_{5,6})$ because the changes (which are common) in $F_{1,2} \cup F_{5,6}$ and $F_{3,4} \cup F_{5,6}$ may not be the same. This makes $BEM_I(F_{1,2} \circ F_{5,6}) \neq BEM_I(F_{3,4} \circ F_{5,6})$. Similarly, $BEM_I(F_{5,6} \circ F_{1,2}) \neq BEM_I(F_{5,6} \circ F_{3,4})$. Therefore, BEM_I does not fulfill this condition.

C2: When a metric does not accomplish C1, it will also not fulfill C2. The metric BEM_I does not fulfill C1 and therefore does not fulfill C2.

C3: When a metric does not accomplish fifth axiom of the modified extensive structure, it will also not fulfill this condition which is the case with BEM_I.

C4: A metric not satisfying the condition C3 cannot accomplish the condition C4. Hence, BEM_I does not accomplish C4.

It can be concluded that BEM_I does not fulfill the independence conditions.

5.4   BEM$_I$ and the Modified Relation of Belief

MRB1: When Axiom 1 of modified extensive structure is fulfilled by a metric, then it also satisfies MRB1. BEM_I fulfills Axiom 1 of modified extensive structure (proved above) and therefore, BEM_I satisfies MRB1.

MRB2: If Axiom 1 of modified extensive structure is satisfied by a metric then that metric satisfies MRB2. BEM_I fulfills Axiom 1 of modified extensive structure and therefore, it satisfies MRB2.

MRB3: In order to prove MRB3, let all the changes of the change file $F_{3,4}$ are included in $F_{1,2}$, then it is obvious that $BEM_I(F_{1,2}) >= BEM_I(F_{3,4})$. Therefore, this axiom is satisfied.

MRB4: In order to prove MRB4, let all the changes of the file $F_{3,4}$ are included in $F_{1,2}$ and $F_{1,2} \cap F_{5,6} = \varnothing$. Then it needs to be proved that $BEM_I(T_{3,4}) >= BEM_I(T_{1,2}) \Rightarrow BEM_I(T_{3,4} \cup T_{5,6}) >= BEM_I(T_{1,2} \cup T_{5,6})$. Due to the fact that $BEM_I(F_{3,4}) >= BEM_I(F_{1,2})$ and that there are no common changes between $F_{3,4}$ and $F_{5,6}$, the value of $BEM_I(F_{3,4} \cup F_{5,6})$ will be more than $BEM_I(F_{1,2} \cup F_{5,6})$. This proves that the metric BEM_I satisfies MRB4.

MRB5: This axiom is satisfied because changes in a process cannot be less than 0.

Therefore, modified relation of belief is fulfilled by BEM_I. Thus, BEM_I is a measure above the level of the ordinal scale.

BEM_E has also been validated using Zuse framework. Both metrics fulfill Axiom 1 to Axiom 4 but do not fulfill Axiom 4 and Axiom 5 of the modified extensive structure. They do not fulfill any of the independence conditions C1 to C5 but fulfill all the axioms MRB 1 to MRB5. Therefore, as per [19], both metrics are above ordinal scale.

6 Metrics Computation Algorithm

All the evolution data between any two versions of a process is stored in database table. It lists the activities, number of changes and change category. Whenever a process evolves, a new table is created and the evolution data is recorded. The metrics are computed using Algorithm 1.

Algorithm 1: Input: Table from the Database Output: Metric Values
Read each row from the table
Compute BEM_I and BEM_E using each row's data
Time complexity of metric's computation is $O(n)$ when number of rows are n.

7 Cohesiveness of Changes in a BPEL Process Vis-à-Vis Changes in a Service

A web service is invoked using client code. When a service undergoes changes, its corresponding client code may also undergo changes. In [11], changes in a web service have been classified into three categories as explained below.

1. Mandatory changes: The client has to include these changes in the code. E.g., corresponding invocation for the deleted operation needs to be removed from client.
2. Optional changes: The client may opt to include them in the code. An example may be addition of service operation.
3. Trivial changes: Changes which are immaterial to the client are known as trivial changes. E.g., addition of documentation which has no effect on client code. This is because while writing client code, this is not used.

Corresponding to these categories, $SCEM_M$ (Mandatory changes), $SCEM_O$ (Optional changes) and $SCEM_T$ (Trivial changes) metrics have been proposed in [11].

A BPEL Process is the consumer of web services. When a service changes, the BPEL process may have to accommodate the corresponding changes—depending upon the type of changes. The metrics proposed in this paper and in [11] are shown to be cohesive. For example, when service client code metrics reflect the mandatory changes [11] in service, then BEM_I and BEM_E metrics of a BPEL process must exhibit a value indicating that a change has occurred for the successful execution of the BPEL process. This cohesiveness is demonstrated with the help of an example.

The example of Booking BPEL Process from Oracle Technology Networks http://www.oracle.com/technetwork/articles/matjaz-bpel1-090575.html is taken. It has two partner services: Airline and Employee services. Employee service is used to give travel status of employee to the process and then based on this status Airline service returns airline booking details to process. Service and BPEL process code

Table 3 Description of the changes in the service and BPEL process

Versions	Changes in the airline service	Service version	Changes in the BPEL process	BPEL version
1	Addition: Travel update & cancel and refund functionality	Airline WSDL Version 1.wsdl	Addition of activities for travel update & cancel and refund functionality	Travel BPEL Version 1.bpel
2	Deletion: Travel update & cancel and refund functionality	Airline WSDL Version 2.wsdl	Deletion of activities for travel update functionality Deletion of activities for cancel and refund functionality	Travel BPEL Version 2.bpel
3	Addition: Client privilege functionality	Airline WSDL Version 3.wsdl	Addition of activities for client privilege functionality	Travel BPEL Version 3.bpel
4	Addition of flight schedule functionality	Airline WSDL Version 4.wsdl	No change	Travel BPEL Version 4.bpel
5	Addition of documentation	Airline WSDL Version 5.wsdl	No change	Travel BPEL Version 5.bpel

Table 4 Metrics for the airline service and the travel booking process

S. No.	Service table	Service metrics	BPEL process table	BPEL process metrics
1	$SV_{1,2}$	$SCEM_M = 2.44$ $SCEM_O = 0.00$ $SCEM_T = 0.00$	$BV_{1,2}$	$BEM_I = 3.00$ $BEM_E = 3.33$
2	$SV_{1,3}$	$SCEM_M = 0.00$ $SCEM_O = 3.31$ $SCEM_T = 0.00$	$BV_{1,3}$	$BEM_I = 1.80$ $BEM_E = 0.80$
3	$SV_{1,4}$	$SCEM_M = 0.00$ $SCEM_O = 2.44$ $SCEM_T = 0.00$	$BV_{1,4}$	$BEM_I = 0.00$ $BEM_E = 0.00$
4	$SV_{1,5}$	$SCEM_M = 0.00$ $SCEM_O = 0.00$ $SCEM_T = 1.86$	$BV_{1,5}$	$BEM_I = 0.00$ $BEM_E = 0.00$

taken from the reference cited above are modified. The modified versions are shown in Table 3.

Service version V_i ($i \geq 2$) is compared with service version V_1 and the changes are stored in the table $SV_{1,i}$ in the database. The second column of Table 4 lists these tables. Similarly, tables for BPEL process are listed in the fourth column of Table 4.

Metrics as given in [11] and the metrics proposed in this paper are shown in Table 4. Next, we analyze the metrics.

(1) Mandatory changes: $SCEM_M > 0$ and BEM_I and BEM_E have positive values for $SV_{1,2}$ and $BV_{1,2}$. Therefore, there is a clear synchronization between mandatory changes in service client code vis-à-vis changes in the process.

(2) Optional changes: Metrics for $SV_{1,3}$ and $SV_{1,4}$ show optional changes. The business process may accommodate (as in $BV_{1,3}$) or may not (as in $BV_{1,4}$). Again, clearly, the changes are synchronized.

(3) Trivial changes: The last row shows that the BPEL process is unaffected by the changes in $SV_{1,5}$.

8 Conclusion

A BPEL process may evolve over time. Evolution is analyzed in terms of what types of changes have occurred and by how much. In order to understand the type of changes, two categories of changes are proposed: Internal and External changes. Subsequently to estimate the amount of changes, metrics are defined for each of these categories. The corresponding metrics are Internal Evolution Metric (BEM_I) and External Evolution Metric (BEM_E). A high BEM_E could indicate that services used by a BPEL process are not available anymore or the services themselves are evolving. A low BEM_I could indicate that the BPEL process doesn't need any more correction and is mature. Complexity of the proposed metrics is linear. Theoretical validation of the metrics is done using Zuse framework. The proposed metrics are found to be above the ordinal scale.

A BPEL process is composed of web services. When a web service evolves, BPEL process may also evolve. Proposed metrics truly reflect the cohesiveness of changes in a process vis-a-vis changes in services.

As future work, our plan is to propose evolution metrics for choreography (via WS-CDL Process) of services when they evolve.

References

1. Mockus, A., Votta, L. G.: Identifying reasons for software changes using historic databases. In: International Conference on Software Maintenance, IEEE, pp. 120–130 (2000)
2. Lehman, M. M., Ramil, J. F., Wernick, P. D., Perry, D. E., Turski, W. M.: Metrics and laws of software evolution-the nineties view, In: Fourth International Software Metrics Symposium, IEEE, pp. 20–32 (1997)
3. Fdhila, W., Rinderle-Ma, S., Reichert, M.: Change propagation in collaborative processes scenarios, In: 8th International Conference on Collaborative Computing: Networking, Applications and Worksharing (CollaborateCom), IEEE, pp. 452–461 (2012)
4. Dongsoo, K., Minsoo, K., Hoontae, K.: Dynamic business process management based on process change patterns, In: Convergence Information Technology, International Conference on, IEEE, pp. 1154–1161 (2007)

5. Fdhila, W., Baouab, A.., Dahman, K., Godart, C., Perrin, O., Charoy, F.: Change propagation in decentralized composite web services. Collaborative Computing: Networking, Applications and Worksharing, 7th International Conference on. IEEE, 508–511 (2011)
6. Slominski A.: Adapting BPEL to scientific workflows. In: Workflows for e-Science, Springer London, pp. 208–226 (2007)
7. Fenton, N., James, B.: Software metrics: a rigorous and practical approach, CRC Press, (2014)
8. Curtis, B.: Measurement and experimentation in software engineering, Proceedings of the IEEE 68, no. 9, 1144–1157 (1980)
9. Mens, T., Demeyer, S.: Future trends in software evolution metrics, In: 4th international workshop on Principles of software evolution ACM: 83–86 (2001)
10. Parimala N., Rachna Kohar: A Quality Metric for a BPEL Process under Evolution. Eleventh International Conference on Digital Information Management (ICDIM), pp. 197–202 (2016)
11. Kohar, R., Parimala N. A Metrics Framework for measuring Quality of a Web Service as it Evolves. International Journal of System Assurance Engineering and Management, Springer, doi:https://doi.org/10.1007/s13198-017-0591-y, pp. 1–15 (2017)
12. Zuse, H.: A framework of software measurement. Walter de Gruyter. (1998)
13. Choi, S. W., Her, J. S., Kim, S. D.: QoS metrics for evaluating services from the perspective of service providers. e-Business Engineering, ICEBE IEEE International Conference on. IEEE, (2007)
14. Rud, D., Schmietendorf, A., Dumke, R.: Resource metrics for service-oriented infrastructures. In: Proc. SEMSOA, pp. 90–98 (2007)
15. Wetzstein, B., Strauch, S., Leymann, F.: Measuring performance metrics of WS-BPEL service compositions, In: 5th IEEE International Conference on Networking and Services, ICNS'09, IEEE, pp. 49–56 (2009)
16. Dyachuk, D., Deters, R.: Using SLA context to ensure evolution of service for composite services. In: IEEE International Conference on Pervasive Services, IEEE, pp. 64–67 (2007)
17. Khoshkbarforoushha, A., Tabein, R., Jamshidi, P., Shams, F.: Towards a metrics suite for measuring composite service granularity level appropriateness. In: 6th World Congress on Services. IEEE, (2010)
18. Qian, K., Jigang, L., Frank, T.: Decoupling metrics for services composition. In: 5th IEEE/ACIS International Conference on Computer and Information Science and 1st IEEE/ACIS International Workshop on Component-Based Software Engineering, Software Architecture and Reuse (ICIS-COMSAR'06), IEEE, pp. 44–47 (2006)
19. Calero, C., Piattini, M., Pascual, C., Serrano, M. A.: Towards Data Warehouse Quality Metrics. In: International Workshop on Design and Management of Data Warehouses. pp. 2.1–2.10 Switzerland (2001)

Development of Performance Testing Suite Using Apache JMeter

Jidnyasa Agnihotri and Rashmi Phalnikar

Abstract Testing a product has become one of the most important tasks for any organization (Be it small scale or large scale). Without testing the product, it is not delivered to the customer. Testing is an ongoing activity from the beginning of a product's development. A performance testing suite shall be developed using Apache JMeter for the purpose of testing a product. To perform performance testing on client- and server-type softwares, a 100% pure Java application named Apache JMeter is used. Apache JMeter is not a browser, it works at protocol level. Static and dynamic resources performance testing can be done using JMeter. A high level performance testing suite will be developed in capturing aspects of performance at UI and System level. Developing the testing suite helps in saving the time and cost of the organization. The discussion follows and describes benefits of performance testing and the performance testing suite.

Keywords Silk performer · Performance testing · JMeter, etc.

1 Introduction

Testing of any application, product, web applications, etc., has become very important to ensure that the product works efficiently as per the requirement of the customer. Testing department has gained a lot of importance for this reason. Every organization and company (large scale or small scale) does have testing department.

The fast-growing world is leading to the demands of updating and implementing their software's as soon as possible to meet the requirements of the market and also

J. Agnihotri (✉) · R. Phalnikar
Department of IT, MITCOE, Pune, India
e-mail: jigyasathakurdas@gmail.com

R. Phalnikar
e-mail: rashmiphalnikar@yahoo.co.in

S. Bhalla et al. (eds.), *Intelligent Computing and Information and Communication*,
Advances in Intelligent Systems and Computing 673,
https://doi.org/10.1007/978-981-10-7245-1_32

317

to stay at lead in the race. Software should run correctly even if hundreds of people are trying to access it at one time. To stay away from bugs, software must undergo software testing. SDLC has software testing as one of its phase. Testing in itself is very vast and lengthy process. Testing ensures that the software will work properly as per the demands of the customer. Testing leads to successful completion and working of a product.

In this paper, performance testing suite shall be developed using Apache JMeter. The tool shall be developed for a particular product, describing in detail the method to build the testing suite for any product. Also, the importance of testing and the testing suite for a particular product shall be discussed in detail.

Rest of the paper is divided into the following section. Section 2 presents existing system. Section 3 Literature Survey Sect. 4 Architecture Sect. 5 Results and Analysis and Sect. 6 Conclusion.

2 Existing System

Performance testing is a nonfunctional testing and is carried out to test the system under load conditions. A user can do performance testing on a product using various tools. Many tools are available online/in market for the purpose of performance testing. A tool which is easy to handle and is available will always be preferred by any organization.

Silk performer tool [1] helps creating practical load and tests all the environments by creating huge stress. Performance testing on internet applications can be done using WebLOAD [2], a commercial tool for performance testing. A user can create practical load on websites and applications on web using LoadComplete [3]. A demo of original transactions is created for load testing by Rational Performance Tester [4]. Performance of your web application can be improvised using NeoLOAD [5]. Apache JMeter [6] is a Java platform application and can also create a functional test plan.

Silk performer [1] can create huge load and requires very less size, because silk performer is a UI-based performance testing. WebLOAD [2] is the generated load from cloud and the machine present in the environment. Virtual users are created by LoadComplete [3] by first recording the actions performed by the user. These virtual users are created from the cloud or the machines. Thousands of users can be created and handled by HP LoadRunner [7]. It is a combination of more than 2–3 different machines. So, it is complex to handle. When user interaction is involved, only then Rational Performance Tester [8] can be used. NeoLOAD [5] can be of use only for testing web applications.

Apache JMeter is a tool widely used for performance testing. Apart from testing the web applications, JMeter can also be used for testing the Servlets, JMS, LDAP, Web Services, Java classes, FTP, Pearl Scripts, and JAVA objects.

3 Literature Survey

Software testing of a product is done to make sure that the software is of a good quality. Software testing also makes sure that the software is reliable. Performance testing is performed on any application especially web application to understand how it behaves under heavy load. Cloud computing can be used in the field of testing for the purpose of overcoming the space issues [9].

Web services require testing on large scale. Extensive Human Intervention is required for calculating the performance of web service. Enhancing the components and creating new components also needs the obvious human interference. In paper [10], JMeter is used for web services testing which can be done without human intervention and dependencies are reduced so, in return, the speed of execution is increased. For measuring the reasons for the decline in the performance, in the paper [11], JMeter-based distributed test structure and point to point test structure are used to set up a computing environment. A fully automated testing service is made available recently for the ease of the testers. This service is available commercially. This service, known as Testing as a Service (TaaS), reduces the time and increases the efficiency. The paper [12] proposes a framework for performance testing. Problems generally faced by the testers are addressed quite neatly in this framework. Additionally generating and executing the test cases is completely automated. In this paper [12], JMeter is also one of the TaaS moreover it is easily available (open source).

In [11], point to point test structure is developed, it helps in leading to the conclusion that the main reason of decline of performance is memory loss. This loss happens because of Tomcat server. JMeter is a very flexible tool and a tool with large number of possible variations which helps in testing of data from various different sectors [13].

Apache JMeter is a freely available feature rich tool and easy to set up and use. Flexibility of JMeter makes it use more compatible (Fig. 1).

Apache JMeter executes its actions as shown in the flow above. Apache JMeter initially generates load on the server under test. It then collects the responses from these servers. The obtained responses are then analyzed and calculated. Report generation is the last step in the workflow which takes place after the response data is ready. Looking at this workflow and all the other factors form installing JMeter to generating the reports, it is understood that JMeter is the best tool to test the load as it is free, great load generation, and easy user interface.

4 Architecture

Performance testing or load testing is a process of evaluating the quality or capability of a product (Fig. 2).

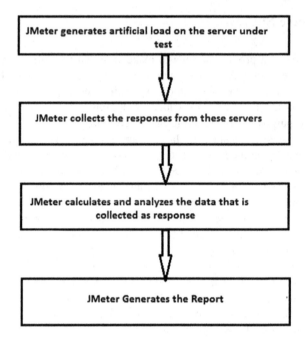

Fig. 1 Workflow for JMeter (*Source* [14])

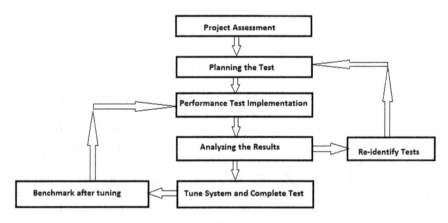

Fig. 2 Performance testing process (*Source* [15])

Figure above describes how the performance testing process is actually carried out. The tests are identified again and again to ensure that the system under test is giving appropriate output every time. After the test is complete, benchmark is noted and the same procedure is carried out again to note all the possible results. Also, no action should be left which does not undergo testing.

Doing performance testing is a tedious task if it has to be done in the traditional way. Developing a testing suite will lead to time deduction. As in, the tester will have the user interface ready, only the test should be started. So, from the point of view of an organization, it is a necessity to have a performance testing suite. The performance testing suite is developed in five modules. Each module is explained in detail ahead.

Module 1: Learning the product
Input—Setup of the product
Output—Understanding the product
Procedure—In this module, the expert of the product will give the training. Understanding the product will lead to the clear idea of what has to be done. The detailed study of the product and all the flows in it are necessary to prepare the performance testing suite. This module will lead to finalization of the product for which the performance testing suite has to be developed.

Module 2: Understanding the Scenarios for Performance Testing
Input—Scripts if available from previous test and different scenarios for Testing
Output—Various scenarios studied
Procedure—In this module, the possible scenarios will be studied. The performance suite developer will understand all the possible scenarios that are available and necessary at the moment. The scenarios which make any difference to the product's operations should be considered for script preparation. All the flows must be thoroughly studied by the developer to record the scripts correctly.

Module 3: JMeter
Input—The product setup and the JMeter Setup
Output—Scripts for the Performance Testing Suite
Procedure—In this module, the possible scenarios are taken into consideration and those scenarios are recorded in JMeter. These scripts will be the important portion of the performance testing suite. Plugins required for the testing are in the development phase. The actual operations performed for the development of scripts include the following steps.

(1) Initially, change the port number of the browser to be used.
(2) Using the JMeter GUI's test script recorder start the recording.
(3) Stop the recorder of JMeter when the necessary navigation is completed.
(4) Change the recorded script and remove all the random generated values and hard coded values.
(5) To check the results, add the appropriate listener.
(6) Now, the initial script should be checked.
(7) Enhancement of script wherever possible shall be done in this step.
(8) Start the performance test, evaluate the results, repeat these actions for the application.

The above steps are performed for every script/action of the product. The script with all the changes and modification which gives expected results is included in the suite.

Module 4: Performance Testing
Input—Performance Testing Suite and the product setup
Output—Performance Test results
Procedure—At the end of this module, we will have performance results of how the system is performing under load conditions. The performance runs are usually carried out for 1 h, 4 h, 1 day, 4 days, 10 days, etc. The performance tests are performed on the product according to the requirement. The load generation is specified accordingly.

Module 5: Verifying the Performance Results
Input—Results of the Performance tests
Output—Check the achieved accuracy
Procedure—At the end of this module, the results achieved are verified for their correctness. It is checked whether the system is behaving according to the expectations. If yes how to enhance is the next step. If there are certain problems in the system behavior, the changes to be done in the system are reported. For example, if a page should load in 5 s and if it taking 55 s, the issue should be reported and the necessary actions should be taken.

Figures 3 and 4 are the visual description of the steps that are to be carried out in the development of performance testing suite. Various steps are combined together to put up in two figures.

Figure 5 represents the graphical user interface of the performance testing suite developed. The load can be managed changing the parameters which appear in the thread group section. The graph represents the presence of threads on the particular timings. The scripts are present on the left-hand side. The performance run must be done through the non-gui mode for the run to be completed without causing any failure or problem to the system. Nineteen scripts are developed so far that are included in the suite.

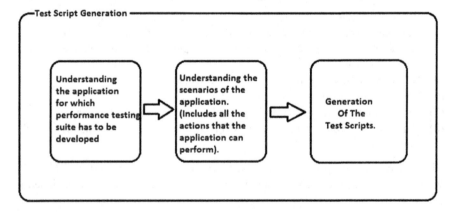

Fig. 3 Test script generation

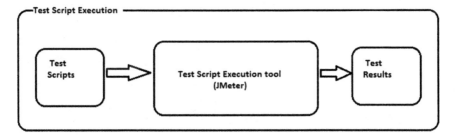

Fig. 4 Test script execution

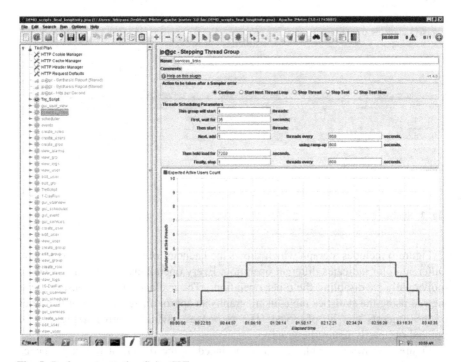

Fig. 5 Performance testing Suite GUI

5 Results and Analysis

Results obtained by the JMeter performance run are very simplified and can be viewed as a CSV file or statistics (in table format) or even as separate graphs for every operation performed. JMeter provides results by writing the results to the CSV file and to the folder which are provided in the command line of the non-gui mode.

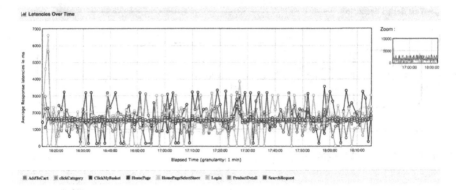

Fig. 6 Latencies over time

Label	#Samples	KO	Error %	90th pct	95th pct	99th pct	Throughput	KB/sec	Min	Max
Total	1126806	1355	0.12%	3503.00	6235.00	36167.93	152.85	12318.23	0	85224
AddToCart	1041	2	0.19%	3066.60	3224.30	3940.92	0.15	0.29	15	35005
clickCategory	104490	54	0.05%	3659.90	3835.00	4758.00	15.12	628.06	114	40741
ClickMyBasket	365	2	0.55%	3487.40	3571.40	3952.20	0.05	1.90	202	4729
HomePage	17619	1	0.01%	3114.00	3141.00	4039.60	2.45	60.74	77	35004
HomePageSelectStore	2100	2	0.10%	3209.00	3412.75	11628.84	0.29	10.22	100	43985
Login	385	0	0.00%	3143.40	3183.50	3512.60	0.05	1.96	59	4232
ProductDetail	203510	110	0.05%	3262.00	3348.00	3770.99	28.56	1184.20	64	39583
SearchRequest	109545	17	0.02%	3260.00	3352.00	3850.00	15.31	505.75	78	38776

Fig. 7 Statistics

Figure 6 includes a graph which shows the latencies of the operations over time. Different color indicates different operation. Every operations graph can be viewed individually by disabling the other operations. The average time responses can be viewed using the statistics table, using graphs an approximate estimate of the times of the requests can be observed. The responses having incorrect reply can be pointed out easily using the graphs.

Figure 7 shows the table which is produced as the output of the performance testing. The statistics are very precise, accurate, and give detailed information about the error, minimum time, maximum time, average time, throughput, etc. The graphs and the statistics tables are equally important to monitor the system.

6 Conclusion

The performance testing is an important task to be done for every product or application. Having convenient means for this testing provides various benefits to the organization. Time and cost savings are the two most important benefits that can be noted for having performance testing suite. Here, the tool used is also free so the cost reduction is to a high level.

In this paper, we discuss the performance testing suite developed for SMGR (System Manager). For the testing of this product, initially silk performer was used which has high cost. The suite developed, gives accurate results when compared with the results of the silk performer. It is noticeable that the results are accurate up to 90%. The reports generated are also very detailed, precise, and give information about every single action performed during the run. Apache JMeter performance testing suite has a GUI which is more user friendly when compared to silk performer. The suite developed in here is specific to SMGR similarly it can be developed for any application (static or dynamic).

Acknowledgements The work of researchers and publishers is meant a lot to the author during the whole process. Author expresses regards for the same. Also the help by the professors proved to be of great use. Author would also like to thank Avaya India Pvt. Ltd. for giving this opportunity of working for Avaya. The suite developed during this project is specific to SMGR (product of Avaya India Pvt. Ltd.) as the project is sponsored by Avaya.

References

1. http://community.microfocus.com/borland/test/silk_performer__application_performance_testing/w/silk_performer_trial/23695.tutorial-1-part-1-getting-started-with-silk-performer.aspx.
2. 2016, Sharma, M., Iyer, V. S., Subramanian, S., & Shetty, A. Comparison of Load Testing Tools.
3. Vinayak Hegde, Pallavi. "Web Performance Testing: Methodologies, Tools and Challenges."
4. 2006, Vol. 45(3), pp. 463–480, Brown AW, Iyengar S, Johnston S. A Rational approach to model-driven development. IBM Systems Journal. 2006.
5. 2016, Vol. 6(5), Bhatia R, Ganpati A. In Depth Analysis of Web Performance Testing Tools. 2016 Sep.
6. http://jmeter.apache.org/usermanual/get-started.html.
7. 2013, pp. 429–434, Gao Q, Wang W, Wu G, Li X, Wei J, Zhong H. Migrating load testing to the cloud: A case study. InService Oriented System Engineering (SOSE), 2013 IEEE 7th International Symposium on 2013 Mar 25. IEEE.
8. 2003, Vol. 26, Issue 8, pp. 888–898, Apte V, Hansen T, Reeser P. Performance comparison of dynamic web platforms. Computer Communications. 2003 May 20.
9. http://www.softwaretestinghelp.com/getting-started-withcloud-testing/.
10. 2014, pp. 314–318, Shenoy S, Bakar NA, Swamy R. An adaptive framework for web services testing automation using JMeter. InService-Oriented Computing and Applications (SOCA), 2014 IEEE 7th International Conference on 2014 Nov 17. IEEE.
11. 2010, Vol. 5, pp. 282–285, Jing Y, Lan Z, Hongyuan W, Yuqiang S, Guizhen C. JMeter-based aging simulation of computing system. In Computer, Mechatronics, Control and Electronic Engineering (CMCE), 2010 International Conference on 2010 Aug 24. IEEE.
12. 2015, pp. 356–361, Ali A, Badr N. Performance testing as a service for web applications. InIntelligent Computing and Information Systems (ICICIS), 2015 IEEE Seventh International Conference on 2015 Dec 12. IEEE.

13. 2016, pp. 1–5, Harikrishna P, Amuthan A. A survey of testing as a service in cloud computing. InComputer Communication and Informatics (ICCCI), 2016 International Conference on 2016 Jan 7. IEEE.
14. http://www.testingjournals.com/jmeter-introduction/.
15. https://www.tutorialspoint.com/software_testing_dictionary/performance_testing.htm.

Characterizing Network Flows for Detecting DNS, NTP, and SNMP Anomalies

Rohini Sharma, Ajay Guleria and R. K. Singla

Abstract Network security can never be assured fully as new attacks are reported every day. Characterizing such new attacks is a challenging task. For detecting anomalies based on specific services, it is desirable to find characteristic features for those service specific anomalies. In this paper, real-time flow-based network traffic captured from a university campus is studied to find if the traditional volume-based analysis of aggregated flows and service specific aggregated flows is useful in detecting service specific anomalies or not. Two existing techniques are also evaluated to find characteristic features of these anomalies. The service specific anomalies: DNS, NTP, and SNMP are considered for study in this paper.

Keywords Network flows · DNS tunnel · DNS amplification reflection
NTP · SNMP

1 Introduction

Network security is an area which never becomes old. Everyday forth, new techniques and attacks are invented by attackers to breach the security of a network. Tackling the security issue is of utmost importance as hardware and software resources are kept on network for sharing. Intrusion detection is an important aspect of network security where intrusions from unauthorized sources are avoided, prevented, and detected. Intrusion Detection Systems (IDS) are of two types: Network IDS and Host IDS. In network IDS, network data is taken by the system and analyzed whereas in Host IDS, resources local to host like logs are taken for

R. Sharma (✉) · A. Guleria · R. K. Singla
Department of Computer Science and Applications, Panjab University, Chandigarh, India
e-mail: rohini@pu.ac.in

A. Guleria
e-mail: ag@pu.ac.in

R. K. Singla
e-mail: rksingla@pu.ac.in

© Springer Nature Singapore Pte Ltd. 2018
S. Bhalla et al. (eds.), *Intelligent Computing and Information and Communication*,
Advances in Intelligent Systems and Computing 673,
https://doi.org/10.1007/978-981-10-7245-1_33

analysis and detection. Host IDS are effective in securing host but its effect is not much as other hosts on the network are not involved. Network IDS (NIDS) is further divided into signature-based NIDS and anomaly-based NIDS. In signature-based NIDS, signatures of known attacks are stored in a database which are used to detect attacks in real time. It is a widely used NIDS but the main problem is not being able to detect zero day attacks. In anomaly-based NIDS, the normal behavior of the system is modeled and deviation from normal behavior is termed as anomalous. Anomaly-based NIDS are able to detect zero day attacks. In the current scenario, every day a new kind of attack is generated and a new service is target of an attack in a new way, therefore, detecting zero day attacks has become very important. Anomaly-based NIDS can use Deep Packet Inspection for detecting anomalies in the network which is a reliable approach as large amount of detailed data is available but at the same time it is quite time consuming. For detection in real time, deep packet inspection is a costly and time consuming solution.

Using Network Flows for detecting anomalies is less costly than deep packet inspection as these flows are based on packet headers only. A number of packets are combined into a flow if they share five attributes: SourceIP, DestinationIP, SourcePort, DestinationPort, and Protocol. It decreases the amount of data to be analyzed and helps in speeding up the process of detection as the flow-based data is less than 1% of the packet data [1]. With many advantages, flows have certain limitations also. As only header information is available in flows, many kinds of anomalies are not detectable using this. Therefore, it cannot be used as an alternative to Deep Packet Inspection [2]. However, it can be used as first step in detection process followed by Deep Packet Inspection wherever required. Anomaly Detection Mechanism can be implemented in supervised mode where labeled dataset is there to train the system or unsupervised mode where labeled dataset is not available. In the absence of labeled dataset, data can be categorized into certain classes such that anomalous data gets confined to these class(es).

2 Problem Definition

Generally for flows based detection, the volume of traffic is taken into consideration. The volume of flows, bytes, and packets per time bin is used in [3–8] for detecting anomalies. Time bin is decided according to the volume of data available. In [4], time bins of 1, 2, and 10 s are used. In [9], time bin of 1 min and in [1] time bin of 10 min are used, respectively. As the size of time bin increases, the information gets hidden in the aggregation of flows. If sum or average of flows, bytes, and packets is taken, a large number of flow records of the order of thousands and lakhs in a single time bin do not provide an accurate picture. Moreover, the problems detected in networks are service specific and can be detected if a service is monitored properly. The research goals of this paper are

- Analyzing time series data for flows, bytes, and packets for different time bin sizes in order to study the effect of size of time bin over anomalous detections.
- Breaking down the flows according to services and analyze the effectiveness of resultant time series data for detecting service specific anomalies.
- Generating time series for DNS service using different attributes in order to study the effect of attribute selection and to characterize flows for effective detection of service specific anomalies.

In this paper, focus is on amplification and tunneling anomalies. The specific services used for analysis are Domain Name System (DNS), Network Time Protocol (NTP), and Simple Network Management Protocol (SNMP) services.

3 Dataset Used

The real-time data is collected from the computer center of Panjab University, Chandigarh which is handling the networking of the whole university.

3.1 Network Architecture

The network is divided into two parts: access network and server network. Dedicated switches are there on both the networks such that the requests for servers and the replies from servers are moved through server network and the other outgoing data moves through access network. UTM is installed on the dedicated switches to check any kind of attack inward or outward. To capture flow data of access network, a machine is attached to the switch installed for outgoing traffic. The port on the dedicated switch is mirrored and the whole data is captured on the machine attached in parallel. The process of data collection is shown in Fig. 1.

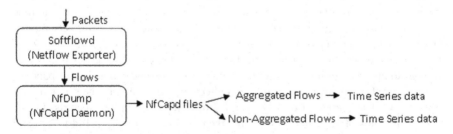

Fig. 1 Data collection architecture

3.2 Data Collection

The machine installed on the network to capture flows is made ready to capture packets on an interface using softflowd that converts the captured packets into flows. These flows are stored in binary nfcapd files by NfDump Collector. The interval for capturing flows in nfcapd files is 5 min by default which was kept unchanged. The captured flows are extracted with and without aggregation and analysis is performed on these flows according to the research goals.

4 Analysis of Flows

Flows are collected for one day, i.e., 24 h from 03-10-2016 1100 to 04-10-2016 1100 h. Non-aggregated flows amount to 453,476,798 and aggregated flows amount to 40,031,076 which is roughly 9% of non-aggregated flows. Hence, if aggregated flows are used for analysis, it will result in huge savings in terms of time and space. However, non-aggregated flows give detailed data as compared to aggregated flows for analysis. Analyzing such a large amount of flows is very much time consuming and requires high power and high speed machine. The machine used for collection and analysis uses i7-5500v CPU@2.4 GHz with 8 GB RAM and with CentOS 7 installed over it.

4.1 Research Goal 1

For generating time series based on number of bytes, packets, and flows, the flows are aggregated using—a option with NfDump that aggregates flows over five parameters SourceIP, DestinationIP, SourcePort, DestinationPort and Protocol. Flows, bytes, and packets are added while aggregating whereas an average is calculated for bytes per packet, bits per second, and packets per second. Aggregation is done over 3 h. In order to get the insight little earlier, we create time series of the aggregated flow records over time bins. There is a tradeoff between the size of the time bin, analysis time, and accuracy of the time series. As the size of time bin increases, the analysis time decreases as there will be lesser number of records to analyze but it may affect the accuracy of the time series. In this paper, it has been studied whether size of time bin affects the accuracy of time series or not. Sperotto [1] used time bin of 10 min for 2 days resulting in 288 records. In this study, time bins of 1, 5, and 10 min are taken for 24 h resulting in 1440, 288, and 144 records, respectively. The graphs generated are shown in Fig. 2. It is quite evident from the graphs that the shapes of the curves are quite similar in all the time bins. A clear pattern which is visible in the curves representing sum of packets, bytes, and flows is the day/night pattern. The volume of packets, bytes, and flows decreases gradually around 2 am till 8 am. These curves also have some spikes which show increase in volume of packets, bytes, and flows on that particular instant of time. As shown in Fig. 2a, spikes show increase in number of packets around 14:00, 17:00, 20:00, and 23:00. If we consider bits per second, packets per second, and bytes per packet, no clear pattern is visible in figures with different time bins but shapes of curves for all the three time bins are same. It has also been shown in Fig. 2d that transfer speed was quite less from 23:00 to 02:00 in the night. The spikes shown in the graphs are very less as compared to the service specific anomalies in the data which may be due to the granularity of the data which is quite low. As the spikes are not coinciding with the anomalies, volume-based analysis cannot detect service specific anomalies. Hence following conclusions can be drawn.

- Size of the time bin does not affect the shape of the curve and hence the accuracy of the time series. For different time bins, same spikes have been shown. Hence, larger time bins can be used to decrease response time. For localization of anomalies, the anomalous interval can be detailed out further.
- Volume-based analysis of flows, packets, bytes, bits per second, packets per second, and packet size cannot detect service specific anomalies.
- Aggregating flows over 3 h results in lesser number of records but do not give clear indications of anomalies. Aggregating over lesser time comparatively may give better results.

4.2   Research Goal 2

For analyzing the data services wise, three services are selected: Domain Name System (DNS) that works on port 53, Network Time Protocol (NTP) that works on port 123, and Simple Network Management Protocol (SNMP) that works on port 161. Most common anomalies for these services/protocols are amplification reflection attacks and data tunneling which are quite common these days. In amplification reflection attacks, an attacker, with spoofed SourceIP address, sends request to legitimate servers and the servers send replies to the target victim as shown in Fig. 3a. Servers involved in the attacks generally send large packets as replies for a small packet of request. For example, a DNS request packets is maximum 80 bytes in size whereas a DNS reply can be as large as 16 K bytes which can be fragmented due to network limitations up to 1500 bytes. As cited by United States Computer Emergency Readiness Team (US-CERT) [10], the amplification factor for DNS protocol is 28–54. If attacker sends requests with spoofed SourceIP address to number of servers and all the servers, behaving as reflectors, send amplified replies to the victim, the victim will get bombarded with data leading to Denial of Service at victim side. NTP and SNMP are other key players because the amplification factor for these two protocols is also high as given in Table 1.

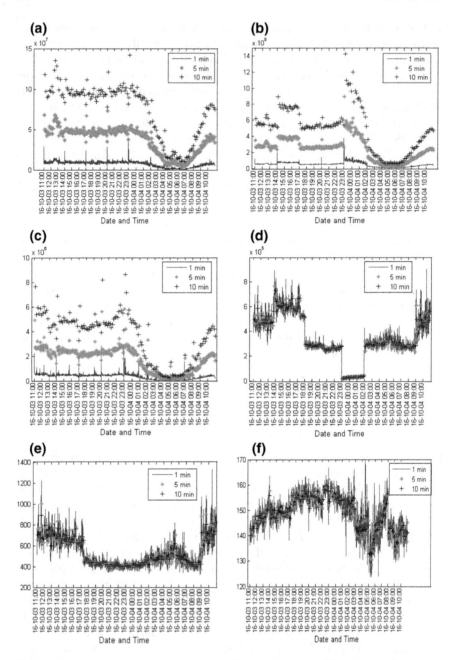

Fig. 2 **a** Sum of packets, **b** Sum of bytes, **c** Sum of flows, **d** Average of bits per second, **e** Average of packets per second and **f** Average of packet size for Time Bin of 1, 5, and 10 min

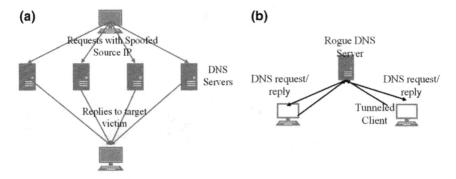

Fig. 3 **a** Reflection amplification attack, **b** DNS Tunneling attack

Table 1 Amplification factors

Protocol	Amplification factor	Protocol	Amplification factor	Protocol	Amplification factor
DNS	28–54	NTP	556.9	SNMPv2	6.3

In DNS Tunneling, data is hidden in the request and reply packets and transmitted through the network without detection. DNS traffic is made to pass through network boundaries as it is meant for name resolution and does not require any checking. Because of this feature, DNS is most exploited for tunneling attacks. Basic working of DNS Tunneling is shown in Fig. 3b.

For analyzing anomalies specific to DNS, again aggregated flows over 3 h have been taken, filtered for destination port number 53,123, and 161 and total number of packets, bytes, and flows are analyzed. The graphs generated for DNS are shown in Fig. 4 for time bin of 1 min. Time bins of 5 and 10 min are not considered as that will also produce same spikes as concluded in Sect. 4.1. As shown in Fig. 4a, b, c, spikes are shown in each figure at different times. Many other spikes are also shown in packets, bytes, and flows time series but are not much prominent. Average packets per second metric is not effective in detecting anomalies as it is displaying a very regular pattern without any spike as shown in Fig. 4e. Bytes per packet and bits per second, however, are able to detect a few anomalies which are shown in the form of spikes clearly distinct from rest of the curve in Fig. 4d, f but these spikes are just 20% of the anomalies present in the data. Hence for DNS flows, bits/s and bytes/packet based time series are also not much effective.

Figures 5 and 6 show the graphs of aggregated flows for destination port 123 and 161, respectively. In Fig. 5, the curves representing total packets, bytes, and flows show spikes indicating anomalous behavior. Bytes per packet and bits/s curves are showing very few spikes as compared to the anomalies present. It can be concluded that leaving packet/s, all the other attributes are able to show some of the anomalies using service specific volume based analysis. In Fig. 6, the anomalies are clearly visible in the curves representing total flows, bytes, and packets. Bits per

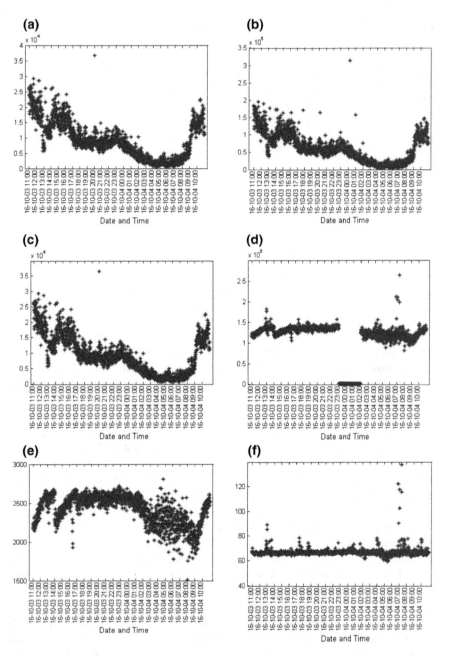

Fig. 4 a Sum of Packets, **b** Sum of bytes, **c** Sum of flows, **d** Average of bits per second, **e** Average of packets per second and **f** Average of bytes per packet for DNS aggregated flows

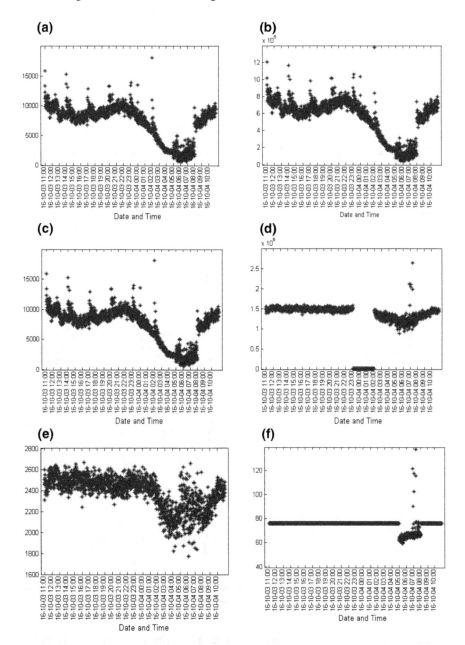

Fig. 5 **a** Sum of Packets, **b** Sum of bytes, **c** Sum of flows, **d** Average of bits per second, **e** Average of packets per second and **f** Average of bytes per packet for NTP aggregated flows

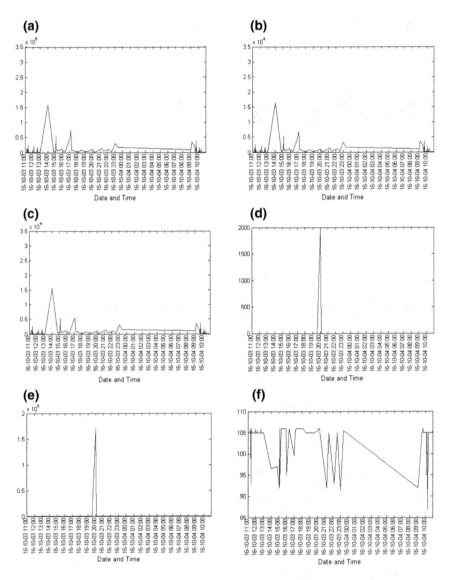

Fig. 6 **a** Sum of Packets, **b** Sum of bytes, **c** Sum of flows, **d** Average of bits per second, **e** Average of packets per second and **f** Average of bytes per packet for SNMP aggregated flows

second and packet per second show a single spike which is quite prominent but many anomalies are left untraced. Hence, volume-based analysis using packets, flows, and bytes can be useful in detecting some SNMP anomalies only.

4.3 Research Goal 3

It is found in Sects. 4.1 and 4.2 that service specific packets, bytes, flows, bits/second, and packet size time series can detect very few DNS anomalies. So it is decided to use specific techniques. In this section, two different techniques have been employed specifically for amplification of reflection and tunneling attacks. For detecting DNS amplification reflection attacks, the technique proposed by Huistra [11], shown in Algorithm 1 has been used. For the study, non-aggregated flow records have been taken which are collected in NfCapd files each of 5 min duration. After applying the technique for DNS, it has been found that on an average 13% source IPs are suspicious in each set of flow records of duration 5 min and the duration during which the suspicious IP addresses are sending request spans complete 5 min approximately. This might be the reason that these anomalies were not much visible in the curves representing total packets, bytes, and flows.

It has been found that out of the listed sourceIPs, 10% of the source IPs are not involved in anomalous activity. Hence, false positive rate is quite high in this technique. For NTP anomalies, same technique is applied with minor modification. As in case of NTP, generally every minute, a host sends requests to all NTP servers for synchronization. So the technique is modified by grouping all the flows in 5 min file by 1 min with same source IP to find out that one IP sends how many requests to NTP servers in a minute. Maximum requests sent by an IP in a minute are 108. As no means are available to find out threshold, top 10% of the sourceIPs can be identified and monitored to find out if they are being used as bots. After applying the technique over NTP flows, it has been found that some hosts are sending only 4 requests per minute while others are sending 108 requests to outside NTP Servers. Top 10% of the sourceIPs are sending approximately 70 or more NTP requests to outside NTP Servers which can be analyzed further using deep packet inspection.

Algorithm 1 Technique for detecting DNS amplification reflection attacks [11]

1. Input: Flow records.
2. Output: Suspicious IP addresses.
3. Filter Flow records with destination port = 53 and combine flow records with same source
4. IPs
5. **if** Total requests generated by an IP > 1000 **then**
6.   IP Address is suspicious.
7. **else if** Total requests generated by an IP > 100 **then**
8.   Calculate Standard Deviation of request size in non-aggregated flows where source ports are different.
9.   **if** Standard Deviation < 1 **then**
10.   IP Address is suspicious.
11.   **end if**
    end if

For detecting SNMP amplification reflection attacks, same technique has been used with threshold set to 1 as in a University network, a single request is also suspicious. If a host is sending an SNMP request outside the enterprise network and the number of requests to a particular IP is large enough, it means that host is compromised and has been directed for initiating SNMP amplification attack. It has been found while analyzing data that all the hosts are involved in sending SNMP requests to other hosts in outer enterprise networks specifically in US.

For Detecting DNS Tunneling attacks, Karasaridis et al. [12] have proposed Tunneling Attack Detector (TUNAD) for detecting DNS Tunnels in near real time. Working of TUNAD is shown in Algorithm 2. The technique has been applied over the non-aggregated flow records. Relative entropy converges to Chi-Square distribution under null hypothesis [12], so threshold was taken as the table value of Chi-Square with degree of freedom equal to number of time bins and false positive rate taken as 10^{-3} as the level of significance. However, the results are very sensitive to the threshold taken.

The graph in Fig. 7 shows the results for 6 h from 1100 to 1800 h on 03-10-2016 and 04-10-2016. It shows the presence of DNS Tunneling anomalies but the results are quite sensitive to the selection of threshold. The UTM installed in the university network also shows that anomalous requests are there in the network in each hour. Hence, the results shown are somewhat in conformance with what the existing security measures show.

Algorithm 2 Technique for detecting DNS Tunneling [12]

1. Input: Flow records
2. Output: Suspicious Flows
3. Combine all 5 min NfCapd files to form hourly flows.
4. For each request packet size, calculate request size frequency.
5. **If** Request packet size > 300 **then**
6.   Mark flows suspicious.
7. **else**

Fig. 7 Relative entropy of hourly request size histograms

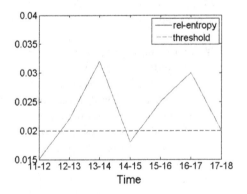

8.   Apply Cross Entropy on packet size histograms.
9.   **If** Relative Entropy > Threshold **then**
10.   Mark flow suspicious.
11.   **end if**
12. **end if**

5   Conclusion and Future Work

From the above study, it is concluded that not all types of anomalies can be detected using volume-based analysis for packets, bytes and flows. For detecting service specific anomalies like amplification reflection attacks and tunneling of data, specific attributes are required to be identified in the given network. Normal profile of the network can be generated based on the identified attributes that can be compared with real-time traffic characteristics to detect anomalies in real time. Detecting service specific anomalies with 100% accuracy and localization of the root cause are challenging tasks to accomplish.

References

1. Sperotto, A., Sadre, R., Pras, A.: Anomaly characterization in flow-based traffic time series. In: International Workshop on IP Operations and Management, pp. 15–27. Springer, Heidelberg. (2008).
2. Sperotto, A., Pras, A.: Flow-based intrusion detection. In: IFIP/IEEE International Symposium on Integrated Network Management (IM), pp. 958–963. (2011).
3. Fernandes, G., Rodrigues, J. J., Proena, M. L.: Autonomous profile-based anomaly detection system using principal component analysis and flow analysis. Applied Soft Computing, 34, 513–525. (2015).
4. Ellens, W., Żuraniewski, P., Sperotto, A., Schotanus, H., Mandjes, M., Meeuwissen, E.: Flow based detection of DNS tunnels. In: IFIP International Conference on Autonomous Infrastructure, Management and Security, pp. 124–135. Springer, Heidelberg. (2013).
5. Johnson, T., Lazos, L.: Network anomaly detection using autonomous system flow aggregates. In: IEEE Global Communications Conference (GLOBECOM), pp. 544–550. (2014).
6. Pena, E. H., Barbon, S., Rodrigues, J. J., Proena, M. L.: Anomaly detection using digital signature of network segment with adaptive ARIMA model and Paraconsistent Logic. In: IEEE Symposium on Computers and Communication (ISCC), pp. 1–6. (2014).
7. Carvalho, L. F., Rodrigues, J. J., Barbon, S., Proenca, M. L.: Using ant colony optimization metaheuristic and dynamic time warping for anomaly detection. In: International Conference on Software, Telecommunications and Computer Networks (SoftCOM), pp. 1–5. (2017).
8. Li, Y., Luo, X., Li, B.: Detecting network-wide traffic anomalies based on robust multivariate probabilistic calibration model. In: IEEE Military Communications Conference, MILCOM, pp. 1323–1328. (2015).

9. Hellemons, L., Hendriks, L., Hofstede, R., Sperotto, A., Sadre, R., Pras, A.: SSHCure: a flowbased SSH intrusion detection system. In: IFIP International Conference on Autonomous Infrastructure, Management and Security, pp. 86–97. Springer, Heidelberg. (2012).
10. US-CERT United States Computers Emergency Readiness Team, https://www.uscert.gov/ncas/alerts/TA14-017A.
11. Huistra, D.: Detecting reflection attacks in DNS flows. In: 19th Twente Student Conference on IT. (2013).
12. Karasaridis, A., Meier-Hellstern, K., Hoeflin, D.: Nis04-2: Detection of dns anomalies using flow data analysis. In: IEEE Global Telecommunications Conference, GLOBECOM'06, pp. 1–6. (2006).

Periocular Region Based Biometric Identification Using the Local Descriptors

K. Kishore Kumar and P. Trinatha Rao

Abstract Biometric systems have become a vital part of our present day automated systems. Every individual has its unique biometric features in terms of face, iris and periocular regions. Identification/recognition of a person by using these biometric features is significantly studied over the last decade to build robust systems. The periocular region has become the powerful alternative for unconstrained biometrics with better robustness and high discrimination ability. In the proposed paper, various local descriptors are used for the feature extraction of discriminative features from the regions of full face, periocular and city block distance is used as a classifier. Local descriptors used in the present work are Local Binary Patterns (LBP), Local Phase Quantization (LPQ) and Histogram of Oriented Gradients (HOG) and Weber Local Descriptor (WLD). FRGC database is used for the experimentation to compare the performance of both periocular and face biometric modalities and it showed that the periocular region has a similar level of performance of the face region using only 25% data of the complete face.

Keywords Periocular region · Local feature extraction methods
Local appearance-based approaches · Equal Error Rate · Rank-1 recognition rates

1 Introduction

The traditional biometric systems [1], such as face and iris systems reaching the state of maturity with almost having high performances of 100% accuracy. These systems work very well under controlled and ideal circumstances like high-resolution images taken from the short distance, well-illuminated conditions

K. Kishore Kumar (✉)
Department of ECE, IcfaiTech School, IFHE University, Hyderabad, India
e-mail: kkishore@ifheindia.org

P. Trinatha Rao
Department of ECE, GITAM School of Technology, GITAM University, Hyderabad, India
e-mail: trinath@gitam.in

© Springer Nature Singapore Pte Ltd. 2018   341
S. Bhalla et al. (eds.), *Intelligent Computing and Information and Communication*,
Advances in Intelligent Systems and Computing 673,
https://doi.org/10.1007/978-981-10-7245-1_34

and from a cooperative subject. The performance of these systems will degrade when they operate in nonideal conditions, i.e. when the images are captured from longer distances when the subject is on the move or from the non-cooperative subject. Now, the research is concentrated on the design of robust biometric systems under nonideal condition with better accuracies. The periocular region has become the best alternative for unconstrained biometrics with better robustness and high discrimination ability.

The periocular region is the sub-portion of the face in the vicinity of an eye with eyebrows, eyelids and eye folds as shown in Fig. 1.

Periocular region [2] is the most discriminating portion of the face, gaining significance as a useful biometric modality under the nonideal conditions, supplementing to the traditional face and iris biometric systems. Periocular region based recognition is still a comparatively new area which is cost-effective [3, 4] and does not require any extra storage space. The periocular region is captured along with the face and iris during the biometric acquisition process. For obtaining the better recognition accuracies, the periocular region can be used as in fusion with face/iris modalities. In the design of age-invariant face recognition systems [5, 6], the periocular region is less affected by the ageing process than compared to the face. Periocular biometrics uses small periocular region [7] templates which use only 25% of the full face image when compared with the large face models makes them faster in the recognition process.

Sato et al. [8] presented the first partial face recognition using the subregions of the face which showed that the eye region achieved better recognition rates compared to others regions suggesting us that the periocular region could be a major area of the face with better discriminative features. Savvides et al. [9] performed partial face recognition on the Facial Recognition Grand Challenge (FRGC) data set and the results showed the eye region achieved better recognition rates compared to others. Teo et al. [10] compared the performance of full face recognition with the eye-based partial face recognition. Park et al. [2] presented the periocular-based recognition system using the local appearance-based approaches for the feature extraction of the periocular region and the performance is compared to the face. Lyle et al. and Merkow et al. presented the use of the periocular biometric for the

Fig. 1 Periocular region

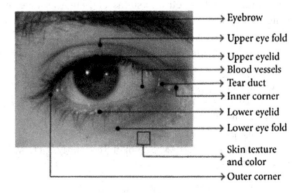

gender classification [7]. Santos and Hoyle [11] showed the fusion of iris and periocular modalities. Miller et al. explored the effects of data quality in a periocular biometric system [12].

2 Proposed Methodology

Proposed methodology for the periocular biometric system [13] is depicted in Fig. 2 which consists of the following steps: Data pre-processing, partitioning testing/training sets, feature extraction and comparison methods.

2.1 Face Recognition Grand Challenge (FRGC)

Face Recognition Grand Challenge (FRGC) database consists of large face images, and it is significantly used for extracting the periocular regions from the face images. It consists of 16,029 still frontal face images, high-quality resolution faces of size 1200 × 1400 with different sessions and variable expressions.

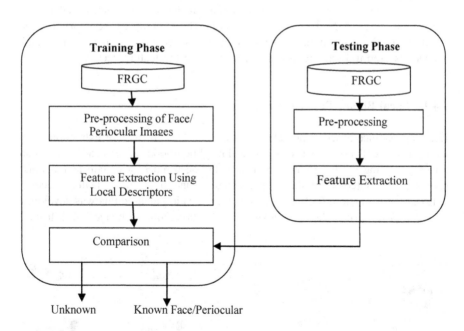

Fig. 2 Block diagram of proposed approach

2.2 Pre-processing of Face/Periocular Images

Pre-processing of the face/periocular images involves the following steps as shown in Fig. 3. (i) Converting a raw colour, facial image into the preprocessed particular images (ii) Geometric normalization (iii) Histogram equalization.

2.3 Periocular Region Extraction

Periocular region is extracted from normalized and equalized facial images, which is accomplished by placing a square bounding box around each eye, centred on the post-geometric normalization eye centre locations.

2.4 Feature Extraction

Each feature extraction technique transforms a two-dimensional image into a one-dimensional feature vector through its unique process. Feature extraction techniques employed in the periocular region are local appearance-based feature representations are LBP, HOG, LPQ and WLD. These multiple descriptors are used for deriving the discriminating features of the periocular region and city block distance is used to compute the similarity between the feature vectors.

2.4.1 Local Binary Pattern (LBP)

Local appearance-based approaches [14] are the class of feature extraction techniques which collects statistics within local neighbourhoods around each pixel of an image providing the information related to the occurrence of certain textures and patterns. The outcome of these approaches is one-dimensional feature vectors. A Local Binary Pattern (LBP) is a texture classification method that was developed by Ojala [15]. LBP collects the texture information from an image into a feature

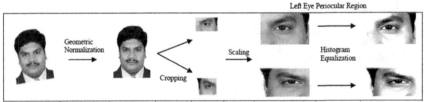

Fig. 3 Process flow for the periocular region extraction

vector by labelling pixels with a binary number by placing a threshold on the neighbourhood around each pixel. A histogram of these values forms the output feature vector. LBP is used extensively for both facial recognition [16] and periocular recognition [17].

The LBP value of pixel of concern P_k, is a function of intensity changes in the neighbourhood of M sampling points on a circle of radius r, then the LBP operator is given by

$$\text{LBP}_{M,r} = \sum_{n=0}^{M-1} s(g_n - g_c)2^n, \tag{1}$$

where

$$S(p) = 1 \quad \text{if } p > 0$$
$$= 0 \quad \text{if } p < 0$$

g_c intensity of the pixel of concern at the centre of the pixels on the circumference of a circle with values of g_n, where $n = 0, ..., M - 1$. In the proposed work, all LBP calculations are made from a circle of radius 1 pixel with 8 pixels along the circumference of the circle.

2.4.2 Local Phase Quantization

LPQ proposed by Ojansivu et al. [18] is a texture descriptor which quantizes the phase information of a discrete Fourier transform (DFT) in patch-sized neighbourhoods of an image. LPQ is robust to image blurring which has been used for face recognition [8]. Like LBP, the resulting LPQ codes are formed into a histogram.

In LPQ, the local spectra at a pixel p is calculated from a short-term Fourier transform and is given by

$$F(u,p) = \sum_{l \in p_x} f(p - l)e^{-j2\pi uTl}, \tag{2}$$

where Pu is a pixel in a $M \times M$ neighbourhood around u. At frequency points $u_1 = [a, 0]^T$, $u_2 = [0, a]^T$, $u_3 = [a, a]^T$, $u_4 = [a, -a]^T$, local Fourier coefficients are computed where a is $1/M$. The phase portion of the Fourier coefficients is defined as the sign of the real and imaginary components of $F(u, x)$ given by

$$q_j(u) = 1, \quad \text{if} \quad q_j(u) > 0$$
$$= 0, \quad \text{otherwise} \tag{3}$$

The LPQ score is the binary coding of the eight binary coefficients $q_j(u)$ and all LPQ calculations were made on a 9×9 pixel window.

2.4.3 Histogram of Oriented Gradients (HOG)

Histogram of Oriented Gradients (HOG) is an edge- and gradient-based feature descriptor [19] originally developed by Dalal and Triggs to detect humans in images. HOG is a local appearance-based approach that counts the occurrences of different gradient orientations in localized portions of a picture. Even though HOG was originally intended for object detection, it has been used for both facial and periocular recognition. HOG is a simple technique which is invariant to geometric and photometric transformation. A modified HOG algorithm is used for extracting features from the periocular region. The first step is computing the gradient of the image using a Prewitt convolution kernel. The gradient magnitude, G_{mag} and gradient angle, G_{angle}, are calculated from the image gradient, G_x in the horizontal direction and G_y in the vertical direction, as defined by

$$G_{mag} = \sqrt{G_x^2 + G_y^2} \tag{4}$$

$$G_{angle} = \text{atan2}(G_x, G_y) \tag{5}$$

The values of G_{mag} and G_{angle} at each pixel location are accumulated into a histogram.

2.4.4 Weber Local Descriptor (WLD)

Weber Local Descriptor (WLD) is a texture descriptor developed by Chen et al. The law states that the change in a signal that will be just noticeable is proportional to the magnitude of the original signal. WLD is concerned with the ratio between the intensity value of a pixel and the relative intensity differences of the pixel to its neighbours, also called the differential excitation and the gradient orientation of a pixel. The WLD feature vector is a histogram of the occurrences of each excitation and direction.

2.5 Classification Using City Block Distance Metric

City Block Distance Metric is used to determine the closeness between two feature vectors obtained from the feature extraction methods and City Block Distance Metric is given by

$$d(k, l) = \sum_{j=0}^{n} |k_i - l_i|, \tag{6}$$

where k and l are feature vectors extracted from two images that have a feature dimensionality of n.

3 Performance Measures

Following performance measures define the robustness of the designed system using face/iris modalities:

Rank 1 recognition rates illustrate the successfulness of the scheme in identifying the best match for a subject.

The rate at which the False Rejection Ratio (FRR) and False Acceptance Ratio (FAR) are equal is defined as Equal Error Rate (EER).

FRGC protocol advocates the verification rate at 0.1% false accept rate be used to compare the performance of two methods.

D shows the separability between the similarity score distributions of the set of true matches and false matches.

4 Results

In the proposed paper, four local feature extraction methods are applied to three image regions, i.e. left eye, right eye and full face images of the FRGC Experiment 1 dataset. Tables 1, 2 and 3 show the performance statistics [20] for the four local feature extraction methods in terms of Rank-1 Accuracy, Equal Error Rate (EER), VR at 0.1% FAR, and D. In most of the cases, face region performs better when compared with the periocular regions as it has more information in terms of nodal points. Particular region showed significant accuracies compared to face with only using 25% of the full face information. LPQ scheme produced the best performance results when extracted from face images (Figs. 4, 5, 6, 7 and Table 4).

Table 1 Results obtained from Local Binary Pattern

Region	Local Binary Pattern (LBP)			
	Rank-1	EER	VR @ 0.1% FAR	D
Left eye	99.7057	8.8312	64.8837	2.7330
Right eye	99.7003	8.2005	69.7148	2.8245
Face	99.9178	7.0837	69.9876	2.9423

Table 2 Results obtained from Local Phase Quantization

Region	Local Phase Quantization (LPQ)			
	Rank-1	EER	VR @ 0.1% FAR	D
Left eye	99.7682	7.1179	75.9181	2.8653
Right eye	99.7814	6.7217	76.6564	2.9434
Face	99.9424	5.4145	79.9201	2.9911

Table 3 Results obtained from Histogram of Oriented Gradients

Region	Histogram of Oriented Gradients (HOG)			
	Rank-1	EER	VR @ 0.1% FAR	D
Left eye	99.6069	8.0829	69.6951	2.8350
Right eye	99.6444	7.5245	72.2473	2.9378
Face	99.8815	6.7912	70.3877	2.9692

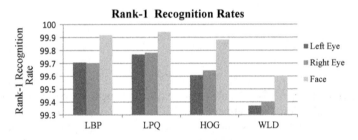

Fig. 4 Rank-1 recognition rates for the LBP, LPQ, HOG and WLD methods

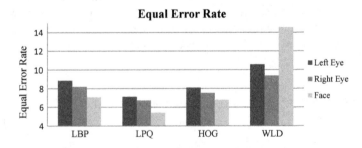

Fig. 5 Equal Error Rates (EER) for LBP, LPQ, HOG and WLD methods

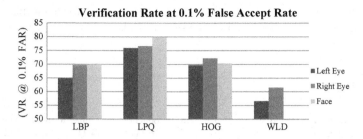

Fig. 6 Verification rate at 0.1% false accept rate (VR @ 0.1% FAR) for the LBP, LPQ, HOG and WLD methods

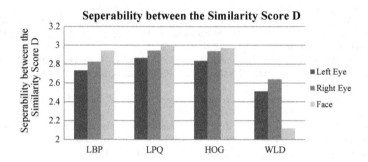

Fig. 7 D shows the separability between the similarity score for the LBP, LPQ, HOG and WLD methods

Table 4 Results obtained from Weber Local Descriptor

Region	Weber Local Descriptor (WLD)			
	Rank-1	EER	VR @ 0.1% FAR	D
Left eye	99.3699	10.5697	56.5907	2.5122
Right eye	99.4010	9.3778	61.5161	2.6390
Face	99.6007	14.5565	41.9101	2.1189

5 Conclusion

In the proposed paper, various local descriptors are used for the feature extraction of discriminative features from the regions of the full face, periocular regions and city block distance is used as a classifier. Local descriptors used in the present work are Local Binary Patterns (LBP) [16], Local Phase Quantization (LPQ) and Histogram of Oriented Gradients (HOG) and Weber Local Descriptor (WLD). FRGC database is used for the experimentation to compare the performance of both periocular and face biometric modalities and it showed that the periocular region has a similar level of performance of the face region using only 25% data of the complete face.

References

1. A. Jain, A. Ross, and S. Prabhakar. An introduction to biometric recognition. IEEE Transactions on Circuits and Systems for Video Technology, 14(1):4–20, January 2004.
2. U. Park, R. Jillela, A. Ross, and A. Jain. Periocular Biometrics in the visible spectrum. *IEEE Transactions on Information Forensics and Security,* 6(1):96–106, March 2011.
3. F. Juefei-Xu and M. Savvides. Unconstrained periocular biometric acquisition and recognition using COTS PTZ camera for uncooperative and non-cooperative subjects. In Proceedings of the IEEE Workshop on ACV, pages 20 1–208, January 2012.
4. F. Juefei-Xu, K. Luu, M. Savvides, T. Bui, and C. Suen. Investigating age invariant face recognition based on periocular biometrics. In Proceedings of the International Joint Conference on Biometrics, pages 1–7, October 2011.
5. Haibin Ling, S. Soatto, N. Ramanathan, and D.W. Jacobs. "Face Verification Across Age Progression Using Discriminative Methods", IEEE Transactions on Information Forensics and Security, 2010.
6. N. Ramanathan. "Face Verification Across Age Progression", IEEE Transactions on Image Processing, Nov 2006.
7. J. Merkow, B. Jou, and M. Savvides. An exploration of gender identification using only the periocular region. In Proceedings of the IEEE International Conference on Biometrics: Theory, Applications, and Systems, pages 1–5, September 2010.
8. K. Sato, S. Shah, and J. Aggarwal. Partial face recognition using radial basis function networks. In Proceedings of the IEEE International Conference on Automatic Face and Gesture Recognition, pages 288–293, April 1998.
9. M. Savvides, R. Abiantun, J. Heo, S. Park, C. Xie, and B. Vijayakumar. Partial holistic face recognition on FRGC-II data using support vector machine. In Proceedings of the IEEE Conference on Computer Vision & Pattern Recognition, page 48, June 2006.
10. C. Teo, H. Neo, and A. Teoh. A study on partial face recognition of eye region. In Proceedings of the International Conference on MV pages 46–49, December 2007.
11. G. Santos and E. Hoyle. A fusion approach to unconstrained iris recognition. Pattern Recognition Letters, 33(8):984–990, 2012.
12. P. Miller, J. Lyle, S. Pundlik, and D. Woodard. Performance evaluation of local appearance based periocular recognition. In Proceedings of the IEEE International Conference on Biometrics: Theory, Applications, and Systems pages 1–6, Sept 2010.
13. K. Hollingsworth, K. Bowyer, and P. Flynn. Identifying useful features for recognition in near-infrared periocular images. In Proceedings of the IEEE International Conference on Biometrics: Theory Applications and Systems, pages 1–8, September 2010.
14. A. Joshi, A. Gangwar, R. Sharma, and Z. Saquib. Periocular feature extraction based on LBP and DLDA. In Advances in Computer Science, Engineering & Applications, Volume 166 of Advances in Intelligent and Soft Computing, pages 1023–1033. Springer, 2012.
15. T. Ojala, M. Pietikainen, and T. Maenpaa. A generalised local binary pattern operator for Multiresolution gray-scale and rotation invariant texture classification. Second International Conference on Advances in Pattern Recognition, pages 397–406, 2001.
16. Kishore K Kumar and P. Trinatha Rao "Face Verification across Ages using the Discriminative Methods and See 5 Classifier". In Proceedings of First International Conference on ICTIS: Volume 2, Springer Smart Innovation, Systems and Technologies 51, Pages 439–448.

17. Mahalingam, Gayathri, and Chandra Kambhamettu. "Face verification with ageing using AdaBoost and local binary patterns", Proceedings of the Seventh Indian Conference on Computer Vision Graphics and Image Processing - ICVGIP 10 ICVGIP 10, 2010.

18. T. Ahonen, E. Rahtu, V. Ojansivu, and J. Heikkilä. Recognition of blurred faces using local phase quantization. In Proceedings of the International Conference on Pattern Recognition, pages 1–4, December 2008.

19. Ramanathan. N "Computational methods for modelling facial ageing: A survey", Journal of Visual Languages and Computing, June 2009.

20. S. Bharadwaj, H. Bhatt, M. Vatsa, and R. Singh. Periocular Biometrics: When iris recognition fails. In Proceedings of the IEEE International Conference on Biometrics: Theory, Applications, and Systems, pages 1–6, September 2010.

Model-Based Design Approach for Software Verification Using Hardware-in-Loop Simulation

Pranoti Joshi and N. B. Chopade

Abstract Increasing demand in high quality products with high safety requirements and reduced time-to-market are the challenges faced during the development of embedded products. These products irrespective of different domains (consumer, automotive, medical, aerospace) incorporate multidisciplinary (electrical, mechanical, electronic) systems as a part of hardware along with complex software that controls the hardware. Late integration of these multidisciplinary systems in the development cycle followed by the software verification and validation may lead to expensive redesigns and delayed time-to-market. Model-based design (MBD) approach can be used to overcome these challenges. Hardware-in-loop (HIL) verification is an effective method that can be used to verify the control software. Plant modeling is the crucial part for HIL verification. This paper will provide a review of one of the steps of model-based design (plant modeling) that can be used for software testing along with the impact of fidelity of model on the verification.

Keywords Model-based design (MBD) · Plant modeling · Hardware in loop (HIL) · Model fidelity · Software verification

1 Introduction

Embedded systems are nothing but control systems that include sensors to sense the parameter of interest. These inputs are taken by the control system, processing is done and the loads are driven accordingly. In modern systems (e.g., cell phones,

P. Joshi (✉) · N. B. Chopade
Pimpri Chinchwad College of Engineering, Akurdi, Pune, India
e-mail: jpranu1234@gmail.com

N. B. Chopade
e-mail: nbchopade@gmail.com

© Springer Nature Singapore Pte Ltd. 2018   353
S. Bhalla et al. (eds.), *Intelligent Computing and Information and Communication*,
Advances in Intelligent Systems and Computing 673,
https://doi.org/10.1007/978-981-10-7245-1_35

ovens, cars, etc.), the control part is taken care of by the software. As the users demand more and smarter systems with new features, the software complexity has increased to about millions of lines. This makes the testing process more challenging. Stringent time-to-market requirements with no compromise in quality add to the testing challenges. In such a scenario, manual testing is obviously not a correct way of testing. Hence, new approaches need to be used to sustain the current demands and scenarios. Socci [1] has mentioned that at a frequency of 1000 lines of code, there are 10–20 defects encountered in an embedded system.

2 Literature Survey

Testing control software in open loop may lead to time as well as cost expenses when it will be integrated with actual plant because of potential errors. Hence, in [2], an approach of closed loop modeling for verification and validation has been suggested. To have an assurance about correctness of the control software, the controller and the plant have to be taken into account that together form a closed-loop system. Plant modeling is one of the phases of model-based design.

A model-based design workflow for embedded control systems can accelerate product development, improve performance, increase reliability, and reduce engineering cost [1].

In [3], authors suggest that data-driven models could serve as a substitution for highly complex physics-based models with an insignificant loss of prediction accuracy for many applications. Thus to assess the quality of models, model fidelity comes into picture.

In [4], analysis of behavioral abstraction relation to the type of behaviors that can be validated and how different model fidelity levels can be related to each other is done. However, several studies have shown that high fidelity simulators may not be necessary to produce effective training results. The fidelity of the model representation also affects the analysis capabilities of the early design method.

Nowadays, many vendors of hardware-in-loop simulation systems are available in the market, but each system has its own pros and cons. A system on chip (SoC) is an integrated circuit (IC) that integrates all components of a computer or other electronic system into a single chip [4]. For power electronics related applications, system-on-chip solution can provide many combinations of digital, analog, or mixed signal architecture inside one die. Having reconfigurable logic along with the processor on a single die provides high processing power in less computation time (up to resolution in nanoseconds). In [5], authors have developed a Xilinx Zynq based HIL platform that can be suitable for power electronics applications. The below sections will include limitations of traditional software testing methods, overview of model-based design, hardware-in-loop simulation, and model fidelity impact on HIL.

3   Traditional Design Approach of Application Control Units (ACU)

In a traditional workflow, requirements analysis and high level design tradeoffs are limited to paper or done using expensive prototype hardware. Control software is written manually based on these framed requirements, as is the testing and verification [6]. Embedded products contain interdisciplinary subsystems whose design and development process is carried out separately using specific tool and methods. After design and development of each subsystem, integration of the subsystems is carried out followed by validation process.

Thus, the errors/faults in the system can be detected very late in the development cycle; after which the entire process needs to be repeated till validation stage which incurs more cost, design time, and efforts. The testing of the embedded system (ACU) is also dependent on the availability of the hardware components even if a prototype or actual system is used. This approach may cause damage to the system hardware affecting the cost of failure.

An embedded system includes an ACU (controller) and the system plant which includes the sensors and the loads are shown in Fig. 1. The "process" part of the control system forms the plant model that is used to replicate the behavior of actual hardware. It involves the system behavior modeling, sensor, and load modeling. In traditional method for ACU testing, actual hardware is used. The sensor outputs are read by the ACU and depending on the control signals sent to the plant, the loads are driven. Using model-based design methodology, this hardware is replaced by the models. The sensor models will sense the system parameters from the plant model which will be sent to ACU. Necessary control signals will be sent from ACU to the model which will drive the corresponding loads.

Fig. 1 Embedded System as a controller and a process

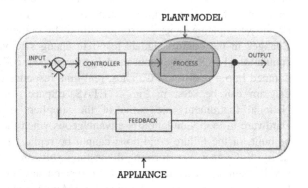

3.1 Concept of Model-Based Design (MBD)

Model-based design is a model-centric approach to the development of control, signal processing, communications, and other dynamic systems. Rather than relying on physical prototypes and textual specifications, model-based design uses a model throughout development [6]. A model includes every component relevant to system behavior: algorithms, control logic, and physical components. Model-based design has an advantage of early error detection in the software during the development cycle even before the hardware is available. Simulation helps designers spot problems that would require hardware changes. It is very important as hardware changes are more expensive than software fixes [6].

Requirements analysis and research is the first and most important phase of MBD. The purpose and application define the scope/boundaries for the plant model development as well as the controller design. Plant modeling can be used for controller design as well as verification.

Once the controller is designed, both the plant model and controller are simulated together. Code generation then takes place that can be dumped into the actual controller [7]. Verification is done after code generation to check that the algorithm still confines to the requirements. Various tools, viz., MATLAB/Simulink, LabVIEW, Dymola are used for modeling [8].

4 Hardware-in-Loop Simulation

Hardware-in-loop simulation is the technique used in the development and testing of embedded systems. An HIL simulation must also include the models of sensors and actuators. These models act as a bridge between the plant model and the control hardware under test. The value of each sensor is controlled by the plant simulation and is read by the embedded system under test. Likewise, the embedded system under test implements its control algorithms by outputting actuator control signals. Changes in the control signals result in changes to variable values in the plant simulation [1]. Once the plant model is developed, code is generated out of it and dumped into a real-time PC. Thus, real-time response can be obtained. The block diagram can be seen in Fig. 2. ETAS, dSpace, OPAL-RT technologies, and National Instruments are some of the suppliers of HIL hardware platforms. Hardware-in-loop simulation is advantageous when the hardware system is yet to be built, or the failure conditions cannot be replicated physically. It also ensures safety of the people performing the testing. Software testing can be started early in development cycle as there remains no dependency on the actual hardware greatly facilitating reduction in development time. Automated tests can be run that increase test coverage and shorten testing times by running complete test suites and overnight tests. HIL systems testing 24 h, 7 days per week are not fiction but reality [9, 10].

Fig. 2 HIL process flow

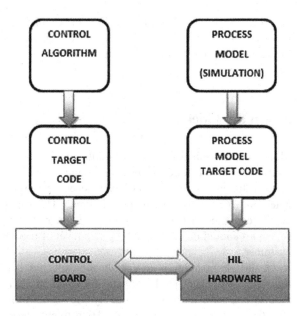

4.1 Plant Modeling

Plant model is a description of the physical hardware that is to be controlled by the embedded system. Plant model may include interdisciplinary systems. A plant model can be developed using experimental data or mathematically where different types of systems whether mechanical, thermal, and electrical are described using differential equations based on physical laws. Such kind of modeling is also known as modeling using first principles. The problem with this type of approach is that certain physical phenomena are very difficult to be described using physics [9].

The values of constants that need to be defined while deriving the equations may not be available in all cases. System identification proves fruitful in such cases. The model and controller forming the loop can be seen in Fig. 3.

Fig. 3 Plant model (includes sensors, actuators, and system behavior)

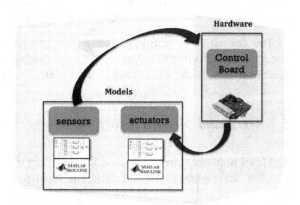

When the input–output relationship of a component or a part of hardware is known, then instead of following a detailed approach system identification or curve fitting can be used. System identification includes black box, gray box, and white box modeling.

5 Model Fidelity

Model fidelity defines the accuracy of the representation of a model when compared to the real world. Fidelity must be quantified with respect to a referent (set of metrics against which a model can be compared), i.e., metrics should exist to determine if a model resembles a referent. The metrics can include the level of detail, resolution, accuracy, precision, and error. Efficiency, cost-effectiveness, and schedules should dictate the appropriate level of modeling to meet the needs of the customer.

Quality of modeling is very application specific. It will not be a case always that the same component model is valid for different applications. In a broader sense, models can have high or low fidelity. Low fidelity models are the ones built on statistical fits on experimental data. High fidelity models are the ones constructed using physics theory.

5.1 Scaling-Based Approach to Determine Model Fidelity

Scaling method can be used to determine fidelity based on the type of method used to construct the model and the model response. Lowest level defined as: no abstraction used high fidelity. Highest level defined as: highly abstract, low fidelity. The more precise description about fidelity can be given as

Case 1: If the model is constructed using the experimental data and the response is as intended, but the model response for different set of data is deviating from its intended response then the fidelity is low.
Case 2: If the model is constructed using the physics theory and the response is as intended, and is the same for different sets of data then the fidelity is high.

The desired level of fidelity is one that provides analysis results which are complete and effective in aiding the decision-making process. Fidelity level of the model impacts on the verification of the system.

5.2 HIL and Model Fidelity

The level of fidelity directly impacts the execution time required for the model and also the development time and cost. The main purpose of HIL is to verify the control algorithm and not the hardware design. The main focus is to get the inputs from the ACU and provide corresponding feedback signals to the ACU. Hence, behavioral kind of modeling can be leveraged instead of doing detailed modeling as far as the desired response is obtained from the model.

6 Case Study: Temperature Sensor Modeling

Consider embedded software has been released that includes a feature of sensing water temperature inside a tank. To test this feature, the controller will require feedback from a temperature sensor. The data flow is described in Fig. 4. If the traditional testing method is followed, a temperature sensor, a tank with water or a prototype including these will be required to proceed with testing. Two types of testing are needed to be carried out to ensure maximum test coverage, test-to-pass, and test-to-fail. Generating failure conditions using real hardware is time consuming as well as cost adding.

Generating fault conditions in software is really easy, less time consuming and can be done using automation as well, avoiding the presence of test engineer to perform the testing.

The boundary conditions for temperature sensor can be checked such as how the software responds if the temperature lies beyond the normal operating range. Conditions such as sensor damage, accurate control of heater according to the sensor feedback can also be tested.

Figure 5 shows the heater control done by ACU based on the feedback obtained from the temperature sensor. When the desired temperature level is met, heater is

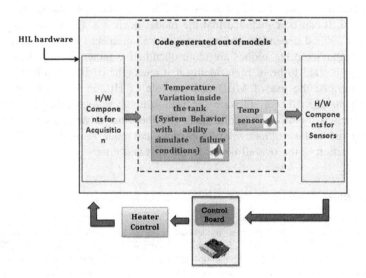

Fig. 4 HIL data flow

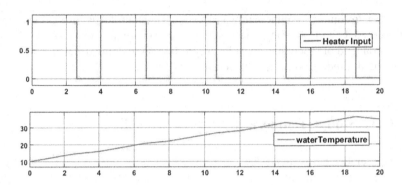

Fig. 5 Heater control using temperature sensor feedback

turned off. If the desired value is less than the required value, then the heater is turned on to maintain the water temperature.

7 Conclusion

Traditional software verification methods that have been employed in organizations are not able to cope up with the current embedded system trends. Reduced time-to-market, high quality, and safety are the major concerns while designing any product. Model-based design that was dominant in automotive domain is also being employed in other domains. In current scenario, MBD can make up to the expectations by the use of models throughout the development cycle and also as a communication between customers and vendors. Plant modeling forms the base for the product development as well as testing. The importance of model fidelity was highlighted. It can be concluded that the model fidelity is application and implementation method dependent and no specific measurement technique exists. Most of the measurement techniques are more quantitative rather than qualitative. The model can be said to be of high fidelity if it meets the desired requirements. The paper described the use of MBD and the role of HIL in software verification. Employment of MBD for software verification involves high initial cost, but can then be reused for different applications and the other advantages of early error detection, reduced use of hardware components and model reuse that it brings, leads to the reduction in the overall development cost and time.

References

1. Socci, Vince. "Implementing a model-based design and test workflow." In *Systems Engineering (ISSE), 2015 IEEE International Symposium on*, pp. 130–134. IEEE, 2015.
2. Preuße, Sebastian, Hans-Christian Lapp, and Hans-Michael Hanisch. "Closed-loop system modeling, validation, and verification." *Emerging Technologies & Factory Automation (ETFA), 2012 IEEE 17th Conference on*. IEEE, 2012.

3. Zhou, Datong, Qie Hu, and Claire J. Tomlin. "Quantitative Comparison of Data-Driven and Physics-Based Models for Commercial Building HVAC Systems."
4. Kelemenová, Tatiana, et al. "Model based design and HIL simulations." *American Journal of Mechanical Engineering* 1.7 (2013): 276–281.
5. Kaczmarek, Michał, and Przemysław Koralewicz. "Hardware in the loop simulations of industrial application using system on the chip architecture." *Signals and Electronic Systems (ICSES), 2016 International Conference on.* IEEE, 2016.
6. Kelemenová, Tatiana, et al. "Model based design and HIL simulations." *American Journal of Mechanical Engineering* 1.7 (2013): 276–281.
7. http://in.mathworks.com/solutions/model-based-design.
8. Köhl, Susanne, and Dirk Jegminat. *"How to do hardware-in-the-loop simulation right."* No. 2005-01-1657. SAE Technical Paper, 2005.
9. http://www.manmonthly.com.au/features/plant-modelling-a-first-step-to-early-verification-of-control-systems/WP_Plant_Moeling.pdf.
10. Murphy, Brett, Amory Wakefield, and Jon Friedman. *"Best practices for verification, validation, and test in model-based design."* No. 2008-01-1469. SAE Technical Paper, 2008.

Silhouette-Based Human Action Recognition by Embedding HOG and PCA Features

A. S. Jahagirdar and M. S. Nagmode

Abstract Human action recognition has become vital aspect of video analytics. This study explores methods for the classification of human actions by extracting silhouette of object and then applying feature extraction. The method proposes to integrate HOG feature and PCA feature effectively to form a feature descriptor which is used further to train KNN classifier. HOG gives local shape-oriented variations of the object while PCA gives global information about frequently moving parts of human body. Experiments conducted on Weizmann and KTH datasets show results comparable with existing methods.

Keywords Action recognition · PCA · HOG · Silhouette image

1 Introduction

In the recent days, there has been huge swell in video data as social media has become popular for education, information, and entertainment. Video surveillance has increased considerably in last decade which produces huge video data. Developing an intelligent video surveillance system in which human actions and events can be detected automatically is a challenging task taken up by researchers. This paper tries to find solution for automatic human action recognition. Human action recognition is a procedure of detecting and labeling the actions. Human action recognition has applications [1] in Intelligent Video Surveillance Systems, Human–Computer Interaction, Content-Based Video Retrieval, Robotics, Sports Video Analytics, etc.

A. S. Jahagirdar (✉)
MIT College Engineering, Pune, India
e-mail: aditi.jahagirdar@mitcoe.edu.in

M. S. Nagmode
Government College of Engineering, Avasari (Kh), Pune, India
e-mail: manoj.nagmode@gmail.com

© Springer Nature Singapore Pte Ltd. 2018
S. Bhalla et al. (eds.), *Intelligent Computing and Information and Communication*,
Advances in Intelligent Systems and Computing 673,
https://doi.org/10.1007/978-981-10-7245-1_36

The various methods proposed in literature basically defer in type of feature selected and classification algorithm used. The features can be broadly divided as shape features, texture features, motion features, or combination of these. In this paper, we have used Histogram of Oriented Gradients and principal component Analysis as features and proposed a system where these two features are concatenated to form a feature vector. Classification of test data is done using Nearest Neighbor Classifier. The algorithm is tested on two well-known datasets, viz., Weizmann and KTH and show satisfactory results.

Organization of remaining paper is as follows: In Sect. 2, review of earlier work is done. In Sect. 3, proposed framework is discussed. Sections 4 and 5 give experimental setup and results. Section 6 gives conclusion of the work.

2 Previous Work

The methods proposed for human action recognition by various authors can be broadly classified on basis of features used for representing the videos. Bag of Words and Spatio Temporal Interest Points (STIP) are most widely used local representations [2]. In contrast to this, holistic approaches extract features directly from raw videos. Local feature methods do not require steps like background subtraction and tracking which are to be used in holistic approaches. The low level features used in both local and holistic approaches are mostly appearance based or motion features.

Recent trend in action recognition shows use of fusion of shape, motion, and texture features for constructing a feature vector and using various methods like covariance matrices, Fisher vector or GMM, etc., to form a feature descriptor [3–6].

Also various machine learning methods like neural network and convolutional neural network [7, 8] are explored for classification of actions. A genetic programming method is proposed in [8, 9] for extracting and fusing spatio-temporal feature descriptor. It simultaneously extracts and fuses color and motion information in one feature.

Another important question is what should be the size of feature vector. The local features which are used by researchers have proposed dimensionality reduction techniques. Dimensionality reduction is achieved in [10, 11] using PCA and Kernel PCA. In [12], Robust PCA is applied before extracting the features. PCA is used in [13] along with HOG feature to achieve tracking and recognition simultaneously. Authors of [14, 15] propose tracking of individual body parts to form a motion curve. Basic feature vector is formed by segmentation and curve fitting of motion curve. Authors have used PCA for dimensionality reduction and classification is done in eigenspace.

Use of Histogram of Oriented Gradients (HOG) descriptor is explored by many researchers. HOG was first introduced by Dalal and Triggs in 2005 for Human Detection in images [16]. For detecting humans in a video use of Histogram of

Optical Flow (HOF), a motion descriptor along with HOG was proposed in [17]. Many of the methods proposed in literature make use of HOG, HOF features along with some modifications for detecting human action.

3 Proposed Framework

The aim of this work is to identify human actions from Weizmann dataset and KTH dataset. The proposed method is divided in four major steps as shown in Fig. 1: Foreground segmentation, Silhouette Extraction, HOG & PCA feature extraction and classification. K-NN classifier is used for classifying the actions.

3.1 Foreground Segmentation and Silhouette Extraction

In first stage, video sequence is converted to frames and foreground segmentation is done using background subtraction method to find Region of Interest. Here, adaptive background subtraction method is used where background is modeled and then progressively updated with every incoming frame.

In Eq. 1, It denotes current frame, It(x) denotes pixel value at location (x), and Lt denotes binary motion label for that frame then

$$Lt(x) = \begin{cases} 1, & \text{if } |It(x) - Bt(x)| \geq \tau \\ 0, & \text{if } |It(x) - Bt(x)| \leq \tau \end{cases} \tag{1}$$

where Bt is the background for the current frame and τ is the threshold. In this work, running Gaussian average model is used for computing background image. The background model is updated for each incoming frame using Eq. 2 [18]:

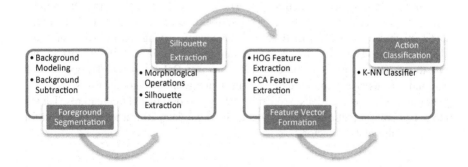

Fig. 1 Overview of proposed action recognition scheme

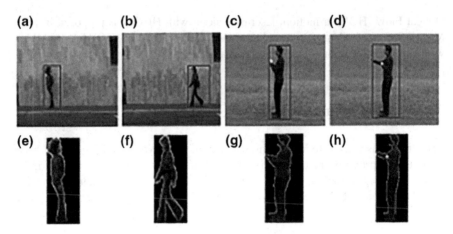

Fig. 2 Human detection and extraction from sample frames. (**a–d**) show detection of moving human in sample frames from Weizmann and KTH dataset respectively. (**e**), (**f**), (**g**), and (**h**) show object extracted from frame

$$Bt(x) = \alpha It(x) + (1 - \alpha) * Bt - 1(x). \tag{2}$$

where α is updating coefficient. Value of α is between 0 and 1. Value of α is determined empirically. The background subtraction process results in creating a foreground blob, which is the required silhouette. Morphological operations, erosion, and dilations are further applied to obtain robust silhouette of the object. Feature extraction process is performed on this extracted silhouette of the object which reduces computational complexity significantly. Figure 2 shows sample frames with foreground blobs and corresponding object extracted.

3.2 HOG Feature Vector

Histogram of Gradient Orientation (HOG) counts occurrences of gradient orientations in a localized part of image. HOG describes appearance as well as shape of the object in image in terms of distribution of intensity, giving the information about edge directions. HOG descriptor is invariant to geometric and photometric transformations. HOG was first introduced by Navneet Dalal and Bill Triggs in 2005 [16] and was used for human detection in image with very good results.

In this work, specifications defined in [19] are used to find HOG features. Taking into consideration that there is redundancy in video data, 20 frames are selected for generating the feature vector.

3.3 PCA Feature Vector

Principal component analysis is a second-order statistical measure to represent data with maximum possible variance values and lower dimensions. Principal components (PCs) give global information of extracted silhouettes [20].

Eigenvectors and eigenvalues are computed for the covariance matrix of zero mean data. Eigenvectors having higher eigenvalues are selected for representing the data. A feature vector is formed by converting diagonal of eigenvalues to row matrix.

A matrix of eigenvalues computed for each frame represents a video. Choice of number of frames used to represent one video is not defined and is a tradeoff between computational complexity and accuracy of recognition.

3.4 HOG-PCA Feature Vector

HOG feature and PCA features extracted from silhouette of object are embedded together to form HOG-PCA feature vector. HOG gives appearance information in form of shape and PCA gives motion information in form of principal component.

3.5 KNN Classifier

KNN is a simple classification algorithm which classifies data based on selected similarity measure. It is basically a lazy learning algorithm which is memory based rather than learning based. Even if it requires storing all training data and does almost all computations at the time of classification, it gives good results when number of classes is more in number.

4 Experimental Setup

In this element, we explain the data sets used for evaluating the algorithm. The algorithms are tested on two well-known publically available and widely used action dataset, viz., Weizmann and KTH.

4.1 Weizmann Dataset

Weizmann Dataset [21] used for testing the algorithm is specifically recorded for action recognition. Database consists of 90 low-resolution video sequences.

9 different have performed 10 natural actions. Some actions are periodic and others are nonperiodic. Videos are recorded in controlled environment using static camera. Only one moving object is present in frame. Background is uniform and uncluttered.

4.2 KTH Dataset

KTH Dataset [22–25] contains six natural human actions like walking, jogging, running, etc. Actions are performed by 25 different people in 4 different situations. All videos are recorded with uniform backgrounds with 25 fps frame rate. Videos are downsampled to the spatial resolution of 160 × 120 pixels. Total 600 videos are available in this dataset.

5 Experimental Results

5.1 Results with Weizmann Dataset

Figure 3 show results obtained with Weizmann dataset in form of confusion matrix. Threefold cross validation was used for evaluating the performance.

It is observed that average recognition accuracy achieved with PCA-HOG embedded feature is much higher than any one of these features used individually.

In Weizmann dataset, action of waving hand is confused with jumping jack where hands are moved along with jumping. Similarly, running action is confused with walking action because of similar characteristics of frames. Also, it is seen that recognition rate is higher for actions which are performed multiple times in a video (i.e., jack, jump) than actions performed once in full video (i.e., bend). Table 1 shows comparison of results obtained with our method.

(a)

	Bend	Jump	Jack	Run	Walk	Wave
Bend	0.9	0	0	0	0.1	0
Jump	0.05	0.85	0	0	0.1	0
Jack	0	0	1	0	0	0
Run	0	0	0	0.9	0.1	0
Walk	0	0	0	0.1	0.9	0
Wave	0	0	0.1	0	0	0.9

(b)

	Bend	Jump	Jack	Run	Walk	Wave
Bend	0.8	0.1	0	0	0.1	0
Jump	0	0.9	0.1	0	0	0
Jack	0	0	0.9	0	0	0.1
Run	0	0	0	0.8	0.2	0
Walk	0	0	0	0.1	0.9	0
Wave	0	0	0.4	0	0	0.6

(c)

	Bend	Jump	Jack	Run	Walk	Wavel
Bend	0.91	0	0	0	0.04	0.05
Jump	0.03	0.92	0	0	0.05	0
Jack	0	0	1	0	0	0
Run	0	0	0	0.94	0.06	0
Walk	0	0	0	0.05	0.95	0
Wavel	0	0	0.08	0	0	0.92

Fig. 3 Confusion Matrix on the Weizmann Dataset. **a** HOG Feature, **b** PCA Feature, **c** HOG +PCA Feature

Table 1 Comparison of results obtained with Weizmann and KTH Dataset

Method used	Recognition accuracy	
	Weizmann (%)	KTH (%)
PCA+KNN	76.66	81.66
HOG+KNN	85.83	90.83
HOG+PCA+KNN	94	91.83

(a)

	Box	Clap	Jog	Wave	Run	Walk
Box	0.85	0.1	0	0	0.05	0
Clap	0.15	0.85	0	0	0	0
Jog	0	0	0.9	0	0	0.1
Wave	0	0.1	0	0.9	0	0
Run	0	0	0.1	0	0.8	0.1
Walk	0	0	0.1	0	0.05	0.85

(b)

	Box	Clap	Jog	Wave	Run	Walk
Box	0.75	0.2	0	0	0.05	0
Clap	0.1	0.8	0.1	0	0	0
Jog	0	0	0.7	0	0.2	0.1
Wave	0.1	0.1	0	0.8	0	0
Run	0	0	0.1	0.1	0.8	0
Walk	0	0	0.2	0	0.05	0.75

(c)

	Box	Clap	Jog	Wave	Run	Walk
Box	0.94	0.06	0	0	0	0
Clap	0.05	0.95	0	0	0	0
Jog	0	0	0.9	0	0.05	0.05
Wave	0.1	0	0	0.9	0	0
Run	0	0	0.1	0	0.9	0
Walk	0	0	0.05	0	0.03	0.92

Fig. 4 Confusion Matrix on the KTH Dataset. a HOG Feature, b PCA Feature, c HOG+PCA Feature

5.2 Results with KTH Dataset

Figure 4 show results obtained with KTH dataset in the form of confusion matrix. Threefold cross validation was used for evaluating the performance.

It is observed that recognition accuracy achieved with PCA-HOG embedded feature is much higher than any one of these features used individually for KTH dataset. It is seen that action of boxing is confused with action of clapping as in many frames the human pose (shape) is similar. Also jogging action is confused with running and walking action for same reason. Table 1 shows comparison of results obtained with our method.

6 Conclusion

In this paper, we propose a method to use HOG and PCA as features for action recognition from video data. The essential thought behind using these features together was that the local object appearance and shape of an object within an image can be described using HOG and motion parameter can be represented by PCA. Since HOG operates on local cells, it is invariant to geometric and photometric transformations. As feature extraction is applied on extract silhouette image, computational cost is reduced considerably.

The classifier used here is simple NN classifier with Euclidian distance measure. It simply computes the nearest neighbor of a test feature and sets its neighbor's label as a result. Number of frames to be used for detecting any action depends on type of action. Selecting the frames from which features are to be extracted plays an important role.

In future work, nonlinear Kernel PCA (KPCA) can be explored as a feature for representing an action.

References

1. Ahmad, Mohiuddin, Irine Parvin, and Seong-Whan Lee: Silhouette History and Energy Image Information for Human Movement Recognition, Journal of Multimedia, 2010.
2. Zhen X, Shao L: Action recognition via spatio-temporal local features: A comprehensive study, Image and Vision Computing. 2016 Jun 30, 50:1–3.
3. Cheng, Shilei, et al.: Action Recognition Based on Spatio-temporal Log-Euclidean Covariance Matrix, International Journal of Signal Processing, Image Processing and Pattern Recognition 9.2 (2016): 95–106.
4. Carvajal, Johanna, et al.: Comparative Evaluation of Action Recognition Methods via Riemannian Manifolds, Fisher Vectors and GMMs: Ideal and Challenging Conditions, arXiv preprint arXiv:1602.01599 (2016).
5. Patel C, Garg S, Zaveri T, Banerjee A, Patel R: Human action recognition using fusion of features for unconstrained video sequences, Computers & Electrical Engineering, 2016 Jun 18.
6. Zhen, Xiantong et al.: Embedding motion and structure features for action recognition, IEEE transaction on Circuits and Systems for Video Technology 23.7 (2013): 1182–1190.
7. Feichtenhofer C, Pinz A, Zisserman A: Convolutional two stream network fusion for video action recognition, in IEEE conference on Computer Vision and Pattern Recognition 2016 (pp. 1933–1941).
8. Ijjina EP, Chalavadi KM: Human action recognition using genetic algorithms and convolutional neural networks, Pattern Recognition, 2016 Nov 30: 59: 199–212.
9. Liu L, Shao L, Li X: Learning spatio-temporal representations for action recognition: A genetic programming approach, IEEE Transaction on cybernetics, 2016 Jan;46(1):158–70.
10. Kuehne H, Gall J, Serre T: An end to end generative framework for video segmentation and recognition, In Applications of Computer Vision (WACV), 2016 IEEE Winter Conference on 2016 Mar 7 (pp. 1–8).
11. Oruganti VR, Goecke R: On the Dimensionality reduction of Fisher vectors for human action recognition, IET Computer Vision. 2016 Feb 26;10(5):392–7.
12. Huang S, Ye J, Wang T, Jiang L, Wu X, Li Y: Extracting refined low rank features of robust PCA for human action recognition, Arabian Journal for Science and Engineering. 2015 May 1;40(5):1427–41.
13. Lu WL, Little JJ: Simultaneous tracking and action recognition using the PCA-HOG descriptor, in Computer and Robot Vision, 2006, The 3rd Canadian Conference on 2006 Jun 7 (pp. 6–6) IEEE.
14. Chivers, Daniel Stephen: Human Action Recognition by Principal Component Analysis of Motion Curves, Diss. Wright State University, 2012.
15. Vrigkas, Michalis et al.: Matching Mixtures of curves for human action recognition, Computer Vision and Image Understanding 119(2014):27–40.

16. Dalal Nvneet, Bill Triggs: Histogram of oriented gradients for human detection, 2005 IEEE Computer Society Conference on Computer Vision and Pattern Recognition (CVPR'05). Vol. 1 IEEE, 2005.
17. Dalal Navneet, Bill Triggs and Cordelia Schmid: Human detection using oriented histograms of flow and appearance, European conference on computer vision, Springer Berlin, Heidelberg, 2006.
18. Bouwmans, Thierry: Traditional and recent approaches in background modeling for foreground detection: An overview, Computer Science Review 11 (2014): 31–66.
19. Chris McCormick: HOG Person Detector Tutorial 09 May 2013, http://mccormickml.com/2013/05/09/hog-person-detector-tutorial.
20. Kim, Tae-Seong and Zia Uddin: Silhouette-based Human Activity Recognition Using Independent Component Analysis, Linear Discriminant Analysis and Hidden Markov Model, New Developments in Biomedical Engineering. InTech, 2010.
21. Blank, Moshe, et al.: Actions as space-time shapes, Computer Vision, 2005. ICCV 2005. Tenth IEEE International Conference on. Vol. 2. IEEE, 2005.
22. Schuldt, Christian, Ivan Laptev, and Barbara Caputo: Recognizing human actions: a local SVM approach, Pattern Recognition, 2004. ICPR 2004. Proceedings of the 17th International Conference on. Vol. 3. IEEE, 2004.
23. Siddiqi, Muhammad Hameed, et al.: Video based Human activity recognition using multilevel wavelet decomposition and stepwise linear discriminant analysis, Sensors 14.4 (2014):6370–6392.
24. Junejo, Imran N., Khurrum Nazir Junejo, Zaher Al Aghbari: Silhouette-based human action recognition using SAX-Shapes, The Visual Computer 30.3 (2014):259–269.
25. Chaaraoui, Alexandros Andre, Pau Climenr_Perez, Francisco Florez-Revuelta: Silhouette-based human action recognition using sequences of key poses, Pattern Recognition Letters 34.15 (2013): 1799–1807.

Unified Algorithm for Melodic Music Similarity and Retrieval in Query by Humming

Velankar Makarand and Kulkarni Parag

Abstract Query by humming (QBH) is an active research area since a decade with limited commercial success. Challenges include partial imperfect queries from users, query representation and matching, fast, and accurate generation of results. Our work focus is on query presentation and matching algorithms to reduce the effective computational time and improve accuracy. We have proposed a unified algorithm for measuring melodic music similarity in QBH. It involves two different approaches for similarity measurement. They are novel mode normalized frequency algorithm using edit distance and n-gram precomputed inverted index method. This proposed algorithm is based on the study of melody representation in the form of note string and user query variations. Queries from four non-singers with no formal training of singing are used for initial testing. The preliminary results with 60 queries for 50 songs database are encouraging for the further research.

Keywords QBH · Music similarity · Pattern matching · Information retrieval

1 Introduction

Query by humming (QBH) is one of the most natural ways of expressing query and search a song from the musical database. Researchers in this domain of computational musicology are working on QBH since last decade or so. Content-based retrieval with the melody of the song as input with QBH is a challenge and has very limited success in commercial applications so far. The interfaces for song search used and available today are mainly based on keywords on metadata associated with the song.

V. Makarand (✉)
Cummins College of Engineering, Pune, India
e-mail: makarand.velankar@cumminscollege.in

K. Parag
Iknowlation Research Labs Pvt. Ltd, Pune, India
e-mail: paragindia@gmail.com

© Springer Nature Singapore Pte Ltd. 2018   373
S. Bhalla et al. (eds.), *Intelligent Computing and Information and Communication*,
Advances in Intelligent Systems and Computing 673,
https://doi.org/10.1007/978-981-10-7245-1_37

A typical query by humming system can be represented with block diagram as shown in Fig. 1. Users submit the query in the form of hummed query. At data cleaning stage, different signal processing techniques are applied, such as background noise removal, melody segmentation, etc. to make query ready for further processing. The query is processed to identify and represent an underlying pattern which is matched with song patterns already stored in the database to find best matches. Depending on the best possible matches results are displayed with ranking.

One of the major challenges in QBH is different users have different tonic or fundamental frequency and identifying tonic becomes an initial challenge. The query can be presented by the same user using different tonic at different instances and at a different tempo. The dynamic time warping algorithm is used for normalizing queries on timescale. Query transposes are used for same query representation considering possible tonic information. This leads to many possible representations of the same query and increases the query processing time. Since queries are likely to be run for millions of songs in the commercial applications, acceptable time complexity, and results are necessary for any commercial application.

A novel mode normalized frequency (MNF) algorithm for edit distance measure is proposed which eliminates the need of query transposes. N-gram inverted index method uses precomputed n-grams of songs. Inverted index of each n-gram includes a list of songs having these patterns. It is useful for the efficient matching of the identified n-grams from the QBH. The performed experiment uses the raw query data and presented results for different algorithms and unified algorithm. The novel unified algorithm narrow downs the possible results using n-gram approach and then apply MNF algorithm for edit distance on potential shortlisted songs.

The paper is organized in the following manner. Section 2 gives necessary musical background and considerations used in the computation. Section 3 refers to

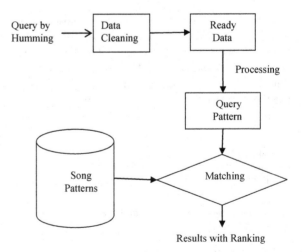

Fig. 1 Query by humming system

related work, methodologies, and different approaches used by the researchers. Section 4 detailed basic framework and approximate pattern matching algorithms used for the study. Section 5 explains computational approach using MNF algorithm. Section 6 covers details of results obtained using different algorithms. The contribution of the paper is summarized with future directions in Sect. 7.

2 Music Background and Contemplations

Musical notations are represented on the logarithmic scale. The sample octave is represented with 12 notes having frequencies F_1 to F_{12}. The equal tempered scale is one of the most widely used musical scales. General formula for the calculation of frequencies F_k from base reference frequency F_1 for an octave is

$$F_k = F_1 \cdot 2^{k/12} \quad \text{where } k = 2 \text{ to } 12.$$
$$F_{12} = F_1 \cdot 2^{12/12} = F_1 \cdot 2 = 2 F_1$$

$$(2.1)$$

The 12th note as F_{12} has a frequency double of F_1 and it marks the end of the current octave and beginning of next octave. For processing of the audio query, 100 samples are considered per second. In order to perceive any note, general minimum duration of note to be sung is about 30–40 ms. Observations of the query by untrained singers and general song patterns revealed 100 ms as the minimum duration of any note as a rough estimate., i.e., 10 consecutive samples of the same frequency are required to decode a specific note. These estimate found to be appropriate for QBH.

Notes sung during query by users are not likely to be perfect as singers. They are likely to deviate from the original note frequency. For such imperfect queries to accommodate possible errors in the query, the algorithm design has considered a range of frequency keeping actual frequency as a center to represent the specific note. It has covered entire frequency spectrum using the range such that each frequency will be associated with some or the other note. Another important observation is users can have different pitch range and it reflects in the queries with different frequencies for same song rendition by different users. The same person can also submit the query in different scales and octaves. This leads to the concept of frequency transpose considerations. As note sequence, $F_1 F_3 F_2$ will be perceived similar to all its aliases as $F_2 F_4 F_3$ or $F_9 F_{11} F_{10}$, etc. These sequences represent same relative difference as $+2$, -1 in subsequent notes as compare to previous note. Relative differences are used for precalculation of n-grams for songs which eliminates the need of any transposes according to the tonic.

3 Related Work

Research on QBH has been prevailing since last two decades with many researchers attempting to get better results. These results are influenced by two major factors as quality of query itself and the algorithms used for search and comparison. Algorithms not only need to be efficient in terms of accuracy but also needs to have better time complexity considering millions of songs to be searched for the query in the application. Queries are represented in the form of a string of notes in the majority of the work with some exceptions as additional information as rhythm or duration of notes with a pitch. Results are evaluated as top 10 hit ratios or mean reciprocal rank. Top 10 hit ratios give a percentage of queries for which intended song appears in top 10. Mean reciprocal rank (MRR) is calculated using the following formula. In case of four queries for the algorithm, the intended song appear at say position 1, 5, 2, and 10, then MRR will be $(1/1 + 1/5 + 1/2 + 1/10)/4 = 0.45$. MRR value is having a range between 0 and 1 with more value indicating better results and lower value represents poor results.

Some standard datasets are available for western music but as per our knowledge, no standard dataset is available for the queries and Hindi songs to compare algorithms and test accuracies. Tansen [1] QBH for Hindi songs tested queries using different syllables, such as "Ta", "La", "Da", and evaluated the impact of different syllable on the performance of the system. Tonal representation of melody with semitone and octave abstraction along with Q max algorithm is used to compute similarity [2]. Authors emphasized the need of query self-tuning to improve results considering the results for the subjects with good tuning the top 10 hit ratio was 64%. Existing QBH systems have slow searching speeds and lack practical applications and proposed GPU-accelerated algorithm with parallel processing to improve search efficiency by 10 times [3]. Considering wide use of mobile for QBH, the query needs to be small in duration and testing of commercial applications revealed the time required is a crucial factor [4].

Importance of tracking fundamental frequency and its impact on the performance of the system is studied and proposed further research for deeper insight into different methods [5]. Different approaches for melodic similarity for Indian art music are evaluated and it is observed that results are sensitive to distance measures used by the algorithms [6]. The results are not comparable, as in the majority of the cases datasets used are different by the different researchers.

Recently, new approaches have been applied to improve the accuracy of QBH system by researchers. In addition to the conventional method, the results are ranked according to user preferences or history to identify user taste and the songs most likely queried at top positions [7]. Use of combining lyrics information available with the user in melody achieved 51.19% error reduction for the top 10 results [8]. Segment-based melody retrieval approach can solve the problem of noise and tempo variation in the query better than global linear scaling [9]. Despite extensive research on the QBH, present systems for song search mainly relies on keyword-based search with limited use in the commercial applications. This is our main motivation for the research work in QBH domain of computational musicology. It has revealed many further challenges and opportunities for the research with various directions.

4 Basic Framework and Pattern Matching Algorithms

The humming query is processed using open-source tool PRAAT for experiments. The default settings of PRAAT are used for query data processing [10]. Framework for content-based music retrieval is proposed using human perception [11] and this work fits as a module in it. The sampling rate for pitch samples used is 100 samples per second. No data cleaning signal processing methods used for our experimentation, however, removal of noise can improve accuracies further.

Pitch listing information with samples is further converted to note string with the novel approach Mode Frequency Normalization. This algorithm is the outcome of study and experimentation with queries and song patterns. Approximate pattern matching algorithms are necessary to match the humming query patterns as exact matches are very unlikely for the non-singers queries. Observations about hummed queries revealed that some queries are time stretched, whereas some of them are compressed on the timescale. In some cases, few notes are prolonged as compared to others. It prompted us to consider compressed queries on the timeline with removing duplicate notes in succession.

For query pattern matching, edit distance and n-gram methods are used. Use of Euclidian distance and n-grams for pattern matching on compressed and non-compressed queries generated four possible alternatives as (1) Simple Euclidian distance. (2) Compressed Euclidian distance. (3) N-gram method. (4) Compressed n-gram method. For compressed queries, the original song patterns are also compressed for calculating distance measure. The cost of calculation for compressed queries is less but it eliminates the comparison of pattern with same successive notes.

Edit distance method uses Euclidian distance measure with distance as a minimum number of insertions, deletions, or replacements required to match two strings. The less the edit distance more similar the song will be for query. Dynamic programming approach is used to find edit distance. For n-gram approach, the calculations are restricted till 4-grams as it was observed unnecessary to calculate further grams as occasionally matches are found for higher grams. In case of 2-gram approach, all note patterns with two successive notes with relative distance are identified from the query and compared with song patterns to an identified number of occurrences of 2-grams. 2-gram approach is like substring matching algorithm. The more number of n-gram matches indicate more likely the pattern matching for the given query.

The inverted index structure is used with pre-calculated relative pitch 2-grams (RP2G), 3-grams (RP3G) and 4-grams (RP4G) from the song patterns. This approach eliminates the need for substring matching of the query with each song pattern. To find matching n-grams of the query, pre-calculated values of n-grams are used. For query example string with relative note distances as +1 +1 +4, the 2-grams are +1 and +4. An inverted index for string +1 will contain a list of songs containing string +1 with in sorted order of a number of occurrences of +1. For each song, 2-gram measure will be the sum of all matching 2-grams from the query. The more the value of 2-grams, the more the song is likely to similar to the query. Similarly 3-grams (+1 +1 and +1 +3 in the above example) and 4-grams (+1 +1 +4 in the above example) are also pre-calculated for song list. Results are evaluated for different pattern matching algorithms using n-grams and edit distance methods.

The unified algorithm uses four individual algorithms in combination as per the need. It advances to the next algorithm if desired results are not obtained in the previous algorithm. At first, it uses RP4G then RP3G and RP2G followed by MNF algorithm. It uses minimum cutoff value for n-grams. Present cutoff values are decided considering our observations. Cutoff is minimum grams matched of the query with the song patterns. Cutoff values considered are 2, 5, and 8, respectively, for RP4G, RP3G, and RP2G at present, however, further fine tuning is possible. The implementation of individual algorithms and unified algorithms is done using C programming language and the results are tested for queries from different users. The proposal is to submit the tested code to PRAAT open-source tool at the later stage and make it available to all.

5   Mode Normalized Frequency (MNF) Algorithm

The note patterns for about 50 Hindi Popular songs studied which are selected randomly from a different era. It was observed that few notes are repeated in the majority of the note sequence pattern. These notes are different in different song patterns, however, the trend prevails. Further introspection revealed that one note is appearing more number of times compared to others. These observations lead us to consider mode concept in statistics which is maximum occurrence symbol or value in the set. Assuming this note as a reference, the song sequences are normalized and converted them to mode normalized frequency form.

For normalized notes pattern, notes are represented using A–Z alphabetic sequence with N as middle alphabet considered as a reference. All notes are considered relative to it, as a note with distance +2 as P or −2 as L with respect to N and so on. The original song notes pattern is converted to the normalized pattern by keeping a relative distance of notes same.

MNF algorithm for query processing:

1. Read pitch listing for the Query
2. Find maximum occurring frequency in the pitch listing (N)
3. Find other notes and their frequencies using the formula 2.1
4. Find range of each note to cover entire frequency range

5. Convert pitch listing of Query to notes representation
6. Read converted QBH for notes
7. If note appears 10 times in a consecutive manner add to output sequence
8. Continue reading the sequence till end
9. Output mode normalized sequence.

MNF algorithm eliminates the need for computation for transposes in Euclidian distance measure; however, some exceptions for songs or queries do need transpose calculations considering the pattern with non-repeating note for better results. Following sample for song and query with appropriate representations gives an idea about the conversion and representations used. * followed by note represents next octave in the notations. Sa* appearing having more occurrences in compressed note pattern is replaced by N in the normalized query pattern as shown in the example below. Pattern matching is done using edit distance method in MNF algorithm.

Compressed song note sequence pattern: g Sa* ni Sa* re* ni Dh ni Sa* Dh dh Dh ni
Relative note sequence pattern: +8 −1 +1 +2 −3 −2 +2 +1 −3 −1 +1 +2
Normalized   Query   pattern   for   MNF   edit   distance   algorithm:
FNMNPMKMNKJKM

6   Analysis of Results

Our experiments include humming samples recorded using ordinary recording mike of laptop and mobile devices. High-quality recording system was not used purposely as it would not be available with common listeners to generate queries. For the humming queries, the samples are generated from people with no formal training of singing. This has helped us to test our algorithms for imperfect samples from the unprofessional users. In the experiments, 60 humming queries are used with a rendition by four amateur youngsters. The queries are supposed to retrieve one of the songs from the list of 50 sample songs as a part of the top songs of the ranked list.

Analysis of results is done for compressed queries and compressed song patterns using RP4G, RP3G, RP2G, MNF algorithm, and Unified approach. Present experiments show unified algorithm is promising as it is time efficient for the majority of queries to reduce the average computational time and overall accuracy is also improved. It was observed in the run stage of a unified algorithm that out of 60 queries, for 24 queries only 4-grams got results, for 21 queries 3-grams got results, for 8 queries 2G used and for 7 queries need to be occurred to use MNF algorithm for top 10 results.

Individual user query and overall results for top 3 hit ratios are shown in Table 1 for all algorithms. Considering the mobile query interface, top 3 results carry more importance than top 10, thus top 3 results are shown. It can be observed from the table that for individual algorithms, RP3G is better compare to rest followed by

Table 1 Top 3 hit ratios in percentage

Queries	RP4G	RP3G	RP2G	MNF	Unified
User 1	54.5	81.8	43.4	54.5	81.8
User 2	36.4	63.6	30.0	60.0	81.8
User 3	0	53.3	20.1	46.1	46.6
User 4	30.8	79.9	36.1	46.1	76.9
Overall	30.4	68.9	33.9	51.7	71.9

Table 2 MRR for different algorithms

Algorithms	RP4G	RP3G	RP2G	MNF	Unified
MRR	0.2473	0.5076	0.232	0.45	0.59

MNF algorithm. The proposed unified algorithm has the best accuracy out of all algorithms. Another observation is that results vary according to the user's query quality. It can be seen from the table that user 3 generated the poor-quality queries among all others.

The mean reciprocal ratio or MRR gives an accuracy of methods. More the MRR value, better the algorithm. Table 2 shows MRRs for different algorithms. The unified algorithm has obtained better accuracy as compared to others for the test data used. These preliminary results are encouraging for further experimentations.

7 Summary and Future Directions

The unified algorithm a novel approach proposed in this paper is promising as it reduces the cost of computation time with precomputed the inverted index for fast look up. For 75% of the queries, the results obtained using RP4G and RP3G in the unified algorithm and for the rest of the queries, RP2G and MNF algorithm were used. The initial results show that the unified algorithm is time efficient considering the average time and the results are also improved with MRR as 0.59 as compared to 0.5076 of the best individual algorithm of RP3G and 0.45 of the developed MNF algorithm.

Since users normally refer to the first page of results, if the intended song appears quite below in the list it is not likely to be found. Thus, top 3/5/10 hit ratios become more important than MRR from user's point of view, however, MRR is a better major for the accuracy of the algorithm. Further experimentation with more queries from different users will help to fine-tune the unified algorithm for better accuracy. Query and song segmentation approach needs to be evaluated to see further precision. This will help us to fine-tune the unified algorithm and build a robust system.

Acknowledgements The authors gratefully acknowledge the support by MKSSS's Cummins College of Engineering for providing experimental setup and the efforts by our UG students Aditi Pawle, Snehal Jain, Sonal Gawande, and Sonal Avhad for the active help in preparing query samples and experiments. Volunteer singer's contribution for generating queries is highly appreciable. We would like to thank Dr. Sahasrabuddhe H. V. for his valuable suggestions and inputs related to musical knowledge and experiments.

References

1. Kumar P, Joshi M, Hariharan S, Rao P: Sung note segmentation for a query-by-humming system. Intl Joint Conferences on Artificial Intelligence IJCAI (2007).
2. Salamon J, Serra J, Gómez E: Tonal representations for music retrieval: from version identification to query-by-humming. International Journal of Multimedia Information Retrieval. 2(1) pp 45–58, (2013).
3. Ruan L, Wang L, Xiao L, Zhu M, Wu Y: A Query-by-Humming System based on Marsyas Framework and GPU Acceleration Algorithms. Appl. Math. pp 261–72, Feb (2013).
4. Chandrasekhar V, Sharifi M, Ross DA: Survey and Evaluation of Audio Fingerprinting Schemes for Mobile Query-by-Example Applications. ISMIR Vol. 20, pp. 801–806 (2011).
5. Molina E, Tardón LJ, Barbancho I, Barbancho AM: The Importance of F0 Tracking in Query-by-singing-humming. In ISMIR pp. 277–282, Nov (2014).
6. Gulati S, Serra J, Serra X.: An evaluation of methodologies for melodic similarity in audio recordings of indian art music. In Acoustics, Speech and Signal Processing (ICASSP), IEEE International Conference pp. 678–682 Apr, (2015).
7. Liu NH: Effective Results Ranking for Mobile Query by Singing/Humming Using a Hybrid Recommendation Mechanism. IEEE Transactions on Multimedia. pp 1407–20, (2014).
8. Wang CC, Jang JS: Improving query-by-singing/humming by combining melody and lyric information. IEEE/ACM Transactions on Audio, Speech, and Language Processing. pp 798–806 (2015).
9. Liu, N. H: Effective Results Ranking for Mobile Query by Singing/Humming Using a Hybrid Recommendation Mechanism. IEEE Transactions on Multimedia, 1407–1420(2014).
10. Ramesh V: Exploring Data Analysis in music using tool praat. ICETET. IEEE International Conference, pp. 508–509, (2008).
11. Makarand, Velankar, and H. V. Sahasrabuddhe: Novel Approach for Music Search Using Music Contents and Human Perception, IEEE International Conference on Electronic Systems, Signal Processing and Computing Technologies ICESC, (2014).

Predict Stock Market Behavior: Role of Machine Learning Algorithms

Uma Gurav and Nandini Sidnal

Abstract The prediction of a dynamic, volatile, and unpredictable stock market has been a challenging issue for the researchers over the past few years. This paper discusses stock market-related technical indicators, mathematical models, most preferred algorithms used in data science industries, analysis of various types of machine learning algorithms, and an overall summary of solutions. This paper is an attempt to perform the analysis of various issues pertaining to dynamic stock market prediction, based on the fact that minimization of stock market investment risk is strongly correlated to minimization of forecasting errors.

Keywords Machine learning algorithms · Stock market prediction
Efficient market hypothesis · Ensemble machine learning

1 About Stock Markets

Stock exchanges are considered as a backbone of financial sectors of any country. Most of the stock brokers, as shown in Fig. 1, dealing with sellers, buyers and trading in the stock market (SM) use technical, fundamental, or time series analysis for stock market (SM) prediction purpose. These methods do not ensure correct SM prediction results, as these are based on the past stock market trends and not on stock prices [1, 2]. It is, therefore, necessary to explore some improved methods, such as ensemble machine learning (EML) algorithms of prediction [3]. Basically, ML is the ability of a program to learn, adapt, and take intuitive decisions in the face of different real-life applications which processes nonlinear, dynamic big

U. Gurav (✉)
K.I.T's College of Engineering, Kolhapur, India
e-mail: umabgurav@gmail.com

N. Sidnal
K.L.E's College of Engineering, Belgaum, India
e-mail: sidnal.nandini@gmail.com

© Springer Nature Singapore Pte Ltd. 2018
S. Bhalla et al. (eds.), *Intelligent Computing and Information and Communication*,
Advances in Intelligent Systems and Computing 673,
https://doi.org/10.1007/978-981-10-7245-1_38

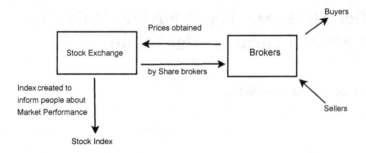

Fig. 1 How stock market works

datasets. EML works on the principal, various algorithms may overfit the data in some area, but the correct combination of their predictions will have better predictive power.

The Efficient Market Hypothesis (EMH): EMH is a popular SM hypothesis [2] that states the future stock price is completely unpredictable knowing the past trading history of the stock. In EMH, any information acquired from examining the history of the stock is immediately reflected in the price of the stock. The fundamental problem of this "hypothesis", is that any market strategy will be unable to survive large swings in the market.

The Random Walk Hypothesis: This hypothesis [2] states that stock prices do not depend on the past stock prices, so patterns/trends cannot be explicitly determined. Further, it adds, with the advancement of more powerful computing infrastructure, efficient ML algorithms and underlying pricing patterns analytics optimal solution can be found [2]. Hence, ML Techniques can seriously challenge traditional EMH like hypothesis [4].

1.1 Stock Market Prediction Methods

1. **Technical Analysis**: This method determines the stock prices based on the past performance value of the stock (using time series analysis) [2]. Technical Indicators used for market analysis are Fibonacci retracements and extensions, Moving Averages, and Momentum indicators, like RSI (Relative Strength Index), volume analysis, Mean reversion and co-integration, Bollinger bands, MACD, EMA (Exponential Moving Average), Stochastic Oscillators, etc. The main problem with these technical indicators is that these are applicable for a specific time frame. There are different models for HFT (High-frequency trading), either intra-day or long-term trading. As per the historical observations, long-term trading seems more beneficial in terms of profit making [5]. In some cases, reverse engineering strategy works well, once graph reverses, needs to take another path or follow another strategy. For the statistical models, time

series analysis, autoregression, logistic regression strategies work well [6, 7]. To solve the probability distribution problem of datasets, outliers, or extreme values are considered and mean reversion is calculated. All above-mentioned technical indicators are based on the price, which always reflects what has already happened in the market. Thus, the conclusion is that technical analysis is reactive not truly predictive of what will happen [1] and stock prices are a result of multiple variables.

2. **Fundamental Analysis**: This method deals with the past performance of the company rather than the actual stock. For more complex indicators to capture the price in a more precise manner, indicators used are Regression analysis, Autocorrelation function, GARCH Models, Regime shifting models like ANOVA, Markov Chains, or any other Stochastic model [1, 2]. Hence, the stock price is a function of multiple factors which are volatile in nature.

3. **Discounted Cash Flow analysis**: known to evaluate the intrinsic value of a stock but it is not applicable for cyclical. From Price to Sales judgments can be done only whether the stock is cheap on a certain sector or on the overall market [8, 9]. The problem with this is that experts in SM are calculating stock values manually.

4. **Sentiment Analysis (SA)**: **eToro** platform allows the user to capture the activity of traders on the social network that are successful in their trades [1] based on the SA. Silicon Market is a combination of advanced AI with ML algorithms that learns from past trades, to inform future trading decisions and utilizes adaptive and flexible ML to create an automated trading strategy [1]. Social media can predict stock up to 80% revenue correlations with Google trends. Reference [10] uses data from Google trends to predict stock market movements, using a GARCH model to replicate. StockTwits which uses Twitter mentions/searches to discover emotions of people. Twitter trends are better than Google trends because twitter news is quicker, and accessible. Gaining profit from Market Mood or Market Sentiment, works on the principle be greedy when people are fearful and be fearful when people are greedy [10]. Ergo, DeepStreetEdge are some more predictive methods.

1.2 Models Used for Stock Market Prediction

1. Six sigma (SS): SS method has its strength in the elimination of variations, fluctuations from repeatable processes. The main issues in trading are when to buy and sell shares? If these issues are addressed then the final outcomes are better. SS can help refine, enhance the performance, improve a system that predicts the direction of a stock, for example, managing a hedge fund [11].

2. Time series-based mathematical models: In this model shown in Fig. 2, the correlation between datasets can be used with the help of statistical tools. Time series models are built to study how a time series moves over time and

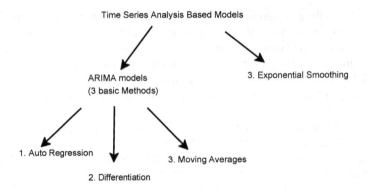

Fig. 2 Time series analysis-based models

application of volatility forecasting models to predict changes in volatility. High-frequency data is prone to all sorts of biases and there has to be correction mechanisms. Hence for the perfect dataset, even a simple linear model or logistic regressions can predict something as complex as stock market prices [7]. Two types of time series analysis are ARIMA and Exponential Smoothing [12]. ARIMA (Autoregressive integrated moving average): The concept of moving averages is incorporated in ARIMA models. Integrated moving average detects movement, speed, and acceleration. The drawback of standard ARIMA modeling is that it only considers proximal entries in the time series to forecast subsequent entries [12]. A more advanced hybrid version of ARIMA models with the combination of EML algorithms can consider cyclicality problems of stock markets. Further performance enhancements can be done by including data as external regressions which are outliers error value between actual and forecast values.

Exponential smoothing: A second popular technique in which previous time series entries contribute as a weighted moving average to forecast the next term. Mean reversion and random walk theory, earning momentum are powerful long-term forecasting tools in finance [12].

2 Methodologies: Machine Learning (ML) Algorithms

Improving on trading strategy iteratively with the help of different types of ML algorithms as shown in Fig. 3 is the solution to this problem [13–15]. This is a continuous process, capturing exact point in time snapshot of the fluctuating stock value and then using it for further iterations is the challenge [16].

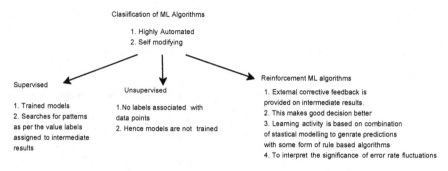

Fig. 3 Classification and types of ML algorithms

2.1 Analysis of Various Machine Learning Algorithms Used by Data Scientists

See Table 1.

Table 1 Analysis of ML algorithms

Analysis		
ML algorithms	When and How to use for stock Market prediction [14]	Advantages/disadvantages/applications [14]
1. **Linear Regression** (LR)	1. LR used for continuous values (Time series analysis) and multidimensional problems like SM for predicting exact point in time value and not a category forecasting	**Advantages**: It is one of the most interpretable ML algorithms, as it requires minimal tuning, it runs very fast. Generally, used in prediction problems. In LR mathematical forms, SM estimate E_t for the next iteration is calculated based on previous iteration's values. For given instance x, at time t, for k number of iterations, considering error ε, LR equation, is formulated as follows: $E_t = \alpha_o + \alpha_{x_1}, t_1 - 1 + \ldots \alpha_{xk}, t_k - 1 + \varepsilon$ **Applications**: Estimating Sales, Risk Assessment, Regression Trees
2. **Logistic Regression** (LogR)	1. LogR are used with manual discretization which is very important considering dynamicity to introduce nonlinearity 2. LogR applies the logistic function to linear	**Advantages**: 1. It is very easy to debug and tune 2. It extracts as much signal as possible and leaves out noise. The fundamental problem is to have the data that maximizes the signal. Stock market firms are in HFT. In low to medium

(continued)

Table 1 (continued)

Analysis		
ML algorithms	When and How to use for stock Market prediction [14]	Advantages/disadvantages/applications [14]
	combination of features for prediction of a dependent variable, which analyze individual components of the markets such as the dependency of certain stocks on the basis of news [14] 3. LR and LogR are the most utilized technique in High Frequency Trading (HFT)	frequency space, this technique gives accurate results **Disadvantages**: Over fitting **Applications**: Credit Scoring, predicting the revenues of a certain product, earthquake prediction
3. **Decision Trees** (DT)	DT are of 2 types depending on outcome type. 1. Continuous Variables 2. Binary Variables DT creates a model of optimal decisions and their possible consequences, including chance event outcomes. DT are used in circumstances where parameters are nonlinearly related, in data exploration and for feature selection in predictive analytics	**Advantages**: DTs are self explanatory. DT efficiently works with both categorical and numerical variables. Even in case of missing data and outliers it works very well, which in turn reduces data preparation time **Disadvantages**: Accuracy of expected outcome is directly proportional to number of decisions in DT. Sometimes, unrealistic DT leads to bad decision making. Not efficient for continuous variables, creating large DT with more number of branches leads to higher time complexity which is a major drawback **Applications**: In finance for option pricing, in remote sensing applications for extraction of similar patterns [14]
4. **LDA/QDA Gaussian-based classification**	LDA/QDA Gaussian-based classification are used for samples within the same distribution	**Applications**: For complex prediction problems
5. **K-Means Clustering**	1. For Unsupervised Learning/Classification 2. Routinely used in applications like Google news to cluster similar stories, can be applied to group the similar web pages. 3. Unsupervised learning for outlier data in SM prediction	**Advantages**: In case of global clusters, K-means results in tighter clusters than hierarchical clustering algorithms. Time complexity is less, hence faster than hierarchical clustering for large number of variables [14] **Applications**: Google, yahoo uses for clustering web pages, estimates the predictability of the direction of movement of stock prices correctly

(continued)

Table 1 (continued)

Analysis		
ML algorithms	When and How to use for stock Market prediction [14]	Advantages/disadvantages/applications [14]
6. **Support Vector Machine** (SVM)	SVM classified into two categories: Linear and Nonlinear 1. In nonlinear SVMs it is very difficult to classify training data using a hyper plane because of nonlinearity [14] 2. If scalability is main criteria then SVM works well 3. In SVM each iteration causes the "error/residue" to become 1/2 of the previous iteration, leading to very rapid convergence resulting in faster computations 4. Dynamic/nonlinear classification problems 5. High-dimensional data like text [17]	**Advantages**: Gives best classification performance in terms of time complexity and accurate results, on the training data, overfitting does not occur **Applications**: SM predictions used by finance institutions. Enhancements can be done using a stock-screener (classification), rather than trying to predict exact stock price (nondeterministic regression) problem, SVMs with reinforcement learning achieves better results to compare relative performance of stocks, which helps in SM investment related decisions
7. **Neural Networks** (NN)	1. Used for prediction and classification, where manipulating/weighting features can give more precise results than simple classification 2. Can be used for stock market predictions, prices fed into the network, with several technical indicators as input features, performs adaptive filtering of the raw price data, feed filtered data into the Recurrent NN (RNN) model, train it, and predict next week's move, learning nonlinear models [18]	**Advantages**: NN is a "black box" method, complex part of NN is Gaussian mixture model or deep learning algorithm which is more advanced **Disadvantages**: The nature of the predictions is unclear, reusability of the same NN model is questionable. Out of sample overfitting is the measure issue, NN training is tiresome and output is unstable, hence NN is poor for investment purpose problems like SM
8. **Deep Learning** (DL)	1. DL Convolutional or Recurrent Networks are specifically used for real-life dynamic problems 2. Learning nonlinear models	**Advantage and Application**: SM prediction, other complex problems 2. Drawback of NN is autoencoder (adjusting of weights) which requires frequent setup, for all intermediate layers but advantage of DL is, to

(continued)

Table 1 (continued)

Analysis		
ML algorithms	When and How to use for stock Market prediction [14]	Advantages/disadvantages/applications [14]
	3. Drawbacks of NN are overcome in DL	memorize single node which captures the information of almost all of the intermediate nodes. Hence, it's always good to innovate on existing NN models rather than using them as it is [19]
9. **Random Forest** (RF)	1. When data is limited for classification/decision problems 2. RF handles different types of features like numerical, binary, or categorical without any transformation. It focuses on the important variables estimation for classification purpose by doing implicit feature selection 3. Unlike Random walk, RF for regression tasks does not predict beyond the given range of target values in training dataset [14]. Mathematical Random Walk equation is as shown: $R(t + 1) = R(t) + e$ 4. Hybrid RF with clustering and for feature selection works well for stock predictions. It is fast, requires nearly no parameter tweaking to get some results, very powerful technique that performs well in hybrid models combined with other ML techniques	**Advantages**: Unlike DT, no overfitting issue, hence pruning RF is not always necessary. Time complexity of Hybrid RF is less. Algorithms are fast but with some exceptions in case of noise. Parallel RF processing makes it run faster on distributed databases as accuracy is much more compared to DT since no over fitting of data, works well in case of missing data, resistant to outliers, data scientists save data preparation time, input data preparation not needed **Disadvantage**: Analysis of RF becomes difficult if large number of DT are included which takes lot of time to process the data, performance decreases in real-time predictions. If the data consists of multilevel categorical variables, then, there are chances of algorithm bias in favor of those multilevel attributes which in turn increases time complexity compared to DT. **Applications**: All sorts of predictions in banking sectors, the automobile industry, health care industry used for regression tasks like prediction of average number of social media shares and likes. Pattern matching in speech recognition, classification of images and texts [14]
10. **Feature Selection** (FS)	1. Variation of FS is (chi-squared feature selection) 2. Helps in model size reduction	**Advantages**: Features are added/removed recursively **Disadvantages**: It does not give accuracy benefits in discriminative models
11. **Ensemble Machine learning**	1. EML (Tree Ensembles including RF and Gradient Boosted DT) is nothing but	**Advantages**: EML are accurate than single ML algorithms. EML are combination of different predictive

(continued)

Table 1 (continued)

Analysis		
ML algorithms	When and How to use for stock Market prediction [14]	Advantages/disadvantages/applications [14]
Techniques (EML)	correct combinations of all predictions blended with statistics gives good results [3, 20] 2. For time series prediction, algorithm like a NN, SVM, or DT are used in ensemble ML [20] 3. Unlike traditional regression approach, provides with a way to take multiple ML algorithms, models, and combine their predictions. Classifiers that make them up are complex, nonlinear, multivariate problems like stock predictions [3]	models which reduces error variance hence accuracy of automated system improves [21] **Applications**: EML are used in variety of ML problems, such as feature selection, confidence estimation, extraction of missing features, incremental reinforcement learning, handling imbalanced data [3]
12. **Naive Bayes** (NB)	Based on the Bayes theorem 1. NB works effectively with categorical input variables 2. Requires relatively little training data than other discriminative models like logistic regression, provided NB conditional independence assumption holds 3. Good for multi-class predictions	**Advantages**: NB classifier converges faster [17] **Applications**: To classify a news article, sentiment analysis, document categorization, used for classifying news articles, Spam email filtering
13. **Conditional Random Fields** (CRF)	1. CRF are of 2 Types: Markov random fields, Maximum Entropy Markov Models 2. CRF are used in models where we have to predict multiple unknowns and there exists some special structure of dependencies between the unknowns	**Advantages**: Extremely helpful when dealing with graphical models. Because of conditional nature of CRF, independent assumptions by hidden markov model (HMM) are not required to be done **Disadvantage**: CRF avoids label bias problem which is a drawback of Maximum entropy Morkov models (MEMMs) **Applications**: Bio-informatics, computational linguistics and speech recognition

(continued)

Table 1 (continued)

Analysis		
ML algorithms	When and How to use for stock Market prediction [14]	Advantages/disadvantages/applications [14]
14. **Regression trees**(RT)	RT Are of two types: Multivariate RT, Multivariate Adaptive Regression Splines in which multiple random variables can be used as parameters 2. RT are used to predict a score rather than a category	**Advantages**: Adaptive, complexity can be increased **Applications**: Regression is most widely used in stock market prediction
15. **A/B testing and data analytics**	Removes some of the issues with confounding variables and allows separation of correlation	**Applications**: 95% of the stuff done in the industry with big data is A/B testing and analytics
16. **Apriori ML Algorithm**	A priori ML makes use of large item set properties	**Advantages**: Easy to implement, can be processed in parallel **Applications**: Detecting adverse drug reactions, used for association analysis, Market basket analysis, Google autocomplete forms
17. **Reservoir sampling** (Weighted Distributed Reservoir sampling)	useful in a number of applications that require us to analyze very large data sets	**Advantages**: A variation of this algorithm, a weighted distributed reservoir sample, where every item in the set has an associated weight, with sample data such that the probability of item selection is proportional to it's weight. Hence, number of passes are reduced over the i/p data **Applications**: CloudEraML [22] has a sample command that can be used to create stratified samples for text files and Hive tables (the HCatalog interface to the Hive Metastore)

2.2   The Challenge

The problem here is that prediction of price movement is not enough for successful application in trading. ML prediction model tells us the move will be +1% or −1% tomorrow. Because this 1% might mean −10% with some probability or +20% with some other much smaller probability. SPX is always going up almost all the time for long-term investments, but every 1% of precision improvement in prediction is very difficult to predict. Hence, instead of focusing on the stock prices, focusing on a probability distribution, is a better solution.

3 Conclusion

ML algorithms form the basis for big data analysis related complex problems like SM prediction. In this paper, various technical indicators, challenges in SM prediction, classification types, analysis of ML algorithms as shown in Table 1 are discussed. The analysis part can be used to design EML which is a correct combination of all predictive models in order to make a good decision better. This forms the basis for subsequent research. Further, research stages would be the design of the functional prototype, the design of EML algorithms, comparative analysis of EML algorithms on the common benchmark platform based on the certain technical indicators and performance evaluation, which can be done in near future.

References

1. www.StocksNeural.net.
2. www.vatsals.com.
3. Robi Polikar, "Ensemble Learning, Methods and applications", Date: 19 January 2012, Springer Verlag.
4. Machine learning algorithms, http://www-formal.stanford.edu/jmc/whatisai/node2.html.
5. Robert Nau, http:blog.optiontradingpedia.com, besttradingplatformfordaytraders.blogspot.com.
6. Rupinder kaur, Ms. Vidhu Kiran, "Time Series based Accuracy Stock Market Forecasting using Artificial Neural Network", IJARCCE. doi:https://doi.org/10.17148/IJARCCE, 2015.
7. AI Stock Market Forum: http://www.ai-stockmarketforum.com/.
8. How to Value Stocks using DCF and the Dangers of Doing, Do Fundamentals Really Drive The stock market? www.mckinsey.com, A Mckinsey report.
9. Ben McClure, Discounted Cash Flow Analysis, Investopedia, Investopedia.com, 2010 http://www.investopedia.com/university/dcf/.
10. https://www.udacity.com/course/machine-learning-for-trading-ud501.
11. http://www.angoss.com, Key Performance Indicators Six Sigma and Data Mining.pdf, 2011.
12. Forecasting home page (Introduction to ARIMA models), price Prediction Using the ARIMA Model-IEEE Xplore, UKSim-AMSS, 16th ICCMS, 2014.
13. www.kdnuggets.com.
14. www.dezyre.com.
15. Paliouras, Georgios, Karkaletsis, Vangelis, Spyropoulos, Machine Learning and Its Applications: advanced lectures, Springer-Verlag New York, Inc. New York, NY, USA 2001, table of contents ISBN-978-3-540-44673-6.
16. Alexandra L'Heureux; Katarina Grolinger, Hany F. El Yamany; Miriam Capretz, Machine Learning with Big Data: Challenges and Approaches, IEEE Access, 2017, doi:https://doi.org/10.1109/ACCESS.2017.2696365.
17. J. Huang, J. Lu and C. X. Ling, "Comparing naive Bayes, decision trees, and SVM with AUC and accuracy," Third IEEE International Conference on Data Mining, 2003, pp. 553–556. doi: https://doi.org/10.1109/ICDM.2003.1250975.
18. Neelima Budhani, Dr. C.K. Jha, Sandeep K. Budhani, Application of Neural Network in Analysis of Stock Market Prediction, www.ijcset.com, IJCSET, 2012.
19. Maryam M. Najafabadi, Flavio Villanustre, Taghi M. Khoshgoftaar, Naeem Seliya, Randall Wald and Edin Muharemagic, Deep learning applications and challenges in big data analytics, Journal of Big Data, SpringerOpen Journal, 2015.

20. L. K. Hansen and P. Salamon, Neural network ensembles, IEEE Transactions on Pattern Analysis and Machine Intelligence, vol. 12, no. 10, pp. 9931001, 1990.
21. W. N. Street and Y. Kim, A streaming ensemble algorithm (SEA) for large-scale classification, Seventh ACM SIGKDD International Conference on Knowledge Discovery Data Mining (KDD-01), pp. 377382, 2001.
22. blog.cloudera.com.

Stability of Local Information-Based Centrality Measurements Under Degree Preserving Randomizations

Chandni Saxena, M. N. Doja and Tanvir Ahmad

Abstract Node centrality is one of the integral measures in network analysis with wide range of applications from socioeconomic to personalized recommendation. We argue that an effective centrality measure should undertake stability even under information loss or noise introduced in the network. With six local information-based centrality metric, we investigate the effect of varying assortativity while keeping degree distribution unchanged, using networks with scale free and exponential degree distribution. This model provides a novel scope to analyze the stability of centrality metric which can further find many applications in social science, biology, information science, community detection and so on.

Keywords Centrality · Local information · Stability · Assortativity
Degree distribution

1 Introduction

The relationship among nodes in the network has varied meaning and analogy. These relations are observed in different domains and fields, such as; the relationship among neurons in the neural network, among web-pages in web graph, among online users in social relationship graph, and among authors in scientific

C. Saxena (✉) · M. N. Doja · T. Ahmad
Department of Computer Engineering, Jamia Millia Islamia, New Delhi, India
e-mail: cmooncs@gmail.com

M. N. Doja
e-mail: ndoja@yahoo.com

T. Ahmad
e-mail: tahmad2@jmi.ac.in

© Springer Nature Singapore Pte Ltd. 2018   395
S. Bhalla et al. (eds.), *Intelligent Computing and Information and Communication*,
Advances in Intelligent Systems and Computing 673,
https://doi.org/10.1007/978-981-10-7245-1_39

collaborations network. Complex networks have been explored to study these relations at the microscopic level of elements like nodes and their relations as links. Structural centralities are such measures which uncover the explicit roles played by the nodes and the links. The concept of centrality is a characterization of a node, subject to its importance according to the structural information of the network. Centrality has become an important measurement with vast implications in theoretical research such as physical science [1] or biological science [2], and practical significance in applications, such as e-commerce [3] and social networks [4]. However, the node centrality concept is vast and can be furnished in various ways, putting forward multiple coexisting centrality measures. Among various aspects, centrality based on the local information is of greater importance due to its low computation complexity and simplicity. Also, in the evident scenario of real complex networks having incomplete information or loss of information and dynamically changing the topological behavior of the network, local information-based centralities provide best solutions. The important issue here is to have the ability of this metric to be stable and robust to noise in the network or to any randomization, such created under real situations. It has been proven that different network topologies, such as scale free, exponential, and heavy-tailed degree distribution [5] can affect network-based operations, such as robustness of network [6] and epidemic spreading [7]. So far the issue of the topological structure has not been investigated for different centralities aiming its stability issue. This has motivated us to explore the stability of local information-based centrality with different topologies of the networks. We studied these centralities under scale-free and exponential distribution of networks and examined the effect of network randomization (perturbation) when the degree of the nodes was kept preserved. To examine the stability problem in local information-based centrality, we have analyzed six centrality measurements namely, *H-index* [8] (*h*), *leverage centrality* [9] (*lc*), *local structural entropy* [10] (*lse*), *local clustering coefficient* [11] (*lcc*), *topological coefficient* [12] (*tc*), and *local average connectivity* [13] (*lac*). We provide a framework to categorize these metrics in a novel dimension. In our experiments, we vary assortativity of the network leaving the crucial property of degree distribution unchanged according to **noise model 1**. We evaluate these metrics on the basis of average rank difference and standard deviation of average rank difference for original and randomized networks. We also compare overlaps of sets of top-ranked nodes, varying range of perturbation with the original network as explained in **noise model 2** defined by Yang [21]. We evaluate this model empirically using real and synthetic networks under scale free and exponential degree distribution.

The rest of the paper is organized as follows. Section 2 presents a framework of the stability issues covering theoretical details and related works in this direction. In Sect. 3 results and experiments evaluate the different stability performance metric on different datasets, and finally, the paper is concluded in Sect. 4.

2 Background

2.1 Local Information-Based Centrality Measure

There have been dozens of centralities based on the global and local topological information of a network. A local network around each node in terms of its connectedness, topology, and type form a network which can formulate its influence. We have examined six local information-based measures, which require only a few pieces of information around the node, defined in this section. *H-index* of a node v_i is the largest value h, such that v_i has at least h neighbors of degree no less than h. *Leverage centrality* finds nodes in the network which are connected to more nodes than their neighbors and determines central nodes. *The local structural entropy* of a node is based on Shannon entropy of local structure which depends on the degree of its neighboring nodes in the local network of the target node. *The local clustering coefficient* of a node is the ratio of a number of links among its neighbors to possible numbers of links that could formulate. *The topological coefficient* of a node is the sum ratio of a number of neighbors common to a pair of nodes among directly linked neighbors to the number neighboring nodes of target node. *Local average connectivity* of a node is the sum of local average connectivity of its neighbors in the subnetwork induced by the target node.

2.2 Degree Preserving Randomization

Rewiring network randomly while keeping its degree distribution, and hence degree of each node constant is to enact real-time perturbation of the network that mapped incompleteness or state of loss of information. As local information-based centrality measurements are degree dependent, therefore, preserving degree of each node while perturbation is an essential step. Primarily, this leads to change in mixing pattern of a network, and hence changing the assortativity of a node. A pair of edges are picked randomly and rewired as illustrated in Fig. 1, ensuring degree preserving randomization, e.g., edges AC and BD are picked as candidates, now the connections between end points are exchanged to form new edges AD and BC. To study the effect of randomization on centrality stability of the network nodes, we use **noise model 1** and **noise model 2** on six different datasets in this paper. These datasets differ both in size and subject to type as the scale free and the random exponential networks as shown in Table 1. The *facebook* dataset consist of nodes as people with edges as friendship ties extracted from social network facebook. The *brain-network* is medulla neuron network of fly-drosophilla. The *foodweb* is the food web network data of Little Rock Lake, Wisconsin USA. The *c. elegans-metabolic* is a metabolic network data. Synthetic scale free and synthetic exponential distribution datasets are also considered for the validation of real data. The basic topological features of the networks are listed in Table 1.

Fig. 1 Network perturbation
preserving degree of nodes

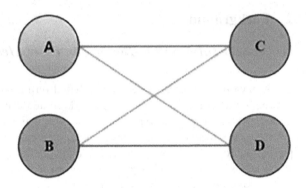

Table 1 Topological statistics of datasets: number of nodes (n), number of edges (m), maximum
degree (k_{max}), best estimate of the degree exponent of the degree distribution (γ), assortativity of
unperturbed network (r), and assortativity of perturbed network (r_p)

Network	n	m	k_{max}	γ	(r)	(r_p)
Scale-free						
Synthetic	2000	5742	1378	2.1	−0.4157	−0.4179
Facebook	6621	249,959	840	2.3	0.1225	0.1149
Brain	1781	33,641	16,224	1.9	−0.3235	−0.3252
Exponential						
Synthetic	1000	9945	495	–	−0.1252	−0.1174
C. elegans	453	4596	644	–	−0.0625	−0.2276
FoodWeb	183	2494	108	–	−0.2374	−0.1585

2.3   Related Work

There are a few numbers of studies on finding stability of pagerank algorithm under
randomization [14–17]. Goshal and Barabasi [18] have proposed stability criteria for
scale free and exponential network when degree preserving random perturbation and
incompleteness in network are observed. Authors have studied the ranking stability
of top nodes and manifested the initial success of pagerank algorithm to scale-free
nature of the underlying WWW topology. Lempel and Moran [14] have examined
the stability of three ranking algorithms—pagerank, HITS, and SALSA—to
perturbation, while Senanayake et al. [16] have studied the page rank algorithm
performance with underlying network topology. Andrew et al. [17] have analyzed
the stability of HITS and PageRank to small perturbations. Sarkar et al. [19]
investigated community scoring and centrality metric in terms of different noise
levels. So far no work has been observed carrying stability study of local
information-based centralities.

2.4 Stability

To examine the stability of centrality measurements we furnish the data with degree preserving randomization according to two noise models as explained. **Noise Model 1**: We implement number of perturbations in the network of size more than 2000 nodes equal to the size of network and for other networks it is 10 times the size of the network. The reason lies on the fact to ensure the range of perturbation depending upon the size of network and to realize meaningful change in assortativity of the network. Where **assortativity** is defined as a measure that evaluate the tendency of a node to be connected to similar nodes in the network. Figure 2 reports the distribution of ranking scores for perturbed and unperturbed networks for three centrality measures—*h-index*, *leverage centrality,* and *local structure entropy*—on synthetic scale-free and exponential network data, real scale-free facebook, and exponential *c. elegans* network data. The *h-index* confronts the most concentrated distribution as compared with other two measures, which are less concentrated. This indicates that when data is exposed to degree preserving perturbations, the node ranking may also change and stability problem exists to diffscore of various centrality indices,erent ranges. To characterize the stability of centrality measure, we define three metrics:

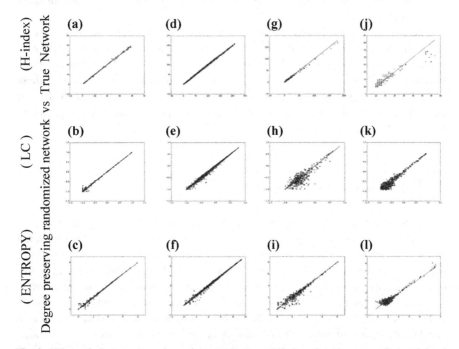

Fig. 2 Effect of degree preserving randomizations on *h*-index, leverage centrality and local structure entropy. Synthetic Scale free and exponential distribution also real facebook and *c. elegans* networks are examined. Figure shows the scatter plot of ranks in both original network and degree preserved randomized network for the same. Synthetic scale-free network (**a**), (**b**), (**c**), Facebook (**d**), (**e**), (**f**), *C. elegans* (**g**), (**h**), (**i**), and synthetic exponential (**j**), (**k**), (**l**)

Mean Bias, μ calculates the average of rank difference of same node from perturbed and unperturbed network for some centrality measurement and is defined as:

$$\mu = \frac{\sum_{xy} \delta_{xy}}{n(n-1)}, \tag{1}$$

where n is the size of the network and δ_{xy} is the difference between true rank of a node calculated from centrality measure on original network (x) and randomized network (y) as shown in Fig. 2. Lower the values of μ, stable the centrality of nodes would be.

Standard Deviation of Mean Bias, σ measures the susceptibility of ranks against the change in data occurred due to randomization. It is defined as:

$$\sigma = \sqrt{\frac{\sum_{xy} (\delta_{xy} - \mu)^2}{n(n-1)}} \tag{2}$$

High value of σ indicates that ranks of a few nodes are quite unstable and low value indicates the similar unstable level of ranks for all nodes due to randomization or unstable centrality.

Jaccard Similarity Index [20], measures the overlap between two rank vectors. In order to check how top ranking nodes change under different proportion of perturbation, we compare top 25 nodes according to rank given by each centrality measures for different networks. Highest value (1) indicates that the set of these nodes is not changed and lowest value (0) indicates that it has changed totally due to randomization offered to the network when compared with original vector of rank for top nodes.

3 Experiments and Results

For experimentation, we have two base topological configurations for network; scale free and exponential and generated synthetic data for the same also considered two real networks for each configuration as mentioned in the Table 1. We perform degree preserving randomization on empirical datasets. For each mentioned centrality measure, we calculate the ranking vectors for perturbed and unperturbed networks. The results of mean bias μ, standard deviation σ of mean bias are shown in Fig. 3. It is well noticeable from the histograms for inverse mean bias and inverse standard deviation of bias score of various centrality indices, that overall h-index outperforms other centrality measurements for each topological configuration. However, all other centrality measures show up higher stability for the scale-free topology of underlying network. Apparently higher values for centrality stability score of scale-free configuration suggests the role of network characteristic in determining its performance under mentioned stability metrics. This fact is also supported by the work of Ghoshal and Barabasi [18], that the accomplishment achieved by page rank algorithm in ranking the web contents is equally credited to the scale-free characteristic of www. For investigating configuration of top ranked nodes in randomized networks according to different centrality metrics we use **noise model 2** as introduced by Yang [21]. Jaccard overlap of top rank nodes for all centrality measures on different datasets are evaluated with varying assortativity of the network as shown in Fig. 4. In majority of the cases, h-index performs better than other metrics retaining top ranked node under varied level of randomization. It can be realized from the experimental evaluations that the scale-free networks are more stable for local information-based centrality stability performance under perturbation of network and h-index is the most stable metric when compared with the present set of centrality metrics considered for evaluation.

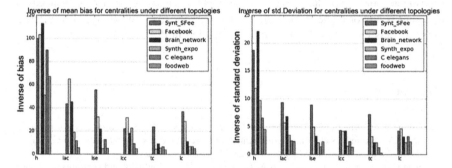

Fig. 3 Inverse of mean bias $(1/\mu)$ and inverse of standard deviation of mean bias $(1/\sigma)$ for different centrality measures under different topology networks

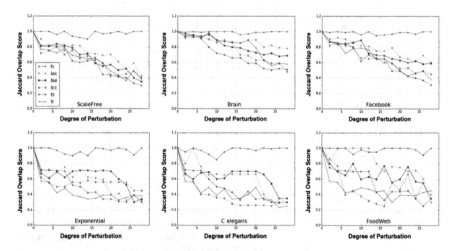

Fig. 4 The Jaccard similarity index between the top vertices of the original and the perturbed networks for varying levels of assortativity changes with increasing number (degree) of perturbations (**noise model 2**)

4 Conclusion

In this work, we investigate how changing assortativity affect different centrality metrics under given topologies of the network. We find that h-index centrality outperforms other benchmark centrality metrics based on the local information in the network, when network is perturbed keeping its degree distribution constant according to noise model 1. To further explore stability notion, we use Jaccard similarity index for different centrality rank vectors of top nodes when network is randomized according to noise model 2 with original top nodes vector. We find that the top rank nodes also called stable nodes are more prevalent to networks with scale-free degree distribution and h-index shows high value for Jaccard overlap on majority of datasets. Our method introduces contemporary measures to analyze and evaluate stability of different centrality metrics which can be further explored to investigate other important networks parameters. This work can be enhanced by studying behavior of network with other properties like transitivity, also different noise models to effect networks could be studied in this regard.

References

1. Guimera, R., Mossa, S., Turtschi, A., & Amaral, L. N.: The worldwide air transportation network: Anomalous centrality, community structure, and cities' global roles. *Proceedings of the National Academy of Sciences, 102*(22), 7794–7799 (2005).
2. Jeong, H., Mason, S. P., Barabási, A. L., & Oltvai, Z. N.: Lethality and centrality in protein networks. *Nature, 411*(6833), 41–42 (2001).
3. Lamberti, F., Sanna, A., & Demartini, C.: A relation-based page rank algorithm for semantic web search engines. *IEEE Transactions on Knowledge and Data Engineering, 21*(1), 123–136 (2009).
4. Malliaros, F. D., Rossi, M. E. G., & Vazirgiannis, M.: Locating influential nodes in complex networks. *Scientific reports, 6* (2016).
5. Strogatz, S. H.: Exploring complex networks. *Nature, 410*(6825), 268–276 (2001).
6. Borge-Holthoefer, J., Rivero, A., García, I., Cauhé, E., Ferrer, A., Ferrer, D., & Sanz, F.: Structural and dynamical patterns on online social networks: the spanish may 15th movement as a case study. *PloS one, 6*(8), e23883 (2011).
7. Barthelemy, M., Barrat, A., & Vespignani, A.: Dynamical processes on complex networks (2008).
8. Hirsch, J. E.: An index to quantify an individual's scientific research output. *Proceedings of the National academy of Sciences of the United States of America*, 16569–16572 (2005).
9. Joyce, K. E., Laurienti, P. J., Burdette, J. H., & Hayasaka, S.: A new measure of centrality for brain networks. *PLoS One, 5*(8), e12200 (2010).
10. Q. Zhang, M. Li, Y. Du, Y. Deng, "Local structure entropy of complex networks", arXiv preprint arXiv:1412.3910. 2014 December.
11. Zhang, Q., Li, M., Du, Y., & Deng, Y. Local structure entropy of complex networks. *arXiv preprint* arXiv:1412.3910 (2014).
12. Xu, J., & Li, Y.: Discovering disease-genes by topological features in human protein–protein interaction network. *Bioinformatics, 22*(22), 2800–2805(2006).

13. Li, M., Wang, J., Chen, X., Wang, H., & Pan, Y.: A local average connectivity-based method for identifying essential proteins from the network level. *Computational Biology and Chemistry*, *35*(3), 143–150 (2011).
14. Lempel, R., & Moran, S.: Rank-stability and rank-similarity of link-based web ranking algorithms in authority-connected graphs. *Information Retrieval*, *8*(2), 245–264 (2005).
15. de Kerchove, C., Ninove, L., & Van Dooren, P.: Maximizing PageRank via outlinks. *Linear Algebra and its Applications*, *429*(5–6), 1254–1276 (2008).
16. Senanayake, U., Szot, P., Piraveenan, M., & Kasthurirathna, D.: The performance of page rank algorithm under degree preserving perturbations. In *Foundations of Computational Intelligence (FOCI), 2014 IEEE Symposium on* (pp. 24–29). IEEE (2014, December).
17. Ng, A. Y., Zheng, A. X., & Jordan, M. I.: Link analysis, eigenvectors and stability. In *International Joint Conference on Artificial Intelligence* (Vol. 17, No. 1, pp. 903–910). LAWRENCE ERLBAUM ASSOCIATES LTD (2001, August).
18. Ghoshal, G., & Barabási, A. L.: Ranking stability and super-stable nodes in complex networks. *Nature Communications*, *2*, 394 (2011).
19. Sarkar, S., Kumar, S., Bhowmick, S., & Mukherjee, A.: Sensitivity and reliability in incomplete networks: Centrality metrics to community scoring functions. In *Advances in Social Networks Analysis and Mining (ASONAM), 2016 IEEE/ACM International Conference on* (pp. 69–72). IEEE (2016, August).
20. Gower, J. C.: Similarity, dissimilarity, and distance measure. *Encyclopedia of Biostatistics* (2005).
21. Yang, R.: Adjusting assortativity in complex networks. In *Proceedings of the 2014 ACM Southeast Regional Conference* (p. 2). ACM (2014, March).

Hybrid Solution for E-Toll Payment

Ajinkya R. Algonda, Rewati R. Sonar and Saranga N. Bhutada

Abstract E-toll system has been huge improvement in decreasing the over traffic jams that have become a big problem in metro cities nowadays. It is the best method to handle the huge traffic. The traveler passing through the traditional type of transport have to pay toll bill by waiting in the long queue, so it results in loss of petrol, loss of time, pollution, and tension of carrying cash with us. So by using our system of E-Toll payment the traveler revokes the tension of waiting in the queue to make the payment, which decreases the fuel consumption and taking cash with them can be avoided.

Keywords RFID · FASTag · OCR · OBU · IU · AVC · AVI
ETC · CCH · Gantry

1 Introduction

Nowadays the amount of traffic has been increasing due to increase in vehicles. Everybody prefers their own vehicles as compare to public transport. It increases traffic and the solution over this is wide roads, hence tooling system is used to collect tolls. The traditional toll collection system is like vehicles come at toll gate and owner of that corresponding vehicle has to pay money according to the toll charged onto his vehicle. This process of toll payment is so time-consuming because manual processing requires more time often leads to delay and customer has to wait. So one-step toward making India a digital India is making traditional toll payment system as a E-toll payment system, in which a user has to buy FASTag just at Rs. 250 only from

A. R. Algonda (✉) · R. R. Sonar · S. N. Bhutada
Department of Information Technology, MIT College of Engineering, Pune, India
e-mail: ajinkyaalgonda449@gmail.com

R. R. Sonar
e-mail: revasonar9@gmail.com

S. N. Bhutada
e-mail: saranga.bhadade@mitcoe.edu.in

© Springer Nature Singapore Pte Ltd. 2018   405
S. Bhalla et al. (eds.), *Intelligent Computing and Information and Communication*,
Advances in Intelligent Systems and Computing 673,
https://doi.org/10.1007/978-981-10-7245-1_40

Axis or from ICICI bank and he has to recharge it for payment for toll. The FASTag will work as a prepaid card. The amount of toll will be debited from the account of owner of the respective vehicle. If any vehicle is not having FASTag then he can be identified by its registered vehicle number which can be sensed by OCR and for users registered nothing, bill will be delivered to his postal address with validity of 80 days. These approach increases speed of toll collection process.

2 Related Work

1. **South Korea**. Motorway tolls can be paid using cash, major credit cards, or a Hi-Pass card [1]. The Hi-Pass system permits owner to pay tolls without any interference at the tollbooth. To use it, owner must have both device name as OBU which is installed on the windscreen of the vehicle and a Hi-Pass card which is inserted into the OBU. The OBU, sometimes marketed as a "Hi-Pass device" or ETC unit (Electronic Toll Collection), is produced by a variety of manufacturers. It can be purchased from any store which sells automobile electronics. A Hi-Pass card can be topped up in advance (Hi-Pass Plus card) or used in conjunction with credit cards issued specifically for the toll system. The card can often be found in stores that sell the OBU.

2. **Italy**. Telepass is the name of the system used to collect the toll on a vehicle in Italy. The Telepass is used for all the vehicles which are traveling on Italian roadways. Telepass comprise a device named as OBU which is placed on the windscreen. The OBU is operated on power supply like battery [2]. The communication between OBUs and the E-Toll booths can be done through dedicated short-range communication. Telepass can be used in the opened and the closed system on the roadways in both system, i.e., the open or closed system the toll which will be charged on the vehicles depends on the type of vehicles (car, bus, auto, etc.). The speed of the vehicle of Telepass users should not exceed 30 km/h

Fig. 1 Telepass in Italy [2]

during it is in Telepass lane when OBU has been checked then the OBU gives a high beep and barricade lifts. When Telepass users leave the toll root, the OBU gives another a high beep. Then a number plate of the vehicle is captured and the vehicle is allowed to pass through the tollbooth. OCR detects vehicle's number plate and the user of the respective vehicle gets a bill statement (Fig. 1).

3. **Canada.** The Electronic toll Collection (ETC) is used in Canada is known as Canada 407 Express toll route [3]. There are no toll gate hence the name Express toll route (ETR). This system uses Transponder and Cameras to toll collection automatically. It is primitive example of Highway that exclusively uses Open road toll collection. A radio antenna recognized when vehicle having transponder has arrived and leaved the road and calculates total fare. For vehicle without having transponder, OCR is used for number plate reorganization and the toll fare sent to user. A small electronic "transponder" is placed on the windscreen of the vehicle. The system tests transactions automatically at arrival and end of the toll booths for supporting a distance-based tolling system. The sensors placed on upper side of gantry log the 407 ETR arrival and exit point. On successful transaction, a system beeps at four times and indicates green light. The optical character recognition (OCR) uses for capture a no. plate of vehicle without having transponder. The statement will be send to enrolled **user's** address (Fig. 2).

4. **Singapore.** The ERP (Electronic Road Pricing) is a E-Toll collection system to manage traffic. Singapore is the first city in all over the world who have implemented electronic tolling system to address road blockage [4]. The cars in Singapore city are having a special gadget placed inside the car which is called as the IU (In Vehicle unit) gadget. This gadget is generally placed at the windscreen of the vehicle. A card is inserted into the IU called as cash card or value card. Each and every IU has its own unique ID for every vehicle (Fig. 3). **Electronic gantry.** The gantry recognizes the vehicle with its unique IU. When a vehicle having the unique IU passes through the gantry then the gantry

Fig. 2 Canada 407 ETR for ETC [2]

Fig. 3 IU gadget

Fig. 4 Electronic gantry in Singapore

recognizes the vehicle and its IU and debit amount of the appropriate toll for the associated vehicle. Sensors placed over the gantry communicate with IU through the short-range communication system and the deducted amount is shown on the LCD screen on the IU. The gantry is electrically controlled and it will generate the amount of the toll as per time of day thus effectively adjusting and moderating the traffic (Fig. 4).

5. **United States**. E-ZPass is an E-Toll collection system used on most of the toll roads, bridges, tunnels in the Midwestern and the northwestern US [5]. The E-ZPass communicates with the reader equipments built into the lane of tollways by transmitting a unique radio signals. The **E-ZPass is mounted on the vehicle's** windscreen of the vehicle. Some of the vehicles has the windscreen by which the radio signal gets blocked; for such type of vehicles or for the historical type of vehicles, an external tag is provided, which is designed to attach

Fig. 5 E-ZPass in US

the vehicle's front license plate. The E-ZPass lanes having a low-speed limit of 8–24 km\h so that it would be easy to sense the E-ZPass on the vehicle (Fig. 5). Each E-ZPass tag is programmed for a specific type of the vehicle, so that the toll charges for each vehicle is also different from others. It is possible to charge wrong toll amount by any vehicle by interchanging the tag over it; means if the tags are **programmed vehicle's class does not match the vehicle**. This will be considered as a violation of E-Toll payment rules and fine according to the toll can be charged to the user having a tag on their vehicle. To eliminate this, the commercial vehicle uses E-ZPass, which are blue in color and the white tag is given for standard passenger's vehicles. These E-ZPass are also used by government servant for their vehicles. In New York, the E-ZPass tags which are orange in color are given to emergency vehicles or to the employee of the metropolitan Transportation authority, an employee of Port authority, employees of New Jersey and New York State Thruway Authority. All these agencies are interconnected to each other for exchanging the tag data and for processing toll transactions.

3 Proposed Idea

All above-mentioned countries have been adopted same technology for E-toll collection. But in India, NHAI ties up with Axis and ICICI bank for E-toll collection. They use FASTag which is easily available in these banks at very low cost.

The FASTag is available just at Rs. 200 only. But the thing is that the Axis and ICICI bank are the private sector banks, so each and every person cannot afford taxes charged by these kinds of banks. So, we can propose our idea with the help of all other nationalized as well as private sector banks. So that each and every person can be able to get the benefit of the E-toll payment system.

4 How Will ETC Work in India

Traffic jam is a headache at the tollways. Sometimes, it occurs due to lack of change, sometimes due to heavy vehicles and sometimes while collecting information about the vehicles. As a solution over it, our system will provide separate lanes for the E-Toll payment users, so that the traffic can be managed very easily.

The FASTag E-toll payment system has been adopted by India over national highways only, but our proposed idea will make it for all the tollways in India.

This idea is very suitable for the Indian system. There are 350 toll plazas in all over the India. NHAI is making cashless payment mechanism (FASTag) working at 123 toll plazas. NHAI will give 10% cash back on toll payments for FASTag users. The cash back amount for a particular month will be credited into the FASTag account at the start of the next month [6].

Ministry of Road Transport and Highways has decided to make all toll plazas on national highways across India ETC enabled. As we know, generally people keep their vehicle on while waiting in toll payment queue. This results in more fuel consumption. If E-toll payment comes into the picture, then this will save fuel around Rs. 86,000 crore [7].

5 Strategy Used for Govt. Vehicles and Ambulance

In India, some of the vehicles do not have any kind of toll on all tollways. Like Govt. vehicles and the ambulance. So we can do one thing for avoiding the ambiguity regarding this is, we will register these vehicles as noncharged vehicles in the database. Means whenever the Govt. vehicle or the ambulance will arrive at toll gate the system will not charge any kind of toll to the corresponding vehicle and these type of vehicles are allowed to go without waiting in a queue.

6 How the Payment Can Be Made

(1) **For users registered with FASTag**

- The FASTag can be recharged from Rs. 100–100,000.

- So when the FASTag will be sensed by the sensors present at the tollways the amount charged by the particular vehicle will be deducted from the bank account linked to that particular FASTag user.
- We can promote the **BHIM** app for payment purpose.

(2) **For users registered with number plate**

- For non-FASTag users, the OCRs will be used.
- These OCRs will sense the number plate of the vehicle and through the number plate information of the particular owner will be taken and the toll will be deducted from the registered user of the given number plate.

(3) **For users registered with nothing**

- For these types of users, the bill will be delivered to his postal address with the validity of 80 days from his journey.
- If the particular user will not be able to pay his toll till due date, he will be charged some fine according to the rules.

7 Violation Handling

Vehicles which are Stolen, vehicles with a class mismatch, vehicles with insufficient balance and those that run through without payment are blacklisted. List of blacklisted vehicles is shared with all toll plazas. These vehicles are stopped when they cross the next toll plaza and appropriate actions are taken.

8 Implementation Procedure of the Idea

1. Start
2. User with registered FASTag arrives at the toll gate, FASTag will be sensed and the toll will be calculated. Go to step 5.
3. User with registered number plate arrives at the toll gate, the info of the owner will be fetched and the toll will be calculated. Go to step 5.
4. User arrives at the toll gate who registered nothing, the toll will be calculated and send to his postal address.
5. Toll paid
6. End.

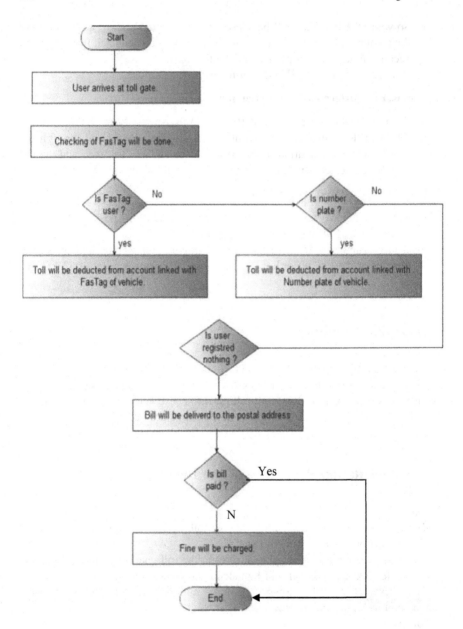

9 Documents Needed to Apply for FASTag

- Customer will need to carry original as well as copy of your KYC document.
- Customer needs to submit the following documents along with the application for FASTag:
- Registration Certificate (RC) of the vehicle.
- Passport size photograph of the vehicle owner.
- KYC documents as per the category of the vehicle owner.

10 Procedure of Getting the FASTag

- Customer can avail ICICI Bank FASTag just by sending an SMS and their executive will call customer within 2 working days.
- SMS—ITOLL(Space)Pincode(Space)Name to 5676766.
- E.g., ITOLL 400051 Roshan Kumar.
- Alternatively, customer can call them at their helpline number 1860-267-0104.

11 Advantages

- By adopting this system, we can take a step toward digital India.
- The corruption can be avoided.
- Using FASTag will manage the traffic over the tollways.
- The problem of change would not make any trouble.
- Time of waiting in a queue at tollway can be saved.
- Saves the fuel wastage at the toll gate.
- Man power reduced at the toll gate.
- Highly secured and reliable.

12 Conclusion

In this system, we have constructed the E-toll payment system with the help of FASTag to offer an efficient way to pay the tolls to enhance the user service quality. It enables the system to classify vehicles for toll payment according to the type of payment method.

The proposed system will clear the problem of traffic jams, lack of change at the money counters at tollways, etc. E-toll users can thus appreciate the high quality of service provided which in turns highly promotes **Digital India**.

References

1. Evaluations and improvements of the Korean highway electronic toll system M.-S. Chang, K.-W. Kang, Y.-T. Oh, H.-W. Jung.
2. Study of Different Electronic toll Collection Systems and Proposed toll Snapping and Processing System. Apurva Hemant Kulkarni. 2014.
3. KHADIJAH KAMARULAZIZI, DR. WIDID ISMAIL electronic toll collection system using RFID technologies Journal of Theoretical and Applied information technology © 2005–2010 JATIT & LLS.
4. Electronic Road Pricing: Experience & Lessons from Singapore. Prof. Gopinath Menon Dr. Sarath Guttikunda 2010.
5. Electronic toll collection syetm using passive RFID technology, Khadijah Kamarulazizi, Dr. Widad ismail.
6. Cashless Payment, http://economictimes.indiatimes.com/news/economy/infrastructure/e-toll-starts-from-monday-now-make-cashless-payments-at-275-toll-plazas/articleshow/51909252.cms.
7. How Indian System work in India, https://www.quora.com/How-will-electronic-toll-collection-work-in-India.

Enhancing Distributed Three Hop Routing Protocol in Hybrid Wireless Network Through Data Weight-Based Scheme

Neha S. Rathod and Poonam Gupta

Abstract The reason for improvising of our concept is to involve multiple nodes in data transmission that diminishes delay in a routing of a network. In hybrid wireless networking, most of the existing routing performs in a linear way. In such possibility sometimes a maximum number of nodes are in idle state. So for that, there is a concept of distributed three hop routing that depreciates the burden of a load on the network. Dividing data and load on a node is an emphatic method of transmission and pronouncement the shortest path using a fuzzy classifier that makes the faster routing approach. But one of the biggest flaws of the existing scheme is that it does not perform as per exceptions due to which it has higher overhead and fails to achieve better congestion control. This paper promoted weight-based assignment technique by using distributed three hop routing protocol. By considering immediate one hop that evolved in shortest path and analyzing a response delay on two nodes on that basis data is fragmented by using weight-based assignment technique. To intensify the efficiency of a routing protocol in the hybrid wireless networks a weight-based data assignment technique is used for data allocation in distributed routing protocol using the artistry of least delay detection to maintain less data congestion in the network.

Keywords Response delay · Data fragmentation · Data assignment
Shortest path identification · Three hop routing

N. S. Rathod (✉) · P. Gupta
Computer Engineering Department, G. H. Raisoni College of Engineering and Management,
Wagholi, Pune, India
e-mail: neharathod73@gmail.com

P. Gupta
e-mail: Poonam77gupta@gmail.com

© Springer Nature Singapore Pte Ltd. 2018
S. Bhalla et al. (eds.), *Intelligent Computing and Information and Communication*,
Advances in Intelligent Systems and Computing 673,
https://doi.org/10.1007/978-981-10-7245-1_41

1   Introduction

Hybrid wireless network assimilates infrastructure network and ad hoc network. It has been proven the best substitute for the future generation wireless network will add infrastructure network to ad hoc network. By the hybrid wireless network, deficiency of ad hoc network and infrastructure network can be overcome, such as weak connection, poor communication. In hybrid wireless networks, the routing protocol is a conspicuous part of the stable designing. One of the biggest drawbacks face by the network is, when a maximum number of nodes are in an idle state that aims to delay in routing process as well as it increases the load on the network due to which concept of DTR with weight-based assignment technique arises. DTR represent as distributed three hop routing protocol in which data is fragmented at the source node and transmitted parallelly in a distributed manner. To increase network amplitude for high-performance applications, the system has to work on hybrid networks [1, 2].

In a wireless network, shortest path problem is one of the basic and valuable for network optimization obstacle and it is used for many real applications. The motive behind the shortest path is to find out the shortest way between the two nodes in a graph or network. The shortest path is been calculated based on the fuzzy classifier. In the fuzzy classifier, crisp inputs are taken, which are evolved due to the result of response delay extraction. After fuzzification, the inference engine refers to the rule base containing fuzzy IF-THEN rules. If-then rules are been set to check the possibilities of the shortest path with defined rules which are normalized in between 0 and 1. The final path score and the essential paths will be extracted according to the rules of features criteria. Now based on the if-then rules, the paths which are criticized under the lowest ranks are considered for the data transmission and ultimately they are treated in the route, so they form the shortest path.

In networking, data sends from source to destination, the immediate nodes are selected by the source node. For that purpose sender sending a "Hello" message to all the nodes in the network and on the basis of route reply of the immediate node, it

Fig. 1 Analyze response delay

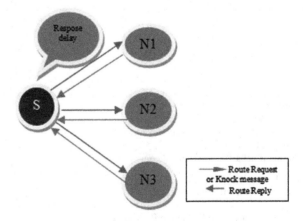

classifies the response delay. Sorting all the delays and select the merest delay which has to be given as a first priority among all the delays (Fig. 1).

Once analyzing response delay of each node in a network, all routing delays are been gathered then based on the weight of the routing delays ratios are being evaluated for the data fragmentation process. Based on the response delay ratios of the nodes, splinter ratios are being decided and then data is divided in the form of bit heap for the communication process. By using some formulas data is segmented these are as follows:

$$\text{Total response time (Rs)} = \text{Rd of N1} + \text{Rd of N2} + \text{Rd of N3} \qquad (1)$$

$$\text{Rd \% Per Node} = \frac{(\text{Rd of node Nn})}{(\text{Total response time})} \qquad (2)$$

Here, Rs is a total response time, Rd is response delay and N1, N2, N3...Nn are nodes.

Hop is one portion of the path between source and destination. Based on the response delay ratios of the three nodes data is segmented into three parts, the three fragments are been delivered at the receiver end. In the two hop routing protocol, the node transmission occurs within a single cell. It takes only a single path transmission and produces high overhead and low reliability [3]. By the distributed three hop routing protocol can improve efficiency and reduce the Overhead. As network consists of three hops in DTR, where first two hops are in the ad hoc transmission mode and one hop was in the cellular transmission mode (Fig. 2).

Most of the methodologies in wireless paradigm are to participate the number of nodes in the network. Most of those are assigning this task to the routing nodes which can create the more havoc at the node end to decrease the routing performance of the nodes. By keeping this issue as of priority proposed system fixed to data weight-based assignment scheme.

Fig. 2 Three hop routing

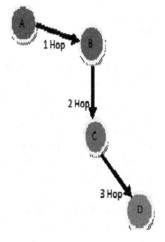

This whole paper is segmented into five sections. Section 2 describes past work as related work. Section 3 reveals the detailing of proposing methodology. The evolution of our methodology is performed in Sect. 4. Finally, Sect. 5 concludes the paper with some scope for future extension.

2 Related Work

Various numbers of methods were discovered to fasten the routing process and involving a number of nodes in the network.

2.1 Shortest Path Identification

K. Sasikala, V. Rajamani et al. reveal about the concept of neuro-fuzzy logic which helps network to reduce delay in the network by performing high-level data to reach the destination in less time and increase time efficiency, network throughput by choosing shortest path [4].

Dimitris Sacharidis, Panagiotis Bouros et al. narrates [5] that fastest route is not considered and examine the simplest route with longer distance. By this, they solve the address issue by the simplest route which will be shortest among all analysis. Sujata V. Mallapur and Siddarama R Patil et al. introduce the FLSBMRP in which by the help of candidate node network can find the most stable path between nodes. In this candidate node used a fuzzy logic technique for nodes collect performance following parameters are considered like enduring bandwidth, enduring power, link quality, node mobility, and reputation index. Other multiple paths can be entrenched through candidate nodes between source and destination [6].

2.2 Analyze Response Delay

Qin Zhu, Li Yu, and Sangsha Fang et al. propose a stochastic network calculus (SNC) scheme which helps to analyze end-to-end delay of a data flow through the stochastic end. For the characterize of data flow it is done on the basis of Linear Fractional Stable Noise (LFSN)-based traffic model [7]. Xianghui Cao, Lu Liu, Wenlong Shen and Yu Cheng et al. propose a distributed delay aware multipath routing method that helps to minimize end-to-end delay of all the commodity nodes of a network [8]. Ze Li and Haiying Shen et al. narrate for the transmission delay represent, another design QoS guaranteed neighbor selection algorithm. For traffic transmission, a short delay is the crucial real-time QoS requirement, the primitive deadline first scheduling algorithm is consolidates by QOD which is a deadline-driven scheduling algorithm for data traffic scheduling in intermediate nodes [9].

2.3 Weight-Based Data Assignment Technique

D. Sumathi, T. Velmurugan, S. Nandakumar, and S. Renugadevi et al. presented a strong load balancing algorithm M-OPTF that work on the basis of dynamic weight assignment scheme, which is for fresh calls and handoff calls. In this, the stray nodes are distributed as per the previously loaded on access points of the base stations. Suppose load is more, then height factor will be small [10].

Haiying Shen, Ze Li, and Chenxi Qiu et al. give the idea how to choose neighbors for data forwarding, for that node require capacity information such as bandwidth and queue size of nodes. In this, an elected neighbors consist of sufficient storage space for each segment. All times current capacity and storage information was updated by exchanging its data to each neighbor [11]. Weiyi Zhang, Benjamin Bengfort, Xiaojiang Du et al. proposed a various decisive bandwidth allotment techniques for the hybrid network like auction based, top down, bottom up allotment techniques. The top- down is lightweight and minimum computation radical among the other techniques [12].

2.4 Hop Routing

Y. Wei and D. Gitlin et al. depict in terms of an omission of route maintenance and a limited number of hops in data routing, it behaves like the two hop transmission. In this, when the bandwidth of node is greater than the neighbor, it sends a message to base station directly. If it not happens it will select a neighbor which has a higher channel and then send it [13].

Nithin Michael and Ao TangHALO et al. ensure to minimize the cost of carrying traffic through packet-switched networks, first link state routing protocol was there in packet forwarding of hop to hop [14]. Jaeweon Choand Zygmunt J. Haas et al. suggested how throughput gain can be achieved in two and three hops relaying scheme. For this technique of multihop relaying technology were used to mitigating unfairness of QOS that totally depends on location wise signal quality [15].

3 Proposed Methodology

In this section, it is discussed how the system works regarding provisions of the perfect network in which data are transmitted by using the weight-based assignment technique with distributed three hop routing protocol. By using this artistry, the maximum number of nodes is involved in a network that diminishes delay in routing. To understand this following are the diagram which demonstrates our proposed methodology (Fig. 3).

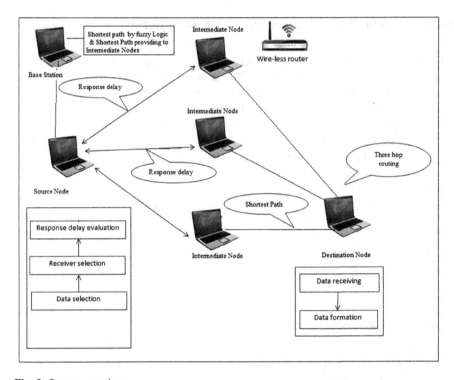

Fig. 3 System overview

Following are the steps which were used in the weight-based assignment with distributed three hop routing method for the wireless communication by that our system is more effective for its working.

3.1 Create Base Station

This is the initial step of the proposed, where every node in a network is registered with a base station on their activation. The base station provides the shortest path to source node on its request for 3 hop routing process. And instance routing path is also provided by the base station on request of instance source node in the process of 3 hop routing protocol.

3.2 Shortest Path Identification Using Fuzzy Logic

In this step, the model user selects the data that need to be delivered to the receiver. Once the receiver and data is selected for routing then system evaluates the shortest path using fuzzy logic theory. According to this theory, all the nodes will assign random integers and these random values are forming a matrix of node weight. This node weight matrix subjected to the evaluation of the shortest path, where five fuzzy crisp values are being created like low, very low, medium, high, and very high based on the node weight. These fuzzy crisp values are forming fuzzy rules, which have eventually given rise to the shortest path based on the IF-THEN rules. Whatever shortest path obtained that considered as immediate one hop from the source node. This can be summarized along with the following algorithm.

Algorithm 1 For Shortest Path Calculation using Fuzzy Logic

//Input: Node Weight as N_{wi}
//Output: Shortest path
Step 0: Start
Step 1: Node Weight Matrix as MAT
Step 2: FOR i=0 TO row
Step 3: FOR j=0 TO column
Step 4: Fuzzy Crisp values F_c
Step 5: Fuzzy Rules as F_r
Step 6: Fuzzy IF-THEN rules
Step 7: Shortest path $S_p = \int F_c \rightarrow F_r$
Step 8: END Inner FOR
Step 9: END Outer FOR
Step 10: Return S_p
Step 11: Stop

3.3 Analyze Response Delay

Once the shortest path is calculated then system identifies all other nodes that are not involved in the shortest path. And then system automatically sends a handshaking message to all these nodes and receives the reply from all of them, thereby records the time delay of all nodes. Now based on the best top two time delay nodes are considered along with the next immediate node from the source node which is evolved in shortest path. So these three nodes are finalized for data fragmentation process and for three HOP process.

3.4 Data Fragmentation Using Weight-Based Assignment Scheme

After analyzing response delay, least delays of two nodes along with the immediate one hop which is ripen through shortest path selects for a data fragmentation purpose. The data are being fragmented based on the ratio of the time delay got from the three nodes. The node which is having highest time delay receives the chunk with lowest size and vice versa. The process of data partition and assignment can be clearly shown below with the algorithm.

Algorithm 2 For Weight-based Data Allocation based on the Response Delay

//Input: User data as D
//Output: Assignment of D_1, D_2, D_3,...Dn to N_1, N_2, N_3...N_n
Where,
D_n Represents Data Segment
N_n Represents Node
Step 0: Start
Step 1: Ping String S to N_1, N_2, N_3
Step 2: Response Delay Collection set as $R_d = \{R_{d1}, R_{d2}, R_{d3}\}$
Step 3: Rs $= \sum Rd$
Step 4: $D_w = \int R_{di}/R_s$ (Data Weight Ratio)
Step 5: $\int D_w \rightarrow D_i$
Step 6: $D_i \rightarrow N_i$
Step 7: Stop

3.5 Data Routing

This step involves the mechanism of data routing to the receiver. Where parallel data routing process triggers with the three data chunks along with the shortest path. In the process, all the data chunks are labeled and handled according to the label. Once all the chunks reaches to the receiver then based on the labels and data is been merged at the receiver end to form the original data with the desired extensions.

4 Results and Discussions

An explicit experiment is conducted for the performance evaluation of the system is estimated using average end-to-end delay of the transmission of data from the sender node to receiver node. So to ensure that proposed model method is of 3 HOP routing protocol average end-to-end delay is better than other systems. Our model compares with the method mentioned in [16] that deals with the technique of iETT (Improved Expected Transmission Time). iETT is an improved version of the ETT (Expected Transmission Time), a routing sequence selection method which selects the routing path based on the availability of the wireless links and other link quality conditions of the network nodes. Whereas proposed system applies and analyzes the availability of the all other nodes in the network based on the DTR. The observed end-to-end delay is recorded in Table 1.

Table 1 Average latency (in seconds)

Packet size (Bytes)	ETT metric	iETT metric	DTR
500	1.25	0.8	0.65
1000	1.5	1.2	1
1500	1.7	1.4	1.1
2000	1.8	1.5	1.18

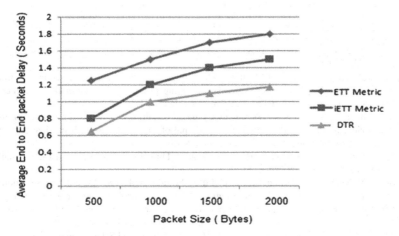

Fig. 4 Average time for different data size

On observing and analyzing above plot some facts are revealed DTR protocols having the excellent hold on transmission mode to select the strong link wireless nodes. Due to this, the end-to-end delay can be diminished reasonably as shown in Fig. 4.

5 Conclusion and Future Scope

As we are appraised of the fact that in almost all the routing protocols in hybrid wireless network routing used to happen based on the routing sequence provided by shortest path algorithms. This technique always utilizes the wireless nodes that are part of the shortest path sequence for data routing, this makes all other nodes to stay idle. Due to this a considerable delay in data transmission is always existed. So the proposed system uses the distributed three hop routing protocol technique, where it involves maximum possible nodes from the network for routing process to decrease the end-to-end transmission delay. This process is powered with the theme of weight-based data assignment technique based on the neighbor node response delay. This empowers the system to perform better to select the best shortest and fastest path of routing.

The proposed system can be enhanced in the future by considering multiple hop routing protocol to get the best end-to-end transmission delay on increasing the number of nodes in the network.

References

1. H Luo, R. Ramjee, P. Sinha, L. Li, and S. Lu, "Ucan: A unified cell and ad-hoc network architecture", In Proc. of MOBICOM, Volume 1, September 2003.
2. P. T. Oliver, Dousse, and M. Hasler, "Connectivity in ad hoc and hybrid networks", IEEE In Proc. of INFOCOM, 2002.
3. Ms. MALARVIZHI, Mrs. P.RAJESWARI, "Improving the Capacity of Hybrid Wireless Networks by Using Distributed Three-Hop Routing Protocol", International Conference on Engineering Trends and Science & Humanities (ICETSH), 2015.
4. K. Sasikala, V. Rajamani, "A Neuro Fuzzy based conditional shortest path routing protocol for wireless mesh network", International Journal of Enhanced Research in Management & Computer Applications, Vol. 2 Issue 5, May-2013.
5. Dimitris Sacharidis, Panagiotis Bouros, "Routing Directions: Keeping it Fast and Simple", ACM, November 05–08-2013.
6. Sujata V. Mallapur, Siddarama R Patil, "Fuzzy Logic Based Trusted Candidate Selection for Stable Multipath Routing", *I.J.* Information Technology and Computer Science, 2015.
7. Qin Zhu, Li Yu and Sangsha Fang, "Per-Flow End-to-End Delay Bounds in Hybrid Wireless Network", IEEE transaction on Wireless communication and Networking, 2013.
8. Xianghui Cao, Lu Liu, Wenlong Shen, Yu Cheng, "Distributed Scheduling and Delay-Aware Routing in Multi-Hop MR-MC Wireless Networks", IEEE transaction on Vehicular technology, 2015.
9. Ze Li, Haiying Shen, "A QoS-Oriented Distributed Routing Protocol for Hybrid Wireless Networks", IEEE transaction on Mobile Computing, 2014.
10. D. Sumathi, T. Velmurugan, S. Nandakumar, S. Renugadevi, "Dynamic Weight Assignment based Vertical Handoff Algorithm for Load Optimization", Indian Journal of Science and Technology, *Vol 9, 2016*.
11. Haiying Shen, Senior Member, Ze Li and Chenx Qiu, "A Distributed Three-hop Routing Protocol to Increase the Capacity of Hybrid Wireless Networks," IEEE Transactions on Mobile Computing, 2015.
12. Benjamin Bengfort, Weiyi Zhang, Xiaojiang Du, "Efficient Resource Allocation in Hybrid Wireless Networks", IEEE transaction on WCNC, 2011.
13. Y. Wei and D. Gitlin, "Two-hop-relay architecture for next generation WWAN/WLAN integration", IEEE Wireless Communication, April 2004.
14. Nithin Michael, Ao Tang, "HALO: Hop-by-Hop Adaptive Link-State Optimal Routing", IEEE/ACM TRANSACTIONS ON NETWORKING, VOL. 23, NO. 6, DECEMBER 2015.
15. Jaeweon Cho, Zygmunt J. Haas, "On the Throughput Enhancement of the Downstream Channel in Cellular Radio Networks Through Multihop Relaying", IEEE JOURNAL ON SELECTED AREAS IN COMMUNICATIONS, VOL. 22, NO. 7, SEPTEMBER 2004.
16. Saad Biaz, Bing Qi, Yiming Ji, "Improving Expected Transmission Time Metric in Multi-rate Multi-hop Networks", IEEE CCNC, 2008.

VLSI-Based Data Hiding with Transform Domain Module Using FPGA

Latika R. Desai and Suresh N. Mali

Abstract In this rapidly growing internet era, researchers are giving more and more attention toward robust, secure, and fast communication channels while hiding sensitive data. The concealment steps can be done through a spatial domain or the transform domain. This paper proposes a data hiding system with an adaptive Very Large-Scale Integration (VLSI) module to enhance the security and robustness of embedded data. The Field Programmable Gate Arrays (FPGA) implementation approach of data hiding technique provides not only pipelined and parallel operations, but also gives the perfect guard against malicious attacks. The proposed algorithm is implemented on a Xilinx Virtex 5 FPGA board. Further, the transform domain technique also optimizes memory space and reduces the execution time through pipelining. The performance of the implemented system is measured using different parameters like resource utilization, Mean Squared Error (MSE), and Peak Signal-to-Noise Ratio (PSNR).

Keywords Security · Data hiding · Field programmable gate arrays
DCT · PSNR

1 Introduction

The demand for robust, secure data hiding techniques in today's real-time applications are increasing. Spatial domain and frequency domain approaches were widely used previously. In the spatial domain, different types of least significant bit (LSB) techniques are used to manipulate pixel values to hide a data. These are simple methods and can be easily detected by human eye, as the converted image gets

L. R. Desai (✉)
Dr. D.Y. Patil Institute of Technology, Pimpri, Pune, Maharashtra, India
e-mail: latikadesai@gmail.com1

S. N. Mali
Sinhgad Institute of Technology and Science, Narhe, Pune, Maharashtra, India
e-mail: snmali@rediffmail.com2

© Springer Nature Singapore Pte Ltd. 2018
S. Bhalla et al. (eds.), *Intelligent Computing and Information and Communication*,
Advances in Intelligent Systems and Computing 673,
https://doi.org/10.1007/978-981-10-7245-1_42

significantly and noticeably deteriorated, even though by taking care of intensity, hiding in the edge area, or corner or dark area of the image. Further, as you increase the data density in the cover image, the security gets compromised accordingly. Embedding and extracting time becomes challenging. To overcome these traditional challenges, frequency domain can be a perfect alternative over spatial domain. Discrete Cosine Transforms (DCT) and Discrete Wavelet Transform (DWT) methods are more preferable in the frequency domain. The authors have implemented the data hiding technique using transform domain through DCT, where the adaptive hardwired mechanism for embedding the sensitive data is proposed. Authors are working on a generic data which is today's demand for the file transfer. Section 2 covers related work in the field. Section 3 throws light on proposed research work by the authors. Section 4 discusses results obtained through experimentation. Section 5 concludes the paper with future scope and directions.

2 Related Work

The data hiding is basically divided into spatial domain and frequency domain. Bassam et al. research work in [1] proposed a novel algorithm to embed and extract the entire secret data in the Haar wavelet-based transform. Special clipping mechanism enhances the PSNR of the stego-image. Authors in [2] introduced high speed, flexible linear feedback shift register (LFSR)-based approach with the reconfigurable hardware implementation. Research work in [3] demonstrated an image Steganography technique to hide the audio signal in the image using transform domain. Authors in [4] give a hardware realization of new algorithm with two stego-keys helping random selection of pixels. Researchers in [5] introduced a novel and fully reversible data embedding algorithm for digital images. Simplicity and efficiency of reversible data embedding algorithms enhance their payload capacity. Research work in [6] implemented a mapping-based hiding technique which involves the different value of X box for encoding 4:2 cover with a message which is difficult to decode. Authors in [7] introduced embedding process in data hiding techniques are classified into spatial domain and frequency domain. Hardware Security modules provide a way for shielding the information from harm in the memory of the tamper proof security service which is the key feature of hardware technique. Research work in [8] compared spatial domain techniques versus transform domain and they demonstrated that spatial domain is simple, but can embed high-quality image while transform were robust in nature. Researchers in [9] proposed improved genetic algorithm getting PSNR upto 38.47 while scope suggested other algorithms such as cuckoo search algorithm and bacteria foraging optimization algorithm is implemented to obtain accurate modeling of Steganography. Authors in [10] demonstrated a fast and real-time, secure and safe communications over networks with a multipoint algorithm for text Steganography hardware implementation. Scope of future work is parallel processing design to optimize the system encryption speed and power consumption. Research work in

[11] concluded a novel methodology behaving like invisible in real-time data transfer. This can be done through the trusted key exchange; it is an extension of existing informed embedding technology. Research work in [12] introduced multilevel embedding technique; they worked on different types of multiplexing and modulation to hide multiple bits. Research work in [13] specified encryption which is one of the solutions for information security, but encrypted messages once intercepted, can easily provide a clue to the adversary or attacker that some message of importance is being communicated. Data hiding, on the other hand, takes the opposite approach and attempts to hide all evidence that communication is taking place. Authors in [14–20] presented information hiding through different techniques like LSB, DCT, DWT, simulation, and implementation using FPGA and microarchitecture and its comparison also information hiding without loss of file formatting, a combination of cryptography and Steganography. Research work in [21] gives an analysis of different method under LSB while authors in [22] used MSB approach in the spatial domain. Researchers in [23] proposed spatial domain technique using LSB through the pipeline. As seen from related work, the authors have observed that the existing most of the available algorithms are available in spatial domain so to improve security, authors need to follow the frequency domain and try to keep tradeoff between capacity, security, and robustness.

3   Proposed Research Work

Data hiding System consists of two main parts, i.e.,

1. Embedder
2. Extractor

To design Embedder in the transform domain (i.e., using DCT) following challenges were introduced while embedding the secret message:

- The random number generation can greatly have a high bandwidth requirement due to the huge number of iterations in embedding process (16 kb × 4 kb READ operations from memory) is getting reduced to only read/write where the message is embedded.
- The 64-bit external memory bus width adds bottleneck for embedding process. It is resolved by LFSR used for address translation.
- By comparing fractions and fixed-point numbers for proposed work, efficient representation for fraction number mechanism is selected because the floating point method has an extremely slow operation. A number of iterations required in floating point is a lengthy while for fixed point gives us low or moderate accuracy.
- The computation overheads for DCT and IDCT are more, but it gets solved by using symmetry for DCT and IDCT computations design structure. That is why authors are using core reusability.

3.1 Embedder

Embedder which is used to form stego is as shown in Fig. 1. We have considered generic data which are in the form of single dimension for cover as well as the message. The host PC provides us a cover data and message data. By dividing cover data into 1×8 windows, calculated DCT through Multiply and Accumulate (MAC) till the end of the cover. In the DCT form cover message bits get embedded through the adaptive embedding technique with LFSR than calculating IDCT it forms the Stego.

3.2 Extractor

In the proposed system, part two of data hiding module is used as a message extractor from stego which is shown in Fig. 2. The host PC provides us stego data by calculating DCT through MAC till the end of the cover. In the DCT form, stego gets extracted through adaptive embedding techniques using LFSR and then the message data gets extracted from the cover data. As we mentioned earlier, we are operating generic data which are in the form of single dimension for cover as well as the message. A Message is sent to the host PC. Here, the process of data hiding is completed successfully.

3.3 Embedding Algorithm

- Data is considered for cover as well as information in the form of a single dimension.

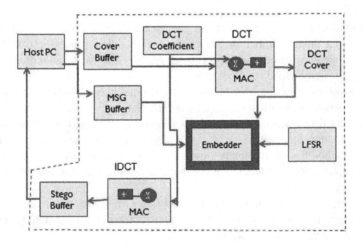

Fig. 1 Proposed system embedder (part1)

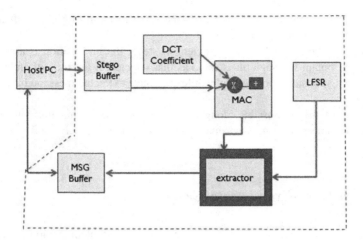

Fig. 2 Proposed system extractor (part2)

- Obtain the number of bits of the message that need to be embedded into the cover.
- Read 8 values of cover (Time/Spatial domain), i.e., 1×8 at one time.
- Perform DCT transform of each of them one by one.
- Store the DCT coefficients of the cover into storage buffer/memory.
- Assign instantaneous LFSR value for n-bit embedding.
- For every 1×8 coefficient, apply the embedding of message bit in DCT coefficient of cover.
- LSFR is used in order to point a cover location for data hiding, i.e., An adaptive approach of stego.
- Perform the IDCT after the stage, where Message is embedded into DCT coefficients of cover. This gives the stego output.
- Repeat the process till the last bit of the message that is to be strategically embedded into the cover.
- Store the result as stego data.
- Stego data are transferred to host personal computer.
- There may be the possibility of an attack at this end.

3.4 Extractor Algorithm

- From host personal computer, communicate the Stego to extraction at receiving the end of the hardware.
- However, first, the Stego undergoes the DCT transform in order to transform from time domain/spatial domain into the frequency domain.

- The message bits are extracted from the same location of DCT coefficient of Stego which is pointed by LFSR.
- Hence, in the pointed DCT coefficient of stego, the bit position is assigned in order to obtain message bit is same as that selected at the time of embedding.
- By taking consideration of applicability we have extracted message bit from the cover.
- Any noise that is possible along the channel might corrupt this bit and it is possible that this bit may not match to bit from message expected.

3.5 Hardware-Experimental Setup

This section demonstrates the FPGA-based hardware implementation of the proposed DCT-based data hiding system. Prior to implementation, a system model was developed to validate the algorithm functional correctness and examine its performance. Next, the resistor-transistor logic (RTL) was developed, simulated, and verified using VHDL language and Modelsim software tool 10.4 through synthesis and analysis summarized in Table 1.

Figure 3 depicts the proposed system, where the cover data (C) and data to be embedded (M) are transferred to the FPGA RAM and there after it is processed block-wise as explained above as per the logic downloaded from the PC to Configurable FPGA board. Then, the bits of M eventually get embedded into various blocks. All the blocks are then reconfigured to reconstruct the stego data (C'). The process of embedding can be validated for various attacks by transferring C' back to PC. The data hiding system is proposed to be implemented initially using Virtex 5, XC5VLX50T, FFG1136C Board.

Figure 4 shows hardware implementation of the data hiding system. On board, FPGA uses the memory of 256 MB × 64 bit. Authors are handling the cover of 16 kb × 8 data of window size for 8 values from the cover and data 4096 (512 × 8 bit). So to map data line of memory with cover size, we need to have memory mapping. 2 kb × 64 memory segmentation of the cover and data is modified like 64 × 64 to match the address range for input and output. The location

Table 1 Design parameters

Parameter's	Value
Target FPGA device	XilinxVirtex 5 XC5VLX50T
Frequency	100 MHz
Input/output signal timing constraint	15 ns
Signal activity source	Simulation file

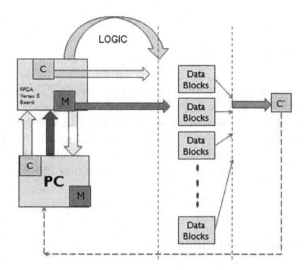

Fig. 3 Proposed system

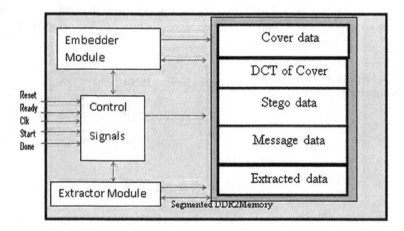

Fig. 4 Hardware implementation of the data hiding system

which is used for DCT computes the cover and after embedding cover with the message is stored which is the same. Here, we optimized the use of memory space. The performance of DCT data hiding FPGA designs helps us to get good PSNR as compared with other methodologies mentioned in the paper. Furthermore, it is desirable to estimate the resource utilization and performance in the data hiding system. We are trying to achieve our goal by using efficient use of FPGA for pipelining processing, optimizing memory utilization, and time.

4 Results

4.1 Experimental Results

The results regarding attack (noise) on the stego are provided in Table 2. Authors demonstrated the stego in terms of PSNR value for various levels of noise as shown in Table 2. The system without noise has PSNR of 60 dB with MSE of 0.06433. And with the noise of 10%, the PSNR value reached 39 dB with MSE of 8.0817260. As we know better PSNR gives robust system. Table 2 and Fig. 5 shown below give our experimental results in the form of PSNR and MSE with noise variation.

4.2 Performance Evaluation

Figure 6 shown below gives a performance evaluation by considering the parameter PSNR of our result with previous researcher's results referred in this paper. Where different methodologies of data hiding are considered like LSB, DWT, MSB, and DCT. It shows that using proposed system gives us better PSNR using DCT.

Table 2 Proposed results for different parameter with noise variation

Sr. No.	NOISE	PSNR	MSE
1	0%	60.04659689	0.064331055
2	5%	42.12002246	3.990966797
3	10%	39.05576235	8.081726074

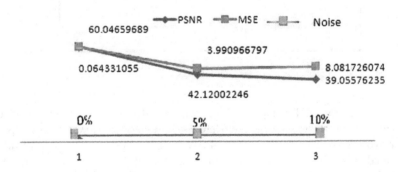

Fig. 5 Proposed PSNR and MSE value with noise variation

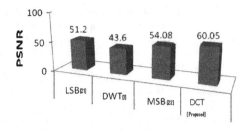

Fig. 6 PSNR comparative evaluation with different methodologies

5 Conclusions and Future Scope

By considering the limitation of the spatial domain, authors have implemented data hiding technique in DCT transform domain. Data hiding using DCT gives us better PSNR as compared with previous other methodologies used by the researchers. The algorithm is implemented in Xilinx Virtex 5, XC5VLX50T, FFG1136C FPGA-based hardware and its performance measures are examined including PSNR and MSE. After adding noise how PSNR gets affected is shown by the results. This research work can be extended to analyze additional parameters like area, power dissipation, and time complexity in the future.

References

1. Bassam Jamil Mohd, Thaier Hayajneh, et al "Wavelet-transform steganography: Algorithm and hardware implementation", International journal of electronic security and digital forensics January 2013.
2. R. Sundararaman and Har Narayan Upadhyay, "Stego System on Chip with LFSR based Information Hiding Approach," International Journal of Computer Application, Vol. 18, No.2, pp. 24–31, 2011.
3. Hemalatha Sa,1, U. Dinesh Acharyaa, "Wavelet transform based steganography technique to hide audio signals in image. Procedia Computer Science 47 (2015) 272–281. © 2015 The Authors. Published by Elsevier B.V.
4. Saeed Mahmoudpour, Sattar Mirzakuchaki, "Hardware Architecture for a Message Hiding Algorithm with Novel Randomizers", International Journal of Computer Applications, Vol 37, No7, Jan 2012.
5. S. Al-Fahoum and M. Yaser Reversible Data Hiding Using Contrast Enhancement Approach International Journal of Image Processing (IJIP), Volume 7: Issue 3: 2013.
6. Mr. Jagadeesha, Mrs. Manjula, "FPGA Implementation X-Box Mapping for an Image Steganography Technique", International Journal of Advanced Research in Electrical, Electronics, Instrumentation Engg. Vol.2 Issue 6, June 2013.
7. Hao Chen, Yu Chen, "A Survey on the Application of FPGA's for Network Infrastructure Security", IEEE Communications Surveys & Tutorials 2010.
8. Satwinder Singh and Varinder Kaur Attri, "State-of-the-art Review on Steganographic Techniques", International Journal of Signal Processing, Image Processing and Pattern Recognition Vol.8, No.7 (2015), pp. 161–170.
9. Rinita Roy, SumitLaha, "Optimization Stego Image rataining secret information using Genetic Algorith with 8-connected PSNR", Published by Elsevier B. V. 2015.

10. Ammar Odeh, Khaled Elleithy, Miad Faezipour, "A Reliable and Fast Real-Time Hardware Engine for Text Steganography" from https://www.researchgate.net/publication/262013703.
11. Dr. S Sivasubramanian, Dr, Janardhana Raju, "Advanced Embedding of Information by Secure Key Exchange via Trusted Third Party using Steganography", International Journal of Latest Research in Science and Technology Vol., 2 Issue 1 Page No. 536–540, 2013.
12. XzMin Wu, et al. "Data Hiding in Image and Video: Part I—Fundamental Issues and Solutions", IEEE Transactions on Image Processing, vol. 12, no. 6, June 2003.
13. Y. K. Lin, "High capacity reversible data hiding scheme based upon discrete cosine transformation," Journal of Systems and Software, vol. 85, no. 10, pp. 2395–2404, 2012.
14. Shabir A. Parah, Javaid A. Sheikh and G.M. Bhat, "Data Hiding in Intermediate Significant Bit Planes, A High Capacity Blind Steganographic Technique," IEEE International Conference on Emerging Trends in Science, Engineering and Technology, pp. 192–197, 2012.
15. Edgar Gomez-Hernandez, Claudia Feregrino-Uribe, Rene Cumplido, "FPGA Hardware Architecture of the SteganographicConText Technique", 18th ICECC,0–7695-3120-2/ 2008IEEE.
16. Hala A. Faroul, MagdiSaeb, "An Improved FPGA Implementation of the MHHEA for Data Communication Security", proceeding of the Design, 1530–1591/05, IEEE.
17. Po-Yueh Chen, Hung-Ju Lin, "A DWT based Approach for Image Steganography", International Journal of Applied Science and Engineering 2006, Vol. 4 No-3, Page No 275–290, 2006.
18. H.B. Kekre, A. Athawale, "Performance Comparison of DCT and Walsh Transform for Steganography", International Conf& Workshop on Emerging Trends in Technology, 2010.
19. Maninder Singh Rana, Bhupender Singh Sangwan, "Art of Hiding: An Introduction to Steganography", International Journal Of Engineering And Computer Science Vol. 1, Issue 1, Page No. 11–12, Oct 2012.
20. Dr. S Sivasubramanian, Dr, JanardhanaRaju, "Advanced Embedding of Information by Secure Key Exchange via Trusted Third Party using Steganography", International Journal of Latest Research in Science and Technology Vol., 2 Issue 1 Page No. 536–540, 2013.
21. BassamJamilMohd, Sa'ed Abed, et. at., "Analysis and Modeling of FPGA Implementations of Spatial Steganography Methods" article in journal of circuits system and computers. February 2014.
22. Multiplexed Stego path on reconfigurable hardware: A novel random approach, BalakrishnanRamalingam, Rengarajan Amitharajan, John Bosco Balaguru Rayappan, Elsevier (science direct) 2016.
23. S. Raveendra Reddy and S. M. Sakthivel, "A FPGA Implementation of Dual Images based Reversible Data Hiding Technique using LSB Matching with Pipelining", Indian Journal of Science and Technology, Vol 8(25) IJST/2015/v8i25/80980, October 2015.

A Novel Approach of Frequent Itemset Mining Using HDFS Framework

Prajakta G. Kulkarni and S. R. Khonde

Abstract Frequent itemset extraction is a very important task in data mining applications. This is useful in applications like Association rule mining and co-relations. They are using some algorithms to extract the frequent itemsets, like Apriori and FP-Growth. The algorithms used by these applications are inefficient to support balancing, distributing the load, and automatic parallelization with good speed. Data partitioning and fault tolerance is also not possible because of excessive data. Hence, there is a need to develop algorithms which will remove these issues. Here, a novel approach is used to work on the extracting the frequent itemsets using MapReduce. This system is based on the Modified Apriori, called as Frequent Itemset Mining using Modified Apriori(FIMMA). To automate the data parallelization, well balance the load and to reduce the execution time FIMMA works concurrently and independently using three mappers. It uses decomposing strategy to work concurrently.

Keywords Association rules · Frequent itemsets · Data partitioning
Load balancing · MapReduce · Hadoop · FIMMA

1 Introduction

Frequent itemset extraction is the basic problem in data mining applications, such as association rule, correlations, sequences, and many more data mining tasks. Hence, this becomes an important research topic to extract the frequently used itemsets. These frequent patterns are useful to take decisions in product marketing, sales, etc. [1]. Association rule mining is popular in data mining [2]. The main goal of Association rule is to find all the rules that fulfill a user-defined threshold. The first

P. G. Kulkarni (✉) · S. R. Khonde
Modern Education Society's College of Engineering, Pune 411001, India
e-mail: prajakta.r999@gmail.com

S. R. Khonde
e-mail: shraddha.khonde@mescoepune.org

© Springer Nature Singapore Pte Ltd. 2018   435
S. Bhalla et al. (eds.), *Intelligent Computing and Information and Communication*,
Advances in Intelligent Systems and Computing 673,
https://doi.org/10.1007/978-981-10-7245-1_43

phase of association rule is to identify frequent itemsets whose support is greater than the threshold and the second phase is to form conditional implication rules, among the frequent itemsets. Frequent itemsets generation defines the two similar itemsets. The first itemset is similar to another. Now a day, there are enormous data generated from different areas such as IT companies and web applications. Existing data mining applications are unable to handle vast data and are only suited for a typical database. Thus, to extract the frequent itemsets from the excessive database is a very critical task [3]. For better utilization of frequent itemsets using large size database, speed is very important. Speeding up the process of FIM is very complex because it consumes most of the time to calculate the input/output intensity. In this modern era, datasets are excessively large and sequential FIM algorithms are unable to compute large database. They, however, failed to analyze data accurately and they suffer from performance degradation. To solve these problems, MapReduce is used to calculate frequent itemsets. Using this approach, the data will not only be distributed in an efficient way but also balanced in the cluster. Hence, the performance of finding frequent itemsets will be optimized [3].

This MapReduce is using the FIM which is based on the Modified Apriori, called FIMMA. In this strategy, we are focusing the data partitioning method, load balancing of data with a parallel approach. FIMMA consumes less time compared with the traditional Apriori. The working of mappers and reducers is done concurrently to optimize the speed, well balancing the load across various clusters [3].

The rest of this paper is partitioned as follows. Section 2 gives the review of the literature. Section 3 defines the problem statement. Section 4 gives the present system architecture. Section 5 explains the algorithms and methodology for the system and discussed the expected results in Section 6. Section 7 concludes this paper.

2 Related Work

The authors of "Association Rule mining extracting frequent itemsets from the large database" have presented a problem of finding the frequent items from the excessive database. The authors have developed the rules that have minimum transactional support and minimum confidence. For this, an algorithm is used that carefully estimates the itemsets for one pass. It adjusts the data between the number of passes and itemsets that are measured in a pass. This process uses pruning system for avoiding certain itemsets. Hence, this gives exact frequent itemsets from excessive databases [4, 5]. A number of parallelization procedures is used to increase the performance of Apriori-like algorithms to find frequent itemsets. MapReduce has not only created but also exceeds in the mining of datasets of gigabyte scale or greater in either homogeneous or heterogeneous groups. The authors have implemented three algorithms, DPC, FPC, and SPC [6]. SPC has straightforward functions and the FPC has static passes merged checking capacities. DPC consolidates the dataset of various lengths by utilizing dynamic strategy and it gives good performance over the other two calculations. Accordingly, these three calculations will scale up to the expanded dataset [6].

When dataset gets larger the mining algorithms becomes inefficient to deal with such excessive databases. The authors have presented a balanced parallel FP-Growth algorithm BPFP [7], a revised version of PFP algorithm [4]. FP-growth algorithm is utilized with the MapReduce approach named as Parallel FP-growth algorithm. BPFP balances the load in PFP, which boosts parallelization and automatically enhances execution. BPFP gives a good performance by utilizing PFP's grouping system [7].

FIUT suggests a new technique for mining frequent itemsets called as Frequent Itemset Ultrametric Tree(FIUT) [8]. It is a sequential algorithm. It consist two main stages to scan the database. First-stage calculates the support count for all itemsets in a large database. The second stage uses pruning method and gives only frequent itemsets. While calculating frequent one itemsets, stage two will construct small ultrametric trees. These results will be shown by constructing small ultrametric trees [8]. Dist-Eclat, BigFIM are two FIM algorithms used with MapReduce Framework. Dist-Eclat focuses on speed by load balancing procedure using k-FIS. BigFIM concentrates on hybrid approach for mining excessive data [9]. Apriori algorithm is additionally used to create kth FIS itemsets. The Kth FIS is used to search frequent itemsets based on the Eclat system. These three algorithms are used with round-robin technique which achieves a better data distribution [9].

PARMA uses parallel mining approach with the benefits of Randomization for extracting frequent itemsets from a vast number of databases. This divides the functionality into two parts, gathering the arbitrary data samples and secondly it uses parallel computing method that is utilized to increase the mining speed. This method avoids the replication that is very expensive. A mining algorithm applies to every segment individually with parallel approach [10]. K-Nearest Neighbor Joins utilizes MapReduce and distributes the excessive information on the number of machines. This is done by the mappers and the reducers give the results in terms of the KNN join. KNN Join is the key component to search the kth-nearest neighbor. MapReduce is utilized for effective computing the data to obtain the best performance result [11, 12]. To diagnose the Heterogeneous Hadoop Cluster and to search primary faults, this paper is used Hadoop schedulers to produce efficient Hadoop clusters even if they are in heterogeneous clusters [13]. It proposes the DHP algorithm (direct hashing and pruning) which is used for minimized candidate set generation for large itemsets. It solves performance degradation problem of large dataset mining. It minimizes candidate itemsets.

FIUT is used with MapReduce to find frequent itemsets. MapReduce is a popular programming approach used for computing massive datasets [14]. It divides into three MapReduce phases. The database is divided into number of input files and given to each mapper. The first MapReduce phase finds out frequent-1 itemset and Second MapReduce phase scans the frequent one itemsets and generates k-frequent itemsets. Third MapReduce phase uses FIUT algorithm and it will create ultrametric tree [3]. FiDoop-DP Data Partitioning uses the Map Reduce programming and gives the effective data partitioning technique for frequent itemset mining. This increases the performance by using the data partitioning technique, which is based on the Voronoi diagram. It extracts the correlations between the transactions. By consolidating the similarity and the Locality-Sensitive Hashing strategy, FiDoop-DP puts most similar records in data partition to increase locality and this is done without repeating records [5, 15]. To differentiate and extract frequent and infrequent itemsets from the massive database two-phase scanning will be done here. In the first scan, it accepts input and distributes it into mappers and finds out infrequent itemsets using minimum support. The reducer combines the result and sends it to the second phase. In this phase, it scans first phase output and gives the final result [16].

3 Problem Statement

Problem statement concentrates on the investigation of Frequent Itemset Ultrametric Tree (FIUT) and to find the efficient way for its execution in HDFS framework Implementation on FIMMA. To show that the proposed algorithm on the cluster is sensitive to data distribution and dimensions, as itemsets with different lengths have different decomposition and construction costs. Improving energy efficiency of FIMMA running on the Hadoop clusters. To improve the performance, a workload balance metric to measure across the clusters computing nodes is developed.

4 System Architecture

FIMMA suggests parallel frequent itemset mining algorithm which uses MapReduce programming technique for development. This removes the issues of existing system and applies automatic parallelization, balancing the load of the excessive database, and well distribution of given data. FIMMA is based on Modified Apriori algorithm to overcome the issues of FIUT algorithm with reduced time. It uses hash-based technique [17].

4.1 Objectives of Proposed System

- Better performance and improved accuracy using automatic parallel processing. It performs with less time execution to scan the database.
- Keeping the cost constraint as it is and dealing with load balancing with automatic parallelization.
- Hashing technique is used to differentiate the traditional Apriori.

The proposed system consists three MapReduce phases. The user is responsible to give the input. This input will be accepted by the job manager as shown in Fig. 1. Job manager splits the input into a number of blocks, processes it and gives the output to the reducer. This output is in the form of frequent one itemset.

The output of the first MapReduce is applied to the mapper of the second phase. It scans all the data to give frequent-k itemsets. It uses pruning method to find out frequent and infrequent itemsets. The result is obtained from reducer, in the form of k-frequent itemsets. The system architecture is as shown in Fig. 1. This is distributed into two main parts.

1. HDFS framework
2. MapReduce Approach.

HDFS is a Hadoop distributed file system and used to store the log files. HDFS framework accepts the data from the user. The user gives the input using SQL queries or in the form structured data and uploads it. This uploaded dataset accessed by the Job manager. Job manager is responsible to distribute the dataset to the available mappers of each data node.

Here, the transactions from input dataset are distributed to the mappers. Each mapper access the input scans it and generates the results in terms of key and values. Key is the item-name and value is the item-count. Using the minimum support the mappers gives the result to the reducers in the form of key and values. Each mapper calculates frequent-1 itemset. Reducer combines the result of each

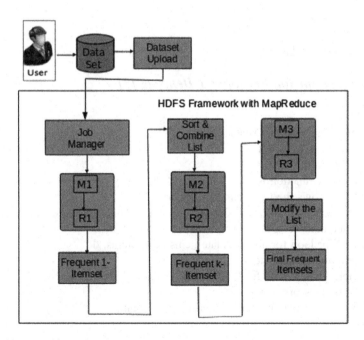

Fig. 1 System architecture

mapper, sorts it and generate a final list. This is frequent-1 itemset list. This output gives to the next MapReduce. It applies another round of scan. In this stage, it accepts the frequent-1 items and compares it with the minimum support and removes the infrequent items. This is called pruning system. Depending upon the users given threshold value of k this MapReduce generates k-frequent itemsets. (where k-itemsets < n number of dataset) It makes the possible combination of frequent itemset from each mapper and gives to the reducer. Second MapReduce updates the list and gives k-frequent itemset. This frequent-k-itemsets applies to the third MapReduce. In this MapReduce, it accepts all k- itemsets and gives the result in terms of top most k-frequent itemset.

FIMMA uses hash-based technique. There is a hash table used after each result of MapReduce. These hash tables are used to store the result generated from frequent-1 itemset to the k-itemset. These hash tables give a unique value to the stored frequent items. Unique value is obtained by calculating mod hash formula. Whenever there comes new input then hash table compares that items with stored one. When new input from first mappers matches with stored one then hash table sets a bit to 1 otherwise 0. This same procedure applies to the result of second MapReduce (i.e., k-frequent itemsets) and on the third MapReduce. In last MapReduce, it checks all set bit and stores into a new list, update it and gives the final result. FIMMA helps to reduce the candidate generation of items by using hash tables, therefore automatically it reduces the time. It controls the huge generation of candidates with minimum support. It avoids the transaction record which does not have any frequent items by comparing hash tables.

5 System Analysis

5.1 Algorithm for Frequent 1 Itemsets [3]

Input: minimum-support, Database D
Output: Frequent-1 itemset
Mapper Algorithm
Step 1: Mapper function is used with-MAPPPER (key offset, values Database D)
Step 2: //TR shows the transaction in Database D
Step 3: for loop is used for all Transactions TR in Database D do
Step 4: Candidate-items < – Splited each transaction TR.
Step 5: Use for loop for all candidate-items in all items, do
Step 6: output(candidate-item, 1)
Step 7: Here ends second for loop
Step 8: Here ends first for loop
Step 9: Mapper Function ends with each items count.

Step 10: The reducer takes input from mappers as input = (candidate-item, 1)
        Reducer Algorithm
Step 11: REDUCE function starts with key, value (key candidate-item, values 1)
Step 12: take a variable total to store output, i.e., total = 0;
Step 13: Use for loop to calculate all candidate-item do
Step 14: Add new frequent item in total +=1 // Here ends for loop
Step 15: Output (frequent1-itemset, total).

5.2   Algorithm for Frequent-K-Itemsets

Input: minimum-support, Database D
output: frequent-k-itemsets.
Mapper Algorithm:
Step 1: In step 1 mapper function is used i.e. MAPPER (key offset, values
       Database)
Step 2: //TR shows the transaction in Database D
Step 3: for loop is used for all Transactions TR in Database D do
Step 4: Candidate-items ← Splited each transaction TR. //Here input database D is
       frequent-1 itemset
Step 5: Use for loop for all candidate-items in all item, do
Step 6: Step 6 applies pruning system using if condition for infrequent items
       if Candidate-item = infrequent item then
Step 7: Remove the Candidate-item which is infrequent from the Transaction TR;
Step 8: If conditions ends here
Step 9: variable Frequent-k-itemset is used to store all k-frequent items and shown
       by-Frequent-k-itemset ← (frequent-k, fr-set)
       //After applying pruning system fr-set is the result with frequent-k items
Step 10: output (Frequent– itemset, 1);
Step 11: Here ends second for loop
Step 12: Here ends first for loop
Step 13: Mapper function is ends here by giving all mappers output.
       Reducer Algorithm
Step 14: Reducer starts from this step using function REDUCER (key k-itemset,
       values 1)
Step 15: Total = 0;
Step 16: used for a loop to count all items from mapper as-for all (k-itemset): do
Step 17: Total += 1;//For loop ends here
Step 18: output = (frequent-k items + total)
Step 19: Reducer function ends here with final k output.

5.3 Algorithm for FIMMA

Step 1: Consider C be the variable for selection of one cluster at a time

Step 2: Here, the database will be scanned using minimum support and it will generate frequent items. It will combine all the possible combinations of frequent itemsets

Step 3: Function Fre1 stores the frequent itemsets → Fre1 = find-freq-1itemset(T)

Step 4: for k = 2 to $f_{k-1} \neq \emptyset$; generate $\emptyset$ from f_{k-1} items

Step 5: Consider H1 is the hash table of size 8. B1 is buckets in the hash table and A1 is Unique value to the frequent itemsets. V1 is the bit vector

Step 6: Calculate the items I up to user threshold value w from ck with min support; ft($1 \leq w \leq k$); end for

Step 7: Calculate frequent items with minimum support

Step 8: y variable to store the result of minimum support, i.e., y = min support (c_k, f_t)

Step 9: get transaction id in variable → target = get-trans id (y)

Step 10: compare the target values with the hash tables

Step 11: for each transaction id in target increment count of all ck

Step 12: if b1 >=min support; then bit vector v1 = 1; otherwise v1 = 0

Step 13: prune the 1 = 0 itemsets and modify the list

Step 14: f_k = items in c_k (min support)\\ end for

5.4 Mathematical Model

Let S be the system which do analysis and read documents; such that:

S = {S1, S2, S3, S4, S5} where—S1 represents a query requesting by the user; S2 represents authentication; S3 represents MapReduce module; S4 denote the sql injection techniques; S5 gives the graphical presentation.

S1 = {U1, U2, U3........Un}; Where, S1 contains SQL query–If S1 is valid then proceed, Else discard.

S2 define user is authenticated or not; Where, Ui = {UI1, UI2.UI3......UIn}

- Ui is the master node which having different storage nodes as clusters.

S3 = Functionality for three MapReduce phases:

- Input: database DI,min sup; Output: Frequent itemsets
- Let DI = {DI1, DI2, ...Din}; where, DI is Input Database

Applying algorithms to Find 1 and k-frequent itemsets, Hash-based Algorithm
$F1$(freq 1-itemset) = scan $\sum 0^{DI}$
; Fk (freq k-itemset) = scan $\sum$ (F1)

$FI = \{FI1, FI2, \ldots .FIK\}$; Where FI is the final frequent itemsets
$$S4 = \{\text{patten 1, pattern 2}\ldots\ldots\text{pattern n}\}$$

Each pattern checks the behavior of query created by end user's S1 (query module). $S1 = \text{avg} + \text{min} + \text{max} + \text{round} + \text{floor} + \text{todate} // \text{possible queries}$

$$S5 = \text{graphical representation for time comparison graph}$$

6   Experimental Results

The experimental results evaluated with the minimum support (sometimes larger data sizes). This upgradation comes at no execution cost, as they prove the way that this implementation, achieves the good performance, compared to other techniques with reduced time. By examining with this work, it demonstrates that, the execution of FIUT is slow. It results that, whenever increased the minimum support and dataset, it gives the balanced output. FIMMA gives an improved performance by using hash table concept. When the threshold value decreases, other methods occupy more memory as well as consume more time.

Table 1 shows the time required to extract frequent itemsets. As shown in table I, first column shows the Size of Dataset. Other columns show the methodologies to find frequent itemset. Proposed system requires less time compared to existing systems. It shows the time in seconds. The performance of this system against the Frequent Itemset Ultrametric Tree (FIUT) method is as shown in Fig. 2. Modified Apriori algorithm is a good algorithm to give the correct results as compared to existing systems. Also, this algorithm shows the faster execution even for a large database. As synthetic dataset is used here, the user can make their own dataset to run the tests.

Table 1 Time required for finding the frequent itemset from dataset

Size of dataset	FIUT	Modified apriori (FIMMA)
1000	139	113
2000	165	144
3000	218	196
5000	277	234
10,000	317	293

Figure 2 shows the graphical representation of the methodologies and datasets. These are the results of two methodologies when the dataset is increased. X-axis shows the methodologies and y-axis shows the time in seconds. Figures 3 and 4 shows the Hadoop implementation results. Figure 3 shows the FIUT implementation with required time in ms. In this, it takes 43,780 ms (shown in red color rectangle) to execute a job. Figure 4 shows the FIMMA implementation. Here it shows the time required to execute the same job. FIMMA takes 21,090 ms to execute a job. Hence, it shows that FIUT takes more time than the FIMMA method. FIMMA gives better performance than the FIUT with reduced time.

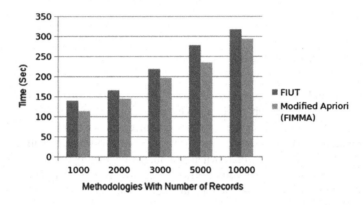

Fig. 2 Comparison graph

```
                GC time elapsed (ms)=8338
                CPU time spent (ms)=43780
                Physical memory (bytes) snapshot=1100861440
                Virtual memory (bytes) snapshot=3386929152
                Total committed heap usage (bytes)=758906880
        Shuffle Errors
                BAD_ID=0
                CONNECTION=0
                IO_ERROR=0
                WRONG_LENGTH=0
                WRONG_MAP=0
                WRONG_REDUCE=0
        File Input Format Counters
                Bytes Read=898560
        File Output Format Counters
                Bytes Written=5900164
Copying files to /output/part-00000 ==> /usr/local/market_hadoop_op/AGED gt 35.txt
Copying files to /output/part-00001 ==> /usr/local/market_hadoop_op/AGED gt 35.txt
Copying files to /output/part-00002 ==> /usr/local/market_hadoop_op/AGED gt 35.txt
Copying files to /output/part-00003 ==> /usr/local/market_hadoop_op/AGED gt 35.txt
```

Fig. 3 Hadoop implementation using FIUT

```
                    GC time elapsed (ms)=2515
                    CPU time spent (ms)=21890
                    Physical memory (bytes) snapshot=1060442112
                    Virtual memory (bytes) snapshot=3379830784
                    Total committed heap usage (bytes)=773062656
         Shuffle Errors
                    BAD_ID=0
                    CONNECTION=0
                    IO_ERROR=0
                    WRONG_LENGTH=0
                    WRONG_MAP=0
                    WRONG_REDUCE=0
         File Input Format Counters
                    Bytes Read=384096
         File Output Format Counters
                    Bytes Written=2708751
Copying files to /output/part-00000 ==> /usr/local/market_hadoop_op/AGED gt 35.t
xt
Copying files to /output/part-00001 ==> /usr/local/market_hadoop_op/AGED gt 35.t
xt
Copying files to /output/part-00002 ==> /usr/local/market_hadoop_op/AGED gt 35.t
xt
Copying files to /output/part-00003 ==> /usr/local/market_hadoop_op/AGED gt 35.t
xt
```

Fig. 4 Hadoop implementation using FIMMA

7 Conclusion

FIMMA technique is used to defeat the issues which are available in existing methods like parallel mining and load adjusting algorithms. In this approach, FIMMA algorithm (Hash-based) is proposed using MapReduce. The Comparison between existing method and proposed methods shows that there is up to 60% reduction in time. Hash-based Apriori is most efficient for generating the frequent itemset than existing methods. Data partitioning and data distribution is done by using this proposed system. FIMMA stores the previous results in hash tables and therefore it reduces time. At each result stage, the new input will be compared with the hash table and if matches with hash table's stored result then vector bit is set to 1 and named as a frequent item. All the set vector's list will be combined by the reducer, modify it and gives the final result. The proposed system works efficiently on MapReduce stages and gives the final result, rather than using the FIUT algorithm and increases the performance.

References

1. Bechini, Alessio, Francesco Marcelloni, and Armando Segatori. "A MapReduce solution for associative classification of big data", Information Sciences, 2016.
2. X Zhou, Y Huang - Fuzzy Systems and Knowledge Discovery. An Improved Parallel Association Rules Algorithm Based on MapReduce Framework for Big Data", pp. 284–288, 2014 11th International Conference on Fuzzy Systems and Knowledge Discovery.

3. Yaling Xun, Jifu Zhang, and Xiao Qin, FiDoop: Parallel Mining of Frequent Itemsets Using MapReduce" IEEE TRANSACTIONS ON SYSTEMS, MAN, AND CYBERNETICS: SYSTEMS, VOL. 46, NO. 3, pp. 313–325, MARCH 2016.
4. R. Agrawal, T. Imieli nski, and A. Swami, "Mining association rules between sets of items in large databases," ACM SIGMOD Rec., vol.22, no. 2, pp. 207–216, 1993.
5. S Deshpande, H Pawar, A Chandras, A Langhe, Data Partitioning in Frequent Itemset Mining on Hadoop Clusters"- 2016 – irjet.net, https://irjet.net/archives/V3/i11/IRJET-V3I11229.pdf.
6. M.-Y. Lin, P.-Y. Lee, and S.-C. Hsueh, "Apriori-based frequent itemset mining algorithms on MapReduce," in Proc. 6th Int. Conf. Ubiquit. Inf. Manage. Common. (ICUIMC), Danang, Vietnam, 2012, pp. 76:1–76:8.
7. L. Zhou et al., "Balanced parallel FP-growth with MapReduce," in Proc. IEEE Youth Conf. Inf. Compute. Telecommun. (YC-ICT), Beijing, China, 2010, pp. 243–246.
8. Y.-J. Tsay, T.-J. Hsu, and J.-R. Yu, "FIUT: A new method for mining frequent itemsets," Inf. Sci., vol. 179, no. 11, pp. 1724–1737, 2009.
9. Kiran Chavan, Priyanka Kulkarni, Pooja Ghodekar, S. N. Patil, Frequent itemset mining for Big data", IEEE, Green Computing and Internet of Things (ICGCIoT), pp. 1365–1368, 2015.
10. M. Riondato, J. A. DeBrabant, R. Fonseca, and E. Upfal, "PARMA: A parallel randomized algorithm for approximate association rules mining in MapReduce," in Proc. 21st ACM Int. Conf. Inf. Knowl. Manage.,Maui, HI, USA, pp. 85–94, 2012.
11. Wei Lu, Yanyan Shen, Su Chen, Beng Chin Ooi, "Efficient Processing of kNearest Neighbor Joins using MapReduce" 2012 VLDB Endowment 2150-8097/12/06, Vol. 5, No. 10, pp. 1016–1027.
12. [Online]. Available:www.vldb.org.
13. Shekhar Gupta, Christian Fritz, Johan de Kleer, and Cees Witteveen, "Diagnosing Heterogeneous Hadoop Clusters", 2012, 23 rd International Workshop on Principles of Diagnosis.
14. J. Dean and S. Ghemawat, "MapReduce: A flexible data processing tool," Commun. ACM, vol. 53, no. 1, pp. 72–77, Jan. 2010.
15. Yaling Xun, Jifu Zhang, Xiao Qin and Xujun Zhao, "FiDoop-Dp Data Partitioning in Frequent Itemset Mining on Hadoop clusters", VOL. 28, NO. 1, pp. 101–113, 2017.
16. Ramakrishnudu, T, and R B V Subramanyam. Mining Interesting Infrequent Itemsets from Very Large Data based on MapReduce Framework", International Journal of Intelligent Systems and Applications, Vol. 7, No. 7, pp. 44–49, 2015.
17. Jong So Park, Ming Syan Chen, Philip S, "An Effective Hash based Algorithm for mining Association rule", '95 Proceedings of the 1995 ACM SIGMOD international conference on Management of data, held on May 22–25, 1995, Pages 175–186.

Issues of Cryptographic Performance in Resource-Constrained Devices: An Experimental Study

Balaso Jagdale and Jagdish Bakal

Abstract Many gazettes, toys, sensors, instruments, etc., devices are mushrooming in the electronics market these days which are running with different operating systems and application frameworks. In Mobile phones also many platforms are available with its own stack, right from hardware level to application level. So it is difficult to all the mobile phones or devices with same characteristics, performance, and features. The conclusive success of a platform entirely depends on its security to the user data ultimately, it constructs the global market. We present here a comparative study of four Mobile Operating Systems architecture and security perspectives of each. Moreover, experiments are carried out with Windows and android smartphone to study the cryptographic performance in these devices which plays as decisive parameters for future applications and system developers to implement secured devices and applications. We show that although algorithms characteristics are fixed, due to different architectures and dependencies in different devices with different operating systems, same efficiency, and performance is not guaranteed by cryptographic as well as other data structure algorithms.

Keywords Security · Cryptography · Mobile platform · Performance

1 Introduction

Users prefer storing data on the mobile device so that they have access to data as and when they need it. The data stored is private and sensitive in nature and hence it is desirable to have an operating system on these devices that secure the data from malicious applications. Here, we look at how the leading mobile operating systems

B. Jagdale (✉)
G. H. Raisoni College of Engineering, Nagpur, India
e-mail: bjagdale@gmail.com

B. Jagdale · J. Bakal
MIT College of Engineering, Pune, India
e-mail: bakaljw@gmail.com

© Springer Nature Singapore Pte Ltd. 2018 447
S. Bhalla et al. (eds.), *Intelligent Computing and Information and Communication*,
Advances in Intelligent Systems and Computing 673,
https://doi.org/10.1007/978-981-10-7245-1_44

implement the security. The operating systems under consideration are Android, iPhone, Symbian (being absolute), and Windows mobile.

Naseer Ahmed and other [1], have compared different operating system used in mobile devices and shown how to utilize resources efficiently. Ahmed and others demonstrated the security algorithms performance in RFID resource-constrained devices [2], where authors have discussed energy and performance challenge. Authors N. R. Potlapally [3] and other have worked with SSL and energy and performance efficient cryptography for mobile platforms. Daniel [4] has worked with Texas and light microcontrollers and studied the performance of ultra-lightweight cryptography. Payal patel [5] and other have worked with ECC implementation for resource-constrained devices. Mathias and other have discussed flexibility and strength of a Mobile OS. Viewpoints can be the manufacturer and end user [6]. In another study [7], the author suggests an application-based selection of operating system and based on the technical parameters and features, the user should select mobile devices.

In Fig. 1, mobile operating system layers are shown which are generic in nature. At the bottom, hardware boards with processors are controlled by the operating system and the user has not control. Middle library API can be accessed by user programs. But application framework controls permission during installations. Moreover, communication services are also staked in the form of HTTP and similar.

There are various places to apply the security controls, such as architectures, hardware, etc., cryptography is a prime security control as compared other security controls Many authors have done applied research of cryptographic algorithms in resource-constrained devices. The reason is the feasibility of performance, power consumption, and strength. The research work of S. Prasanna Ganesan and others [8] shows RSA-based asymmetric cryptography is still a challenge in mobile devices due to required computing power, Infrastructure support, and memory capacity. Creighton T. R. Hager and others [9], studied the performance and energy requirements for various cryptographic algorithms in personal digital assistants and tablets.

Fig. 1 Stack of mobile operating system

Mobile Applications
Application Framework
Application Execution Environment
Middleware (Libraries)
Kernel and Drivers
Hardware Platform

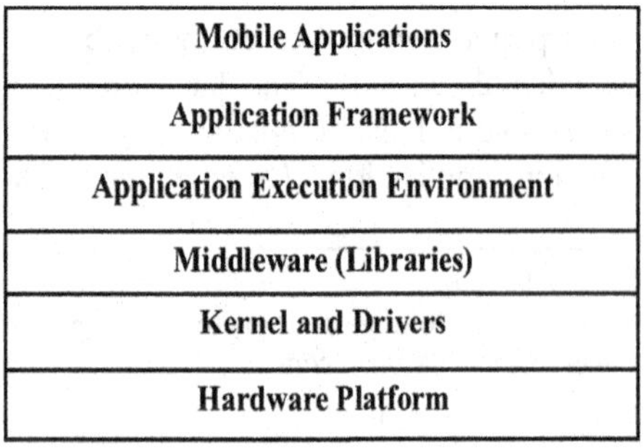

2 Architectures of Mobile Device Operating System

In this section, we are discussing the architecture of the mobile operating systems Android, iPhone, and Windows mobile.

Figure 2 shows android layers. Open Handset Alliance launched android device platform in 2007. The goal was to keep it open, free and diverse for the Internet of things. OS Layers are shown in Fig. 2. Many mobile manufacturers like Motorola, Ericson, HTC, and Samsung have adopted the platform. It usually is loaded on the ARM-Snapdragon processor designed by Qualcomm. It has Good energy monitoring and memory management. Now that android is used in many electronics-enabled appliances, it has become important to study performance and other utilization of resources.

Android implementation is based on Linux kernel [10, 11]. Native libraries are written in C/C ++ to achieve the stable performance of various elements. The main component of Android Runtime, Dalvik Virtual Machine. The architecture of iPhone is shown in Fig. 3. ARM series of processors are used by iPhone. SDKs and drivers are mainly developed in Objective C runtime. Figure 4 shows, windows mobile layers [12], which uses Microsoft Win32 API. Design and applications are similar to the desktop flavor of windows. Windows CE.NET-based platform is used

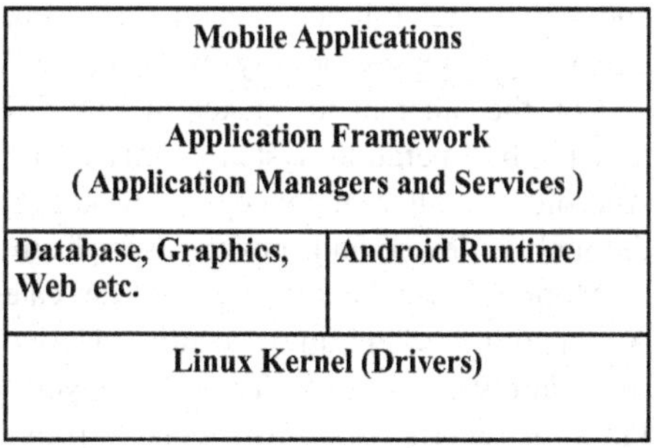

Fig. 2 Architecture of Android

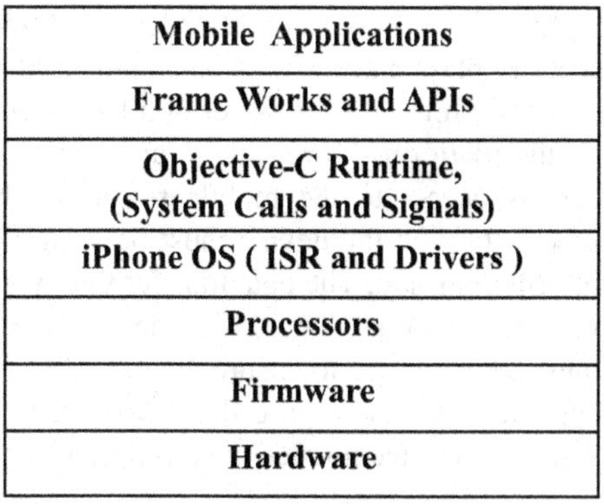

Mobile Applications	
Application Framework (Application Managers and Services)	
Database, Graphics, Web etc.	Android Runtime
Linux Kernel (Drivers)	

Fig. 3 Architecture of iPhone OS

Mobile Applications
Frame Works and APIs
Objective-C Runtime, (System Calls and Signals)
iPhone OS (ISR and Drivers)
Processors
Firmware
Hardware

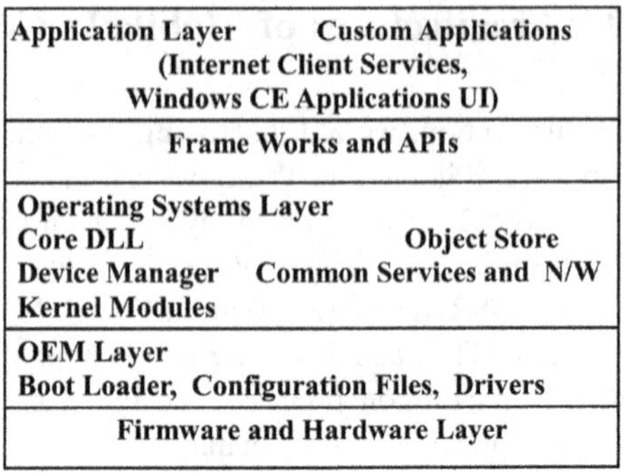

Fig. 4 Architecture of
Windows mobile OS

| Application Layer   Custom Applications |
| (Internet Client Services, |
| Windows CE Applications UI) |
| Frame Works and APIs |
| Operating Systems Layer |
| Core DLL   Object Store |
| Device Manager   Common Services and N/W |
| Kernel Modules |
| OEM Layer |
| Boot Loader, Configuration Files, Drivers |
| Firmware and Hardware Layer |

[12]. Performance is dependent on OEM adaptation layer (OAL) which is placed between the Microsoft Windows CE.NET kernel and target hardware mobile device.

3   Security Aspects and Analysis in Mobile Devices

Users increasingly use their smartphones for Internet access, accessing their email, and store more and more sensitive information on their mobile phones. This, on the one hand, increases the ways in which mobile phones can be infected with malware, and on the other hand, makes mobile devices increasingly attractive as an attack target. The operating systems, [11] Symbian, Windows Mobile, iPhone OS, and Android use different concepts to protect the integrity of Software and Firmware on the mobile device and prevent users from installing malware-infected programs.

Mobile device's security can be categorized as follows: Platform security, Application configuration security, Network services security, Data storage security, Hardware security, Transport layer security, and Application layer security. Platform security: Platform security provides protection to the users' information such as secondary data storage, phone memory data storage, and location information protection, protection to different hardware elements, generating, or storing information, present on the device. Following are the important areas, where security plays a key role. Application configuration security: whereby application to be fixed on the device is verified by the software developer but is tested for safety by the platform. Moreover, in application configuration one can restrict the applications access to the mobile resources. Network services security: the network access through the device must be secure enough to keep the login data, i.e., user information safe. The data transfer that would take place in the sessions must not be vulnerable. Proper encryption-decryption methods must be used. Proper policies must be employed to ensure the integrity, secrecy of the data. Data storage security: applications may require the data stored on the device. Proper authentication and authorization techniques must be used to restrict the open access to the data on the device. Hardware security: security provided for different hardware modules can be categorized here. Locking keypad, passwords set against usage of phone memory, sim card. Transport layer security: can be implemented in the form of secured socket layer which is implemented just above transport layer which provides API to applications. This is implemented during the development of client–server mobile applications with the help of programming libraries. Browser and OS network applications support SSL to access secured sites which is an inherent security provided in the mobile device platform. Application layer security: applications may be installed explicitly or may be installed as a part of other application. Application layer security may be implemented in mobile applications with the help of cryptography libraries provided with the programming SDKs. The authenticity of the application must to be checked. These applications may access data from other applications which may harm the data or functioning of other applications.

(A) Android: Android supports security mode to cater mobile situation. Android has system features similar to shared memory, files access control, user Identifiers and Preemptive multitasking. The Dalvik virtual machine implements main security of the android operating system. The main design aspect of Android is that it has control over applications due to which applications cannot adversely affect system [11]. Android reduces abuse of access of device data from applications by employing permission features statically and dynamically [13]. Rooted phone concept is also implemented in Android, where only advanced system developers can root the device and perform core changes in the system as per the requirements. Through Google play, applications are going through security check before they are deployed by users in the phone.

(B) iPhone: The iPhone OS runs applications with root access. Thus, it makes the operating system vulnerable to security attacks. iPhone has security features which help to protect sensitive data collected by applications. iPhone OS deploys BSD (Berkeley Software Distribution) and CDSA (Common Data Security Architecture) that supports protection of services that are implemented by using CDSA. APIs support send, receive, and store data in the encrypted form [14]. File access permissions are similar to UNIX BSD kernel, as a part of security features.

(C) Symbian: Symbian has several security features for protecting integrity and privacy. Symbian security architecture concentrates mainly on threats of distribution of malicious applications. Applications are signed by users or developers whenever installed on the device. After that access to resources is limited. But there is a risk of using nonsigned applications running on the device. Control of API's is limited for the programs running in the device, thus restricting access of file system for trivial reasons. Each application as public and private file system as a part of security. Moreover, API is provided to implement security features for cryptography, hashing, and random number generation.

(D) Windows mobile: Programmers are allowed to use lower APIs. So programmer has the freedom to control hardware. The build has restricted security features in the operating system. Programs do not have their own boundaries and difficult to protect from each other. Even though, programs cannot access directly, but with DLL one program can access maliciously other user's data. Small process bugs can be tested but this is not enough to protect from the malicious activity of applications. Security policy features are available so that trusted API's can refer before granting permissions to access low-level resources including hardware [12].

4 Cryptographic Experiments

We carried cryptographic experiments in Windows Mobile and Android mobile. Experiments are run for digest algorithm, Symmetric algorithm, Public key algorithm. In digest category, we used the popular SHA1 algorithm. In the same key category, we chose AES (advanced encryption standard). Finally, for Public key cryptography, we experimented RSA algorithm. The experimental setup diagram is as follows.

As shown in the Fig. 5, the experimental setup consists of a personal computer loaded with the three different or possibly a single universal SDK, which could be used to write, compile, and test the application on the emulator. The deployment packages would then be created. These packages will be deployed on the mobile device through a USB cable and will be executed to study their behavior.

The time required to execute a task is measured with timers available in programming languages. Time can be measured in nano or milli seconds resolutions in modern SDKs. Same experimental setup is also applicable for Win mobile instead of android SDK, it is Visual studio 2008, Windows mobile SDK. Cryptographic algorithms were run in both the platforms with 20, 40, 60, 100 K text file samples.

```
Cryptographic task ()//Time measurement in mobile phone
Start
  Note the start time ()
  Open the data file (file1)
```

Fig. 5 Device Development System

```
Output = Encrypt (file1, algorithm, key)
StoreTheCipherOutput (Output, file2)
Note the end time ()
Time required i.e. end time - start time;
End
```

We used HTC touch diamond2 phone to conduct the experiments. This phone is dual OS phone. Android java is used to run these programs in android device. C Sharp is used to win mobile programming. Multiple readings were taken in which minimum time reading is selected as proper cryptographic time. Also, mobile is kept in idle mode while this program is run by killing unwanted processes. Also, SIM card is not installed in mobile to avoid communication events and in the effort to keep the mobile device in silent mode.

5 Results and Discussions

This section shows the results from different experiments in graphical form and security analysis is done based on the performance results and other parameters. As shown in the graph, Fig. 1 shows the performance for the SHA1 algorithm in which time required less means good performance. In SHA1 Android is scoring well as compared to Windows mobile, but in graph 2, we can infer that windows mobile is doing well as compared to the android platform for Advanced encryption standard. In graph 3, again RSA is doing well which is a well-accepted Public key algorithm. Here are some Experimental Development Observations (Fig. 6).

We found that Android has good programming model and many libraries as compared with windows mobile system. The security model of android is better that windows mobile. In windows mobile, live device interface and programming is excellent (plug and play) because of Active Sync utility from Microsoft (Figs. 7, 8).

Moreover, windows have upper hand for GUI design and ease of programming. It is difficult to carry out same experiments of same logic on different platforms and study benchmarking. The main reason is that iPhones, Symbian's, Win mobile,

Fig. 6 SHA1 Algorithm performance

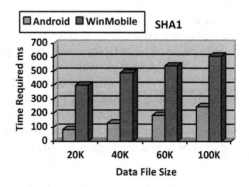

Fig. 7 AES 128 (Rijndel) Algorithm performance

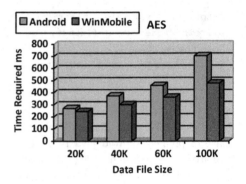

Fig. 8 RSA 1024 Algorithm performance

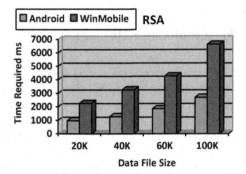

Android platforms are ported on numerous hardware platforms with variable size form factors including Processors, Clock frequencies, RAM, ROM, buses, displays, sensors, etc.

6 Conclusion

Cryptographic results are important to implement security for mobile data while data is being transmitted or received over the network as well as data residing in the mobile phone. As can be seen from experiments, android performs better in public key and hash algorithms while windows perform better in symmetric key algorithms. One important thing to observe is that algorithms are same but OS and lower layers have an impact on performance. There are other ways for security in term of signing the applications or during installations, OS provides security for verification of applications. One can also carry out this kind of applied research in the field of hybrid cryptographic protocols.

The android operating system runs applications in a closeted virtual machine environment. Therefore, the android security architecture is strong. Symbian has a security system in place for authenticating applications and windows mobile also supports this, although the system is less sophisticated. iPhone implements security services at OS level by using BSD and CDSA, leaving less freedom to the developer. It is not possible to point out one as the best OS but we can certainly choose the one over other based on their application development framework and security.

References

1. Naseer Ahmad, Boota, M.W. and Masoom, A.H.: Comparative Analysis of Operating System of Different Smart Phones, Journal of Software Engineering and Applications, 8, 114–126, 2015.
2. Ahmed Khattab, Zahra Jeddi, Esmaeil Amini, Magdy Bayoumi: RBS Performance Evaluation, Springer International Publishing, 10.1007/9783319475455_6, 2016.
3. N. R. Potlapally, S. Ravi, A. Raghunathan, N. K. Jha: A study of the energy consumption characteristics of cryptographic algorithms and security protocols, IEEE Transactions on Mobile Computing (Volume: 5, Issue: 2, Feb. 2006).
4. Daniel Engels, Xinxin Fan, Guang Gong, Honggang Hu, Eric M. Smith: Hummingbird: UltraLightweight Cryptography for Resource Constrained Devices, Springer Lecture Notes in Computer Science, ISBN 9783642149917, 2010.
5. Payal Patel, Rajan Patel, Nimisha Patel: Integrated ECC and Blowfish for Smartphone Security, Elsevier Procedia Computer Science, International Conference on Information Security & Privacy (ICISP2015).
6. C. Mathias: Why mobile operating systems could fade away, Computerworld, 2007.
7. Hielko van der Hoorn: A Survey of Mobile Platforms for Pervasive Computing, University of Groningen Computing Science Master of Science Thesis, Groningen, 2010.
8. S. Prasanna Ganesan, "An Asymmetric Authentication Protocol for Mobile Devices Using Elliptic Curve Cryptography", 978-1-4244-5848-6/10/ IEEE 2010.
9. Creighton T. R. Hager, Scott F. Midkiff, Jung-Min Park, Thomas L. Martin, "Performance and Energy Efficiency of Block Ciphers in Personal Digital Assistants", Proceedings of the 3rd IEEE Int'l Conf. on Pervasive Computing 2005.
10. Black hat report on android, http://www.blackhat.com/presentations/AndroidSurgery-PAPER.pdf, 2009.
11. Google android architecture, http://developer.android.com/guide/basics/ android.html.
12. Microsoft mobile report, http://msdn.microsoft.com/en-us/library/ms920098.aspx.
13. Google Android security report, http://code.google.com/intl/nl/android/devel/security.html.
14. Denmark IT mobile security report, http://itsec.rwth-aachen.de/research/comparative-study-of-protection-echanisms-for-mobile-operating-systems.

Assessment of Object Detection Using Deep Convolutional Neural Networks

Ajeet Ram Pathak, Manjusha Pandey, Siddharth Rautaray and Karishma Pawar

Abstract Detecting the objects from images and videos has always been the point of active research area for the applications of computer vision and artificial intelligence namely robotics, self-driving cars, automated video surveillance, crowd management, home automation and manufacturing industries, activity recognition systems, medical imaging, and biometrics. The recent years witnessed the boom of deep learning technology for its effective performance on image classification and detection challenges in visual recognition competitions like PASCAL VOC, Microsoft COCO, and ImageNet. Deep convolutional neural networks have provided promising results for object detection by alleviating the need for human expertise for manually handcrafting the features for extraction. It allows the model to learn automatically by letting the neural network to be trained on large-scale image data using powerful and robust GPUs in a parallel way, thus, reducing training time. This paper aims to highlight the state-of-the-art approaches based on the deep convolutional neural networks especially designed for object detection from images.

Keywords Computer vision · Deep convolutional neural networks
Deep learning · Object detection

A. R. Pathak (✉) · M. Pandey · S. Rautaray
Data Science Center of Excellence, School of Computer Engineering,
Kalinga Institute of Industrial Technology (KIIT) University, Bhubaneswar, India
e-mail: ajeet.pathak44@gmail.com

M. Pandey
e-mail: manjushapandey82@gmail.com

S. Rautaray
e-mail: sr.rgpv@gmail.com

K. Pawar
Department of Computer Engineering & IT, College of Engineering Pune (COEP),
Pune, India
e-mail: kvppawar@gmail.com

© Springer Nature Singapore Pte Ltd. 2018
S. Bhalla et al. (eds.), *Intelligent Computing and Information and Communication*,
Advances in Intelligent Systems and Computing 673,
https://doi.org/10.1007/978-981-10-7245-1_45

1 Introduction

Computer vision technology has been extensively used in different segments like industry, automation, consumer markets, medical organizations, entertainment sectors, defense, and surveillance, to mention a few. The ubiquitous and wide applications like scene understanding, video surveillance, robotics and self-driving cars triggered vast research in the domain of computer vision during the most recent decade. Visual recognition systems encompassing image classification, localization, and detection have achieved great research momentum due to significant development in neural networks especially deep learning, and attained remarkable performance [1]. The last 4 years witnessed a great improvement in performance of computer vision tasks especially using deep convolution neural networks (DCNNs) [2].

Several factors are responsible for proliferation for DCNNs viz. (i) Availability of large training datasets and fully annotated datasets (ii) Robust GPU to train large-scale neural network models in a parallel way (iii) State-of-the-art training strategies and regularization methods. Object detection is one of the crucial challenges in computer vision and it is efficiently handled by DCNN [3], Restricted Boltzmaan Machine (RBM) [4], autoencoders [5], and sparse coding representation [6]. This paper aims to highlight state-of-the-art approaches for object detection based on the DCNNs.

The contents of the paper are portrayed as follows. Section 2 introduces object detection. The fundamental building blocks of DCNNs from the perspective of object detection are enunciated in Sect. 3. The state-of-the-art DCNN-based approaches for object detection are discussed in Sect. 4. The paper is concluded in Sect. 5.

2 Object Detection

An image or video contains single or more than one classes of real-world objects and abstract things like human, faces, building, scene, etc. The aim of object detection is to determine whether the given instance of the class is present in the image, estimate the location of the instance/instances of the all the classes by outputting the bounding box overlapping the object instance along with obtained accuracy of detection irrespective of partial occlusions, pose, scale, lightening conditions, location, and camera position. It is generally carried out using feature extraction and learning algorithms. Object detection is a preliminary step for various computer vision tasks like object recognition, scene understanding from images and activity recognition, anomalous behavior detection from videos. Detecting instance or instances of the single class of object from image or video is termed as *single object class detection*. *Multi-class object detection* deals with detecting instances of more than one class of objects from the image or video. Following challenges need to be handled while detecting objects from the images.

- Image-based challenges

Many computer vision applications require multiple objects to be detected from the image. Object occlusions (partial/full occlusion), noise, and illumination changes make detection challenging task. Camouflage is a challenge in which object of interest is somewhat similar to the background scene. This challenge needs to be handled in surveillance applications. It is also necessary to detect objects under conditions of multiple views (lateral, front), poses, and resolutions. The object detection should be invariant to scale, lighting conditions, color, viewpoint, and occlusions.

- Processing challenges

Detecting objects at large scale without losing accuracy is a primary requirement of object detection tasks. Some applications require robust and efficient detection approaches, whereas others require real-time object detection. Thanks to specialized hardware like GPUs and deep learning techniques which allow to train multiple neural networks in parallel and distributed way helping to detect objects at real-time.

3 Building Blocks of Convolutional Neural Network

DCNNs was first used for image classification. After achieving state-of-the-art performance in image classification, DCNN has been used for more complex tasks like object detection from images and videos.

3.1 CNN Architecture

Convolutional neural network (CNN) is a kind of feedforward neural network in which the neurons are connected in the same way as the neurons present in the brain of animal's visual cortex area. Figure 1 shows the architecture of CNN. Being hierarchical in nature, CNN encompasses convolution layer with activation function like Rectified Linear Unit (ReLU), followed by pooling layer and eventually fully connected layers. The pattern of CONV-ReLU-POOL is repeated in such a manner that image reduces spatially. The neurons are arranged in the form of three dimensions—width, height, and depth. Depth corresponds to color channels in the image. The image to be recognized is fed as input in terms of [*width* × *height depth*]. For the desired number of filters, the image is convolved with the filter function in order to get the specific feature. This process is repeated for the desired number of filters and accordingly feature map is created. This is done by applying the dot product of the weight assigned to the neuron and the specified region in the image. In this way, the output of a neuron is computed by convolution layer.

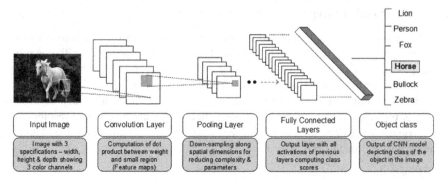

Fig. 1 Architecture of convolutional neural network

Rectified Linear Unit (ReLU) applies activation function at elemental level making it nonlinear. Generally, convolution layer is followed by pooling layer to reduce the complexity of the network and the number of parameters in learning by down-sampling the feature map along spatial dimensions. The last layer in the CNN is a fully connected layer which gives the output of the image recognition task in the form of scores representing the object classes. Highest score represents the presence of a corresponding class of the object in the image.

3.2 Pooling Layers

Addition of pooling layer amidst the consecutive layers of convolutional layer reduces the number of parameters and complexity of the network, and thus, control overfitting. Pooling layer is translation-invariant and it takes activation maps as input and operates on every patch of the selected map. There are various kinds of pooling layers.

- Max pooling: In this pooling, each depth slice of input is operated using pooling. Figure 2 shows working of max pooling where a filter of size 2 × 2 is applied over a patch of an activation map with a stride of 2. The max value among each entry of 4 numbers is chosen and stored into the matrix, getting a spatially resized map. Another pooling approach is average pooling which takes an average of neighborhood pixels. Max pooling gives better results compared to average pooling [7].
- Deformation constrained pooling (Def-pooling): In order to apply deformation of object parts along with geometric constraints and associated penalty, def-pooling is applied [8]. It has the ability to learn deformable properties of object parts and shares visual patterns at any level of information abstraction and composition.

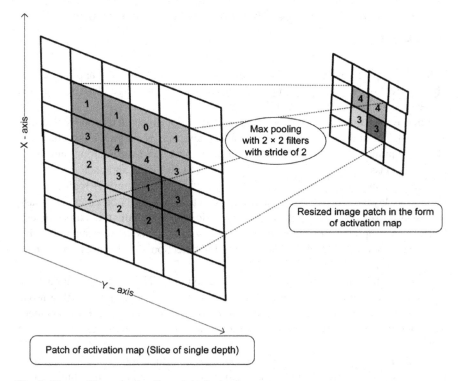

Fig. 2 Max pooling over the slice of single depth

- Fully connected layers: These layers perform high-level reasoning in CNN.
  They exhibit a full connection to all the decision functions in the previous layer
  and convert 2D features into one-dimensional feature vector. Fully connected
  layers possess a large amount of parameters and so require powerful computational resources.

3.3 Regularization

*"Regularization is defined as any modification we make to a learning algorithm
that is intended to reduce its generalization error but not its training error* [9]."
Due to a large amount of parameters for training and ability of architectures to learn
more abstractions using deep learning, there are chances of obtaining negative
performance on test data from the model, i.e., model learns too well such that it
poorly generalizes in case of new data. This is known as overfitting. Regularization
strategies are required in order to stop overfitting.

4 State-of-the-Art Object Detection Approaches Using DCNN

Table 1 compares the state-of-the-art discussion of DCNN-based approaches for object detection. DCNNs have been extensively used for image classification and achieved state-of-the-art results [1].

For the very first time, DCNNs have been used for object detection by Szegedy et al. [10]. The authors have formulated object detection as a regression problem for object bounding box masks and defined object detection as estimation and localization of class and object from the image, respectively. Their approach is known as DetectorNet in which the last layer of AlexNet [1] architecture is replaced with regression layer in order to localize the objects using DNN-based object mask regression. To precisely detect the multiple instances of the same object, DetectorNet applies multi-scale box inference with refinement procedure. But, this approach lacks multiple classes of objects for detection since it uses only single mask regression.

To demystify the working of features extracted in CNN model and diagnose the errors associated with the model, Zeiler and Fergus (ZFNet) [3] put forth a novel visualization technique based on multilayered deconvolution network (deconvnet). This model used deconvnet in order to project features back into pixel space of the image.

Deformable DCNN (DeepID-Net) encompasses feature representation learning, part deformation learning, context modeling, model averaging, and bounding box location refinement [11] and uses cascaded CNN for object detection. It works on deformable part objects. Regions with CNN features (R-CNN) [12] take an input image and evaluate 'n' number of bottom-up region proposals using segmentation. Once region proposals are obtained, it classifies proposals using class-relevant SVMs to get classified regions. This method acts as a baseline model for a large number of approaches put forth for object detection. Fast R-CNN [13] is the extended version of R-CNN to improve the speed of training and testing phase and improve the detection accuracy. Fast R-CNN suffers from the drawback of calculating the proposals for each region in the image, thus, incurring the large cost of computation, this drawback has been removed in Faster R-CNN by Ren et al. [14]. The authors put forth region proposal network (RPN) which is a fully convolution network in which input image is shared with the detector network, this network simultaneously calculates object bounds and object features at each point, thus freeing cost of region proposals.

This method merges RPN with fast R-CNN and creates a unified network. It works on the principle of "attention mechanism" in which RPN guides network where to search for the object bounds. In the paper, by Markus et al. [15], multiple CNN models are used to detect objects at multiple scales. The papers by Lee et al. [16] and Cheng et al. [17] are based on the region-based proposals. Lee et al. [16] handled the issue of intra-class and interclass variability among objects using multi-scale templates of CNN and non-maximum suppression method. On the other

Table 1 Comparative study of State-of-the-approaches for object detection

Paper	Approach	Issues	Features	CNN configuration	Pooling	Classifier	Regularization technique	Deep network training
ImageNet [1]	CNN	Reducing detection errors	Image classification for high-resolution images	5 CNN layers + 3 Fully conn. Layers	Max pooling	Softmax	DropOut	Supervised learning mode, Local response normalization and stochastic gradient descent
ZFNet [3]	CNN and deconv. Neural network	Demystifying the working of intermediate CNN layers and classifier	Visualization of intermediate features in the network	Fully supervised CNN models	Max pooling	Softmax	DropOut, ReLU as activation function	Back-propagating derivative of cross entropy function for training and SGD
DetectorNet [10]	CNN	Object detection and localization	Multi-object detection, DNN-based regression for localization	5 CNN layers + 2 Fully conn. Layers	Max pooling	Softmax	Use of negative classes to regularize the network	Adaptive gradient algorithm based on Stochastic optimization
DeepID-Net [11]	Cascaded CNN	Handling of deformable part objects	Generic object detection	13 CNN layers + 3 Fully conn. Layers (O-Net and T-Net)	Def-pooling	SVM	Pre-training scheme & jointly learning feature representation	Multi-stage deep training model
Faster R-CNN [14]	Region-based proposal network	Computation cost of region-Based proposal handling translation	Cost-free region proposals, Real-time object	VGG-Net (13 CNN & 3 fully conn. layers) + ZF Net	RoI pooling layer	R-CNN acts as a classifier	"Image-centric" sampling strategy based on back-propagation and SGD	"Attention" mechanism of Neural network, back-propagation

(continued)

Table 1 (continued)

Paper	Approach	Issues	Features	CNN configuration	Pooling	Classifier	Regularization technique	Deep network training
Markus et al. [15]	Multi-scale model	Visual Pedestrian detection with high accuracy	Multi-scale person detection	3 stages of CNNs processing image patches	Max pooling	Non-maximum suppression	DropOut	SGD with mini-batch training
RIFD-CNN [17]	Region-based proposal (Fisher discriminative)	Object rotation, intra-class and interclass similarity of objects	Rotation invariant, interclass & intra-class object detection	AlexNet + VGG-Net	Max pooling	Softmax, Linear Support Vector Machine (SVM)	Discrimination regularization	CNN-based pre-training and fine-tuning using SGD
MSS-CNN [18]	CNN working on image pyramids	Handling of scale variation	Multi-scale, context-aware object detection and localization	Image pyramid-based CNN	Max pooling	structured SVM, bounding box regression	Training based on the data augmentation	Multi-scale training, non-max suppression to resolve multiple detection issue

hand, in RIFD-CNN by Cheng et al. [17], the issue of object rotation and intra-class and interclass variability is handled by introducing rotation-based layer and Fisher discriminative layer in the network, respectively. For detecting small objects and localizing them, contextual information based on the multi-scale model of CNN is used in [18], handling the issue of variation in scaling of objects. The aforementioned approaches mainly focus on the specific challenge in object detection viz. multi-scale model, fast and real-time detection, detection accuracy, and localization, interclass and intra-class variation of objects.

It is worth important to amalgamate challenges and address them using unified object detection framework applicable to detect objects in different complex scenarios and thereby enhance the usability of such object detection systems.

5 Conclusion

This paper compares some of the noteworthy approaches to object detection based on DCNNs. DCNN-based approaches are found to be suitable for images and can also be applicable to detect moving objects from the video [19]. The need of the hour is to develop object detection model which can be generalized to work in different application scenarios like face recognition, emotion detection, abandoned object detection (Suspicious object detection), etc. The role of "transfer learning" method for training deep networks would help to cope with the issue [20].

The efficacy of object detection frameworks mainly depends on the learning mode, method of processing the images (parallel programming) and also the platform (CPU, GPU). Continuous change in the scene implies a change in the behavior of objects to be detected, therefore, it is mandatory for such systems to continuously learn the multitude features of objects and detect them despite of a change in their orientation, views, and forms. In addition to this, real-time detection of objects [21] helps to take proactive measures or acts as alarming conditions for effectively monitoring and controlling the public and private places requiring utmost security.

Object detection is a very promising area which can be applied in computer vision and robotics systems, surveillance based on the drone cameras, etc. It is extremely useful in places like deep mines, expeditions to exploring deep ocean floor where human presence is not feasible.

References

1. Alex Krizhevsky, Ilya Sutskever, and Geoff Hinton. Imagenet classification with deep convolutional neural networks. In: Advances in Neural Information Processing Systems. (2012).
2. He, Kaiming, Xiangyu Zhang, Shaoqing Ren, and Jian Sun. Spatial pyramid pooling in deep convolutional networks for visual recognition. In: European Conference on Computer Vision, pp. 346–361. Springer International Publishing. (2014).

3. M.D. Zeiler, R. Fergus.: Visualizing and understanding convolutional neural networks. In: ECCV. (2014).
4. R. Salakhutdinov, G.E. Hinton.: Deep boltzmann machines. In: AISTATS, (2009).
5. S. Rifai, P. Vincent, X. Muller, et al.: Contractive auto-encoders: explicit invariance during feature extraction. In: ICML (2011).
6. Yang, Jianchao, Kai Yu, Yihong Gong, and Thomas Huang.: Linear spatial pyramid matching using sparse coding for image classification. In: Computer Vision and Pattern Recognition, CVPR. pp. 1794–1801. IEEE. (2009).
7. D. Scherer, A. Müller, S. Behnke.: Evaluation of pooling operations in convolutional architectures for object recognition. In: ICANN. (2010).
8. W. Ouyang, P. Luo, X. Zeng, et al.: Deepid-net: Deformable deep convolutional neural networks for object detection. In: Computer Vision and Pattern Recognition, pp. 2403–2412. IEEE. (2015).
9. I. Goodfellow, Y. Bengio, and A. Courville.: Deep Learning. MIT Press. (2016).
10. C. Szegedy, A. Toshev, D. Erhan.: Deep neural networks for object detection. In: Proceedings of the NIPS. (2013).
11. W. Ouyang, P. Luo, X. Zeng, et al.: DeepID-Net: multi-stage and deformable deep convolutional neural networks for object detection. In: Proceedings of the CVPR. (2015).
12. R. Girshick, J. Donahue, T. Darrell, et al.: Rich feature hierarchies for accurate object detection and semantic segmentation. In: Proceedings of the CVPR. (2014).
13. R. Girshick.: Fast R-CNN. In: ICCV. (2015).
14. S. Ren, K. He, R. Girshick, and J. Sun.: Faster R-CNN: Towards Real-Time Object Detection with Region Proposal Networks. In: TPAMI, pp. 91–99. IEEE. (2016).
15. Eisenbach, Markus, Daniel Seichter, Tim Wengefeld, and Horst-Michael Gross.: Cooperative multi-scale Convolutional Neural Networks for person detection. In: International Joint Conference on Neural Networks (IJCNN), pp. 267–276. IEEE. (2016).
16. B. Lee, E. Erdenee, S. Jin, and P. K. Rhee. Efficient object detection using convolutional neural network-based hierarchical feature modeling. In: Signal, Image Video Process. vol. 10, no. 8, pp. 1503–1510, (2016).
17. Cheng, Gong, Peicheng Zhou, and Junwei Han.: RIFD-CNN: Rotation-invariant and fisher discriminative convolutional neural networks for object detection. In: Computer Vision and Pattern Recognition, pp. 2884–2893. IEEE. (2016).
18. E. Ohn-Bar and M. M. Trivedi.: Multi-scale volumes for deep object detection and localization. In: Pattern Recognition, vol. 61, pp. 557–572. Elsevier (2017).
19. S. H. Shaikh, K. Saeed, and N. Chaki.: Moving Object Detection Approaches, Challenges and Object Tracking. In: Moving Object Detection Using Background Subtraction, pp. 5–14. Springer International Publishing (2014).
20. Dauphin, G.M. Yann, X. Glorot, S. Rifai, Y Bengio, I. Goodfellow, E. Lavoie, X. Muller et al.: Unsupervised and transfer learning challenge: a deep learning approach. In: ICML Workshop on Unsupervised and Transfer Learning, pp. 97–110. (2012).
21. P. Viola, M. Jones.: Rapid object detection using a boosted cascade of simple features. In: Computer Vision and Pattern Recognition. pp. I-511-I-518. IEEE. (2001).

Implementation of Credit Card Fraud Detection System with Concept Drifts Adaptation

Anita Jog and Anjali A. Chandavale

Abstract There is a large number of credit card payments take place that is targeted by fraudulent activities. Companies which are responsible for the processing of electronic transactions need to efficiently detect the fraudulent activity to maintain customers' trust and the continuity of their own business. In this paper, the developed algorithm detects credit card fraud. Prediction of any algorithm is based on certain attribute like customer's buying behavior, a network of merchants that customer usually deals with, the location of the transaction, amount of transaction, etc. But these attribute changes over time. So, the algorithmic model needs to be updated periodically to reduce this kind of errors. Proposed System provides two solutions for handling concept drift. One is an Active solution and another one is Passive. Active solution refers to triggering mechanisms by explicitly detecting a change in statistics. Passive solution suggests updating the model continuously in order to consider newly added records. The proposed and developed system filters 80% fraudulent transactions and acts as a support system for the society at a large.

Keywords Credit card · Concept drift adaptation · Credit card fraud detection
Network-based extension

1 Introduction

Credit cards and debit cards are used extensively in today's world. People prefer executing banking transaction online rather than going to the branch. This is due to easy availability of modern resources like laptops, phone, and tablets. Most of the banks switched to centralize processing to support this trend. Another reason for credit card gaining popularity is online shopping trend. Lots of deals, promotions

A. Jog (✉) · A. A. Chandavale
Department of Information Technology, MIT College of Engineering, Pune, India
e-mail: Anita.visal@gmail.com

A. A. Chandavale
e-mail: Anjali.chandavale@mitcoe.edu.in

© Springer Nature Singapore Pte Ltd. 2018

467

S. Bhalla et al. (eds.), *Intelligent Computing and Information and Communication*,
Advances in Intelligent Systems and Computing 673,
https://doi.org/10.1007/978-981-10-7245-1_46

attract customers to buy online. Online shopping allows customers to explore more items in short time. Customers can also compare the prices of the same item from different vendors. All these online activities are more prone to fraudulent cases due to lack of proper training or awareness before using digital transaction. During online transactions, personal data like account number, password, birth date, credit card details, etc. is exposed over the network resulting in financial fraud. Financial fraud causes loss of huge amount worldwide each year. There is a huge amount of loss occurs due to online fraud. As per cybercrime report [1], total card sales volume is $28.844 trillion while the losses occurred from fraud on cards is $16.31 billion. Comparing to past year fraud increased by 19%, while the overall volume of card sales only grew by only 15%. Companies that are responsible for processing electronic transactions need to efficiently detect the fraudulent activity to maintain customers' trust and the continuity of their business. Data mining offers a number of techniques to analyze patterns in data, differentiating genuine transaction from suspicious transactions. The most important aspect of fraud detection is to correctly identify the atypical character of fraud, since the fraudulent transactions are very few as compared to the legitimate transactions. A critical step in an efficient fraud detection process is creating a set of significant characteristics that capture irregular behavior. The detection of fraud is a complex computational task and there is no ideal system that accurately predicts fraudulent transaction. These systems only predict the probability of the transaction to be fraudulent. The paper is divided into following sections. Section 2 presents the related work. Section 3 describes the proposed work. In Sect. 3.3.3 expected result is outlined and Sect. 4 describes the conclusion.

2 Related Work

Various data mining techniques exist for fraud detection [2, 3, 4, 5]. Support Vector machine, genetic programming, Artificial Intelligence, Neural networks, Decision tree are few of them [6, 7, 8, 9]. Using decision tree, the system is described by Yusuf Sahin and Serol Bulkan. In this paper, cost-sensitive decision tree approach is suggested which minimizes the sum of misclassification costs while selecting the splitting attribute at each nonterminal node [10]. Author has compared the performance of this approach with the well-known traditional classification models on a real-world credit card data set. Andrea Dal Pozzolo, Giacomo Boracchi designed two fraud detection systems using ensemble and sliding window approach [11]. Customer feedbacks and alerts provided by the system are used as supervised samples. One more novel approach is described by Véronique Van Vlasselaer and Cristián Bravo [12]. This approach is a combination of the characteristics of incoming transactions along with the customer spending history based on the fundamentals of RFM (Recency–Frequency–Monetary); and characteristics by exploring the network of credit card holders and merchants. A time-dependent suspiciousness score is calculated for each network object. As seen from the literature, it is essential to have a more efficient system capable of detecting the frauds at the very first instance. The following section describes the proposed and developed system based on the concept drift adaption which values the customer feedback. It performs a proactive heuristic search across all the customers.

3 Proposed Work

The proposed and developed system works on the principle of the active solution to detect the fraud. It follows the layered architecture namely data layer, utility layer, manager layer, and controller layer. The data layer is responsible for storing the historical data, transactional data, the model which contains the set of attributes to validate for each customer. Utility layer acts as a support layer. Manager layer is responsible for executing each job separately while controller encapsulates all the flows for every user action. The concept drift adaptation technique is implemented by invoking the scheduler periodically to update the model. Below section covers the overview of how credit card processing happens followed by the technical details of the proposed system.

3.1 Overview of Credit Card Processing

1. *Authorization*: It is the first step in which a merchant swipes the card. Then, the login details are then passed to an acquirer. Acquirer authorizes the transaction. The acquirer is nothing but a bank. It is responsible for processing and settling credit card transactions by consulting a card issuer. The acquirer then sends the request to the card-issuing bank. Card-issuing bank either authorizes or denies the request. Depending upon the decision the merchant is allowed to proceed with the sale.
2. *Batching*: Merchant reviews all the day's sales to make sure all of them were authorized and endorsed by the cardholder. It then submits all the transactions in a batch, to the acquirer to process payment.
3. *Clearing*: Once acquirer receives the batch, it sends it through the card network. Card network is responsible for routing each transaction to the appropriate issuing bank as shown in Fig. 2. The issuing bank then deducts its interchange fees and sends the remaining amount to acquirer through the network. The interchange fee is the amount paid by merchants to a credit card issuer and a card network for accepting credit cards. It usually ranges between 1 and 3%. Card network is Visa, MasterCard, or other networks that act as a mediator between an acquirer and an issuer. They are responsible for authorizing credit card transactions.

A. Jog and A. A. Chandavale

4. *Funding*: This is the last step in processing a credit card. Acquirer deducts its discount charges and sends the remaining amount to the merchant after receiving payment from the issuer. Thus, the merchant got paid for the transaction. Thus, the cardholder is charged. Merchants pay a discount fee to acquirers to incur the processing cost of the credit cards payment.

3.2 Proposed System

Figure 1 shows the block diagram of the proposed system. Using the historical transactional data, the model builder will the model. This model is nothing but the set of differentiating attributes for identifying the fraudulent transaction. This model is used by online fraud detector to decide if each incoming transaction contains any suspicious characteristic. Based on the decision given by online fraud detector, transaction processor either aborts the transaction or proceeds for further processing. Separate scheduler which runs daily triggers the offline calibrator. The offline calibrator is responsible for correcting the model as per customer feedback. This model is updated weekly to consider newly added transactions. The application has following modules starting from user login, viewing account statement, downloading in PDF or excel format to actual transaction execution which covers different validations. Some validations are performed during actual transaction while some heuristic checks are applied across all customers, where the pattern is recognized and suspicious transactions are marked in the database table.

As shown in Fig. 2, the proposed system is broadly divided into four layers

1. *Controller Layer*: It controls the flow of actions. It calls appropriate manager to execute each action. It calls alert utility depending upon the output received from manager layer. For every user action, there is one controller class. It decides different steps that need to be performed to accomplish given task.
2. *Manager Layer*: It executes below described action using database tables. It sends notifications, alerts to the controller.

    (i) *Authentication Manager*: First calls authentication manager to verify the userId, password. If the authentication manager successfully returns as a valid user, then the transaction controller passes the request to the preliminary validator.
    (ii) Fraud detection manager: After passing all the validation checks, the controller calls fraud detection manager to detect any unusual behavior of transaction. It performs few fraud detection checks which include amount, location, valid merchant, last transaction time, etc.
    (iii) Transaction Manager: Finally, the controller calls Transaction manager to execute the transaction dealing with debit/credit, transfer, etc.

3. *Utility Layer*: It takes care of reporting. Notifications, Alerts triggered by manager layer are converted into appropriate Actions.

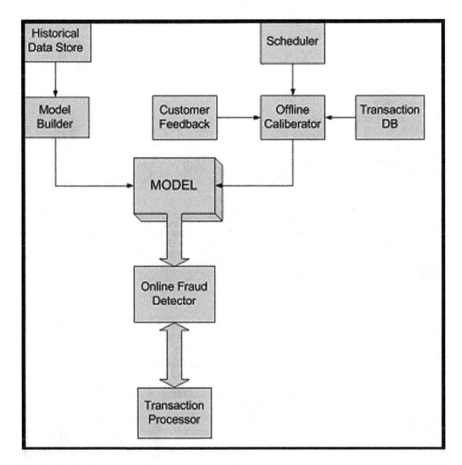

Fig. 1 Block diagram

4. *Database Layer*: It consists of transactional data, historical data, rules model data. Transactions flagged by the algorithm kept in the separate table. Rules Model table is populated based on historical data. And, it gets updated periodically using self-learning algorithm. The following figure shows a quick overview of the database layer.

3.3 Implementation

The system is implemented in JAVA, JSP. The model is determined by WEKA output. SQL server database is used for backend data storage. For download

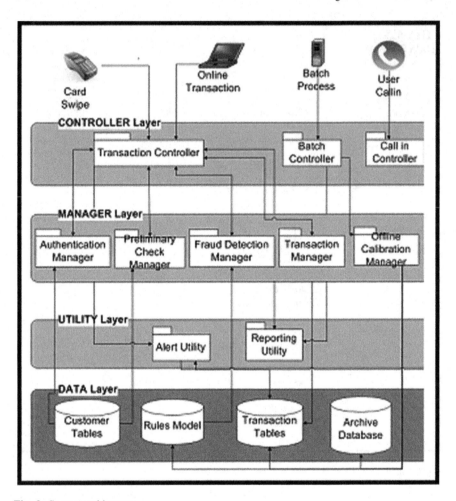

Fig. 2 System architecture

functionality in PDF format, PDFBOX library is used and APACHE POI library is used to support Excel format. After testing the system, it is found that using the fraud detection system, 80% fraudulent transactions are filtered. Heuristics searches also help to detect fraud at the initial instant.

3.3.1 DB Design

The main tables used for fraud detection are:

1. Customer_Vendor_tbl is used to store the RFM attributes, i.e., Recency, frequency, and monetary value for each customer–vendor pair. For each incoming

transaction, the frequency is checked. If it is monthly, then current transaction date and last transaction date for the corresponding vendor is compared. If the difference is less than a month, then after considering certain threshold, the transaction is put on HOLD. Customer care will reach out to the customer to confirm if the transaction is genuine or not.

2. Account_validation_tbl is storing all the different validations for each account. Attribute name in this table represents attribute like location, amount, etc. attribute value represents set of allowed values separated by a comma. For example, it is comma separated list of cities, where the customer can use the card for the transaction. If new city is encountered then the transaction is put on HOLD.

3. Fraud_log_tbl is used to log every transaction that is blocked. The design allows having different attributes for different accounts.

3.3.2 Algorithm

When a customer performs a transaction, then the request comes to credit card fraud detection system and the following algorithm is executed.

Step 1: Perform preliminary checks given below on the incoming transaction.

- *Balance Check*: The credit card balanced is checked to confirm if the there is sufficient balance is the credit card.
- *Validate customer*: Customer is validated to check if he has paid previous bills.
- *Validate card status*: Card status is verified to check if it is in active state.

Step 2: Perform online fraudulent checks given below:

Min/Max amount check: Using the historical data, threshold values for the account are determined and stored in the database. For the incoming transaction, the transaction amount is compared to check if it falls within the account threshold. If transaction amount falls out of threshold, it is marked as suspicious.

- *Customer–Vendor validation*: Using the historical data, the relationship between customer and vendor is determined with respect to frequency and monetary values. For each customer–vendor pair, the minimum amount transacted, max amount transacted, and frequency of the transaction is calculated using historical data and stored in the database. The incoming transaction is verified against corresponding customer–vendor pair for threshold values and frequency. For example, customer ABC transacts with vendor PQR every month between the amounts 1000 to 3000. If the incoming transaction has customer ABC and vendor PQR with amount 7000, the system marks it as suspicious.
- *Location check*: Using historical data, different locations are identified for each customer. These locations are stored as comma separated list in the database. If the incoming transaction has a location which was never found before, it is marked as suspicious.

Step 3: If anyone of above check fails, then transaction is suspicious and is put on HOLD

All transactions that are marked as suspicious then confirmed with the customer. Below algorithm is executed when confirmation is received from the customer.

Step 1: Notify the customer about fraud transaction. Show the validations status to the customer. If the customer confirms the transaction to be fraudulent, abort the transaction. Mark transaction status as "CANCELLED" and no further action is taken

Step 2: If a customer claims that the transactions are genuine, then proceed with transaction successfully and mark as COMPLETED.

Step 3: After the transaction is executed successfully, update the model as given below.

- Recalculate the account threshold values.
- Recalibrate customer–vendor frequency, amount threshold.
- If there is new location found in the current transaction, then update the list of locations.
  With this updated model, the similar transaction will be executed successfully in future.

Step 4: Recalibrate amount threshold, and location list at every end of the day to consider newly added records.

3.3.3 Project Results

Below are the screenshots shown in Figs. 3 and 4 for transaction status, fraud summary, and action taken on suspicious transactions after receiving feedback from the customer.

3.4 Fraud Detection System Evaluation

The data is downloaded from kaggle.com. The datasets contain transactions made by credit cards in September 2013 by European cardholders. Along with this, synthetic data is also added to the database to test different scenarios. The resulted dataset has 152,706 credit card transactions of 120 cardholders. 151,768 transactions legitimate are and 1026 transactions are fraudulent. Criteria used to evaluate the system is alarm rate, sensitivity, false positive rate, precision, negative predictive value, and accuracy as shown in Table 1 [13–15].

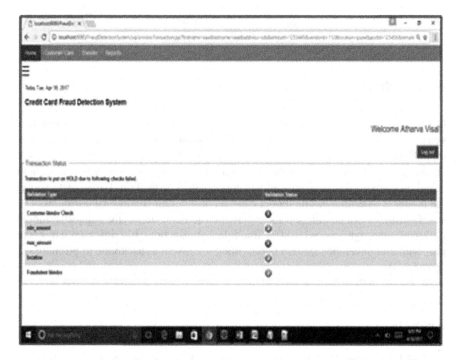

Fig. 3 Transaction status of fraud check

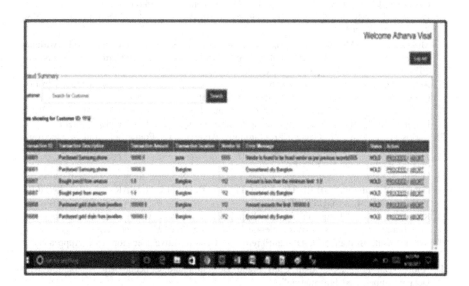

Fig. 4 Fraud summary page

Table 1 Evaluation results

Sensitivity	Precision	Accuracy
43.24%	11.20%	80.12%

4 Conclusion

Credit card fraud detection algorithms have emerged remarkably over last few years. The algorithm used for credit card detection needs to be self-learning and periodically updated. The approach suggested in this paper is most suitable for the changing environment since it changes dynamically. The work is a hybrid system based on Network-based extension and concept drift adaptation. Using past history, the model is determined. The model consists of different conditional attributes. The set of attributes can be different for different customers. Hence, the suggested model is flexible and dynamic with 80% fraudulent transactions filtering rate. Thus, the proposed and developed system is providing the essential backbone to promote the plastic money.

References

1. http://www.pymnts.com/news/2015/global-card-fraud-damages-reach-16b/.
2. Emanuel MinedaCarneiro, "Cluster Analysis and Artificial Neural Networks: A Case Study in Credit Card Fraud Detection," in 2015 IEEE International Conference.
3. Dhiya Al-Jumeily, "Methods and Techniques to Support the Development of Fraud Detection System", IEEE 2015.
4. Mukesh Kumar Mishra, "A Comparative Study of Chebyshev Functional Link Artificial Neural Network, Multi-layer Perceptron and Decision Tree for Credit Card Fraud Detection", 2014 13th International Conference on Information Technology.
5. Andrea Dal Pozzolo, Olivier Caelen," Learned lessons in credit card fraud detection from a practitioner perspective", Expert Systems with Applications 41, 2014.
6. V. Mareeswari, "Prevention of Credit Card Fraud Detection based on HSVM". 2016 IEEE International Conference On Information Communication And Em- bedded System.
7. Carlos A. S. Assis, "A Genetic Programming Approach for Fraud Detection in Electronic Transactions" in Advances in Computing and Communication Engineering (ICACCE), 2015 Second International Conference.
8. Dustin Y. Harvey, "Automated Feature Design for Numeric Sequence Classification by Genetic Programming", IEEE TRANSACTIONS ON EVOLUTIONARY COMPUTATION, VOL. 19, NO. 4, AUGUST 2015.
9. Kang Fu, Dawei Cheng, Yi Tu, and Liqing Zhang, "Credit Card Fraud Detection Using Convolutional Neural Networks", Neural Information Processing, Springer.
10. Sahin Yusuf, BulkanSerol, DumanEkrem," A Cost-Sensitive Decision Tree Approach for Fraud Detection", Expert Systems with Applications,vol.40, pp. 5916–5923, 2013.
11. Andrea Dal Pozzolo, "Credit Card Fraud Detection and Concept-Drift Adaptation with Delayed Supervised Information".
12. Véronique Van Vlasselaer, "APATE: A novel approach for automated credit card transaction fraud detection using network-based extensions" published in Decision Support Systems 2015.

13. Yiğit Kültür, "A Novel Cardholder Behavior Model for Detecting Credit Card Fraud", IEEE international conference on commuting and communication engineering, 2015.
14. https://www.creditcards.com/credit-card-news/assets/HowACreditCardIsProcessed.pdf.
15. https://www.kaggle.com/dalpozz/creditcardfraud.

Intelligent Traffic Control by Multi-agent Cooperative Q Learning (MCQL)

Deepak A. Vidhate and Parag Kulkarni

Abstract Traffic crisis frequently happens because of traffic demands by the large number vehicles on the path. Increasing transportation move and decreasing the average waiting time of each vehicle are the objectives of cooperative intelligent traffic control system. Each signal wishes to catch better travel move. During the course, signals form a strategy of cooperation in addition to restriction for neighboring signals to exploit their individual benefit. A superior traffic signal scheduling strategy is useful to resolve the difficulty. The several parameters may influence the traffic control model. So it is hard to learn the best possible result. The lack of expertise of traffic light controllers to study from previous practice results makes them to be incapable of incorporating uncertain modifications of traffic flow. Defining instantaneous features of the real traffic scenario, reinforcement learning algorithm based traffic control model can be used to obtain fine timing rules. The projected real-time traffic control optimization model is able to continue with the traffic signal scheduling rules successfully. The model expands traffic value of the vehicle, which consists of delay time, the number of vehicles stopped at the signal, and the newly arriving vehicles to learn and establish the optimal actions. The experimentation outcome illustrates a major enhancement in traffic control, demonstrating the projected model is competent of making possible real-time dynamic traffic control.

Keywords Cooperative learning · Multi-agent learning · Q learning

D. A. Vidhate (✉)
Department of Computer Engineering, College of Engineering, Pune, Maharashtra, India
e-mail: dvidhate@yahoo.com

P. Kulkarni
iKnowlation Research Lab. Pvt. Ltd, Pune, Maharashtra, India
e-mail: parag.india@gmail.com

© Springer Nature Singapore Pte Ltd. 2018   479
S. Bhalla et al. (eds.), *Intelligent Computing and Information and Communication*,
Advances in Intelligent Systems and Computing 673,
https://doi.org/10.1007/978-981-10-7245-1_47

1 Introduction

Thousands of vehicles distribute in a large and board urban area. It is a difficult and complicated work to effectively take care of such a large scale, dynamic, and distributed system with a high degree of uncertainty [1]. Though the number of vehicles is getting more and more in major cities, most of the current traffic control methods have not taken benefit of an intelligent control of traffic light [2]. It is observed that sensible traffic control and enhancing the deployment effectiveness of roads is an efficient and cost-effective technique to resolve the urban traffic crisis in majority urban areas [3]. Major vital part of the intelligent transportation system is traffic signal lights control strategy becomes necessary [4]. There are so various parameters that have an effect on the traffic lights control. The static control method is not feasible for rapid and irregular traffic flow. The paper suggests a dynamic traffic control framework which is based on reinforcement learning [5]. The reinforcement learning can present a very crucial move to resolve the above cited problems. It is effectively deployed in resolving various problems [6]. The framework defines different traffic signal control types as action selections; the number of vehicles arriving and density of vehicle at a junction is observed as the context of environment and common signal control indicators, including delay time, the number of stopped vehicles and the total vehicle density are described as received rewards. The paper is divided as Sect. 2 gives the insights about the related work done in the area of traffic signal control. Section 3 describes Multi-agent Cooperative Q learning algorithm (MCQL). Section 4 explains about the system model. Experimental results are given in Sect. 5. The conclusion is presented in the Sect. 6.

2 Related Work

The traffic control systems can be categorized into offline traffic control systems and online traffic control systems. Offline methods make use of theoretical move toward optimizing the controls. Online methods regulate traffic regulator period dynamically as per instantaneous traffic conditions. Many achievements in collaborative traffic flow guidance and control strategy have been made. The approach of [7] the transportation industry. By means of F-B method approach, the traffic jam difficulty was partially resolved [7]. After that, several enhanced methods based on the F-B approach had developed [8]. Driving reimbursement coefficient and delay time was used to estimate the effectiveness of time distribution system given in [9]. The approach reduces the delay of waiting time, making the method appear to be sharp and sensible. There is a need to discover a new proper technique as this method could hardly solve the heavy traffic problem. Traffic congestion situation has been addressed using intelligent traffic control in [10] but congestion problems among neighboring junctions required better technique. The local synchronization demonstrated a fine result to this problem discussed in [11]. Because of complication and unpredictability, it is of limited opportunity to construct a precise mathematical model for traffic system in advance [12]. It has turned out to be a style to resolve traffic problems by taking benefit of computing expertise and machine learning [13]. Among many machine intelligence methods, reinforcement learning is feasible for the finest control of the transport system [14]. The study using the learning algorithm [15] achieved online traffic control. The approach was able to choose the optimal coordination model under different traffic conditions. Some applications [16] that utilize learning algorithm have received much significant effect. A paper implemented an online traffic control through learning algorithm, yielding good effort in the normal state of traffic congestion [17].

2.1 Traffic Estimation Parameters

Signal lights control has a very crucial responsibility in traffic management. Normally applied traffic estimation parameters [18] comprises of delay time, the number of automobiles stopped at a signal, and a number of newly arriving cars.

Delay Time. The delay between the real time and theoretically calculated time for a vehicle to leave a signal is defined as the delay time. In practice, we can get total delay time during a certain period of time and the average delay time of a cross to evaluate the time difference. The more delay time indicates the slower average speed of a vehicle to leave a signal.

Number of Vehicles Stopped. How many vehicles are waiting behind stop line to leave the road signal gives the number of vehicles stopped. The indicator [18] is used to measure the smooth degree of the road as well as the road traffic flow. It is defined as

$$\text{stop} = \text{stopG} + \text{stopR}, \tag{1}$$

where stopG is the number of automobiles stopped before the green light and stopR is the number of vehicles stopped at the red light.

Number of Vehicles Newly Arrived. The ratio of the actual traffic flow to the maximum available traffic flow gives the signal saturation. Newly arrived vehicle is calculated as

$$S = \frac{traffic\,flow}{(dr \times sf)}, \tag{2}$$

where dr is the ratio of red light duration to green light duration and sf is traffic flow of the signal.

Traffic Flow Capacity. The highest number of vehicles crossing through the signal is shown by traffic flow capacity. The result of the signal control strategy is given by the indicator. Traffic signal duration is associated with traffic flow capacity.

2.2   Reinforcement Learning

Reinforcement learning describes about maximize the numerical reward and mapping the state into actions through different way [18, 19]. Signal agents identify situation and responses from traffic scenarios, learn information depend on learning algorithms. Then it makes action choice with respect to its own accumulated information. Increase the traffic flow and decrease the average delay time is the purpose of traffic light control system. In this traffic arrangement, signals at one intersection coordinate with the signals at other intersection for better transport flow. Throughout the process, signals at each intersection develop a strategy of cooperation to maximize their individual benefit. Cooperation between agents is accomplished by distributing partial information of the states with the adjacent agents. In view of changing scenarios of the real traffic situation, multi-agent cooperative Q learning algorithm (MCQL) is developed for intelligent traffic control approach [19–21]. The Q update equation is given as:

$$Q(s, a) \leftarrow Q(s, a) + \alpha(r + \gamma Q(s', a') - Q(s, a)), \tag{3}$$

where the state, action, immediate reward, and cumulative reward at time t correspondingly stands for s_t, a_t, r_t, and Q_t. $Q_t(s, a)$ is called policy function. $\alpha \in [0, 1]$ refers to the learning rate and $\gamma \in [0, 1]$ indicates the discount rate.

3   Multi-agent Cooperative Q Learning (MCQL)

Synchronization in multi-agent reinforcement generates a complex set of presentations achieved from the different agents' actions. A portion of good performing agent group (i.e., a general form) is shared among the different agents via a *specific form*(Q_i) [22]. Such specific forms embrace the limited details about the environment. Such strategies are incorporated to improve the sum of the partial rewards received using satisfactory cooperation prototype. The action plans or forms are created by the way of multi-agent Q learning algorithm by constructing the agents to travel for the most excellent form Q* and accumulating the rewards. When forms Q_1, ..., Q_x are incorporated, it is possible to construct new forms that is *General Form* ($GF = \{GF_1, ..., GF_x\}$), in which GF$_i$ denotes the **outstanding reinforcement** received by agent i all through the knowledge mode [5]. Algorithm 1 expresses *get_form* algorithm that splits the agents' knowledge. The forms are designed by the Q learning used for all prototypes. Outstanding reinforcements are liable for GF which compiles all outstanding rewards. It will be shared by the way of the added agents [21, 22]. Transforming incomplete rewards as *GF* is considered for outstanding reinforcements to achieve the cooperation between the agents. A *status* utility gives the outstanding form among the opening states and closing state for a known form which approximates *GF* with the outstanding reinforcements. The status utility is calculated by summation of steps the agent needed to get to the destination at the closing state and the sum of the received status in the forms among each opening and the closing state [22].

Algorithm 1 *Cooperative Multi-agent Q Learning Algorithm*
Algorithm *get_form* (I, technique)

1. Initialization Q_i(s, a) and GF_i(s, a)
2. Coordination of the agents $i \in I$;
3. Agents collaborate till the target state is found; episode ← episode +1
4. Renewal rule which estimates the reward value;

$$Q(s, a) \leftarrow Q(s, a) + \alpha(r + \gamma Q(s', a') - Q(s, a))$$

5. Fcooperate (episode, tech, s, a, i);
6. Q_i ← GF that is Q_i of agent $i \in I$ is updated by means of GF_i.

Cooperation Models

Various collaboration methods for cooperative reinforcement learning are proposed here:

(i) *Group model*—reinforcements are distributed in a sequence of steps.
(ii) *Dynamic model*—reinforcements are distributed in each action.
(iii) *Goal-oriented model*—distributing the sum of reinforcements when the agent reaches the goal-state (S_{goal}).

Algorithm 2 *Cooperation model*
Fcooperate (episode, tech,s,a,i)/*cooperation between agents as four cases*/
q: count of sequence

1. Switch between cases
2. In case of Group method

   if episode mod q = 0 then
   get_Policy(Q_i, Q^*,GF_i);

3. In case of Dynamic method

   $r \leftarrow \sum_{j=1}^{x} Qj(s,a)$;
   Q_i(s, a) ← r;
   get_Policy(Q_i, Q^*, GF_i);

4. In case of Goal-oriented method

   if S = S_{goal} then
   $r \leftarrow \sum_{j=1}^{x} Qj(s,a)$;
   Q_i(s,a) ← r;
   get_Policy(Q_i,Q^*,GF_i);

Algorithm 3 *get_Policy*

Function get_Policy(Q_i, Q^*, GF_i) /*find out universal agent policy */

1. for loop for each agent i ∈ I
2. for loop for each state s ∈ S
3. if value(Q_i, s) ≤ value(Q^*, s) then

   GF_i(s,a) ← Q_i(s,a);

4. end for loop

Group Model: During the learning process each agents receives reinforcements for their actions. At the last part of the series (step q), each agent gives out the cost of Q_j to GF. If reward value is suitable, that is it improves the usefulness of another agents for given state the agents will afterward donate to these rewards [21–23].

4   Model Design

In a practical environment, traffic flows of four signals with eight flow directions are considered for the development. As shown in Fig. 1, there are altogether four junctions at each signal agent, i.e., agent 1, agent 2, agent 3, and agent 4 for Ja, Jb, Jc, and Jd, respectively.

The control coordination between the intersections can be viewed as a Markov process, denoted by $\langle S, R \rangle$, where represents the state of the intersection, stands for the action for traffic control, and indicates the return attained by the control agent.

Definition of State: Agent receives instantaneous traffic state and then returns traffic control decision by the present state of the road. Essential data such as a number of vehicles newly arriving and number of vehicles currently stopped at signal are used to reflect the state of road traffic.

Fig. 1 Traffic flow and control of four intersections with eight flow directions

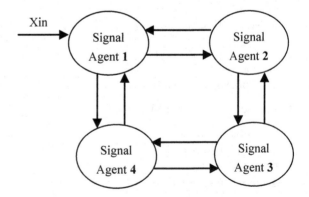

Number of vehicles newly arriving $= X_{max} = x_1, x_2, x_3, x_4 = 10$
Number of vehicles currently stopped at junction $J = I_{max} = i_1, i_2, i_3, i_4 = 20$.
State for agent 1 become (x_1, i_1), e.g., (5,0) that means 5 new vehicles are arriving to agent 1 with 0 vehicles are stopped at junction 1. State for agent 2 become (x_2, i_2), State for agent 3 become (x_3, i_3) and State for agent 4 become (x_4, i_4). State of the system become **Input** as (x_i, i_i). Here, it can get together 200 possible states by combining maximum 10 arriving vehicle and maximum 20 vehicles stopped at signal $(10 * 20 = 200)$.

Definition of Action: In reinforcement learning framework, policy denotes the learning agent activities at a given time. Traffic lights control actions can be categorized to three types: no change in signal duration, increasing signal duration, reducing signal duration.

Value	Action
1	No change in signal duration
2	Increase in signal duration
3	Reduce the signal duration

Action set for signal agent 1 is $A1 = \{1, 2, 3\}$, action set for signal agent 2 is $A2 = \{1, 2, 3\}$ and action set for signal agent 3 is $A3 = \{1, 2, 3\}$.

Each of them is for one of the following actual traffic scenarios. The strategy of no change in signal duration is used in the case of the normal traffic flow when the lights control rules do not change. The strategy increasing the signal duration is mostly used in the case that in one route traffic flow is stopped and the other route is regular. Increasing the signal duration extends the traffic flow while signal lights are still timing. The strategy decreasing signal duration is mostly used in the case that in one route of traffic flow is little while that of the other route is big. Decreasing signal light duration reduces the waiting time of the other route and lets vehicles of that route pass the junction faster, while signal lights keep timing [23, 24].

Definitions of Reward and Return: Agent makes signal control decisions under diverse traffic circumstances and returns an action sequence. We use traffic value display to estimate the traffic flows as [26].

Reward is calculated in the system as given below:

Assume current state $i = (x_i, i_i)$ and next state $j = (x_j, i_j)$.
current state $i \rightarrow$ next state j
Case 1: $[x_i, i_i] \rightarrow [x_i, i_{i-1}]$

$$[X_{max} = 10, I_{max} = 20] \rightarrow [X_{max} = 10, I_{max} = 19]$$

That means: one vehicle from currently stopped vehicle is passing the junction

Case 2: $[x_i, i_i] \rightarrow [x_{i+1}, i_{i-1}]$

$$[X_{max} = 9, I_{max} = 20] \rightarrow [X_{max} = 10, I_{max} = 19]$$

That means: one newly arrived vehicle at junction and one vehicle is passing

Case 3: $[x_i, i_i] \rightarrow [x_i, i_{i-3}]$

$$[X_{max} = 10, I_{max} = 20] \rightarrow [X_{max} = 10, I_{max} = 17]$$

That means: More than one stopped vehicles are passing the junction

Case 4: $[x_i, 0] \rightarrow [x_{i+1}, 0]$

$$[X_{max} = 2, I_{max} = 0] \rightarrow [X_{max} = 3, I_{max} = 0]$$

That means: new one new vehicle is arriving and no stopped vehicle at the junction. Depending on above state transitions from current state to next state, reward is calculated as [24]

$$
\begin{aligned}
\text{Reward is } r_p(i, p, j) &= 1 \quad \text{if } x_1' = x_1 + 1 \ldots\ldots\ldots\text{Case 4} \\
&= 2 \quad \text{if } i_1' = i_1 - 1 \ldots\ldots\ldots\text{Case 1} \\
&= 3 \quad \text{if } i_1' = i_1 - 3 \ldots\ldots\ldots\ldots Case\,2\,\&\,3 \\
&= 0 \quad \text{otherwise.}
\end{aligned}
$$

5 Experimental Results

The study learns a controller with learning rate = 0.5, discount rate = 0.9, and $\lambda = 0.6$. During the learning process, the cost was updated 1000 with 6000 episodes.

Figure 2 shows that delay time versus a number of state given by simple Q learning (without cooperation) and group method (with cooperation). Delay time obtained by cooperative methods, i.e., group method is much less than that of without cooperation method, i.e., simple Q learning for agent 1 in the multi-agent scenario.

Figure 3 shows that delay time versus a number of state given by simple Q learning (without cooperation) and group method (with cooperation). Delay time obtained by cooperative methods, i.e., group method is much less than that of

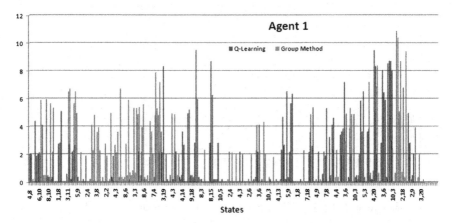

Fig. 2 States versus Delay time for Agent 1 by Q learning and Group Method

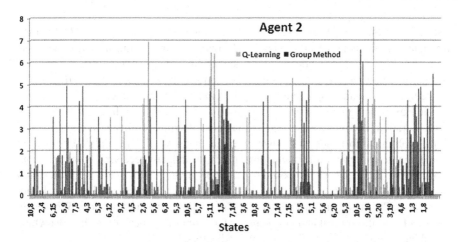

Fig. 3 States versus Delay Time for Agent 2 by Q learning and Group Method

without cooperation method, i.e., simple Q learning for agent 2 in the multi-agent scenario.

Figure 4 shows that delay time vs number of state given by simple Q learning (without cooperation) and group method (with cooperation). Delay time duration obtained by cooperative methods, i.e., group method is much less than that of without cooperation method, i.e., simple Q learning for agent 3 in multi-agent scenario.

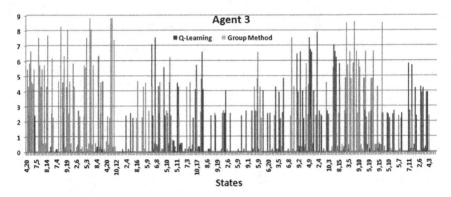

Fig. 4 States versus Delay Time for Agent 3 by Q learning and Group Method

6 Conclusion

Because traffic control system is so complicated and variable that a Q learning model (without cooperation) with defined strategy can rarely manage with the traffic jam and sudden traffic accidents which actually may occur at any time, the demand for combining timely and intelligent traffic control policy with real-time road traffic is getting more and more urgent. Reinforcement learning gathers tests and information by keeping communication with the situation. Although it usually needs a long duration to complete learning, it has good learning ability to a complex system, enabling it to handle unknown complex states well. The application of reinforcement learning in traffic management area is gradually receiving more and more concerns. The paper proposed a cooperative multi-agent reinforcement learning-based models (CMRLM) for traffic control optimization. The actual continuous traffic states are discretized for the purpose of simplification. We design actions for traffic control and define reward and return by mean of traffic cost which combines with multiple traffic capacity indicators.

References

1. F. Zhu, J. Ning, Y. Ren, and J. Peng, "Optimization of image processing in video-based traffic monitoring," *ElektronikairElektrotechnika*, vol.18, no.8, pp. 91–96, 2012.
2. B. de Schutter, "Optimal traffic light control for a single intersection," in *Proceedings of the American Control Conference (ACC '99)*, vol. 3, pp. 2195–2199, June 1999.
3. N. Findler and J. Stapp,"A distributed approach to optimized control of street traffic signals," *Journal of Transportation Engineering*, vol.118, no.1, pp. 99–110, 1992.
4. L. D. Baskar and H. Hellendoorn, "Traffic management for automated highway systems using model-based control,"*IEEE Transactions on Intelligent Transportation Systems*, vol. 3, no. 2, pp. 838–847, 2012.
5. R. S. Sutton and A. G. Barto, *Reinforcement Learning: An Introduction*, MIT Press, Cambridge, Mass, USA, 1998.

6. Artificial Intelligence in Transportation *Information for Application*, Transportation Research CIRCULAR, Number E–C 113, Transportation On Research Board *of the National Academies*, January 2007.
7. Deepak A. Vidhate, Parag Kulkarni "New Approach for Advanced Cooperative Learning Algorithms using RL methods (ACLA)" VisionNet'16 Proceedings of the Third International Symposium on Computer Vision and the Internet, ACM DL pp 12–20, 2016.
8. K. Mase and H. Yamamoto, "Advanced traffic control methods for network management," *IEEE Magazine*, vol. 28, no. 10, pp. 82–88, 1990.
9. Deepak A. Vidhate, Parag Kulkarni "Innovative Approach Towards Cooperation Models for Multi-agent Reinforcement Learning (CMMARL)" in Smart Trends in Information Technology and Computer Communications, Springer Nature, Vol 628, pp 468–478, 2016.
10. L. D. Baskar, B. de Schutter, J. Hellendoorn, and Z. Papp, "Traffic control and intelligent vehicle highway systems: a survey," *IET Intelligent Transport Systems*, vol. 5, no. 1, pp. 38–52, 2011.
11. M. Broucke "A theory of traffic flow in automated highway systems," *Transportation Research C*, vol. 4, no. 4, pp. 181–210, 1996.
12. D. Helbing, A. Hennecke, V. Shvetsov, and M. Treiber, "Micro and macro-simulation of freeway traffic," *Mathematical and Computer Modelling*, vol. 35, no. 5–6, pp. 517–547, 2002.
13. S. Zegeye, B. de Schutter, J. Hellendoorn, E. A. Breunesse, and A. Hegyi, "A predictive traffic controller for sustainable mobility using parameterized control policies," *IEEE Transactions on Intelligent Transportation Systems*, vol. 13, no. 3, pp. 1420–1429, 2012.
14. Deepak A. Vidhate, Parag Kulkarni "Enhancement in Decision Making with Improved Performance by Multiagent Learning Algorithms" IOSR Journal of Computer Engineering, Volume 1, Issue 18, pp 18–25, 2016.
15. A. Bonarini and M. Restelli, "Reinforcement distribution in fuzzy Q-learning," *Fuzzy Sets and Systems*, vol.160, no.10, pp. 1420–1443, 2009.
16. Y. K. Chin, Y. K. Wei, and K. T. K. Teo, "Qlearning traffic signal optimization within multiple intersections traffic network," in *Proceedings of the 6th UKSim/AMSS European Symposium on Computer Modeling and Simulation (EMS '12)*, pp. 343–348, Nov 2012.
17. L.A. Prashanth and S. Bhatnagar, "Reinforcement learning with function approximation for traffic signal control," *IEEE Transactions on Intelligent Transportation Systems*, vol. 12, no. 2, pp. 412–421, 2011.
18. Deepak A. Vidhate, Parag Kulkarni "Multilevel Relationship Algorithm for Association Rule Mining used for Cooperative Learning" in International Journal of Computer Applications (IJCA), Volume 86 Number 4- 2014 pp. 20–27.
19. Y. K. Chin, L. K. Lee, N. Bolong, S. S. Yang, and K. T. K. Teo, "Exploring Q-learning optimization in traffic signal timing plan management," in *Proceedings of the 3rd International Conference on Computational Intelligence, Communication Systems and Networks (CICSyN '11)*, pp. 269–274, July 2011.
20. Deepak A. Vidhate, Parag Kulkarni "Multi-agent Cooperation Methods by Reinforcement Learning (MCMRL)", *Elsevier International Conference on Advanced Material Technologies (ICAMT)-2016]No. SS-LTMLBDA-06-05*, 2016.
21. S. Russell and P. Norvi, *Artificial Intelligence: A Modern Approach*, PHI, 2009.
22. Deepak A. Vidhate, Parag Kulkarni "Performance enhancement of cooperative learning algorithms by improved decision making for context based application", International Conference on Automatic Control and Dynamic Optimization Techniques (ICACDOT) IEEE Xplorer, pp 246–252, 2016.
23. Deepak A. Vidhate, Parag Kulkarni "Improvement In Association Rule Mining By Multilevel Relationship algorithm" in International Journal of Research in Advent Technology (IJRAT), Volume 2 Number 1- 2014 pp. 366–373.
24. Young-Cheol Choi, Student Member, Hyo-Sung Ahn "A Survey on Multi-Agent Reinforcement Learning: Coordination Problems", IEEE/ASME International Conference on Mechatronics Embedded Systems and Applications, pp. 81–86, 2010.

Digital Tokens: A Scheme for Enabling Trust Between Customers and Electronic Marketplaces

Balaji Rajendran, Mohammed Misbahuddin, S. Kaviraj and B. S. Bindhumadhava

Abstract In electronic marketplaces, when the supply of a particular product is limited, and when there is a huge demand for the same, the questions of transparency and integrity prop up. We propose Digital Tokens—defined using proven cryptographic techniques—as a mechanism to assure trust for customers, and issued by a reliable, transparent and third-party intermediary, called digital token service provider (DTSP). The digital tokens are issued to a customer on behalf of a vendor and could be authenticated by both Vendor and the DTSP. This paper details the architecture involving the DTSP, protocols for communication, implementation details, the potential uses and benefits of the system and performance evaluation of such a system.

Keywords Digital token · Timestamp · DTSP · Digital token service provider
Certifying authorities · Electronic market place

1 Introduction

The electronic marketplaces [1] have been witnessing huge demand for select brands of products such as mobile phones from customers. These electronic marketplaces have limited stock of the products, primarily owing to the constraints of

B. Rajendran (✉) · M. Misbahuddin · S. Kaviraj · B. S. Bindhumadhava
Computer Networks and Internet Engineering (CNIE) Division,
Centre for Development of Advanced Computing (C-DAC),
Electronics City, Bangalore 560100, India
e-mail: balaji@cdac.in

M. Misbahuddin
e-mail: misbah@cdac.in

S. Kaviraj
e-mail: skaviraj@cdac.in

B. S. Bindhumadhava
e-mail: bindhu@cdac.in

© Springer Nature Singapore Pte Ltd. 2018
S. Bhalla et al. (eds.), *Intelligent Computing and Information and Communication*,
Advances in Intelligent Systems and Computing 673,
https://doi.org/10.1007/978-981-10-7245-1_48

the supplier, and therefore release the product in batches, which leads to huge competition among potential buyers as they easily outnumber the available quantity, and many buyers return disappointed, as the entire stock goes off in a flash.

We propose a mechanism of Digital Token, wherein a Digital Token is issued to the customer by a trusted intermediary, on behalf of the vendor. The token is then presented by the customer to the vendor, which can be verified for its authenticity and validity by both the intermediary and the vendor. The trusted token can also be validated by the customer if required, by communicating with the intermediary. This mechanism establishes the much-needed trust between the customer and the vendor.

Digital Tokens are useful, whenever the demand is more and supply is less, and when the buyers need a reliable and trustable intermediary. In general, the concept could be applied to any real-life token system, and the potential applications are many as detailed in the later sections of this paper. The benefits of the digital token-based system include: the introduction of a trustable third party that could be relied by the buyer and because of which the vendor relies on it—as it brings in transparency and the vendor can offload the complexities involved in generating the tokens, authenticating and validating them and also will have lesser loads on its servers.

This paper explains the elements of such a Digital Token, the process of authentication and validation of a Digital Token that can be carried out by Vendor and the Intermediary, the protocols for communication between the three parties—Customer, Vendor (eMarketplace) and the Intermediary (referred as the DTSP—Digital Token Service Provider), and finally, the implementation detailing the experimental setup and study of performance issues.

2   Related Work

The concept of Digital Token as a cryptographically encoded message with Digital Time Stamps has not been mentioned elsewhere, although the reference of tokens has been widely used in the literature.

However, the concepts of Digital timestamping [2] as a time validation methodology has been described and used in several PKI-based applications, including e-Tendering, e-Auction, etc. Digital timestamping is comprehensively described in the IETF RFC 3161 standard [3]. The format of timestamping request and the response are detailed. Digital timestamping as a service has been offered by several vendors all over the world, with most of the vendors offering the core CA (Certifying Authority) services.

Digital Watermarking is a technique used for copy-protection of media content, by tracking any unauthorized distribution of the digitally watermarked content. A buyer–seller watermarking protocol [4] was proposed, in such a way that neither the buyer will be able to distribute the watermarked contents received from the seller, nor the seller would be able to create unauthorized copies of the buyer watermarked contents. The proposed technique uses digital signatures and encryption for its protocol. Holmquist et al. [5] proposed the concept of tokens as a physical object required to access digital information. A Credit card company aimed to replace the Credit Cards with a Digital Token [6]. The concept of Token based Secure Electronic Transaction [7] was proposed by Rajdeep et al., which focused on the customers can be sure about the trustworthiness of the Seller before indulging any transaction, they also mentioned that faulty transactions never take place by implementing token-based SET mechanism in electronic commerce sites.

Matsuura et al. [8] proposed digital timestamp mechanism for dispute settlements in electronic commerce by providing long-lived authenticity and archiving by the server. Shazia et al. [9] explained the importance of E-commerce security, implementing PKI, digital signature and certificate-based cryptographic techniques in E-commerce security.

3 Proposed System: Architecture and Processes

3.1 Entities

There are three entities involved in this System: Customer—who is interested in buying a product; Vendor—selling a product through an electronic marketplace; DTSP—Digital Token Service Provider—a trusted intermediary for both the Customer and the Vendor, who issues Digital Tokens, and who can verify the authenticity and time validity of a token.

3.2 Digital Token

A Digital Token is a cryptographic entity derived from customer's identity, time value and a timestamp by the DTSP. The time value indicates the time validity of the token, the timestamp mentions the time the token was generated that will uniquely identify a transaction.

3.3 Architecture

Figure 1 illustrates the architecture containing the overall sequence of communications between Customer, Vendor and DTSP detailing the life cycle of a Digital Token.

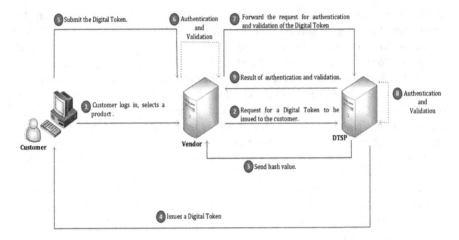

Fig. 1 Digital token life cycle

3.4 Processes

The system is comprised of two main processes: One, issuance of a token to a customer, on behest of the Vendor; Two, authentication and validation of the digital token—which can be either carried out by the Vendor or by the DTSP on request.

Process 1: Digital Token Issuance

Step 1: This is a base operation, wherein the customer visits the vendor website (typically an online marketplace) and selects a particular product, and requests for an advanced registration or booking. The vendor then records the request and assigns a time slot—during which the customer has to appear for buying of the product—and sends a request to DTSP to issue a Digital Token to the customer on its behalf. During the above process, the vendor constructs the Digital Token request—DTRQ, which consists of customer's id (UID), start time (ST) and end time (ET). DTRQ is then digitally signed by the vendor and sent to the DTSP.

Step 2: DTSP, upon receipt of the request—DTRQ, verifies the authenticity and integrity of the request by verifying the digital signature sent along with the request. If authenticated successfully, a timestamp value is obtained and sent as a response to the request in a cryptographically hashed form [10], as follows:

H = Hash (Customers ID + Start Time + End Time + TimeStamp + Nonce);

Here, the nonce represents a random secret value generated by DTSP. Now the DTSP proceeds to generate a digital token, by digitally signing the request and response as

DT = <S, H> where S = Digital Sign (H)

Therefore, a Digital Token is a digital signature of the DTSP along with the generated Hash, which is sent to the customer. Figure 2 illustrates the Token issuance process:

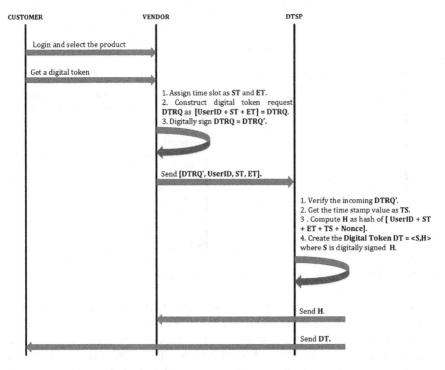

Fig. 2 Digital token issuance process

Process 2: Authentication and Validation of Digital Tokens

The authenticity check will tell whether the token has been really issued by the DTSP to the particular customer in question, on behalf of the said vendor. The validation will have two cases—one, the customer presenting it at the right time slot, two—the customer presenting either before time or after time—the latter case being an expired token, but both invalid.

Step 1: The customer logs into the vendor site (at the specified time slot) and uploads the Digital Token, received from the DTSP corresponding to a particular product. The authenticity and validity of the token can now be carried out either by the vendor or by the DTSP.

Step 2(a): *Authentication and Validation by the Vendor:* The vendor verifies the Digital Signed Hash from Digital Token by using the public key of the DTSP. The vendor then searches his database for the H value from DT using the customer's ID as a filter, and if a match is found then the token is authentic—i.e. the token has been issued by the DTSP, for the particular customers. The vendor will also be able to retrieve the time values—start time and end time and validate the token for the time, and if found to be valid, the vendor may permit the customer to proceed for the subsequent steps involved in buying a said product. Figure 3 illustrates the process of authentication and validation carried out by the vendor.

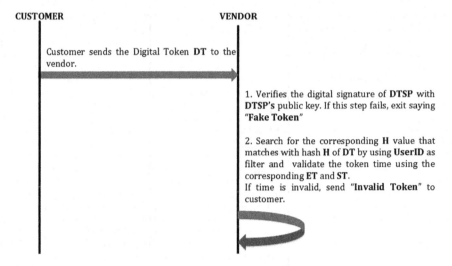

Fig. 3 Authentication and validation of digital token by vender

Step 2(b): Authentication and Validation by the DTSP: The vendor can decide to offload the authentication process to the DTSP, by simply forwarding the digital token. In such a case, the DTSP will receive the digital token presented by the customers, and the customer's id from the vendor. The DTSP will then verify the digital signature from Digital Token using its public key. Now the DTSP will search its database of active tokens issued at behest of the particular vendor, containing H value from DT using the Customers Id as the filter. If found, the token is authenticated successfully, meaning the token has been issued by the DTSP only, and then the time validation can be done by fetching the corresponding start and end time. The result of the validation will then be communicated to the vendor by the DTSP. Figure 4 illustrates the process of authentication and validation carried out by the DTSP.

4   Implementation

A prototype implementation has been developed using Java EE wherein the entities —Vendor and DTSP are modelled as distinct entities as in real situations, and cryptographic operations are carried out with the native Java Crypto libraries.

4.1   Features and Assumptions

1. The DTSP does not store any product information that the customer is buying, therefore helping to protect the privacy.

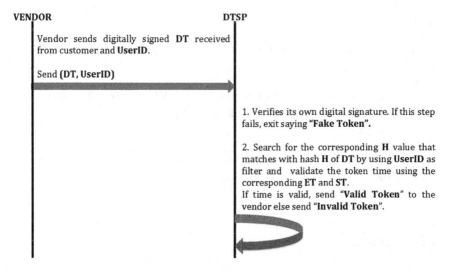

VENDOR   **DTSP**

Vendor sends digitally signed **DT** received from customer and **UserID**.

Send **(DT, UserID)**

1. Verifies its own digital signature. If this step fails, exit saying **"Fake Token"**.

2. Search for the corresponding **H** value that matches with hash **H** of **DT** by using **UserID** as filter and validate the token time using the corresponding **ET** and **ST**.
If time is valid, send **"Valid Token"** to the vendor else send **"Invalid Token"**.

Fig. 4 Authentication and validation of digital token by DTSP

2. The DTSP is running a timestamp server or at least has an access to a timestamp service that will give a reliable time.
3. There is binding between the customer and a Digital token, as DTSP uses customer identification (CustomersID) information supplied by the vendor. The Customers Id is assumed to be an email address as typically is the case with most vendors.
4. The start time could be null by default, which means that the customer can buy the product after it is open for sale at any time, but before the end time.
5. A customer can obtain a digital token only for one quantity of a product at a time. If more quantities are required, the customer has to obtain those many digital tokens.

4.2 Digital Token Encoding

The digital token created by the DTSP on behalf of the vendor is emailed as a file to the customer. The digital token as described above is actually a signature of the DTSP on a message containing a time stamp. The digital signature is encoded in the standard Base64 [11] format, and emailed to the customer. Upon receipt of the token, for authentication and validation, either by the vendor or by the DTSP, both have to decode it back to the standard bytes for further processing. A sample digital token is given below:

–Begin Signature–
ckvZu-
oSlH3+670PLu2QwoXUd2COoKt9GkutCBVi9nmyyeSII/pNDCPLlg8/2U
+WrjfG62
HJatUs-
vaAY2G4UUfvFsipVQFDmH4K4PftSaHdG+/RD+VIG43n5lZxerUy0oB/
QfCKTLAk9aHw/KrF8NxQTdnckw8gkTtzfo/t1LjOw=
–End Signature–
–Begin Hash–
bf65c0043aec52ec26f4c67014e5960d40a4755d
–End Hash—

4.3 *Schema*

The schema of the **DTSP** is given as below:

{ID, VendorID, CustomersID, StartTime, EndTime, Timestamp, Hash, Status}

As it can be observed, the following information comes from the vendor—CustomersID, StartTime, EndTime and VendorID—each vendor is assigned a unique ID by the DTSP. The Hash value computed by the DTSP would be used for searching for a token, in case if the same customers had booked for multiple quantities of the same product or booked for different products from the same vendor having overlapping timeslots.

We use MongoDB [12] for storing the data, as it yields readily itself to JSON format so that it could be used for communications straightaway. The status field is used to store the status of the record/digital token—Active, Expired, etc. All expired tokens are moved out periodically to separate databases, keeping only the active ones, and grouped by vendor to improve the search performance of DTSP. A sample of a fully constructed a document with the DTSP would look as follows:

```
DTSP Document:
{
``_id'' : ObjectId(``55e7f79d8ad5d90404000029''),
``VendorId'':64,
``CustomersId'':``xyz@abc.in'',
``StartTime'':new Date(``3-9-2015 12:30:00''),
``EndTime'':new Date(``3-9-2015 19:30:00''),
``TimeStamp'':new Date(``25-8-2015 16:35:00''),
``Hash'': ``bf65c0043aec52ec26f4c67014e5960d40a4755d'',
``Status'':``Active''
}
```

The schema at the **Vendor** for the purposes of Digital Tokens is given as below:

{ID, CustomersID, ProductID, StartTime, EndTime, Hash, Status}

As it can be observed, the vendor has details of the customers, along with the product information that the customers are seeking to book, and also the time in which the vendor wants to allow the customers to actually buy the product. The vendor can set the 'StartTime' as NULL, if he wants to only set the end time to the customer. The Hash value is obtained from the DTSP, which is a combination of the above information plus the timestamp.

```
Vendor Document:
{
``_id'':ObjectId(``55e7f79d8ad5d90404000029''),
``CustomersId'':``xyz@abc.in'',
``ProductId'':M101256,
``startTime'':new Date(``3-9-2015 12:30:00''),
``EndTime'':new Date(``3-9-2015 19:30:00''),
``Hash'': ``bf65c0043aec52ec26f4c67014e5960d40a4755d'',
``Status'':``Active''
}
```

4.4 Communication Between Vendor and DTSP

The Vendor communicates with DTSP during the process of token issuance and during the process of authentication and validation. During Issuance, the vendor opens up a connection with the DTSP, and communicates its identity, the customer identity information and the time slot, as a key-value pair in JSON format and redirects the customer to the DTSP's page. The Vendor then waits to listen from the DTSP for the corresponding H value and closes or resets its connection with the customer.

In the process of authentication and validation, if the vendor decides to offload the process to the DTSP, the vendor has to communicate its and the customer's identity information, and wait for the result. If a successful result is announced by the DTSP, then the vendor puts the customer through the subsequent process of buying the product. It may be noted that the DTSP communicates the results only with the vendor, which will be useful for the latter to detect any multiple attempts by an attacker to hog the resources by sending fake tokens, or expired tokens.

4.5 Screenshots

A sample of the screenshots depicts the flow of the process as illustrated in Fig. 5.

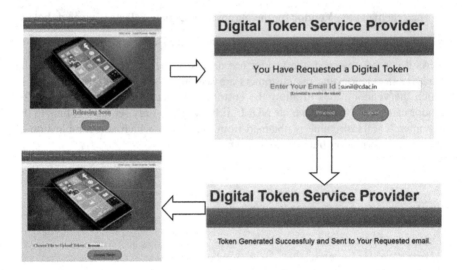

Fig. 5 Sample screenshots

5 Performance Analysis

The performance is analyzed by looking at the time it takes to create a digital token, and the time it takes to authenticate and validate a token.

5.1 Time Taken to Generate a Digital Token

Generating a Digital Token is carried out by the DTSP and involves the following steps, and the average time for each activity is given in Table 1. As it can be seen, approximately 470 digital tokens can be created in less than a minute, by a system that is not fully optimized for this activity. The mailing of digital tokens to the customer is considered as a separate process, outside of the critical sections affecting the performance of the system. Also, it may be noted that the timestamps are obtained from a dedicated timestamping service that implements the RFC 3161, and hence, the time taken to obtain it is higher than obtaining the local server time value.

5.2 Time Taken for Authentication and Validation

The authentication and validation can be carried out by either the Vendor or the DTSP and the logic is same—verification of the presented Digital token with the

Table 1 Average time taken for generation of a digital token

S. No.	Activity	Average duration (in ms)
1	Generating the timestamp—typically done through a timestamping server operated by the DTSP	122
2	Creating the hash of all the values {CustomersID, StartTime, EndTime, TimeStamp, Nonce}	0.75
3	Creating the signature by encrypting the Hash created in step 2 with its private key	4.01
4	Encoding the signature in base64 format and writing into a file	1.66
Total		128.42

Table 2 Average time taken for authentication and validation of a digital token

S. No.	Activity	Avg. time (in ms)	Remarks
1	Reading and decoding the signature given in base64 format	0.55	Failure at this step can mean the token is corrupted
2	Verification of the signature using public key of DTSP	2.56	Failure here means, the customers may be presenting a fake token
3	Searching for the H value (part of DT)	0.26	Failure here means, the customer has presented an invalid token
4	Perform time validation	0.01	Failure here means, the token has been presented before time (as expired tokens have been moved to different database and in that case would be eliminated in previous step)
Total		3.38	

public key of DTSP and search for the record matching the Hash value, and if found then perform the time validation. The above steps and time taken are given below in Table 2.

In the case of authentication and validation, failure is possible at each and every step, and failure at an earlier stage is better than at the later stage—as the latter stages will add to the cost of time. Step 3 and 4 have been optimized using efficient data structures and clean-up techniques.

5.3 Analysis

Following observations can be made from the above experiments:

1. The process of digital token issuance is costlier than the process of authentication and validation.

2. In the event of authentication and validation by DTSP, the overheads involved in communication between the vendor and the DTSP is primarily determined by the latency factor, rather than the process itself.
3. The cryptographic operations–especially the encryption and decryption are relatively faster.

When the vendor is experiencing peak-traffic loads, it may offload the process of authentication and validation of digital tokens to the DTSP, but in such scenarios the vendor may have to take into account the latency delays also, as the time required to do the authentication and validation is meagre 3.38 ms, which means an approximate 17,000 digital tokens could be evaluated in less than a minute. The vendor can use the above value as guidance to timeslot the customers accordingly.

6 Conclusion and Future Work

In this paper, we presented the concept and process of a Digital Token, issued by a trusted intermediary, playing the role of a DTSP—a role that could be played by a Certifying Authorities (CA) which could probably add to another business or revenue line to their existing services—the Digital tokens being issued to the customer who is attempting to buy a high-demand product or service from an electronic marketplace. The DTSP can conduct audit trails of the digital tokens issued, authentication and validation done by it and can publish the same.

The process could have been strengthened for the better by having a tight binding between a digital token and the customer, but then the customer should have a mechanism to apply their digital signature. Server-based digital signing approaches like eSign [13] could be used for high-value transactions. This approach can be extended to create a competitive market place by decoupling the tokens from the market places. Therefore, in our future work, we aim to shift the balance in binding proportions wherein we envisage a strong binding between the customer and the token, while a light binding between the token and the vendor or eMarketplace.

References

1. Sonja Grabner-Kraeuter, The Role of Consumers' Trust in Online-Shopping, July 2002, Journal of Business Ethics. http://link.springer.com/article/10.1023/A:1016323815802
2. S. Haber, WS. Stornetta, "How to Time-Stamp a Digital Document" Springer Berlin Heidelberg, pp. 437–455, 1991
3. C Adams, P Cain, D Pinkas, R Zuccherato, Internet X.509 Public Key Infrastructure Time-Stamp Protocol (TSP), 2001 – IETF RFC 3161 - https://www.ietf.org/rfc/rfc3161.txt
4. "A Buyer-Seller Watermarking Protocol", Nasir Memon, Ping Wah Wong, IEEE Transactions on Image Processing, pp: 643–649, 2001

5. L. Holmquist, J. Redström, and P. Ljungstrand, "TokenBased Access to Digital Information," Proceedings of HUC'99 (1999), pp. 234–245
6. Amex to Implement Digital Tokens to Replace Cards, http://www.infosecurity-magazine. com/news/amex-to-implement-digital-tokens/
7. Rajdeep Borgohain, Chandrakant Sakhatwade, Sugata Sanyal, "TSET: Token based Secure Electronic Transaction", IJCA, Volume 45 – Number 5, 2012.
8. Kanta Matsuura, Hideki Imai, "Digital timestamps for dispute settlement in electronic commerce: Generation, Verification and renewal", CiteSeerx, 2002
9. Shazia Yasin, Khalid Haseeb, Rashid Jalal Qureshi, "Cryptography based E-Commerce security: A Review", Volume 9, Issue 2, IJCSI, March 2012
10. L. Damgard. "Collission-free hash functions and public-key signature schemes." In Advances in Cryptology – Eurocrypt'87, pp. 203–217. Springer-Verlag, LNCS, vol. 304, 1988.
11. S. Josefsson, SJD, "The Base16, Base32, and Base64 Data Encodings", RFC 4648,Network Working Group, October 2006.
12. MongoDB-Documentation. http://docs.mongodb.org/manual/tutorial/
13. eSign – Online Electronic Signature Service, http://www.cca.gov.in/cca/?q=eSign.html, Last accessed May 12, 2017.

BMWA: A Novel Model for Behavior Mapping for Wormhole Adversary Node in MANET

S. B. Geetha and Venkanagouda C. Patil

Abstract Wormhole attack has received very less attention in the research community with respect to mobile ad hoc network (MANET). Majority of the security techniques are toward different forms of wireless network and less in MANET. Therefore, we introduce a model for behavior mapping for Wormhole Attacker considering the unknown and uncertain behavior of a wormhole node. The core idea is to find out the malicious node and statistically confirm if their communication behavior is very discrete from the normal node by formulating a novel strategic approach to construct effective decision. Our study outcome shows enhanced throughput and minimal overhead-latency with increasing number of wormhole node.

Keywords Mobile ad hoc network · Wormhole attack · Adversary
Malicious node behavior · Attack pattern

1 Introduction

Security problems in mobile ad hoc network are always represented as challenging problems owing to it inherent characteristics, i.e., dynamic topology, intermittent link breakage, power consumption, etc. [1, 2]. In past decade, there have been various research work carried out toward improving security system in MANET taking case studies of various forms of adversaries [3, 4]. Although, such studies develop a good baselines of future research but they also suffer from the limitation

S. B. Geetha (✉)
Department of Computer Engineering, MMCOE, Pune, India
e-mail: sb.geetha@gmail.com

V. C. Patil
Department of Electronics & Communication Engineering, BITM, Bellary, India
e-mail: patilvc@rediffmail.com

© Springer Nature Singapore Pte Ltd. 2018
S. Bhalla et al. (eds.), *Intelligent Computing and Information and Communication*,
Advances in Intelligent Systems and Computing 673,
https://doi.org/10.1007/978-981-10-7245-1_49

of their narrowed applicability [5]. Existing routing protocols with security features are quite less and highly symptomatic [6]. There are also good numbers of studies highlighting about node behavior of MANET [7–9]. Even these studies too does not emphasized over the computational complexities associated with the implementations of computationally complex algorithms. Moreover, with a changing time, design aspects of a malicious node are revolutionized for which reason it becomes sometime very hard to differentiate malicious node and regular node. Presence of malicious node does not mean attack implication. Even such nodes could be controlled to assists in packet forwarding. Unfortunately, there were no such research works, where the malicious node does assists in packet forwarding. We comment here that normally in wormhole attack, the malicious node will act as a regular node so that they can divert the incoming route through newly developed wormhole tunnel. This characteristic has not been researched by anyone with respect to wormhole attack and hence the proposed manuscript will brief out the simple mechanism where the malicious node will be used for this reason. This paper presents a novel technique toward leveraging the security features in MANET where both security features and communication capabilities are emphasized in MANET in presence of wormhole adversary. Section 2 discusses about the existing research work followed by problem identification in Sect. 3. Section 4 discusses about proposed methodology followed by an elaborated discussion of algorithm implementation in Sect. 5. Comparative analysis of accomplished result is discussed under Sect. 6 followed by the conclusion in Sect. 7.

2 Related Work

This section discusses about the relevant work being carried out toward wormhole attack in MANET. Our prior work has already reviewed certain techniques of resisting wormhole attack [10]. It was seen that wormhole attack was basically studied with respect to other forms of network, e.g., sensor network [11–14], wireless mesh network [15], and quite a very less in MANET. Singh et al. [16] have presented a technique to countermeasure the adverse effect of wormhole attack over on-demand routing in MANET using case study of misbehaved nodes. The study outcome shows enhanced communication performance. Almost similar direction of study toward wormhole attack was also carried out by Kaur and Chaba [17] to find the technique offer lower delay and throughput. Gupta and Pathak [18] have introduced a technique that explores substitute route for avoiding the packets go in wormhole tunnel. The study outcome shows reduced overhead in routing, reduced packet drop, and increased received packets. Stoleru et al. [19] have presented a technique that emphasizes on the discovery of the neighbor nodes for a secured communication system in MANET. The technique also assists in determining the wormhole node position and explores the neighbor node in much secured manner.

Shastri and Joshi [20] have reviewed the existing techniques toward determining and eradicating the wormhole adversaries in MANET. Similar forms of identifying and localizing wormhole attacks were also carried out by Ojha and Kushwah [21]. The technique has also used threshold-based approach for improving delay and communication performance. Ali and Nand [22] have presented a discussion where two different on-demand routing protocols were investigated in presence of wormhole attack on MANET. In a nutshell, it can be said that there are few techniques to resists the adverse effect of wormhole attacks in MANET. The existing technique has emphasized on detection of the adversaries and presented approaches to identify the tunnel. All the above-mentioned technique has associated advantages as well as pitfalls too. Discussion of the existing research problems is carried out in the next section.

3 Problem Description

From the previous section, it can be seen that there are a various approach that assists in determining the adversary in wormhole where the focus is more on exploring the tunnel and less on finding the exact adversaries. The biggest challenging situation is that exploring exact malicious node is quite a difficult task especially in presence of dynamic topology in MANET. A malicious node can easily act as normal node until it gain complete trust of the network to invoke a collateral damage which is much dangerous than wormhole tunnel. In such situation, none of the existing techniques presented in literature can assist to differentiate malicious or regular node in presence of wormhole attack. This problem is yet to be studied as a solution to this problem will give precise information about the adversaries.

4 Proposed Methodology

The proposed work is a continuation of our prior model of resisting wormhole attack [23]. The present work adopts analytical research methodology, where the focus is laid to find the wormhole adversary node in the simulation area. We consider the complex hiding behavior of wormhole node and chalk out a novel implementation formulation that uses confusing characteristics of a malicious node in order to resist other regular nodes to fall into wormhole tunnel. The schematic architecture of the proposed system is shown in Fig. 1.

The proposed study applies three different strategies motivated from game theory and performs the computation of three different forms of the trust-based coefficient, e.g., positive trust, negative trust, and ambiguity trust. Sequential

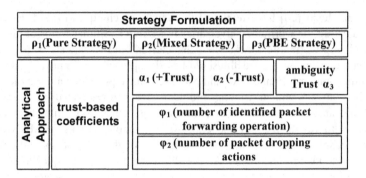

Fig. 1 Schematic architecture of BMWA

rationality is implemented to invoke complex behavior of wormhole adversary in MANET. The prime idea is to ensure that malicious node should be caught at any cost and while doing so the communication performance in MANET should not be affected. The next section briefs about algorithm implementation of BMWA.

5 Algorithm Implementation

This section discusses about the algorithm that is responsible for mapping the malicious behavior of a mobile node in MANET. The algorithm implements a novel logic of strategy formulation for mapping malicious actions of the mobile node and thereby undertakes different steps to mitigate the attack. The algorithm takes the input of the A (simulation area), n (number of the node), m (percentage of the malicious node), T (Threshold), which upon processing leads to the generation of the outcome of a malicious node or regular node. The basis of the algorithm performs both detection and prevention system. The steps of the proposed system are as shown below:

```
Algorithm for BMWA
Input: A, n, m
Output: adversary
Start
1. init A, n, m
2. [R_ns, M_ns]=[ρ₁, ρ₂, ρ₃]
3. If T_θ*(1-T_u)<T
4. If T_θ<val1
5. flag→D_F
6. else
7. flag strategy→P_D
```

```
8. end
9. End
10. compute α₁, α₂, α₃
11. Pcatch=sᵢ*((temp₁-Damf)/statf)
12. Flag n(max(card(α₃))→adversary
End
```

The algorithm implements the potential logic of game theory where three different strategies, i.e., ρ_1 (Pure Strategy), ρ_2 (Mixed Strategy), ρ_3 (PBE Strategy) is being testified. A matrix is constructed for such strategies against regular node (R_{ns}) and malicious node (M_{ns}) (Line-2). We also construct various entities in the proposed model, e.g., (i) Intrusion Profit, (ii) Packet forwarding profit, (iii) Updating Profit, (iv) Resources required for Intrusion, (v) Resources required for packet forwarding, (vi) Resources required for constructing tunneling, (vii) Damage factor due to negative update, and (viii) statistical factor for negative update. The proposed algorithm also consider a threshold T considered for a normal mobile node to perform the update. We also apply probability modeling for constructing this algorithm. Our first probability factor is T_θ that computes the possibility of one mobile node being malicious. Similarly, our second probability factor is also defined for if the malicious node will really perform intrusion or choose to assist in data packet forwarding just to hide its true identity. The proposed system also considers another probability factor T_u to understand the possibility of the uncertainty of one node being a malicious node. We perform empirical formulation to compute the ambiguity factor α_3 (Line-11) as $c.\varphi_1* \varphi_2/((\varphi_1 + \varphi_2)^2*(\varphi1 + \varphi2 + 1))$, where φ_1 and φ_2 are a number of identified packet forwarding operation and pack dropping actions, respectively. A condition is applied for mapping malicious node behavior with Perfect Bayesian Equilibrium (Lines-3 and 4) after computing a decision value *val1* which corresponds to (Packet forwarding profit-Resources required for packet forwarding)/(Packet forwarding profit + Resources required for constructing tunneling). Therefore, the condition shown in Line-3 and Line-4 represents the good state of the node in which case we flag the node as to obtain a strategy for data forwarding (D_F) (Line-5) otherwise, it is flagged as a malicious node with packet dropping (P_D) action (Line-7). However, we also infer that it is highly possible that a malicious node can act as regular node showing D_F, whereas a regular node can also show P_D action. Hence, Algorithm line-3–7 just assists in understanding the possible strategies of a malicious or regular node but it does not confirm the fact that the node is really a malicious or regular node. For this purpose, we consistently monitor three trust-based coefficients ($\alpha_1, \alpha_2, \alpha_3$) (Line-10), The first coefficient α_1 is calculated as $[\varphi_1/(\varphi_1 + \varphi_2)] * \alpha_3$, the second coefficient α_2 is calculated as $[\varphi_2/(\varphi_1 + \varphi_2)] * \alpha_3$.

We compute the probability of detecting malicious node P_{catch} as shown in Line-11. Where s_i is selected strategy, and the variable $temp_1$ is equivalent to

$(T_\theta * (1 - T_u)*$Profit of update$)/(1 - T_\theta*(1 - T_u))$. The other variable Damage factor Dam_f and statistical factor for the negative update, i.e., $stat_f$. It will mean that if the regular node attempts to defame another regular node by offering a false update to its neighbor that it is a malicious node, than it is considered as false updating system. In such case, the regular node is being penalized by offering lower utility value. This process is called as sequential rationality, where both malicious node and the regular node will attempt to obtain the highest utility. For better game strategy, we maintained sequential rationality in such a way that malicious node can get the highest utility only when they attack and in such case regular node will get highest utility if they can catch hold of such attack using updating mechanism. Hence, in such scenario, chances of the malicious node to be get caught is quite high. However, if the malicious node chooses not to perform intrusion but to assists in packet forwarding, a malicious node is discouraged by offering lower utility value. But a closer look in such strategy suggests that by adopting the practice of data packet forwarding by a malicious node, it does not affect any communication performance of MANET due to any security issues as long as there is no intrusion. To further confirm about it, we compute maximum cardinality of the third trust factor, i.e., α_3. If the ambiguity factor is found to be consistently higher that it is assured that the node is malicious and there is no routing being performed through this node once the false reporting of its neighbor regular node is found to be lower. Hence, in this manner, the proposed system not only identifies the malicious node but also discourages them to initiate any forms of attacks.

6 Results Discussion

The assessment of the proposed study is carried out considering $1000 \times 1000 \, m^2$ simulation areas with 100 mobile nodes being randomly deployed. We consider 10% of the malicious node being present during the simulation. We consider all the three strategies of game theory, i.e., pure, mixed, and PBE for both the types of the nodes in order to incorporate sequential rationality. Each mobile node is considered to have a range of 20 m during the assessment of the result. For a better assessment, we compared the outcome of our study with respect to multicast ad hoc on-demand vector (MAODV) routing protocol in MANET (Fig. 2).

The study outcome shows that proposed BMWA offers a better improvement over throughput obtained by existing MAODV. It also offers higher extent of reduction in routing overhead as well as routing latency as compared to existing MAODV. Hence, the security techniques incorporated in BMWA does not seem to negatively affect communication quality. A closer look at the outcome will also show the capability to enhance communication performance increase with an increase in malicious nodes, which means the presence of malicious node does not create security loopholes in BMWA. Hence, the equilibrium between security and communication performance is maintained.

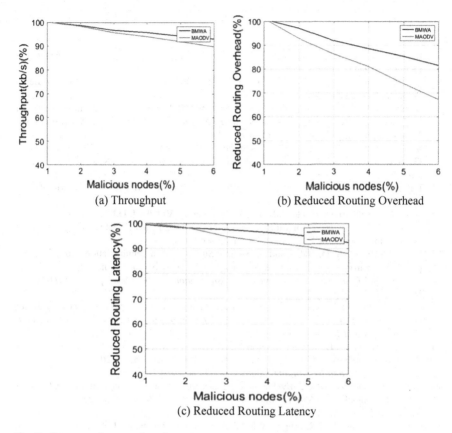

Fig. 2 Outcome of proposed study

7 Conclusion

The prime factor for working on this presented paper is as following: (i) existing research studies toward MANET security considering wormhole attack is quite less, (ii) a wormhole attacker can perform camouflaging by spoofing itself with the identity of the regular node. Therefore, in such scenario, there are no existing techniques that offer comprehensive disclosure of malicious node identity. Hence, this paper presents a probabilistic technique that assists in formulating a decision by a regular node to determine if its neighbor or its other communication node is wormhole adversary. Proper identification of wormhole adversary will protect the regular node from wormhole tunnel. Our outcome shows that proposed system offers a better balance between malicious node identification with communication performance in MANET.

References

1. R. Rajendran, A Survey on Multicast Routing Protocol and Its Challenges in Manet, Lap Lambert Academic Publishing, 2015
2. S. Paul, Introduction to MANET and Clustering in MANET, Anchor Academic Publishing-Computer, 2016
3. A. Nadeem and M. P. Howarth, "A Survey of MANET Intrusion Detection & Prevention Approaches for Network Layer Attacks," in *IEEE Communications Surveys & Tutorials*, vol. 15, no. 4, pp. 2027–2045, Fourth Quarter 2013.
4. S. Das and P. Nand, "Survey of hash security on DSR routing protocol," *2016 International Conference on Computing, Communication and Automation (ICCCA)*, Noida, 2016, pp. 520–524.
5. Rawat, Danda B., Security, Privacy, Trust, and Resource Management in Mobile and Wireless Communications, IGI Global, 2013
6. A. L. Sandoval Orozco, J. GarciaMatesanz, L. J. Garcia Villalba,J. D.M´arquez Diaz, and T.-H. Kim, "Security Issues in Mobile Adhoc Networks", Hindawi Publishing Corporation, International Journal of Distributed Sensor Networks, 2012
7. P. M. Nanaware, S. D. Babar and A. Sinha, "Survey on bias minimization and application performance maximization using trust management in mobile adhoc networks," *2016 3rd International Conference on Computing for Sustainable Global Development (INDIACom)*, New Delhi, 2016, pp. 831–834.
8. P. Ramkumar, V. Vimala and G. S. Sundari, "Homogeneous and hetrogeneous intrusion detection system in mobile ad hoc networks," *2016 International Conference on Computing Technologies and Intelligent Data Engineering (ICCTIDE'16)*, Kovilpatti, 2016, pp. 1–5.
9. [9] A. Lupia and F. De Rango, "A probabilistic energy-efficient approach for monitoring and detecting malicious/selfish nodes in mobile ad-hoc networks," *2016 IEEE Wireless Communications and Networking Conference*, Doha, 2016, pp. 1–6
10. S.B. Geetha and V.C. Patil, "Evaluating the Research Trends and Techniques for Addressing Wormhole Attack in MANET", International Journal of Computer Applications, Vol. 110, No. 6, 2015
11. R. Singh, J. Singh, and R. Singh, "WRHT: A Hybrid Technique for Detection of Wormhole Attack in Wireless Sensor Networks", Mobile Information Systems, pp. 13, 2016
12. L. Lu, M.J. Hussain, G. Luo, and Z. Han, "Pworm: passive and real-time wormhole detection scheme for WSNs", International Journal of Distributed Sensor Networks, pp. 16, 2015
13. S. Mukherjee, M. Chattopadhyay, S. Chattopadhyay, and P. Kar, "Wormhole Detection Based on Ordinal MDS Using RTT in Wireless Sensor Network", Journal of Computer Networks and Communications, pp. 15, 2016
14. H. Chen, W. Chen, Z. Wang, Z. Wang, and Y. Li, "Mobile beacon based wormhole attackers detection and positioning in wireless sensor networks", International Journal of Distributed Sensor Networks, Vol. 10, No. 3, 2014
15. R. Matam and S. Tripathy, "Secure Multicast Routing Algorithm for Wireless Mesh Networks", Journal of Computer Networks and Communications, pp. 11, 2016
16. Y. Singh, A. Khatkar, P. Rani, and D.D. Barak, "Wormhole Attack Avoidance Technique in Mobile Adhoc Networks", In Advanced Computing and Communication Technologies (ACCT), Third International Conference, pp. 283–287, 2013.
17. G. Kaur, V. K. Jain, and Y. Chaba, "Wormhole attacks: Performance evaluation of on demand routing protocols in Mobile Adhoc networks", In Information and Communication Technologies (WICT), 2011 World Congress, pp. 1155–1158, 2011.
18. C. Gupta, and P. Pathak, "Movement based or neighbor based tehnique for preventing wormhole attack in MANET", In Colossal Data Analysis and Networking (CDAN), Symposium, pp. 1–5, 2016.
19. R. Stoleru, H. Wu, and H. Chenji, "Secure neighbor discovery and wormhole localization in mobile ad hoc networks", Ad Hoc Networks, Vol. 10, No. 7, pp. 1179–1190, 2012

20. A. Shastri and J. Joshi, "A Wormhole Attack in Mobile Ad-Hoc Network: Detection and Prevention", In Proceedings of the Second International Conference on Information and Communication Technology for Competitive Strategies, pp. 31, 2016

21. M. Ojha and R.S. Kushwah, "Improving Quality of Service of trust based system against wormhole attack by multi-path routing method", In Soft Computing Techniques and Implementations (ICSCTI), 2015 International Conference, pp. 33–38, 2015

22. S. Ali and P. Nand, "Comparative performance analysis of AODV and DSR routing protocols under wormhole attack in mobile ad hoc network on different node's speeds", In Computing, Communication and Automation (ICCCA), 2016 International Conference, (pp. 641–644), 2016

23. S. B. Geetha, V.C. Patil, "Elimination of Energy and Communication Tradeoff to Resist Wormhole Attack in MANET", International Conference on Emerging Research in Electronics, Computer Science and Technology, pp. 143–148, 2015

AWGN Suppression Algorithm in EMG Signals Using Ensemble Empirical Mode Decomposition

Ashita Srivastava, Vikrant Bhateja, Deepak Kumar Tiwari and Deeksha Anand

Abstract Surface Electromyogram (EMG) signals are often contaminated by background interferences or noises, imposing difficulties for myoelectric control. Among these, a major concern is the effective suppression of Additive White Gaussian Noise (AWGN), whose spectral components coincide with the spectrum of EMG signals; making its analysis problematic. This paper presents an algorithm for the minimization of AWGN from the EMG signal using Ensemble Empirical Mode Decomposition (EEMD). In this methodology, EEMD is first applied on the corrupted EMG signals to decompose them into various Intrinsic Mode Functions (IMFs) followed by Morphological Filtering. Herein, a square-shaped structuring element is employed for requisite filtering of each of the IMFs. The outcomes of the proposed methodology are found improved when compared with those of conventional EMD-and EEMD-based approaches.

Keywords EMG · Additive white gaussian noise (AWGN) · Ensemble empirical mode decomposition (EEMD) · Morphological filtering

A. Srivastava (✉) · V. Bhateja · D. K. Tiwari · D. Anand
Department of Electronics and Communication Engineering, Shri Ramswaroop Memorial Group of Professional Colleges (SRMGPC), Lucknow 226028, Uttar Pradesh, India
e-mail: srivastavaashita95@gmail.com

V. Bhateja
e-mail: bhateja.vikrant@gmail.com

D. K. Tiwari
e-mail: srmdeepaktiwari57@gmail.com

D. Anand
e-mail: deeksharoma2012@gmail.com

© Springer Nature Singapore Pte Ltd. 2018
S. Bhalla et al. (eds.), *Intelligent Computing and Information and Communication*,
Advances in Intelligent Systems and Computing 673,
https://doi.org/10.1007/978-981-10-7245-1_50

1 Introduction

EMG is an electrical demonstration of a shrinking muscle. The acquisition of a pure EMG signal becomes a major constraint in its proper analysis and diagnosis [1]. Surface EMG signals are often corrupted by Baseline Wander and AWGN [2]. AWGN gets superimposed in EMG signal as a result of thermal vibration of atoms in conductors, which is currently a more demanding task [3]. Many filtering approaches have been developed in recent past for the effective suppression of AWGN from EMG signals [4]. Among these, the most simple and the cost-effective ones' deploy Gaussian filtering for AWGN suppression. However, processing via Gaussian filter also attenuates useful diagnostic information from EMG signals. Hence, effective minimization of AWGN without deteriorating the signal quality (upon restoration) is a prominent challenge towards the development of filtering algorithms for EMG signals. Ren et al. [5] introduced an approach of background and gaussian noise suppression in EMG signals based on the combination of Independent Component Analysis (ICA) and Wavelet-based filtering. But, ICA has a disadvantage that the scale of the resulting signal is not same as the original signal. Mello et al. [6] propounded an approach of digital Butterworth filter design to delimitate the frequency band of surface Electromyograms (EMG) and remove the mains noise and its harmonics. Yet, it is highly impractical to design a Butterworth Filter for the elimination of such a high-frequency noise. Phinyomark et al. [7] performed EMG signal denoising using adaptive wavelet shrinkage function. However, proper selection of wavelets is a necessary task which acts as a constraint. In addition, this approach is also computationally intensive. Djellatou et al. [8] carried out a dual-adapted fast block least mean squares algorithm (DA-FBLMS) to remove noise contaminations from Surface Electromyogram (sEMG). However, processing via least mean square (LMS) algorithm for denoising signals leads to a slower convergence. Recently, Harrach et al. [9] proposed a technique of Canonical Correlation Analysis (CCA) followed by noise intensity ratio thresholding for AWGN elimination. But, the technique has been performance limited as it hampered the diagnostic quality/utility of the EMG signal. In this paper, an improved algorithm for preprocessing (AWGN suppression) of EMG signals based on the EEMD followed by morphological filtering is presented [10]. EEMD has a major advantage that this algorithm uses the full statistical properties of white noise and hence works effectively. Further, combinations of morphological operators are used to retain the shape of the original EMG signal upon restoration. After that, the signal fidelity assessment parameters, i.e., signal-to-noise ratio (SNR) and percentage root mean square difference (PRD %) are evaluated for the performance validation of the proposed algorithm. The rest of the paper has been sectioned as follows: Sect. 2 reports the proposed AWGN suppression algorithm. Section 3 comprises of the results section which demonstrates the simulation outcomes along with their analytical discussion; whereas Sect. 4 concludes the work.

2 Proposed AWGN Suppression Algorithm

The proposed AWGN suppression algorithm combines the EEMD algorithm with morphological filtering to preprocess the corrupted EMG signals. Initially, the unprocessed EMG signal is fed to the EEMD module, where the signal decomposition via EEMD takes place which breaks the EMG signal into IMFs. These IMFs are then filtered by morphological operations (opening and closing) to get the denoised signal. At the end, the signal fidelity assessment is performed using SNR and PRD% as the performance metrics.

2.1 EEMD

The EMD, particularly drafted for nonlinear and nonstationary signals, is a local, data-driven technique which dissolves the signal without the use of any precedent premise [11]. EMD decomposes a multifrequency component signal into non-coinciding frequency constituent named intrinsic mode functions (IMFs) and a residual signal [12]. The deduced IMFs are mono-frequency components that fulfill two criterions:

- The total amount of maxima (both the local maxima and minima) and zero crossings in the time series must not differ by more than one.
- The average value of the upper envelope (interpolated through maxima) and lower envelop (interpolated through minima) is zero in the entire time series.

EMD being such an adequate dissolving technique reaps some troublesome crunches. The main problem with the original EMD is the continual appearance of mode mixing, i.e., the detail related to one scale can appear in two different intrinsic modes. To dispatch the mode mixing problem, a new noise-assisted data analysis (NADA) technique brought by Huang et al. [13], the Ensemble EMD (EEMD), that defines the true IMFs as the mean of an ensemble of trials, each residing of the signal plus a white noise of finite amplitude. The procedural steps of EEMD are shown in Algorithm 1. The resultant IMFs, namely $c_{ij}(t)$ are averaged across trials to get the final IMFs, $c_i(t)$ as shown in Eq. No. (1) below:

$$c_i(t) = \frac{1}{N} \sum_{j=1}^{N} c_{ij}(t), \tag{1}$$

where, i is the IMF order, j denotes the trial index, and N is the total number of trials.

Algorithm 1 Procedural Steps for EEMD algorithm

	BEGIN
Step 1:	*Input:* Baseline Corrupted EMG signal
Step 2:	*Process:* Add a white noise series to the corrupted EMG signal
Step 3:	*Compute:* All Local Maxima and Minima
Step 4:	*Compute:* Upper and Lower Envelopes by Cubic Spline function
Step 5:	*Compute:* Average of Upper and Lower Envelopes
Step 6:	*Compute:* Difference by subtracting average from the noisy signal
Step 7:	*Process:* Check if the difference is an IMF
Step 8:	*Compute:* Residue signal by subtracting difference from the noisy signal if it is an IMF otherwise repeat steps 1–7
Step 9:	*Process:* Check if residue signal is a monotonic function
Step 10:	*Process:* Repeat steps 1–9 iteratively with changed versions of noise each time
Step 11:	*Compute:* Mean (Ensemble) of corresponding decomposed IMFs as the final true IMFs
	END

2.2 Morphological Filtering

Morphological filtering is a nonlinear transformation technique mostly used for the sake of regionally altering the geometrical attributes of a signal. It employs the simple applications of set theory. Morphological filters are the type of set functions which adjust the linear representation of the signal to get the desired geometrical shape of the signal [14]. In this technique, the whole examined signal is compared with a small template called structuring element (SE), whose suitable selection plays a very vital role for the effective filtering of the signal. The whole signal and structuring element are the two sets in mathematical morphology; onto which various operations are applied to change the structural shape of the signal. A huge variety of structuring elements are present in the mathematical morphology which enriches its applications. The shape and size of structuring element should be selected literally as its inaccurate choice may distort the adjoining wave of the EMG signal [15]. Linear, Triangular, Octagonal, and Square-shaped structuring elements are usually applied in the morphological filtering and shown in the Fig. 1a–d.

In the proposed approach, square-shaped structuring element of width 4 is chosen as shown in Fig. 1d. Dilation, Erosion, Opening, and Closing are basic operations in the mathematical morphology to alter the signal's geometrical shape. AWGN is suppressed by the usage of morphological operations comprising high pass as well as lowpass filter properties. Hence in the proposed algorithm, opening and closing operations have been used for the adequate suppression of AWGN. The respective expression of opening and closing operations is given by:

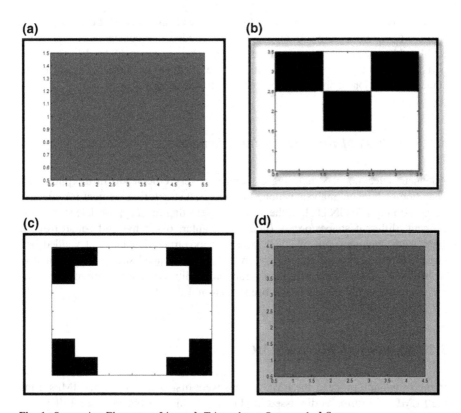

Fig. 1 Structuring Elements **a** Linear, **b** Triangular, **c** Octagonal, **d** Square

$$A \circ B = (A \ominus B) \oplus B \tag{2}$$

$$A \bullet B = (A \oplus B) \ominus B \tag{3}$$

where: A denotes the matrix of decomposed canonical components, B represents the structuring element and $\oplus$ and $\ominus$ denotes the dilation and erosion, respectively.

Opening, i.e., erosion succeeded by dilation and Closing, i.e., dilation succeeded by erosion is deployed on the decomposed IMFs from the AWGN corrupted EMG signals. These operations are carried out on all the IMFs one by one. After that, an average of the outputs obtained from the opening and closing operation performed on the IMFs is calculated to obtain the filtered EMG signal with minimized gaussian noise.

3 Results and Discussions

In the proposed algorithm, the EMG signals have been acquired from The UCI Machine Learning Repository [16] which includes 10 normal and 10 aggressive physical actions measuring the distinct human activities. Herein, for the simulation purpose, five aggressive actions EMG signals have been taken as the test data. The number of iterations and the noise parameter needs to be considered for applying EEMD Algorithm. Herein, ensemble number (NE) is taken to be 10. The noise parameter (Nstd), defined as the ratio of the standard deviation of noise to the standard deviation of the signal power, is selected to be 0.2.

3.1 Analysis of the Structuring Element

Adequate selection of structuring element is a very crucial job in morphological operation so here an analysis is done to select best structuring element for effective suppression of AWGN [17]. In the process of selecting most appropriate structuring element, different shapes/orders (linear, triangular, octagonal and square) of the same were deployed; out of which square shape structuring element yielded best results. In the present simulation setup, a "square"-shaped structuring element of width: 4 has been used. The reconstruction quality comparison of results using different structuring elements has been given in Table 1.

3.2 Analysis of Filtered EMG Signals

In the simulation process, the EMG signals were first decomposed into IMFs using the EEMD algorithm as discussed in [11]. After this, the decomposed IMFs were filtered using morphological filters and the filtered IMFs were reconstructed for the AWGN suppressed signal. Figure 2. demonstrates the results obtained using the proposed algorithm along with the outputs of the other conventional versions (EMD-Gaussian [18] and EEMD-Gaussian [10]).

Figure 2a shows an original EMG signal (Pushing3) in which AWGN is present. From Fig. 2b, it is observed that only a little portion of AWGN component has been suppressed using EMD-Gaussian algorithm although it is present in huge quantity in the original EMG signal. Figure 2c shows the output of the EEMD-Gaussian methodology in which a major portion of the AWGN component has been suppressed. But, there is an improvement in the output of the proposed algorithm over the other two conventional ones. In this, AWGN has been suppressed significantly

Table 1 Comparison of SNR (in dB) values with different structuring elements used in the proposed filtering algorithm

EMG signal Ref. no.	Original	Triangular structuring element	Linear structuring element	Octagon structuring element	Square structuring element
S1	−48.5911	−20.7906	−20.2755	−20.0123	−17.3911
S2	−53.3090	−24.0179	−22.0869	−21.4212	−20.6459

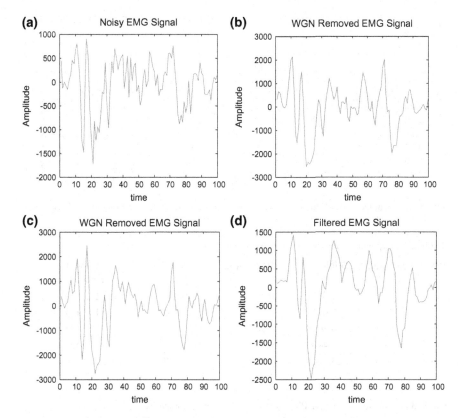

Fig. 2 (**a**) Corrupted EMG Signal (Pushing3), Restored Outputs Generated with (**b**) EMD-Gaussian [18], (**c**) EEMD-Gaussian [10], (**d**) Proposed Algorithm-EEMD-Morphological Filtering

from the original EMG signal which is clearly seen in Fig. 2d. The noisy EMG signal along with the corresponding filtered outputs using the proposed algorithm has been shown in Fig. 3a, b, respectively.

Furthermore, SNR (dB) and PRD (%) have been used as performance metrics for the evaluation and validation of the proposed algorithm [19]. The comparison of SNR (dB) and PRD (%) values of the AWGN suppressed EMG signal using the proposed algorithm with the EMD-Gaussian and EEMD-Gaussian-based approaches have been shown in Tables 2 and 3, respectively.

Signal-to-noise ratio is a parameter which evaluates the improvement in the signal quality relative to noise. The SNR value of the algorithm which has been proposed is reported to be −17.3911 for signal (S1) which is much better as compared to EMD-Gaussian and EEMD-Gaussian. Similar results are obtained during simulations when tested with other signals. Also, the PRD % value for signal (S1) of the proposed algorithm is 0.3110 which is less in comparison to the other methodologies thus signifying minimum distortion in the original EMG signal.

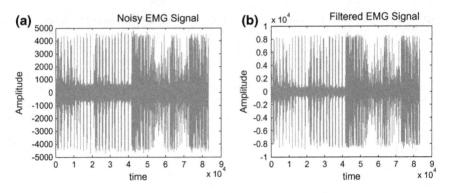

Fig. 3 a Corrupted EMG Signal (Pushing3). **b** Proposed Methodology Filtered EMG Signal

Table 2 Computed values of SNR (in dB) for performance comparison

EMG signal Ref. no.	Original	EMD-Gaussian	EEMD-Gaussian	Proposed methodology
S1	−48.5911	−20.4691	−19.0516	−17.3911
S2	−50.0350	−21.6478	−19.7652	−17.9144
S3	−53.3090	−22.9461	−22.6957	−21.7131
S4	−62.4497	−30.0952	−26.9520	−21.8559
S5	−83.7804	−36.1392	−28.3775	−22.5433

Table 3 Computed Values of PRD (%) for performance comparison

EMG signal Ref. no.	EMD-Gaussian	EEMD-Gaussian	Proposed methodology
S1	0.3451	0.3447	0.3110
S2	0.3580	0.3576	0.3216
S3	0.3525	0.3520	0.3255
S4	0.3514	0.3511	0.3465
S5	0.3583	0.3580	0.3488

Hence these two performance metrics confirm the superior performance of the proposed algorithm over the other two algorithms used. Therefore, the proposed algorithm certainly outshines over the other two methodologies.

4 Conclusion

This paper introduced an improved algorithm for minimization of AWGN from EMG signals based on EEMD algorithm which is followed by morphological filtering [20]. EEMD algorithm effectively decomposes the EMG signal into IMFs while morphological filtering successfully reduces the AWGN present with minimum distortion of the original EMG signal. The performance of this methodology was compared with EMD-and EEMD-based approaches. The proposed algorithm achieved better results than the other two conventional approaches in terms of suppression of AWGN in the EMG signal as validated by values of SNR (dB) and PRD (%).

References

1. Luca, C. D. J., Adam, A., Wotiz, R., Gilmore, L. D., Nawab, S. H.: Decomposition of Surface EMG Signals, Journal of Neurophysiology, vol. 96, pp. 1647–1654, (2006).
2. Ahsan, M. R., Ibrahimy, M., Khalifa, O. O.: EMG Signal Classification for Human Computer Interaction A Review, Journal of Scientific Research, vol. 33, pp. 480–501, (2009).
3. Srivastava, A., Bhateja, V., Tiwari, H., Satapathy, S. C.: Restoration Algorithm for Gaussian Corrupted MRI using Non-Local Average Filtering, 2nd International Conference on Information Systems Design and Intelligent Applications (INDIA-2015), vol. 2, pp. 831–840, (2015).
4. Canal, M. R.: Comparison of Wavelet and Short Time Fourier Transform Methods in the Analysis of EMG Signals, Journal of Medical Systems, Springer, vol. 34, no. 1, pp. 91–94, (2010).
5. Ren, X., Yan, Z., Wang Z., Hu, X.: Noise Reduction based on ICA Decomposition and Wavelet Transform for the Extraction of Motor Unit Action Potentials, Journal of Neuroscience Methods (Elsevier), vol. 158, pp. 313–322, (2006).
6. Mello, R. G. T., Oliveira, L. F., Nadal, J.: Digital Butterworth Filter for Subtracting Noise from Low Magnitude Surface Electromyogram, Journal of Computer Methods and Programs in Biomedicine (Elsevier), vol. 87, pp. 28–35, (2007).
7. Phinyomark, A., Phukpattaranont, P., Limsakul, C.: EMG Signal Denoising via Adaptive Wavelet Shrinkage for Multifunction Upper-Limb Prosthesis, The 3rd International Conference on Biomedical Engineering, pp. 35–41, (2010).
8. Djellatou, M. E. F., Nougarou, F., Massicote, D.: Enhanced FBLMS Algorithm for ECG and Noise Removal from sEMG Signals, IEEE 18th International Conference on Digital Signal Processing, pp. 1–6, (2013).
9. Harrach, M. A., Boudaoud, S., Hassan, M., Ayachi, F. S., Gamet, D., Grosset, J. F., Marin, F.: Denoising of HD-sEMG Signals using Canonical Correlation Analysis, Medical & Biological Engineering & Computing, vol. 55, no.3, pp. 375–388, (2017).
10. Zhang, X., Zhou, P.: Filtering of Surface EMG using Ensemble Empirical Mode Decomposition, Journal of Medical Engineering and Physics, vol. 35, no. 4, pp. 537–542, (2014).
11. Vergallo, P., Lay-Ekuakille, A., Giannoccaro, N. I., Trabacca, A., Morabito, F. C., Urooj, S., Bhateja, V.: Identification of Visual Evoked Potentials in EEG detection by Empirical Mode Decomposition, (IEEE) 11th International Multi-Conference on Systems, Signals and Devices- Conference on Sensors, Circuits & Instrumentation Systems, pp. 1–5, (2014).
12. Tiwari, D. K., Bhateja, V., Anand, D., Srivastava, A.: Combination of EEMD and Morphological Filtering for Baseline Wander Correction in EMG Signals, International Conference on Micro-Electronics, Electromagnetics and Telecommunications (ICMEET 2017), Springer-LNEE Vishakhapatnam (India), pp. 1–8, (2017).
13. Wu, Z., Huang, N. E.: Ensemble Empirical Mode Decomposition: A Noise Assisted Data Analysis Method, Advances in Adaptive Data Analysis, vol. 1, no. 1, pp. 1–51, (2005).

14. Bhateja, V., Verma, R., Mehrotra, R., Urooj, S., Lay-Ekuakille, A., Verma, V. D.: A Composite Wavelets and Morphology Approach for ECG Noise Filtering, 5th International Conference on Pattern Recognition and Machine Intelligence (PReMI2013), vol. 8251, pp. 361–366, (2013).
15. Verma, R., Mehrotra, R., Bhateja, V.: An Improved Algorithm for Noise Suppression and Baseline Correction of ECG Signals, Springer, vol. 199, pp. 733–739, (2013).
16. Frank, A., Asuncion, A.: UCI Machine Learning Repository, University of California, School of Information and Computer Science, Irvine, Calif, USA, (2010).
17. Shrivastava, A., Alankrita, Raj, A., Bhateja, V.: Combination of Wavelet Transform and Morphological Filtering for Enhancement of Magnetic Resonance Images, International Conference on Digital Information Processing and Communications (ICDIPC 2011), pp. 460–474, (2011).
18. Andrade, A. O., Nasuto, S., Kyberd, P., Sweeney-Reed, C. M., Kanijn, V.: EMG Signal Filtering based on Empirical Mode Decomposition, Journal of Biomedical Signal Processing and Control, vol. 1, pp. 44–55, (2006).
19. Verma, R., Mehrotra, R., Bhateja, V.: An Integration of Improved Median and Morphological Filtering Techniques for Electrocardiogram Signal Processing, 3rd IEEE International Advance Computing Conference, Springer, pp. 1212–1217, (2013).
20. Qiang, L., Bo, L.: The Muscle Activity Detection from Surface EMG Signal using the Morphological Filter, Applied Mechanics and Materials, vol. 195, pp. 1137–1141, (2012).

Visible-Infrared Image Fusion Method Using Anisotropic Diffusion

Ashutosh Singhal, Vikrant Bhateja, Anil Singh
and Suresh Chandra Satapathy

Abstract In this paper, Visible and Infrared sensors are used to take complementary images of a targeted scene. Image fusion thus aims to integrate the two
images so that maximum information and fewer artifacts are introduced in the fused
image. The concept of merging two different multisensor images using the combination of Anisotropic Diffusion (AD) and max–min approach is carried out in this
paper. Herein, each of the registered source images are decomposed into approximation and detailed layers using AD filter. Later, max–min fusion rules are applied
on detail and approximate layer, respectively, to preserve both spectral as well as
structural information. Image-quality assessment of the fused images is made using
structural similarity index (SSIM) , fusion factor (FF), and entropy (E) which
justifies the effectiveness of proposed method.

Keywords Multisensor · Entropy · Anisotropic diffusion (AD)

A. Singhal (✉) · V. Bhateja · A. Singh
Department of Electronics and Communication Engineering,
Shri Ramswaroop Memorial Group of Professional Colleges (SRMGPC),
Lucknow 226028, Uttar Pradesh, India
e-mail: ashutoshsinghal71@gmail.com

V. Bhateja
e-mail: bhateja.vikrant@gmail.com

A. Singh
e-mail: lfsanilsinghgkp@gmail.com

S. C. Satapathy
Department of CSE, PVP Siddhartha Institute of Technology,
Vijayawada, Andhra Pradesh, India
e-mail: sureshsatapathy@gmail.com

© Springer Nature Singapore Pte Ltd. 2018
S. Bhalla et al. (eds.), *Intelligent Computing and Information and Communication*,
Advances in Intelligent Systems and Computing 673,
https://doi.org/10.1007/978-981-10-7245-1_51

1 Introduction

Image fusion is a method of synthesizing different registered source images having complementary information; enhancing the capability of human visual perception. The aim is to produce a new image, which carries complementary as well as common features of individual images. Multisensor image fusion which involves visible and infrared imaging sensors can be applied for various applications involving guidance/detection system, navigation, surveillance, and targeting as well as remote sensing images [1]. Visible images are organized in visible spectrum and are formed due to scattering whereas thermal radiation creates infrared images and are related with brightness. In unpropitious weather condition and during night use of infrared over visible sensor is required [2]. There is a need to form a composite image so that there will be no loss of information in order to provide a compact representation of targeted scene with enhanced apprehension capability. Literature in the last two decades provides various studies discussing number of image multiresolution decomposition methodologies, such as Discrete Wavelet Transform (DWT) [3], Laplacian Pyramid (LP) [4] and data driven methods, such as Empirical Mode Decomposition (EMD) [5], Independent Component Analysis (ICA) [6] are employed. But the outcome of these methods employs lack of directional information, anisotropy and contains artifacts in the fused image. To overcome this problem AD filter [7] is used which involve multiple iteration for the purpose of image pixel segregation and processing into approximate and detail layer, respectively. Various fusion mechanisms, such as Principal Component Analysis (PCA) [8], pixel-based image fusion techniques which includes Intensity Hue Saturation (IHS), maximum, minimum, averaging, and weighted averaging image fusion have been popularly used. In addition, several other edge preserving and smoothing schemes, such as Guided and Bilateral filter [9], are widely used for fusion purpose. These methods [10] produce gradient reversal artifacts and halo effects [7]. Godse et al. [11] worked on wavelet fusion technique by using maximum intensity approach but the fused image suffered from contrast reduction and blurring. Sadhasivam et al. [12] performed fusion using PCA by selecting maximum pixel intensity but the obtained fused image has low contrast and illumination. From the above discussion, it can be inferred that there is a need to remove the artifacts and redundant information from the fused image so that it contain the complementary information of both the source images. AD-based approach uses intra-region smoothing; preserving edges and removes artifacts [6]. AD processing of source images result in preserving both high-frequency information and contrast. Max-and Min-based approach improves the redundant information and feature enhancement property in the fused image. Therefore, it is assimilated that combination of AD and max–min can be better modeled to provide fusion of visible and infrared images. In this paper, the decomposition of source images is carried out by using AD and further min rule is applied on approximate layer and max rule is applied on detail layer to yield final fused image. The proposed fusion results are promising when evaluated by using parameters, such as SSIM, FF, and E.

2 Proposed Visible-Infrared Fusion Method

The scope of the problem presented in this paper is to perform Visible and Infrared image fusion by using combination of AD and Max–Min fusion rule. Max and Min fusion rule involves selection of maximum and minimum pixel from every corresponding pixel of input images, which in turns forms resultant pixel of input images. Thus, every pixel of fused image will be pixel having maximum and minimum intensity of corresponding pixel in the input images. It helps to suppress the redundant information and highlights the components having maximum information in the source image.

2.1 Anisotropic Diffusion (AD) Filtering

Perona and Malik [13] performed nonlinear filtering on source images to produce base layer and named it as anisotropic diffusion filter. It is an inhomogeneous process that reduces the diffusivity at that location where there is much likelihood to be edges. AD utilizes flux Eq. (1) in order to control the diffusion of an image I which is governed by Eq. (1).

$$A_t = p(m, n, t)\Delta A + \nabla p . \nabla A \tag{1}$$

Where, $p(m, n, t)$ = Flux function also known as diffusion rate, Δ, ∇ are laplacian and Gradient operator, respectively, t = iteration. Equation (1) is also known as Heat Eq. whose solution is given as:

$$I_{m,n}^{t+1} = I_{m,n}^{t} + \lambda \left[c_N . \overline{\nabla}_N I_{m,n}^{t} + c_S . \overline{\nabla}_S I_{m,n}^{t} + c_E . \overline{\nabla}_E I_{m,n}^{t} + c_W . \overline{\nabla}_W I_{m,n}^{t} \right] \tag{2}$$

In Eq. (2) $I_{m,n}^{t+1}$ represent coarser image which depend on previous coarser image scale. Value of λ lies between $0 \leq \lambda \leq 1/4$. $\overline{\nabla}_N$, $\overline{\nabla}_S$, $\overline{\nabla}_E$ and $\overline{\nabla}_W$ represent nearest neighbor distance in respective direction. They can be defined as:

$$\overline{\nabla}_N I_{m,n} = I_{m-1,n} - I_{m,n} \tag{3}$$

$$\overline{\nabla}_S I_{m,n} = I_{m-1,n} - I_{m,n} \tag{4}$$

$$\overline{\nabla}_E I_{m,n} = I_{m-1,n} - I_{m,n} \tag{5}$$

$$\overline{\nabla}_W I_{m,n} = I_{m-1,n} - I_{m,n} \tag{6}$$

Perona and Malik [13] further proposed two equations as given below as Eqs. (7) and (8).

$$s(\nabla A) = e^{-\left(\frac{\|\nabla A\|}{c}\right)^2} \tag{7}$$

$$s(\nabla A) = \frac{1}{1 + \left(\frac{\|\nabla A\|}{c}\right)^2} \tag{8}$$

Equations (7) and (8) regulates smoothing and edge-preservation. Equation (7) is utilized if image consists of high contrast edges over low contrast edges while Eq. (8) is effective when wider regions dominate over smaller regions in the image [7]. These attributes of the conventional AD filter justifies its usage for multisensor images.

2.2 Image Fusion Algorithm Based on the Combination of AD and Max-Min Rule

The first preprocessing step involves conversion of multisource image from RGB to gray scale. After preprocessing, visible and infrared images (source images) are processed using AD. When AD filter is applied on both the registered multisource images; base layer is obtained. Further, the procedural algorithmic steps for the same are detailed in Table 1.

Table 1 Proposed fusion algorithm using AD, Max–Min fusion rule

BEGIN
Step 1: Input source images A_1 and A_2
Step 2: Decompose source image by AD to obtain base layer B_1 and B_2. This can be achieved via following steps:
2. (a) Determine diffusion rate of source image using Eq. (1)
2. (b) Eqs. (7) and (8) can be utilizes to yield base layer
Step 3: Detail layer D_1 and D_2 can be obtained by subtracting base layer from source image as given in Eq. (9)
$D_k(m,n) = A_k(m,n) - B_k(m,n)$ (9)
Where, $D_k(m,n)$ is detail layer, $A_k(m,n)$ is source image and $B_k(m,n)$ is base layer
Step 4: Fuse detail layer of both input source images using max fusion rule given by Eq. (10)
$D_k(m,n) = \max(D_k^1(m,n), D_k^2(m,n))$ (10)
Step 5: Fuse base layer of both input source images using min fusion rule given by Eq. (11)
$B_k(m,n) = \min(B_k^1(m,n), B_k^2(m,n))$ (11)
Step 6: Base and detail layer are combined using max rule to obtained final result. This is achieved by employing Eq. (12)
$F_k(m,n) = \max(B_k(m,n), D_k(m,n))$ (12)
$F(m, n)$ is resultant fuse image
END

After the generation of final reconstructed image using the proposed fusion algorithm; their performance parameters, such as SSIM, FF and E [14–16] are calculated. Higher values of these parameters justify the worthiness of the proposed fusion algorithm.

3 Results and Discussions

Fusion of different Visible and Infrared images is performed by decomposing it into base layer by applying AD. Further finding detail layer and combining each layer of corresponding source images by utilizing Max and Min fusion rule as fusion algorithms. The proposed fusion method is tested with four different sets of visible and infrared images IMG-1(p1, q1, r1), similarly for IMG-2, IMG-3, IMG-4 to confirm the effectiveness. It can be inferred that in IMG-1 due to image capture from visible sensor house rooftop and greenery in targeted scene is clearly visible whereas roads are not clearly visible as shown in Fig. 1 (p1) whereas in Fig. 1 (q1) due to image captured from infrared sensor roads and person are clearly visible but the house rooftop and greenery are not clearly visible. Figure 1 (r1) represents the fusion of visible and infrared image from the proposed algorithm which shows house rooftop, roads, greenery and person in the fused image. High values of SSIM depict measure similarity in structure between the source and fused images as shown in Table 2. This has been the methodology of utilizing the fused image to improve the loss of information. Similarly, in IMG-2, Fig. 1 (p2) clearly shows alphabets on the roof and street lights whereas cars and people are not clearly visible while in Fig. 1 (q2), the people and car can be clearly visualized while the streetlight and alphabet on the roof is not. The obtained fused images show both the complementary features which inferred the effectiveness of proposed fusion method. Figure 1 also depicts the other test input source images and the obtained fusion results.

The obtained results are evaluated for fusion quality using performance parameters, such as E, FF, and SSIM. The overall impact is that the Entropy (E) of the fused image has been improved which shows a high amount of information content in the fused image. It is further verified through Fusion Factor (FF) shown in Table 2. Large values of E, FF, and SSIM depicts better performance of proposed fusion approach for various source images. Hence, the proposed methodology is validated to show its worth in conserving the spectral as well as spatial features.

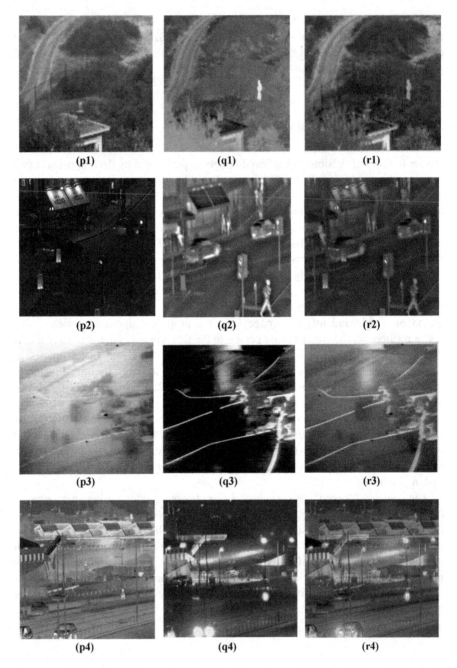

Fig. 1 An illustration of image fusion results for different sets of visible and infrared images using proposed methodology: (**p1–p4**) Input source images (Visible), (**q1–q4**) Input source images (Infrared), (**r1–r4**) Output fused images

Table 2 Image quality computation using various fusion metrics

Images	E	FF	SSIM
IMG-1	6.5518	6.4997	0.7536
IMG-2	4.9323	5.2636	0.9583
IMG-3	7.1145	6.2284	0.9342
IMG-4	6.5427	6.8427	0.8802

4 Conclusion

An AD-based fusion method applicable to visible and infrared images employing Max and Min fusion rule is proposed in this paper. Since, the image acquired by using infrared radiations contains few details with high contrast, whereas visible imaginary provides plenty of high-frequency information but have low contrast. AD processing of source images results in preserving both high-frequency information and contrast. This yields a better result than other multiresolution decomposition techniques [1, 3, 14]. AD and max–min-based image fusion approach leads to minimum loss of information and artifacts. It can be easily observed from the result and the same can be easily validated using different fusion metrics pertaining to E, FF, and SSIM. Further, the AD can be utilized with other fusion algorithm, such as Karhunen–Loeve (KL) Transform [7], Non-Subsampled Contourlet Transform (NSCT) [17], etc. to obtained better results.

References

1. Srivastava, A., Bhateja, V., Moin, A.: Combination of PCA and Contourlets for Multispectral Image Fusion, In: International Conference on Data Engineering and Communication Technology (ICDECT-2016), Pune, India, Vol. 2, pp. 1–8, Springer (March, 2016).
2. Bulanon, D. M., Burks, T. F., Alchanatis, V.: Image fusion of visible and thermal images for fruit detection, In: Biosystems Engineering, Vol. 103, No.1, pp. 12–22—Elsevier (May, 2009).
3. Krishn, A., Bhateja, V., Himanshi, Sahu, A.: Medical Image Fusion Using Combination of PCA and Wavelet Analysis, In: International Conference on Advances in Computing, Communication and Informatics (ICACCI-2014), Gr. Noida (U.P), India, pp. 986–991, IEEE (September, 2014).
4. Sahu, A., Bhateja, V.,Krishn, A., Himanshi: Medical Image Fusion with Laplacian Pyramid, In: International Conference on Medical Imaging, m-Health & Emerging Communication Systems (MEDCom-2014), Gr. Noida (U.P.), pp. 448–453, IEEE (November, 2014).
5. Hui, C. S., Hongbo, S., Renhua, S., Jing, T.: Fusing remote sensing images using a trous wavelet transform and empirical mode decomposition, In: Pattern Recognition Letters, Vol. 29, No.3, pp. 330–342, Elsevier (1 February, 2008).
6. Mitianoudis, N., Stathaki, T.: Pixel-based and region-based image fusion schemes using ICA bases, In: Information Fusion, Vol. 8, No. 2, pp. 131–142, Elsevier (2007).
7. Bavirisetti, D. P., Dhuli, R.: Fusion of Infrared and Visible sensor images based on Anisotropic diffusion and Karhunen-Loeve Transform, In: IEEE Sensor Journal, Vol. 16, No. 1, pp. 203–209, IEEE (14 September 2015).

8. He, C., Liu, Q., Li, H., Wang, H.: Multimodal Medical Image Fusion Based on IHS and PCA, In: Symposium on Security Detection and Information Processing, Vol. 7, pp. 280–285, Elsevier (2010).

9. He, K., Sun, J., Tang, X.: Guided Image Filtering, In: IEEE Transaction on Pattern Analysis and Machine Intelligence, Vol. 35, No. 6, pp. 1397–1408, IEEE (June 2013).

10. Li, S., Kang, X., Hu, J.: Image Fusion with Guided Filtering, In: IEEE Transactions on Image Processing, Vol. 22, No. 7, pp. 2864–2875, IEEE (July 2013).

11. Godse, D. A., Bormane, D. S.: Wavelet based Image Fusion using Pixel based Maximum Selection rule, In: International Journal of Engineering Science and Technology, vol. 3, no. 7, pp. 5572–5578, (2011).

12. Sadhasivam, S. K., Keerthivasan, M. K., Muttan, S.: Implementation of Max Principle with PCA in Image Fusion for Surveillance and Navigation Application, In: Electronic Letters on Computer Vision and Image Analysis, Vol. 10, No. 1, pp. 1–10, (2011).

13. Perona, P., Malik, J.: Scale-Space and Edge Detection Using Anisotropic Diffusion, In: IEEE Transactions on Pattern Analysis and Machine Intelligence, Vol. 12, No. 7, pp. 629–639, (July 1990).

14. Himanshi, Bhateja, V., Krishn, A., Sahu, A.: Medical Image Fusion in Curvelet Domain Employing PCA and Maximum Selection Rule, In: International Conference on Computers and Communication Technologies (IC3T-2015), Hyderabad, India, vol. 1, pp. 1–9, IEEE (July, 2015).

15. Moin, A., Bhateja, V., Srivastava, A.: Multispectral Medical Image Fusion using PCA in Wavelet Domain, In: Proc. (ACM-ICPS) Second International Conference on Information and Communication Technology for Competitive Strategies (ICTCS-2016), Udaipur, India, pp. 1–6, (March, 2016).

16. Krishn, A., Bhateja, V., Himanshi, Sahu, A.: PCA based Medical Image Fusion in Ridgelet Domain, In: International Conference on Frontiers in Intelligent Computing Theory and Applications (FICTA-2014), Bhubaneswar, India, vol. 328, pp. 475–482, Springer (November 2014).

17. Xiao-bo, Q., Wen, Y. J., Zhi, X. H., Zi-Qian, z.: Image Fusion Algorithm Based on Spatial Frequency-Motivated Pulse Coupled Neural Networks in Nonsubsampled Contourlet Transform Domain, In: Acta Automatica Sinica, vol. 34, no.12, pp. 1508–1514, Elsevier (2008).

Fast Radial Harmonic Moments
for Invariant Image Representation

Shabana Urooj, Satya P. Singh, Shevet Kamal Maurya
and Mayank Priyadarshi

Abstract The main objective of this paper is to reduce the reconstruction error and fast calculation of Radial Harmonic Fourier Moments (RHFM). In the proposed work, the fast RHFM has been applied on the original gray image for reconstruction. Before applying RHFM on grayscale image, the image in portioned into radial and angular sectors. Results are compared with traditional methods. The proposed approach results in better reconstruction error. Also, moments can be calculated at high speed using proposed approach.

Keywords Radial harmonic Fourier moments · Reconstruction error
Computational speed

1 Introduction

In the image processing technology, the immunity of an Image is a very violent problem and quality of image also one problem. The improvement of the quality of an image one fast technology applied by author fast Radial Harmonic Fourier Moments (RHFM), and give the batter results by the new fast RHFMs. The author will apply the proposed approach to the gray image for reconstruction. This will reduce the consume time and error of an image in the reconstruction performance. The proposed approach of the radial harmonic Fourier moments are the regenerated techniques of the traditional approach. The Fourier Moments, the linear combination of sine and cosine wave are the Fourier series. The application of Fourier moments is in image processing, signal processing, and quantum mechanics. First reconstruction of an image by Zernike moments/pseudo-Zernike moments into

S. Urooj (✉) · S. P. Singh · S. K. Maurya · M. Priyadarshi
School of Engineering, Gautam Buddha University, Gr. Noida, UP, India
e-mail: shabanaurooj@ieee.org

© Springer Nature Singapore Pte Ltd. 2018
S. Bhalla et al. (eds.), *Intelligent Computing and Information and Communication*,
Advances in Intelligent Systems and Computing 673,
https://doi.org/10.1007/978-981-10-7245-1_52

image watermarking conducted by Shirani et al. [1–3]. Invariant extraction has been executed by Tchebichef moments in Deng et al. for the reconstruction of an image by radial exponent moments [4]. The proposed reconstruction of an image through Radial Harmonic Fourier Moments [5, 6].

2 Proposed Work

2.1 Radial Harmonic Fourier Moments

In the polar coordinate system, RHFM of an image $f(r, \theta)$, $0 \leq r \leq 1$, and $0 \leq \theta \leq 2\pi$, are describe on unit disk. Let ϕ_{nm} are the RHFMs and order and repetition of function is $n, m, n \geq 0, m \geq 0$ then RHFMs are defined as follows:

$$M_{nm} = \frac{1}{\pi} \int \int f(r, \theta) P^*_{nm}(r, \theta) r dr d\theta \tag{1}$$

where, $P*_{nm}(r, \theta)$ are the conjugate of $P_{nm}(r, \theta)$ and defined as follows:

$$P_{nm}(r, \theta) = T_n(r) \exp(jm\,\theta) \tag{2}$$

$$T_n(r) = \begin{cases} \sqrt{1/r} & \text{for } n = 0 \\ \sqrt{2/r} \sin(n+1)\pi r & \text{for } n = \text{odd} \\ \sqrt{2/r} \cos n\pi r & \text{for } n = \text{even} \end{cases} \tag{3}$$

The radial basis function $P_{nm}(r, \theta)$ also satisfies orthogonal conditions as follows,

$$\int_0^{2\pi} \int_0^1 P_{nm}(r, \theta) P^*_{k\ell}(r, \theta) r dr d\theta = 2\pi \delta_{nk} \delta_{ml} \tag{4}$$

where δ is Kronecker delta and 2π is the normalization factor.

The image $f(r, \theta)$ can be reconstructed using RHFMs as

$$\hat{f}(r, \theta) = \sum_{n\ell \in S} E_{n\ell} H_{n\ell}(r, \theta) \tag{5}$$

In the rectangular coordinate system [7] reconstruction of RHFM two types by this method circular inscribed mapping and on the circle circumscribed mapping. In which circular circumscribed mapping are not suitable for the pattern recognition image watermarking and orthogonal moments. Then we will use only circular inscribed mapping.

The image function $'f'$ is mapped into polar coordinated to calculate the coefficient of Eq. 1 [8, 9]. We divide the whole image in radial sector N_r and angular size of N_φ into polar coordinates so that the grid density is high enough as per sampling theorem, i.e., $N_\varphi \geq 2\ell\text{max}$. In this way, we partitioned the angular and radial components of Eq. 1 into M equal parts as:

$$N_r = \frac{1}{M}\sum_{u=0}^{M-1} u \quad \text{and} \quad N_\theta = \frac{2\pi}{M}\sum_{v=0}^{M-1} v \tag{6}$$

Therefore, we obtained M^2 subregions inside the unit disk (Fig. 1). Now, the coordinates (N_r, N_θ) are changed to rectangular coordinates as $x = (N_r \times \frac{N}{2} \times \cos N_\theta)$, and $y = \left(N_r \times \frac{N}{2} \times \sin N_\theta\right)$. Therefore, the polar coordinate can be written in terms of portioned data as follows;

$$f_p(N_r, N_\theta) = f(i, j) \tag{7}$$

Let us define $\psi_{n\ell}$ as partitioned RHFMs and Eq. 1 can be rewritten as follows:

$$\psi_{n\ell} = \frac{1}{2\pi}\int_0^{2\pi}\int_0^1 f(N_r, N_\theta)[T_n(N_r)]^* \exp(-i\ell N_\theta)N_r dN_r dN_\theta \tag{8}$$

Fig. 1 Angular radial partitioning of an image to N angular and M radial sectors

Substituting Eq. 6 into Eq. 8 we get:

$$\psi_{n\ell} = \frac{1}{2\pi} \int_0^{2\pi} \int_0^1 f(N_r, N_\theta)[T_n(N_r)]^* \exp(-i\ell N_\theta) N_r \frac{1}{M} du \frac{2\pi}{M} dv$$

$$\psi_{n\ell} = \frac{1}{M^2} \sum_{u=0}^{M-1} \sum_{v=0}^{M-1} f(N_r, N_\theta) H_n(N_r) \exp(-i\ell N_\theta) \tag{9}$$

where, the term $H_n(N_r)$ is the radial kernel of the proposed method partitioned RHFMs and Eq. 3 can be rewritten as follows:

$$T_n(r) = \begin{cases} \sqrt{1/r} & \text{for } n = 0 \\ \sqrt{2/r} \sin(n+1)\pi r & \text{for } n = \text{odd} \\ \sqrt{2/r} \cos n\pi r & \text{for } n = \text{even} \end{cases} \tag{10}$$

3 Reconstruction of Image and Error in Reconstruction

The image $f(r, \theta)$ reconstructed approximately by order and repetition of fast RHFMs ($n \leq N, m \leq M$). Then, the reconstruction performance of an image are denoted by $\hat{f}(r, \theta)$. The orthogonal function will gives the independence beneficiation for the image reconstruction.

Then normalized reconstruction error of an Image (NIRE)

$$\varepsilon^2 = \frac{\int \int_{-\infty}^{\infty} \left[f(r, \theta) - \hat{f}(r, \theta)\right]^2 dr d\theta}{\int \int_{-\infty}^{\infty} f^2(r, \theta) dr d\theta} \tag{11}$$

4 Experiment Result

In the first experiment, the proposed method is tested and validated for reconstruction capabilities. The results are also compared with the traditional approach. Standard grayscaled Lena image is used to validate the proposed approach. The standard Lena image of size 512×512 is down sampled to the size of 128×128. For this experiment a set of moments/coefficients are calculated using Eq. 7 and again image reconstruction process is performed from these sets of moments/coefficients. The mean square error is calculated for measuring the reconstruction performance using the following equation:

$$\text{MSRE} = \frac{\sum_{x^2+y^2 \leq 1} |f(x, y) - \hat{f}(x, y)|}{\sum_{x^2+y^2 \leq 1} f^2(x, y)}$$

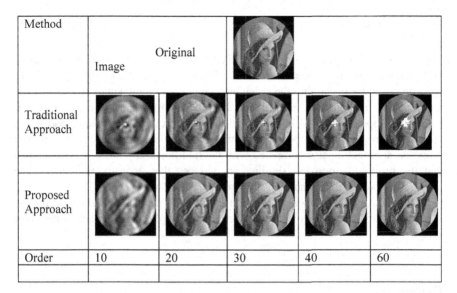

Method	Image	Original		
Traditional Approach				
Proposed Approach				
Order	10	20	30	40	60

Fig. 2 Reconstruction performance of image by fast RHFMs

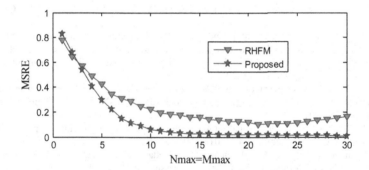

Fig. 3 Comparative analysis of reconstruction error using proposed method and traditional method

The results are shown in Fig. 2 for moments order and repetition 10, 20, 30, and 40. If we take a closer look at the center of reconstructed images, we can observe that there is information loss using the traditional method as we increases the order of moments. On the other hand, the reconstruction error continuously decreases using proposed method. In continuation, we also plot behavior of reconstruction error using the proposed method and traditional method. The results are shown in Fig. 3. It can be seen that after certain degree of moments, the reconstruction error of the traditional method starts increasing while this error continuously decrease using the proposed method.

5 Conclusions

The traditional computing approach of RHFM suffers from numerical integration error which results in reconstruction error. To tackle this issue, a new approach has been proposed by portioning the radial and angular parts into small sectors. To test the proposed methods, two experiments have been carried out. In the first experiment, the proposed method is tested for its reconstruction performance and in the second experiment; the proposed approach is tested for computational speed. The proposed approach is also compared with the traditional approach. It is concluded that the proposed approach significantly reduces the reconstruction error. The proposed approach also results in high-computational speed.

In future work, the proposed approach for RHFMs can be applied on the pattern recognition, and image segmentation. We can also apply this approach to the color Image.

References

1. S. Shirani, M. Farzam, A robust multimedia watermarking technique using Zernike transform, in 2001 IEEE 4th Workshop on Multimedia Signal Processing ((MMSP 01), 2001, CANNES, pp. 529–534).
2. Singh, Satya P., Shabana Urooj, and Aime Lay Ekuakille. "Rotational-invariant texture analysis using radon and polar complex exponential transform." Proceedings of the 3rd International Conference on Frontiers of Intelligent Computing: Theory and Applications (FICTA) 2014. Springer, Cham, 2015.
3. Singh, Satya P., and Shabana Urooj. "An improved CAD system for breast cancer diagnosis based on generalized pseudo-Zernike moment and Ada-DEWNN classifier." Journal of medical systems 40.4 (2016): 1–13.
4. Y. D. Zhang, C. Shao, Orthogonal moments based on exponent functions: exponent- Fourier moments, Pattern Recognition. 47 (2014) 2596–2606.
5. C. Singh, S. K. Ranade, A high capacity image adaptive watermarking scheme with Radial Harmonic Fourier Moments, Digit. Signal Processing. A Rev. J. 23 (2013) 1470–1482.
6. C. Singh, R, Upneja, A Computational Model of Enhanced accuracy of radial harmonic Fourier moments, in World Congress of Engineering, 2012, London, pp. 1189–1194.
7. H.T. Hu, Y.D. Zhang, C. Shao, Q. Ju Orthogonal moments based on exponent function: Exponent Fourier Moments, Pattern Recognition. 47(2014) 2596–2606.
8. Singh, Satya P., and Shabana Urooj. "Combined rotation-and scale-invariant texture analysis using radon-based polar complex exponential transform." Arabian Journal for Science and Engineering 40.8 (2015): 2309–2322.
9. Singh, Satya P., and Shabana Urooj. "Accurate and Fast Computation of Exponent Fourier Moment." Arabian Journal for Science and Engineering (2017): 1–8.

A Wide-Area Network Protection Method Using PMUs

Namita Chandra and Shabana Urooj

Abstract This paper proposes the idea of utilizing the data of PMU's (Phasor Measurement Unit) to detect different types of faults in transmission lines. Data measured from PMUs is collected at a system control center. Proposed method equates the positive sequence voltages of different buses to detect the faulted bus. With the help of synchronized phasor measurements, bus voltages are estimated by using different paths, furthermore, they are matched for the detection of the faulted line also. The proposed protection scheme is verified for balanced and unbalanced faults. Simulation has been done with MATLAB/SIMULINK for a 230 kV IEEE 9-bus system.

Keywords Wide-area measurement (WAM) · PMU · Backup protection

1 Introduction

In general, the aim of backup protection is to clear the faults that are not cleared by the primary protection. Backup protection operates more slowly than primary protection, to ensure proper coordination, and is less selective [1]. As it has to protect a larger part of a power system, settings become more difficult and dependent on the operating state of the system [2]. The main drawbacks of conventional backup protection are coordination among several backup protections and delay caused due to measuring devices.

With the growth of computer applications and communication networks, wide-area measurement system (WAMS) is one of the most significant innovative developments in the field of existing power system [3]. WAM is able to offer a

N. Chandra (✉) · S. Urooj
Department of Electrical Engineering, School of Engineering, Gautam Buddha University, Greater Noida 201310, Uttar Pradesh, India
e-mail: namitachandra.94@gmail.com

S. Urooj
e-mail: shabanabilal@gmail.com

© Springer Nature Singapore Pte Ltd. 2018
S. Bhalla et al. (eds.), *Intelligent Computing and Information and Communication*,
Advances in Intelligent Systems and Computing 673,
https://doi.org/10.1007/978-981-10-7245-1_53

real-time understanding of the dynamic performance of a power system that updates once per cycle, through inventions in the field of synchronized measurement technology and PMUs [4]. It is a vital resource for constructing new applications that can benefit power system protection and control.

In a recent report [4] 70% of wide-area disturbances were due to relay maloperation. These reported commotions were majorly caused due to poor relay settings or unseen failures in the protection system. Incorrect operation of the relay in wide-area disturbances is a source of concern these days, as these disturbances have played a major part in recent power failures [5]. Considering the shortcomings of backup security it has become essential to develop new protection scheme using WAMS.

Various schemes have been developed in [6–9] for faulted line identification based on the WAMS. In [6] positive sequence voltage magnitudes are paralleled to detect bus nearest to the fault and then positive sequences current angles at each line are associated to identify the faulty line. In [7] PMU MATLAB simulator is created. Its aim is to help in the understanding of the behavior of algorithms internal to the PMU. A wide-area-based backup protection scheme for series compensated line is created. It also works in the case of voltage and current inversions. PMUs were assumed at all the buses [8]. PMUs can also be employed in state estimation of energy management systems [9]. The study is focused on the improving the power systems immunity toward tragic letdowns by using PMUs within the system.

In this paper, a new protection scheme has been proposed using the data collected from time-synchronized phasor measurement units placed at selected buses, resulting in less number of PMUs. This scheme is simple in terms of computation and is cost effective. Simulations have been done with MATLAB/SIMULINK environment on 230 kV, 9-bus system.

In this paper proposed protection scheme is discussed in Sect. 2. Simulation results are detailed in Sects. 3 and 4 conclude the paper.

2 Proposed Protection Scheme

Whenever the system gets unbalanced (i.e., fault condition), its unbalanced phasors can be transformed into three balanced phasors (positive, negative, and zero) known as sequence components. This method matches the sequence voltages of unlike buses to decide the faulted bus. It was found that during the instance of a fault the voltage of the bus closest to the fault will observe maximum swerve [8]. For detection of faulted bus, positive sequence voltages of selected buses are compared. After determining the faulted bus, line at fault is also determined by estimating non-PMU bus voltages using voltages and currents of buses having PMU through different paths. A single line diagram of a 9-bus system with PMUs at selected locations used for the proposed protection scheme is shown in Fig. 1. Locations of PMUs are decided considering that each transmission line is connected to a PMU bus [10–12].

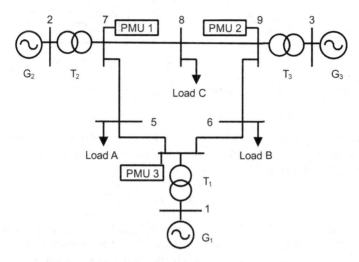

Fig. 1 Single line diagram of a 9-bus system

2.1 Identification of Faulted Bus

For the detection of the faulted bus, positive sequence voltage magnitudes are collected from the PMU buses at a system control center. The basic principle behind the detection of the faulted bus is that at the instance of a fault, voltage magnitude is reduced to a certain extent. The condition used to detect if there is a fault in the system is as follows:

$$|V_{FAULT}| < K \ V_R, \tag{1}$$

where V_{FAULT} is the positive sequence voltage magnitude at the time of the fault, V_R is the rated voltage magnitude, and K is a constant. K is used to calculate the voltage thresholds for detecting a fault in a system. Its value is set at 0.7 for this system and is decided after carrying out some simulations.

If Eq. (1) is true the bus with a minimum value of positive sequence voltage magnitude will be assumed as a faulted bus. In this scheme, if a fault is nearer to a non-PMU bus, it will not be identified as a faulted bus as only PMU buses are compared for detection of the faulted bus. So, until the detection of faulted line, the nearest PMU bus will be assumed as the faulted bus.

2.2 Identification of Faulted Line

After locating faulted bus, lines connected to the faulted bus are tested for the identification of the faulted line. For the detection of faulted line, positive sequence

Fig. 2 Single line diagram of a subnetwork

voltages are estimated from different paths, for the non-PMU buses utilizing the values of PMU buses. For estimating the non-PMU bus voltages, a hyperbolic form of the long transmission line equation is utilized.

$$V_{YX} = V_X \cosh(\gamma L) - I_X Z_C \sinh(\gamma L), \tag{2}$$

where V_{YX} is the estimated positive sequence voltage for a Y (non-PMU) bus using the positive sequence voltages and currents of X (PMU) bus. Z_C is the surge impedance of the line of length L and γ is the propagation constant.

The basic concept for detection of the faulted line is that during normal conditions estimated voltages of a non-PMU bus through different paths would be almost equal. But, if a fault occurs on the line connected to the bus whose voltage is estimated, then the estimated values of voltages through different paths may differ. This logic is used for the detection of the line at fault. This estimation is done only for the buses which are directly connected to the faulted bus. When bus X is the faulted bus as shown in Fig. 2 then ΔV_Y is calculated for all the buses connected to the faulted bus as follows:

$$\Delta V_Y = |V_{YX} - V_{YZ}|. \tag{3}$$

During normal conditions, the value of ΔV will be approximately equal to zero. But in the case of fault, the line with the minimum value of ΔV will be considered as the faulted line.

$$\text{Faulted line} = \text{Min}\{\Delta V\} \tag{4}$$

2.3 Flowchart of the Proposed Protection Scheme

PMUs measure the voltage and current phasors of all the buses. These measurements are transferred to a phasor data concentrator (PDC) at the system control center. The PDC links the data by time tag to build a WAMS. The flowchart of this protection scheme is given in Fig. 3.

Fig. 3 Flowchart of the
proposed scheme

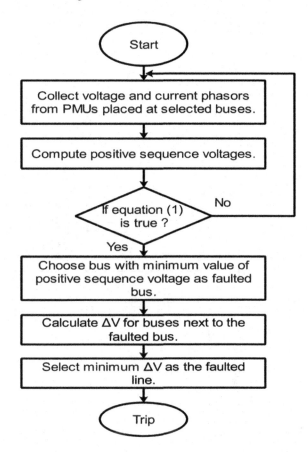

3   Simulation Results

The proposed protection scheme is tested for a 9-bus power system as shown in
Fig. 1 with faults on line 7–8. Using MATLAB R2013a, simulations are carried out
for different faults (balanced and unbalanced), using positive sequence voltages and
currents. The sequence components are calculated using the simPowerSystem tool
box present in MATLAB/SIMULINK. The number of PMUs required for the
protection of transmission lines in this system is 3. PMUs are placed on bus 4, 7,
and 9 so that each line is connected to at least one PMU.

3.1   Results from Identification of Faulted Bus

A single phase, line-to-ground (LG) fault of 5 Ω is created on line 7–8 at 0.4 s,
30 km from bus 7. Figure 4a shows the positive sequence voltage magnitudes of

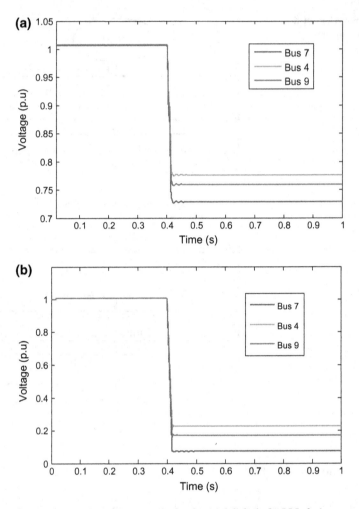

Fig. 4 Positive sequence bus voltage magnitudes for (**a**) LG fault (**b**) LLL fault

buses 7, 9, and 4. The simulated output shows that all the buses have been affected from the LG fault but bus 7 has been affected the most. Bus 7 has minimum positive sequence voltage hence it is assumed as the faulted bus. Figure 4b shows the positive sequence voltage magnitudes for LLL (three phase) fault.

3.2 Results from Identification of Faulted Line

To verify the result of the faulted bus identification, the faulted line is identified. If bus 7 is identified as the faulted bus then either line 7–8 or line 7–5 will be the

Table 1 Results of faulted line identification

Type of fault	Fault resistance (Ω)	Fault location from bus 7 (%)	ΔV_8 (kV)	ΔV_5 (kV)
L-G	5	30	5.8	9
		70	1.9	6.7
	50	30	4.2	6.7
		70	1.7	5.5
L-L-G	5	30	11.95	18.62
		70	3.71	14.14
	50	30	7.6	13.1
		70	3	10.8
L-L-L	5	30	18.7	29.09
		70	5.54	22.04
L-L-L-G	5	30	18.7	29.09

faulted line. Hence, positive sequence voltages of the buses next to the faulted bus 7 are estimated through different paths using Eq. (2). As bus 8 and 5 are the buses next to the faulted bus, V_{87}, V_{89} are calculated for bus 8 and V_{57}, V_{54} are calculated for bus 5. Then ΔV is calculated from Eq. (3) as follows:

$$\Delta V_8 = |V_{87} - V_{89}|. \tag{5}$$

$$\Delta V_5 = |V_{57} - V_{54}|. \tag{6}$$

On comparing the values of ΔV_8 and ΔV_5, ΔV_8 was found the minimum. Hence, confirming the fault on line 7–8 as bus 7 was considered the faulted bus. Table 1 shows the results of faulted line identification with different types of faults, fault locations, and fault resistances. It is apparent from the following table that this protection scheme is capable for the identification faulted line for balanced and unbalanced faults in all cases. Results show the effectiveness of the proposed protection scheme.

4 Conclusion

In this paper, a backup protection scheme was proposed using WAMS (Wide-Area Measurement System). Positive sequence voltage magnitudes of only PMU buses were compared for the identification of faulted bus. Utilizing the measurements of PMUs, the faulted line was identified. Positive sequence voltage of buses next to the faulted bus was estimated using different paths. The proposed protection scheme was tested on IEEE 9-bus system for unbalanced and balanced faults. A major achievement of this protection scheme is the lessening in the number of PMUs without being intricate. The main benefit of this scheme is its simplicity, cost-efficiency, and low computation. In future, this protection scheme can also be tested for non-fault events like load encroachment and power swing.

References

1. Jena, M.K., Samantaray, S.R., Panigrahi, B.K.: A New Wide-Area Backup Protection Scheme for Series-Compensated Transmission System. IEEE Systems Journal https://doi.org/10.1109/JSYST.2015.2467218 (2015).
2. Phadke, A.G., Wall, P., Ding, L. et al.: J. Mod. Power Syst. Clean Energy 4: 319. https://doi.org/10.1007/s40565-016-0211-x (2016).
3. Urooj, S., Sood, V.: Phasor Measurement Unit (PMU) Based Wide Area Protection System. 2nd International Conference on Computing for Sustainable Global Development (INDIACom) (2015).
4. Horowitz, S.H., Phadke, A.G.: Third zone revisited. IEEE Trans Power Deliv 21(1):23–29 (2006).
5. Report of the Enquiry Committee on grid disturbance in northern region on 30th July 2012 and in northern, eastern & north-eastern region on 31st July 2012. The Enquiry Committee, Ministry of Commerce and Industry, Government of India, New Delhi, India, (2012).
6. Eissa, M.M., Masoud, M.E., Elanwar, M.M.M.: A novel backup wide area protection technique for power transmission grids using phasor measurement unit. IEEE Trans. Power Del., vol. 25, no. 1, pp. 270–278, Jan. (2010).
7. Dotta, D., Chow, J.H., Vanfretti, L., Almas, M.S., Agostini, M.N.: A MATLAB-based PMU Simulator. Power and Energy Society General Meeting (PES), IEEE, (2013).
8. Nayak, P.K., Pradhan, A.K., Bajpai, P.: Wide-Area Measurement-Based Backup Protection for Power Network With Series Compensation. IEEE Trans. Power Del., Nov (2013).
9. Ree, J.D., Centeno, V., Thorp, J.S., Phadke, A.G.: Synchronized phasor measurement applications in power systems. IEEE Trans. Smart Grid, vol. 1, no. 1, pp. 20–27, Jun. (2010).
10. Roy, B.K.S., Sharma, R., Pradhan, A.K., Sinha. A.K.: Faulty Line Identification Algorithm for Secured Backup Protection Using PMUs. Electric Power Components and Systems, 45:5, 491–504, https://doi.org/10.1080/15325008.2016.1266417 (2017).
11. Roy, B.K.S., Sinha, A.K., Pradhan, A.K.: An optimal PMU placement technique for power system observability. Int. J. of Elect. Power Energy Syst., Vol. 42, No. 1, pp. 71–77, November (2012).
12. Phadke, A.G., Thorp, J.S., Adamiak, M.G..: A new measurement technique for tracking voltage phasors, local system frequency, and rate of change of frequency. IEEE Trans Power Appar Syst 102(5):1025–1038 (1983).

Analysis and Prediction of the Effect of Surya Namaskar on Pulse of Different Prakruti Using Machine Learning

Jayshree Ghorpade-Aher, Abhishek Girish Patil, Eeshan Phatak, Sumant Gaopande and Yudhishthir Deshpande

Abstract *"Surya Namaskar"* is the key for Good health! Today's social life can be made easier and healthier using the mantra of *"YOGA"*. Nadi Parikshan is a diagnostic technique which is based on the ancient Ayurvedic principles of Wrist Pulse analysis. Nadi describes the mental and physical health of a person in great depth. This information can be used by practitioners to prevent, detect as well as treat any ailment. Surya Namaskar is a Yoga exercise which has multiple health benefits and a direct impact on Pulse. Prakruti of a person is a metaphysical characteristic and a combination of the three doshas in Ayurveda viz. *Vatta, Pitta,* and *Kapha* which remains constant for the lifetime. Experimentation was carried out to analyze the effect of Surya Namaskar exercise on the Pulse of different Prakruti. The Pulse was recorded for a group of young students aged between 19 and 23 years with different Prakruti before Surya Namaskar and after Surya Namaskar for a period of 4 days. This paper analyzes the effect of Surya Namaskar on human Pulse and proposes a framework to predict Pulse after Surya Namaskar. The changes that Surya Namaskar causes in the Pulse are studied and used to predict Pulse after performing Surya Namaskar. This analysis helps understand how Surya Namaskar benefits the health of a person. Performing Surya Namaskar in our daily routine would improve the health of the society as a whole making the subjects energetic and active.

J. Ghorpade-Aher (✉) · A. G. Patil · E. Phatak · S. Gaopande · Y. Deshpande
Department of Computer Engineering, M.I.T. College of Engineering, Pune, India
e-mail: jayshree.aj@gmail.com

A. G. Patil
e-mail: abhilampard@gmail.com

E. Phatak
e-mail: eeshan.phatak@gmail.com

S. Gaopande
e-mail: sumantgaopande@gmail.com

Y. Deshpande
e-mail: udealwithit20@gmail.com

© Springer Nature Singapore Pte Ltd. 2018
S. Bhalla et al. (eds.), *Intelligent Computing and Information and Communication*,
Advances in Intelligent Systems and Computing 673,
https://doi.org/10.1007/978-981-10-7245-1_54

Keywords Nadi Parikshan · Wrist pulse · Pulse analysis · Surya Namaskar Yoga · Machine learning

1 Introduction

Ayurveda, is a traditional effective medical science which was developed in the Indian subcontinent thousands of years ago where "*Ayur*" means life and "*Veda*" means science. '*Nadi Parikshan*' (Wrist Pulse Diagnosis), can accurately diagnose both physical and mental diseases as well as the imbalances in doshas [1]. It is an important diagnostic method used by Ayurvedic practitioners which consists of a various examination and physical tests [2]. Surya Namaskar is a Yoga exercise which has been practiced since ancient times by *Rushimunis*.

Various evidence-based research studies have suggested that Surya Namaskar has a positive impact on the human body and its functionalities such as balancing metabolism function, tuning the central nervous system, supporting urogenital system, boosting gastrointestinal system, etc. In today's busy life, practicing Surya Namaskar regularly revitalizes body, improves energy level, and keeps mind calm [3]. Figure 1 (*src:* http://www.harekrsna.de/surya/surya/surya-names.htm.) depicts

Fig. 1 The 12 Asanas of Surya Namaskar

the various steps in performing Surya Namaskar. The steps include 12 different *asanas* having different benefits. The 12 asanas are as follows:

1. *Pranamasana*
2. *Hasta Uttanasana*
3. *Hasta Padasana*
4. *Aekpadprasarnaasana*
5. *Adhomukhasvanasana*
6. *Ashtanga Namaskar*
7. *Bhujangasana*
8. *Adhomukhasvanasana*
9. *Ashwasanchalanasana*
10. *Uttanasana*
11. *Hasta Uttanasana*
12. *Pranamasana*

Various research studies have proved that regular practice of Surya Namaskar exercise improves cardiopulmonary efficiency in healthy adolescents and is a beneficial exercise for both males and females. The regular practice of Surya Namaskar for a long time, results in reduced Pulse rate and this is attributed to factors like the increased vagal tone and decreased sympathetic activity [4].

2 Proposed Framework

The objective of this paper is to analyze changes in Pulse caused by Surya Namaskar and predict Pulse after Surya Namaskar.

Figure 2 shows the block diagram for the Proposed Framework of System that includes various steps in building the predictive model. The data recorded is split into training data and testing data. The predictive model will be built by using feedforward neural network and backpropogation. Testing data will be used to improve the accuracy of the predictive model using cross-validation. This predictive model will predict Pulse after Surya Namaskar.

2.1 Data Acquisition

'*Nadi Tarangini*' is a noninvasive Pulse-based diagnosis system which is an alternative to the traditional method of "Nadi Pariksha'. It has three pressure sensors to record the Pulse which gives 3 Pulse waveforms at the 3 points on the wrist of the *tridoshas* viz. *Vatta, Kapha,* and *Pitta,* respectively. The pressure is slowly increased on the wrist until one of the 3 Pulse waveforms start to take the form of a typical Pulse signal. The recorded data is obtained in the form of Time Series data

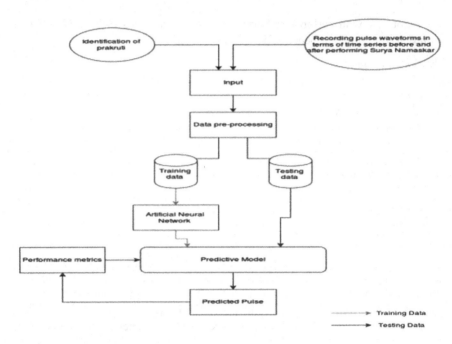

Fig. 2 Block diagram for the proposed framework of system

Fig. 3 *Nadi Tarangini* to record pulse

and is stored in the form of Comma Separated Values (.csv file) where three columns represent the amplitude values of respective Dosha signal in Vatta–Pitta–Kapha sequence. The sampling frequency of the device is around 500 Hz [5, 6]. Figure 3 shows the real-time images of the Surya Namaskar practical sessions that were conducted for data acquisition.

Table 1 Data acquisition information

Sr. No.	Parameter	Description
1	Number of subjects	35 (23 Males, 12 Females)
2	Age group	21 ± 2 years
3	Time	3:00 P.M.–5:00 P.M. (Indian Standard Time)
4	Recording of pulse	Pulse recorded for duration of 2 min, 1. Before Surya Namaskar 2. After Surya Namaskar

The physical and mental constitution of the body; which remains constant during the lifetime of a person is identified by Prakruti of that person. In Ayurveda, the *"Panchamahabhutas"* manifests the features of **Prithvi** (Earth), **Akash** (Space), **Tejas** (Fire), **Vayu** (Wind), and **Jal** (Water). The amalgamation of these elements is responsible for the *tridoshas*—Vatta, Pitta, and Kapha [7]. The Prakruti of a person is identified by using a questionnaire and based on the answers given by the person. The questions are related to the daily lifestyle and habits of the subject which defines the Prakruti of that subject. Table 1 shows the description of the dataset for the data acquisition process.

Disclaimer: The Pulse was recorded using a noninvasive device. No human being was harmed for the study. Only healthy people were considered who were without any apparent medical condition or illness. The procedure for recording Pulse is similar to one followed by many other authors [5, 8].

2.2 Data Preprocessing

Baseline removal is a wave preprocessing technique which arranges the peaks of a wave in a uniform manner about an axis that divides each side into equidistant parts.

As shown in Fig. 4a, all the peaks are seen in a haphazard manner before baseline removal. After applying baseline removal technique the wave is brought about the X-axis in a uniform manner as shown in Fig. 4b. This allows comparing

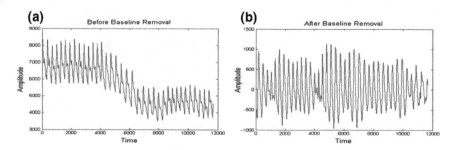

Fig. 4 **a** Before baseline removal. **b** After baseline removal

the change in amplitude and frequency for the Pulse waves before and after performing Surya Namaskar. Denoising a wave is any technique applied to remove noise from a signal and preserve the useful information. The denoising function has been applied that smoothens the shape of the wave such that even intricate curves can be studied to conclude minute changes after performing Surya Namaskar.

2.3 Feed Forward Neural Network

The prediction of the Pulse signal after performing Surya Namaskar will be non-linear. A feedforward neural network (FFNN) is a type of neural network containing many neurons which are arranged in multiple layers viz. input layer, hidden layer, and output layer. Each neuron in a hidden layer joins another respective neuron in the previous layer and its next layer. Each such connection has a weight associated with it which gets adjusted as the training process continues. These are also called as multilayered perceptron (MLP). As there is no feedback involved, they are called as FFNN. A MLP is adopted as a prediction model [9, 10–12]. The features given to the input layer will be:

1. Point in Pulse Before Surya Namaskar
2. P to P time interval between consecutive Pulse as shown in Fig. 5
3. D to D time interval between consecutive Pulse as shown in Fig. 5

These features will be given to each neuron in the input layer for each Dosha (Vatta, Pitta, or Kapha).

Every input has an associated weight which determines its influence on the change in a wave. Each neuron in the network is governed by an activation function given by:

$$\frac{1}{1 + e^{-f(x)}},$$ (1)

Fig. 5 P to P and D to D
interval

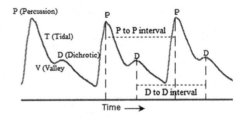

Time ⟶

where,

$$f(x) = \sum_{i=1}^{4} w_i x_i + \text{Bias}$$

W   Weight associated with corresponding
X   Input Parameter
Bias Measure of the threshold of a Neuron.

The FFNN will be trained using backpropagation algorithm. This is based on calculating the error between the predicted and actual values. The change in this error with respect to the weights is calculated. This is called the error derivative. Error derivative will be used to readjust the weights as training progress [13]. Once the training is completed the neural network along with the weights forms the prediction model. A new Pulse signal before Surya Namaskar can be given to this model which predicts the "After Surya Namaskar" Pulse. Once, the network is trained, on providing a new input value (Pulse signal before performing Surya Namaskar), it will predict the corresponding output. As the dataset is increased for training the network, this degree of error can be minimized. Using MATLAB's "ntstool", nonlinear prediction of time series data is possible [14–16]. It uses a FFNN and has two parameters:

1. Input Vector: defines the set of values before Surya Namaskar
2. Target Vector: Expected set of values after Surya Namaskar

This neural network is trained and returns the output vector and the mean error in prediction.

3 Observations and Discussion

Analyzing and observing the Pulse signals recorded under two conditions viz. before Surya Namaskar and after Surya Namaskar; the amplitude of the Pulse signals increased. It was also observed that the frequency increased in the latter case (After Surya Namaskar). Figure 6a, b, c shows one such example of the 3 Pulse signals (Vatta, Pitta and Kapha Dosha signals) recorded for both the conditions at Afternoon Prahar.

In Fig. 6a, b, c, the y-axis represents the Pressure amplitude and x-axis represents the timestamps. In all the three cases after Surya Namaskar, there was an increase in amplitude and frequency. The increase in frequency is a direct indication of the increase in Pulse rate of the Pulse. There was a significant change in all the doshas viz. Vatta, Pitta and Kapha signals as shown in Fig. 6a, b, c, respectively. In all the signals, the Dichrotic Notch appears to be diminished or less prominent and the Pulse becomes smoother after Surya Namaskar. This is because while performing

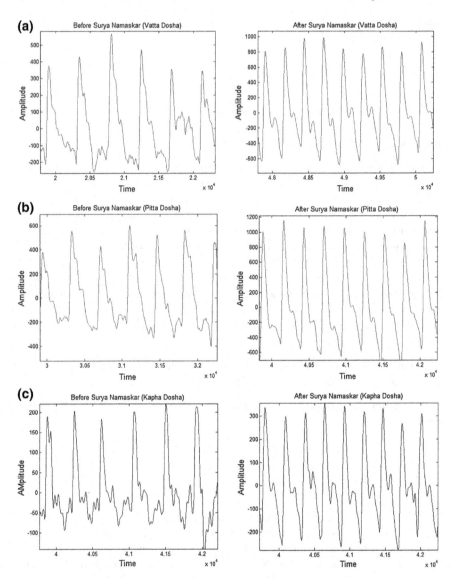

Fig. 6 a Pulse signal (Vatta). **b.** Pulse signal (Pitta). **c.** Pulse signal (Kapha)

Surya Namaskar exercise, blood flow to the muscle is tightly coupled to its metabolic rate. This is due to the increase in the cardiac output and means arterial blood pressure and regional redistribution of blood flow through adjustments in the vascular resistance. The arterial compliance increases and blood flows faster and as a result the Pulse becomes smooth [8].

4   Conclusion

Surya Namaskar has an immediate impact on Pulse and improves the overall health of an individual. The young age group considered in this study show significant changes which signify the usefulness of this exercise and benefits of Yoga. Surya Namaskar has an important role in our day to day life and has a positive impact on the society as a whole, by balancing the three doshas. The changes in Pulse that Surya Namaskar brings about can be further analyzed in detail and can have many applications in healthcare domain. Medical practitioners can use the system and understand the Surya Namaskar benefits on the body of a patient. Ayurveda, despite being an effective medical system, not much technological advancement have been there. This research is an effort toward proving these results using machine learning.

References

1. V. Lad.: Secrets of the pulse: The ancient art of Ayurvedic pulse diagnosis. Motilal Banarasidas. Delhi 2005.
2. Prasad, G. P., Bharati, K., & Swamy, R. K.: Some important aspects of nadipariksha from basavarajiyam. Ancient science of life, 24(1), 27–9. 2004.
3. Amit Vaibhav, Swati Shukla and Om Prakash Singh.: Surya Namaskar (Sun Salutation): A Path to Good Health. International Journal of Pharmacological Research IJPR Volume 6 Issue 07 (2016).
4. Pratima M. Bhutkar, Milind V. Bhutkar, Govind B. Taware, Vinayak Doijad and B.R. Doddamani.: Effect of Surya Namaskar Practice on Cardio-respiratory Fitness Parameters: A Pilot Study. Al Ameen J Med Sci (2008) 1(2).
5. Aniruddha Joshi, Anand Kulkarni, Sharat Chandran, V. K. Jayaraman and B. D. Kulkarni.: Nadi Tarangini: A Pulse Based Diagnostic System. Engineering in Medicine and Biology Society, 2007.
6. Nadi Tarangini. http://www.naditarangini.com. Accessed: 25th January 2017.
7. Dinesh Kumar Meena, Devanand Upadhyay, Rani Singh, B. K. Dwibedy.: A Critical Review of Fundamental Principles of Ayurveda. International Ayurvedic Medical Journal (IAMJ): Volume 3, Issue 7, July 2015.
8. L.S. Xu, K.Q. Wang, L. Wang, Naimin Li.: Pulse Contour Variability Before and After Exercise. Proceedings of the 19th IEEE Symposium on Computer-Based Medical Systems (CBMS'06).
9. Jayshree Ghorpade-Aher, Abhishek Patil, Yudhishthir Deshpande, Sumant Gaopande and Eeshan Phatak.: A Proposed Framework for Prediction of Pulse, based on the Effect of Surya Namaskar on Different Prakruti at Different Prahars of the Day. 2017 International Conference On Big Data Analytics and Computational Intelligence (ICBDACI), Chirala, India, 2017.
10. F. Scarselli and A. C. Tsoi.: Universal approximation using feedforward neural networks: a survey of some existing methods, and some new results. Neural Networks, vol. 11, pp. 15–37, 1998.
11. Ming Yuchi, Jun Jo.: Heart Rate Prediction Based on Physical Activity using Feedforward Neural Network. International Conference on Convergence and Hybrid Information Technology 2008.
12. Priddy K.L., Keller P.E.: Introduction in Artificial Neural Networks: An Introduction. Bellingham Wash. pp. 1–12. St. Bellingham USA, 2005.

13. Gashler, Michael S., and Stephen C. Ashmore.: Training Deep Fourier Neural Networks to Fit Time-Series Data. Intelligent Computing in Bioinformatics. Springer International Publishing, 2014.
14. Neural Network Time-Series Prediction and Modeling—MATLAB & Simulink-MathWorks, UK https://in.mathworks.com/help/nnet/gs/neural-network-time-series-prediction-and-modeling.html. Accessed: 25th Jan 2017.
15. Nitin Sharma, Neha Udainiya.: Mechanism of Chakras in Suryanamaskar and its benefits: A Conceptual Study. International Ayurvedic Medical Journal (IAMJ): Volume 3, Issue 8, August-2015.
16. L. Xu, D. Zhang, K. Wang, N. Li, and X. Wang.: Baseline wander correction in pulse waveforms using wavelet-based cascaded adaptive filter. Computers in Biology and Medicine, vol. 37, pp. 716–731, 2007.

Cognitive Depression Detection Methodology Using EEG Signal Analysis

Sharwin P. Bobde, Shamla T. Mantri, Dipti D. Patil
and Vijay Wadhai

Abstract This paper illustrates a new method for depression detection using EEG recordings of a subject. It is meant to be used as a computerised aid by psychiatrists to provide objective and accurate diagnosis of a patient. First, data from the occipital and parietal regions of the brain is extracted and different channels are fused to form one wave. Then DFT, using FFT, is applied on the occipito-parietal wave to perform spectral analysis and the fundamental is selected from the spectrum. The fundamental is the wave with the maximum amplitude in the spectrum. Then classification of the subject is made based on the frequency of the fundamental using rule-based classifier. Detailed analysis of the output has been carried out. It has been noted that lower frequency of the fundamental tends to show hypoactivation of the lobes. Moreover, low-frequency characteristics have also been observed in depressed subjects. In this research, 37.5% of the subjects showed Major Depressive Disorder (MDD) and in all 80% of the subjects showed some form of depression.

Keywords Depression · Electroencephalography (EEG) · Discrete Fourier Transform (DFT) · Fast Fourier Transform (FFT) · Alpha band
Delta band · Rule-based classifier

S. P. Bobde (✉) · S. T. Mantri
MIT College of Engineering, Pune, India
e-mail: bsharwin@gmail.com

D. D. Patil
MKSSS's Cummins College of Engineering for Women, Pune, India
e-mail: diptivt@gmail.com

V. Wadhai
KJEI's Trinity Academy of Engineering, Pune, India

© Springer Nature Singapore Pte Ltd. 2018
S. Bhalla et al. (eds.), *Intelligent Computing and Information and Communication*,
Advances in Intelligent Systems and Computing 673,
https://doi.org/10.1007/978-981-10-7245-1_55

1 Introduction

According to the WHO report issued in 2016, depression affects more than 350 million people of all ages worldwide, and this makes it the most prevalent global disease. It is also recorded that more women are affected by depression as compared to men. More than 8,00,000 people commit suicide due to depression each year [1]. The global burden of depression is increasing and there is an immediate need for a coordinated response from all health and social sectors at the country level [2]. Depression is characterized by persistent low-mood state and loss of interest and passion. In Major Depressive Disorder (MDD), difficulty with daily functioning in personal, family, social, educational, occupational areas is observed. Many people also show symptoms of anxiety. Depression is managed through psychological intervention and drug prescriptions [3].

There is a need for an objective diagnostic tool, which is technologically advanced and accurate. In this chapter, one such system has been proposed. The input to the system is the patient's Electroencephalography (EEG) recording. The system extracts readings from the occipital and parietal lobes of the brain, which are of main interest in this research. The readings are then fused to create one occipito-parietal wave. Then, the spectral analysis of the wave is carried out using Discrete Fourier Transform (DFT). Here, FFT has been used to perform DFT on the wave. The classification is dependent o the fundamental in the spectrum. After the fundamental has been selected, its frequency is recorded and the subject is classified accordingly.

2 Previous Work

Previous studies have studied different parts of the brain separately along with different frequency bands. There are four main frequency bands in EEG on which there has been extensive research done; they are delta (0.5–4 Hz), theta (4–8 Hz), alpha (8–13 Hz) and beta (13–30 Hz). It has been observed that the alpha band is most informative about a subject's state of depression [4]. Alpha activity mainly corresponds to the occipital and parietal lobes [5]. Alpha band has also been used to distinguish between depressed and normal subjects in the past. Increase in slow alpha activity in depressed subjects in occipital and parietal regions has also been noted in a 2010 study [6]. Relatively decreased alpha activity in patients with (MDD) has also been noted. In fact, decreased parieto-occipital alpha power is also stated as a biomarker of depression and excess alpha is also considered as a favourable response to antidepressive treatments [7]. A 2005 study found abnormal low-frequency activity in the occipital lobe of untreated patients with depression. They also found increased occipital lobe delta dipole density as a risk factor for depression [8]. Another paper stated that alpha and to a lesser extent delta oscillations can distinguish depressed and healthy individuals [9, 10]. Taking these facts into consideration our method focuses on the occipital and parietal regions of the brain. These facts also realise a relation between depression and the occipito-parietal wave. To detect and classify depression, it is checked if the power of the alpha band is low and if it shows low-frequency characteristics. The characteristic frequency is the frequency of the fundamental found in the spectral analysis of the wave. To extract these properties, DFT is performed on the occipito-parietal wave.

3 Data Collection

EEG readings are non-stationary and non-Gaussian signals that arise from the brain. They have high temporal resolution and low spacial resolution [11, 12]. In this research, EEG data from 80 different subjects has been used. Each dataset contains EEG data of 1 min. It has been recorded on a 64-channel EEG headset, which gives high precision in differentiating between different parts of the brain. The dataset has been obtained from PhysioNet and PhysioBank, which are research resources for complex psychologic signals [13]. The particular dataset was recorded for motor/imagery tasks [14].

Disclaimer
The EEG recordings used for research were collected from a reputable source. The data has been used in past research experiments. The resulting product is usually used for forensic experiments. No human was harmed during collection or utilisation of the data.

4 Methodology

In the first stage, the occipital and parietal regions of the brain are selected from the EEG data. Then the data from the channels is fused to produce the occipito-parietal wave. For feature extraction of DFT, using FFT, is performed on the occipito-parietal wave. Then an apt range is defined from which the fundamental of the wave is selected. This fundamental is the required feature for classification.

4.1 Occipito-Parietal Wave

As the occipital and parietal lobes of the brain are of main interest, data from these parts is fused to form one wave. In a 64-channel EEG, there are 17 channels which point to these two parts of the brain. These channels are P7, P5, P3, P1, Pz, P2, P4, P6, P8, PO7, PO3, POz, PO4, PO8, O1, Oz and O2. The fusing of data is done by taking the mean of the channels' readings at every time instant. This results in a new wave with the same number of readings. Let this new occipito-parietal wave be W. The index position of each record in the dataset is i. Therefore, the resulting voltage value at a recorded instant i will be W_i. Let SF be the sampling frequency of the recording headset. Therefore, the index positions are given by Eq. 1 as follows:

$$\{i \in Z^+ \,|\, 1 \le i \le (60 \times \text{SF})\} \tag{1}$$

Let the set of required channel numbers be Ch as in Eq. 2, where $P7, P5 \cdots O2$ represent the selected channel numbers.

$$\text{Ch} = \{P7, P5, P3, \cdots Oz, O2\} \tag{2}$$

$$W_i = \frac{\sum_{\text{Ch}} \text{EEG}_{i,\text{Ch}}}{|\text{Ch}|} \tag{3}$$

The voltage reading of W at index i will be given by Eq. 3. Here, EEG is the data frame containing readings of all 64 channels at all i time instances. Figure 1a shows the EEG readings of selected 17 EEG channels of subject 30. All 17 graphs show similar readings, but actually they all differ in voltage readings. Figure 1b shows the resulting occipito-parietal wave of subject 30.

4.2 Feature Extraction

Fast Fourier Transform is a computational algorithm which facilitates DFT of an input signal [15]. It helps in performing fast spectral analysis. It is a more efficient way of performing DFT of a signal. FFT transforms the input signal into a sequence of complex numbers which represent pure sinusoidal waves. The real part of the number represents the amplitude and the imaginary part represents the phase of the sinusoidal wave. For this system the amplitude of the waves is important. Therefore, the absolute value of each complex number is taken. The feature for

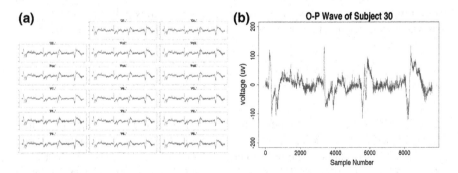

Fig. 1 **a** Plot of selected 17 channels **b** Resulting occipito-parietal wave

classification here is the fundamental obtained after spectral analysis of the occipito-parietal wave W. The FFT algorithm outputs a list of complex numbers. The list will be of length $|i|$, which was used before. First, the amplitude values obtained after FFT are saved. Let this vector of amplitudes be F. Therefore, let each element of the vector, using DFT equation, be given by Eq. 4 as follows:

$$F_k = \left| \sum_{n=0}^{N-1} W_n \times E_N^{kn} \right|, \quad 0 \leq k \leq N-1 \tag{4}$$

where $EN = e - j2\pi/N$, $k_i = i - 1, N = |i|$, $N = 60 \times SF$.

The FFT algorithm gives the output in a logarithmic scaling. Frequency scaling is done to get the correct output. Let the new vector of correct amplitudes be F'. The relation between F' and F is given by Eq. 5. The resulting set F' contains correct values of amplitude in μV for each sinusoidal wave with a frequency equal to the index position at which it occurs. Two such plots of subjects' number 1 and 20 have been shown in Fig. 2 as examples.

$$F_i' = \frac{1}{2} \times \left(\frac{F_i}{SF} \right) \tag{5}$$

To select the fundamental, a frequency range needs to be defined from which the fundamental's frequency can be selected. This range is very important as it forms the basis of the diagnosis. It should be broad enough to cover maximum real-world instances. It should cover only the important. Thus range from 1 Hz to the Nyquist frequency will be apt. The Nyquist frequency in this study's samples is 80 Hz. This range contains all 4 major frequency bands. The Nyquist frequency is given by Eq. 6.

$$Nyq = \frac{SF}{2} \tag{6}$$

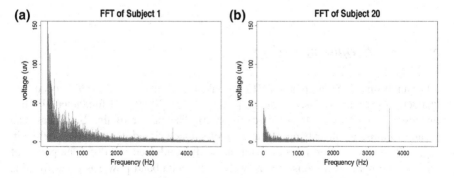

Fig. 2 FFT spectrum of **a** Subject 1 **b** Subject 20

After the range is defined, the fundamental can be selected. The fundamental is that sinusoidal wave which has the maximum amplitude in the defined range. Because of the narrowing of range, the algorithm does not consider high and comparatively lower frequency noise which is present because of the hardware and can affect the output.

5 Analysis

5.1 Square Mean Voltage

To compare the fundamental's frequency with the activation in the occipital and parietal regions, it needs to be related to power. As power is proportional to intensity, which is the square of amplitude (here voltage), it can be related to the occipito-parietal wave. The previous range of frequency is used to extract a part of the spectrum. Then, the mean amplitude in that range is found and its square is computed. The equation for mean voltage is given by Eq. 7.

$$MV = \frac{\sum_n F'_n}{|n|}, \quad 1 \leq n \leq Nyq, \quad n \in i, \ |n| < |i| \tag{7}$$

$$SMV = MV^2 \tag{8}$$

Square mean voltage is derived from Eq. 8. Then, the frequency of the fundamental was graphically compared to Square Mean Voltage (SMV). As power is proportional to SMV the comparison gives an understanding of the relation between the fundamental's frequency and activation of the occipital and parietal lobes of the brain. It is shown in Fig. 3. A regression line was also plot along with it. The line yielded a positive slope.

This means that as the frequency of the fundamental decreases the activation of these lobes tends to decrease. This hypoactivation of the parietal and occipital lobes of the brain is an indication of depression [7].

5.2 Low-Frequency Activity

After analysing all 80 graphs of FFT spectral analysis manually, certain reoccurring characteristics were realised. The subjects with low SMV had fundamental's frequency ranging from 1 to 4 Hz (delta band). The power of this band was also greater to those in other bands. Examples of subjects 5 and 26 are shown below in Fig. 4. This observation lines up with the study conducted in 2005. The power of the occipito-parietal wave is less, as well as the delta band power is high [8]. This further validates the classification of depression in the subjects.

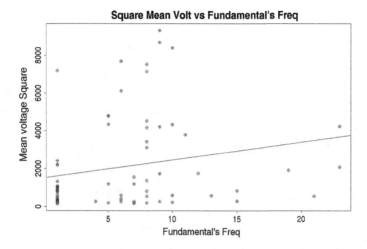

Fig. 3 Plot of square mean voltage versus fundamental's frequency

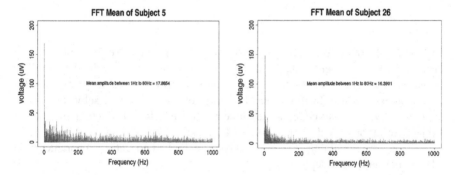

Fig. 4 FFT spectrum of subjects to be classified as depressed

5.3 Classification

The classification bins were defined after performing analysis and inspecting the occipito-parietal waves and their spectrums together. The occipito-parietal waves have been analysed for their intensity, pattern, spectrum patterns, and fundamental's properties. Similar wave patterns have also been seen in other studies which took pre-classified subjects for analysis [16]. The classification bins have been decided as severe, moderate, mild and normal. The classification is done using a rule-based classifier, which selects bins based on the fundamental's frequency. The fundamental's frequency range from 1 to 4 Hz shows server depression of patients. Patients lying in this category showed all signs of major depressive disorder (MDD) considered in this analysis.

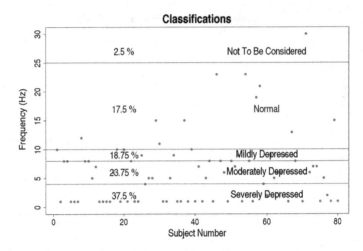

Fig. 5 Plot of fundamental's frequency versus subject number, also showing classification labels and percentage of subjects present

6 Results

EEG data of 80 subjects has been processed and out of these 76 subjects were classified. In the experimental analysis, four subjects were classified as outliers based on their fundamental's frequency. Figure 5 shows the results graphically for clearer understanding. In this research, 37.5% of the considered subjects showed signs of MDD. In particular, these subjects showed clear distinction from other subjects in all statistical parameters. 80% of total subjects showed some level of depression. The remaining 20% of the subjects exhibited no statistical parameters in EEG which indicated depression.

7 Conclusion

Analysis of 80 subjects has been performed and 78 have been classified according to various levels of depression. 61.25% of the subjects indicated moderate or severe depression, which means they need professional help. Out of these 37.5% indicated severe depression. 2.5% of the total subjects were outliers and could not be classified, mainly due to noise in the EEG signal. The main emphasis was given on the occipital and parietal lobes of the brain, which have been studied to be good sources of indicators for depression. The data of these lobes is fused to achieve good data. Analysis has also been done on low-frequency characteristics of brainwave which is also an indicator of depression. Classification bins were defined based on analysis of the occipito-parietal wave and its frequency spectrum. The final classification has been made using a rule-based classifier.

The system can diagnose subjects before their condition can become severe. It is based on an objective understanding of the human brain. This can benefit victims of depression and the society in many ways. Early diagnosis means the patient can live a happy life. It is free from the subjective bias of psychologists and psychiatrists. A better, objective help can improve the communal standard of living and decrease the suicide statistics.

References

1. World Health Organization, revised on November 2016, available at http://www.who.int/topics/depression/en/.
2. WHO: World Health Assembly (WHA) Resolution 65th Report by Secretariat, 2012, pp 1–4. http://apps.who.int/gb/ebwha/pdf_files/WHA65/A65_R4-en.pdf?ua=1 (accessed November 28, 2014).
3. World Health Organization, mhGAP Intervention Guide - Version 2.0 for mental, neurological and substance use disorders in non-specialized health settings, 2016.
4. Behshad Hosseinfard, Mohammad Hassan Moradi, Reza Rostami, "Classifying depression patients and normal subjects using machine learning techniques and nonlinear features from EEG signal," Elsevier, Computer methods and programs in Biomedicine, 2012.
5. David M. Groppe, Stephan Bickel, Corey J. Keller, Sanjay K. Jain, Sean T. Hwang, Cynthia Harden, and Ashesh D. Mehta, "Dominant frequencies of resting human brain activity as measured by the electrocorticogram," Neuroimage; 79: 223–233, 1st Oct 2013.
6. Vera A. Grin-Yatsenko, Inke Bass, Valery A. Ponomarev, Juri D. Kropotov, "Independent Component approach to the analysis of EEG recordings at early stages of depressive disorders," Elsevier, Clinical Neurophysiology 121, 281–289, 2010.
7. Sebastian Olbrich & Martijn Arns, "EEG biomarkers in major depressive disorder: Discriminative power and prediction of treatment response," International Review of Psychiatry; 25(5): 604–618, October 2013.
8. Alberto Fernández, Alfonso Rodriguez-Palancas, María López-Ibor, Pilar Zuluaga, Agustín Turrero, Fernando Maestú, Carlos Amo, Juan José López-Ibor, Jr, and Tomás Ortiz, "Increased occipital delta dipole density in major depressive disorder determined by magnetoencephalography," J Psychiatry Neurosci.; 30(1): 17–23, Jan 2005.
9. Mahdi Mohammadi, Fadwa Al-Azab, Bijan Raahemi, Gregory Richards, Natalia Jaworska, Dylan Smith, Sara de la Salle, Pierre Blier and Verner Knott, "Data mining EEG signals in depression for their diagnostic value," BMC Medical Informatics and Decision Making, 15:108, 2015.
10. SUBHA D. PUTHANKATTIL and PAUL K. JOSEPH, "Classification Of Eeg Signals In Normal And Depression Conditions By Ann Using Rwe And Signal Entropy," Journal of Mechanics in Medicine and Biology Vol. 12, No. 4 1240019 (13 pages), 2012.
11. D. Puthankattil Subha & Paul K. Joseph & Rajendra Acharya U & Choo Min Lim, "EEG Signal Analysis: A Survey," J Med Syst (2010) 34:195–212, 2010.
12. M. Rajya Lakshmi, Dr. T. V. Prasad, Dr. V. Chandra Prakash, "Survey on EEG Signal Processing Methods," International Journal of Advanced Research in Computer Science and Software Engineering, Volume 4, Issue 1, January 2014.
13. Goldberger AL, Amaral LAN, Glass L, Hausdorff JM, Ivanov PCh, Mark RG, Mietus JE, Moody GB, Peng CK, Stanley HE. PhysioBank, PhysioToolkit, and PhysioNet: Components of a New Research Resource for Complex Physiologic Signals. *Circulation* 101(23):e215–e220 [Circulation Electronic Pages; http://circ.ahajournals.org/content/101/23/e215.full]; 2000 (June 13).

14. Schalk G., McFarland D.J., Hinterberger T., Birbaumer N., Wolpaw J.R. "A General-Purpose Brain-Computer Interface (BCI) System," IEEE Transactions on Biomedical Engineering 51 (6):1034–1043, 2004.
15. W.T. Cochran, J.W. Cooley, D.L. Favin, H.D. Helms, R.A. Kaenel, W.W. Lang, G.C. Maling, D.E. Nelson, C.M. Rader, P.D. Welch, "What is the fast Fourier transform?," Proceedings of the IEEE, Volume: 55, Issue: 10, Oct. 1967.
16. Rajendra Acharya, Vidya K. Sudarshan, Hojjat Adeli, Jayasree Santhosh, Joel E.W. Koh, Amir Adeli, "Computer-Aided Diagnosis of Depression Using EEG Signals," European Neurology 2015;73:329–336, 2015.

Biogas Monitoring System Using DS18B20 Temperature Sensor and MQTT Protocol

Suruchi Dedgaonkar, Aakankssha Kaalay, Nitesh Biyani
and Madhuri Mohite

Abstract Nonrenewable energy resources such as coal, petroleum, and natural gas are becoming extinct and thus there is a need for renewable energy resources in the long run. One of the most important renewable energy resources is biogas. Biogas is the gas produced after anaerobic digestion of organic matter by micro-organisms. Biogas mainly contains methane (about 60%). There are a number of factors affecting the production of biogas and one of them is temperature. The temperature of the biogas plant should be held constant with variation <1 °C and within the range 30–55 °C. Thus, there is a need of a proper monitoring system for the biogas plant. In this study, we try to develop a monitoring system for the biogas plant using temperature sensor DS18B20, MQTT protocol, Mosquitto broker, and Raspberry Pi. A web-based system, where the temperature sensor values will be uploaded periodically and an end user will monitor the temperature of biogas plant remotely, is proposed.

Keywords Biogas monitoring system · DS18B20 · Temperature sensor
MQTT protocol · Mosquitto · Raspberry Pi

S. Dedgaonkar (✉) · A. Kaalay · N. Biyani · M. Mohite
Information Technology Department, VIIT, Savitribai Phule Pune University, Kondhwa
(BK), Pune, Maharashtra, India
e-mail: suruchigd@gmail.com

A. Kaalay
e-mail: aakanksshak21@gmail.com

N. Biyani
e-mail: niteshbi294@gmail.com

M. Mohite
e-mail: madhurimohite123@gmail.com

© Springer Nature Singapore Pte Ltd. 2018   567
S. Bhalla et al. (eds.), *Intelligent Computing and Information and Communication*,
Advances in Intelligent Systems and Computing 673,
https://doi.org/10.1007/978-981-10-7245-1_56

1 Introduction

The biogas is in use since a long time in India, but the technology improvement is necessary in order to improve energy outputs. Biogas is one of the most trusted and reliable energy resources on earth. There is a need for an eco-friendly substitute for energy resources that are depleting continuously and will soon be exhausted completely. Biogas serves this. Biogas is the gas produced by the complex breakdown of organic matter in the absence of oxygen, it does not have any geographical limitations and also biogas is very simple to use and apply. Kitchen waste consists of major part of the feed used for the biogas plant as kitchen waste is very high in calorific value and also it is highly accessible and cheap, kitchen waste also helps in increasing the methane content of the gas produced. There is a need for a monitoring system of biogas process, since a monitoring system can help to increase the efficiency, enhance performance, and also get better results at minimal cost. Biogas is the most important and efficient resource of energy at a cheaper price and thus it needs to be monitored for maximum output. There is no proper monitoring system for the biogas plants as such that will govern the working and output of the plant. There is no efficient system to measure the parameters of the gas produced in the plant and take necessary actions according to them Fig. 1 shows homemade biogas plant used for simulation purpose [1].

Fig. 1 Homemade biogas plant used for monitoring

2 Message Queuing Telemetry Transport

MQTT is a protocol, which sits on the top of the TCP/IP protocol that is designed for machine-to-machine communication and for the internet of things connectivity. MQTT is a lightweight protocol, unlike HTTP which is very bulky in comparison. MQTT is a protocol which is designed for publish/subscribe protocol which consists of a broker as a medium. MQTT runs on low bandwidth and is still extremely powerful. It can be used for high-cost connections because it is readily available [2–4, 8–10]. In this study, temperature sensor sends values continuously to MQTT. The broker examines the values as to whom it has to be sent and sends accordingly. On the other hand, the subscribers which are previously subscribed to that topic keep receiving the messages from that particular topic until they cancel the subscription [15, 16].

A. **MQTT client**:

All the components of the IOT environment can act as a client, which sends or receives data in this protocol. The components can be a microcontroller, sensor, or the end user devices. In our study, we will be using a temperature sensor, which will send the values to the broker, which will be uploaded to a web page and also will be stored in a database and will be sent to the user application. A client can be a publisher or a subscriber, it depends upon the role assigned to the device. In both the cases, once the connection is established the device must announce itself as a subscriber or a publisher.

A publisher can publish the temperature values. A subscriber needs to subscribe to the same topic to which the publisher is sending the values to receive the temperature values. For a device to behave as an MQTT client, an MQTT library must be installed on the client and the broker and the client should be on the same network.

B. **MQTT Publisher**:

MQTT publisher is a client, which publishes the data to a particular topic to which the subscriber is subscribed and sends the values to that topic. The data is received by the subscriber through this topic.

C. **MQTT Subscriber**:

MQTT subscriber is a client, which is subscribed to the topic and receives values from that topic, which are published by the publisher. The subscriber can access data on that topic only if he is subscribed to that topic.

D. **MQTT Broker**:

An MQTT broker is a bridge between the publisher and the subscriber. The job of the broker is to establish a communication between a publisher and a subscriber. Broker sends values from a publisher to a subscriber. A broker can handle up to hundreds of clients simultaneously. When the data is incoming, the broker must identify all the clients that are subscribed to the corresponding topic and make sure that each and every one of them receives the data. In this study, the broker receives the values from the temperature sensor connected to Raspberry Pi and sends the values to the subscribed user. Broker also performs the task of authentication and authorization of the clients for security. The clients send the username and password along with the connect message which is verified by the broker. Broker also makes sure that the messages are received by the subscribed users only and no other device can access the data.

There are a number of brokers which can be used for implementation of MQTT protocol.

Mosquitto—It is an open-source MQTT broker. In this study, we will be using Mosquitto broker [11].

Installation and testing of Mosquitto broker on Raspberry Pi:

The commands for installation of mosquito are given below. The following commands are to be entered into the terminal of Raspberry Pi

```
- > sudo apt-get install mosquitto
- > sudo apt-get install mosquito -clients
- > sudo apt-get install mosquito mosquito-clients python-mosquitto
```

Next, we need to edit the configuration file of Mosquitto which is mosquito.conf. We need to stop the Mosquitto service and then edit the file

```
- > sudo /etc./init.d/mosquito stop
- > sudo nano/etc./mosquito/mosquito.conf
```

Enter the following lines into the file

```
Pid_file /var/run/mosquito.pid
persistence true
persistence_location
/var/lib/mosquito
log_dest topic
log_type error
log_type warning
log_type notice
log_type information
connection_messages true
log_timestamp true
include_dir /etc./mosquito/conf.d
```

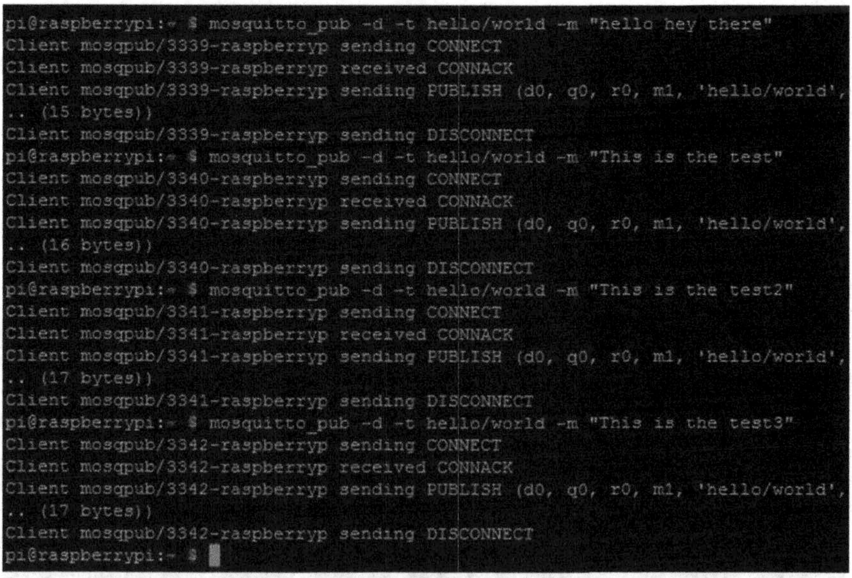

Fig. 2 Publisher window-publishing the text messages

Save the file by pressing CTRL+X and then pressing Y and ENTER
Now, start the Mosquitto service again

```
- > sudo /etc./init.d/mosquito start
```

Now, open two terminals, one for the publisher and the second one for the subscriber.

Enter the following command into subscriber window to subscribe to the topic to receive the messages published by the publisher

```
- > mosquito_sub -d -t hello/world
```

Enter the following command into publisher window (Fig. 2,3)

```
- > mosquito_pub -d -t hello/world -m "this is first test".
```

```
pi@raspberrypi:~ $ mosquitto_sub -d -t hello/world
Client mosqsub/3337-raspberryp sending CONNECT
Client mosqsub/3337-raspberryp received CONNACK
Client mosqsub/3337-raspberryp sending SUBSCRIBE (Mid: 1, Topic: hello/world, Qo
S: 0)
Client mosqsub/3337-raspberryp received SUBACK
Subscribed (mid: 1): 0
Client mosqsub/3337-raspberryp sending PINGREQ
Client mosqsub/3337-raspberryp received PINGRESP
Client mosqsub/3337-raspberryp received PUBLISH (d0, q0, r0, m0, 'hello/world',
... (15 bytes))
hello hey there
Client mosqsub/3337-raspberryp sending PINGREQ
Client mosqsub/3337-raspberryp received PINGRESP
Client mosqsub/3337-raspberryp sending PINGREQ
Client mosqsub/3337-raspberryp received PINGRESP
Client mosqsub/3337-raspberryp received PUBLISH (d0, q0, r0, m0, 'hello/world',
... (16 bytes))
This is the test
Client mosqsub/3337-raspberryp received PUBLISH (d0, q0, r0, m0, 'hello/world',
... (17 bytes))
This is the test2
Client mosqsub/3337-raspberryp sending PINGREQ
Client mosqsub/3337-raspberryp received PINGRESP
Client mosqsub/3337-raspberryp received PUBLISH (d0, q0, r0, m0, 'hello/world',
... (17 bytes))
This is the test3
```

Fig. 3 Subscriber window-receiving the text messages published by the publisher

3 Sensor

DS18B20 (Temperature sensor):

DS18B20 is a 1-wire digital temperature sensor. The DS18B20 digital thermometer provides 9-bit to 12-bit Celsius temperature measurements and has an alarm function with nonvolatile user-programmable upper and lower trigger points. The DS18B20 communicates over a 1-wire bus that by definition requires only one data line (and ground) for communication with a central microprocessor. In addition, the DS18B20 can derive power directly from the data line ("parasite power"), eliminating the need for an external power supply. Each DS18B20 has a unique 64-bit serial code, which allows multiple DS18B20s to function on the same 1-wire bus. Thus, it is simple to use one microprocessor to control many DS18B20s distributed over a large area. Applications that can benefit from this feature include HVAC environmental controls, temperature-monitoring systems inside buildings, equipment, or machinery, and process monitoring and control systems (Fig. 4).

The connections of DS18B20 with Raspberry Pi are as follows:

Connect GND of DS18B20 to pin 6 of Raspberry Pi

Connect DATA of DS18B20 to pin 7 of Raspberry Pi

DS18B20 to pin 1 of Raspberry Pi (3.3 v supply)

Fig. 4 DS18B20
(Temperature sensor)

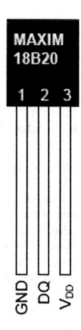

4 Raspberry Pi

Raspberry Pi is a small credit card-sized computer developed for basic programming and learning purposes [5–7, 17]. It has 1 GB of RAM with 1.2 GHZ 64-bit quad-core ARM Cortex-A53 processor. It has 4 USB ports and an HDMI port, also it has 40 GPIO pins and we can install an operating system on it. In this study, we will be installing our MQTT broker on Raspberry Pi, and also the temperature sensor will be connected to Raspberry Pi for taking the values of the biogas plant. To access the Raspberry Pi terminal, we will be using SSH configuration through putty by using the IP address of the Raspberry Pi.

5 Web Interface

In this study, we will be uploading all the values coming from the temperature sensor of the biogas plant through MQTT protocol to a web page. The values will be uploaded continuously on the web page in real time for monitoring. The web page is designed using the functionality of HTML along with the JavaScript and PHP to design a dynamic page [12]. The values received from the sensor are stored in SQL database from where the data is sent to the web page. As soon as the values are received by the sensor, the values are sent to the web page which will be sent the values to the users with the topic concerning the temperature values if the temperature values are below or above the range and also if there is a continuous change in the temperature of the biogas plant. The user can monitor the temperature values of the biogas plant from the web page in real time. The web page consists of graphs and charts of the incoming temperature values for a graphical view of the current temperature of the biogas plant.

6 Android Application

This study consists of an end user android application, which is designed to monitor and analyze the values according to time and date. The values received from the temperature sensor are stored in a database in SQL and all the values are sorted by date, time, and range. The application consists of an interface where the user can select to view the data by time and date. The user can view the temperature of the biogas plant by selecting the date and time. The user has an option of checking what the value of the plant was on a particular date and at a particular time. This will help in understanding the production and output of the biogas plant better. The user can analyze the data over time and can take helpful decisions according to the data.

7 Design and Implementation

In this proposed study, we have created a biogas monitoring system with temperature sensor DS18B20 and MQTT protocol for monitoring the temperature of the biogas plant. The temperature sensor will sense the temperature values of the biogas plant and will send these values to the MQTT protocol. The job of MQTT protocol is to send the alerts to the registered users if the temperature goes below or above the ideal temperature for the biogas plant. Simultaneously all the values are uploaded to a web page for a web-based monitoring. The protocol is installed on Raspberry Pi which acts as a client in this architecture for the protocol and all the installations and working takes place on Raspberry Pi (Fig. 5) [13, 14].

Fig. 5 Overall flow of data

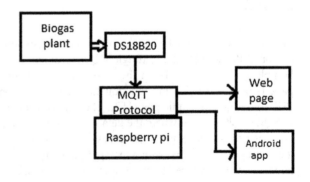

The connection of Raspberry Pi and DS18B20 temperature sensor is shown in (Fig. 6).

The values can be received and observed on the terminal of the Raspberry (Fig. 7).

The system also displays the temperature of the biogas plant on 2X16-bit LCD screen in degrees as well as in Fahrenheit (Fig. 8).

The data can be sent to MQTT broker once the connection is established between the Raspberry Pi and the temperature sensor. For sending the values from sensor to the broker, we use Python language.

The following code helps in sending the data to the MQTT broker

Fig. 6 Connection of Raspberry Pi and DS18B20

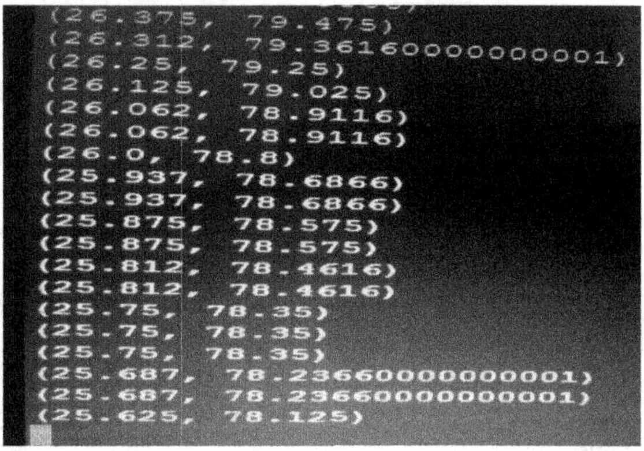

Fig. 7 Raspberry Pi terminal showing sensor values

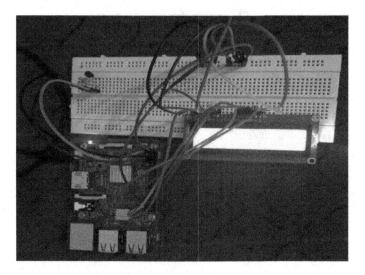

Fig. 8 Temperature values on LCD screen in degrees and Fahrenheit

```
import paho.mqtt.client as paho
import time client = paho.Client()
client.on_publish = on_publish
client.connect("192.168.225.105", 1883)
client.loop_start()
while True:
    temperature = read_temperature()
    (rc, mid) = client.publish
    ("test/temp", str(temperature), qos = 1)
    time.sleep(30)
```

8 Conclusion and Future Scope

This study proposes a proper monitoring system for analyzing real-time temperature values of the biogas plant which helps in maintaining the temperature of the plant according to our needs and getting desired and ideal output from the plant. The system consists of a web interface, which helps users in monitoring the system remotely over the Internet and taking decisions accordingly. The system consists of an end user android application, which allows user to monitor the temperature, and also offers users with options for monitoring the temperature of the biogas plant according to time and date which helps user in building a study or a pattern of the plant's temperature over the time.

The system also facilitates with future advancements in the proposed system to control the temperature of the biogas plant externally using a number of different ways.

The ways by which the temperature can be controlled are as follows:

(1) Energy from cooling water and also exhaust gas can be used to control the temperature of the biogas plant according to our needs.

(2) Insulations, heating elements, heat exchanges, steam injections and water baths, etc., can be used to control the temperature of the biogas plant according to our requirements.

References

1. Mohammad Shariful Islam, Asif Islam, Md. Zakirul Islam "Variation of Biogas Production with Different Factors in Poultry Farms of Bangladesh".
2. Aleksandar Antonić, Martina Marjanović, Pavle Skočir and Ivana Podnar Žarko "Comparison of the CUPUS Middleware and MQTT Protocol for Smart City Services".
3. Mohsen Hallaj Asghar, Nasibeh Mohammadzadeh "Design and Simulation of Energy Efficiency in Node Based on MQTT Protocol in Internet of Things".
4. Krešimir Grgić, Ivan Špeh, Ivan Heđi "A Web-Based IOT Solution for Monitoring Data Using MQTT Protocol".
5. F. Xia, L.T. Yang, L. Wang, and A. Vinel, "Internet of things," International Journal of Communication Systems, vol. 25, pp. 11011102, 2012.
6. S. Charmonman, P. Mongkhonvanit, V. Dieu, and N. Linden "Applications of internet of things in e-learning," International Journal of the Computer, the Internet and Management, vol. 23, no. 3, pp. 1–4, September December 2015.
7. D. Evans, "The internet of things – how the next evolution of the internet is changing everything," white paper, April 2011.
8. L. Dürkop, B. Czybik, and J. Jasperneite, "Performance evaluation of M2 M protocols over cellular networks in a lab environment," Intelligence in Next Generation Networks (ICIN), February 2015, pp. 70–75.
9. HiveMQ Enterprise MQTT Broker. (2016, Feb. 8). [Online]. Available: http://www.hivemq.com/.
10. Scalagent, "JoramMQ, a distributed MQTT broker for the internet of things," white paper, September 2014.
11. Eclipse Mosquitto. (2016, Feb. 12). [Online]. Available: http://mosquitto.org/.
12. Eclipse Paho JavaScript Client. (2016, Feb. 12). [Online]. Available: https://eclipse.org/paho/clients/js/.
13. Maxim Integrated. (2016, Feb. 14). [Online]. Available: www.maximintegrated.com.
14. Smart home implementation based on internet and Wi-fi technology, "YAN Wenbo, WANG Quanyu, GAO Zhenwei Control Conference (CCC), 2015 34th Chinese.
15. MQTT for Sensor Networks (MQTT-SN) Protocol Specification, Andy Stanford-Clark and Hong Linh Truong, Andy Stanford-Clark and Hong Linh Truong (andysc@uk.ibm.com, hlt@zurich.ibm.com) November 14, 2013.
16. MQTT-S A publish/subscribe protocol for Wireless Sensor Networks, Hunkeler, U.; IBM Zurich Res. Lab., Zurich; Hong Linh Truong; Stanford-Clark, A, Communication Systems Software and Middleware and Workshops, 2008. COMSWARE 2008. 3rd International Conference.
17. Raspberry Pi as a Wireless Sensor Node: Performances and Constraints, Vladimir Vujovic and Mirjana Maksimovic, Faculty of Electrical Engineering, East Sarajevo, Bosnia and Herzegovina, IEEE 2014.

Time-Efficient and Attack-Resistant Authentication Schemes in VANET

Sachin Godse and Parikshit Mahalle

Abstract VANET (Vehicular Ad hoc Network) brings evolution in the transportation system. It is vulnerable to different kind of attacks. Authentication is the first line of defense against security problems. The previous researcher comes with some cryptographic, trust-based, ID-based, and signature-based authentication schemes. If we consider the performance of previous schemes, process time and efficiency need to be improved. Faster authentication can help to establish communication in a short time and RSU can serve more number of vehicles in a shorter time span. We presented AECC (Adaptive Elliptic Curve Cryptography) and EECC (Enhanced Elliptic Curve Cryptography) schemes to improve the speed and security of authentication. In AECC key size is adaptive, i.e., different sizes of keys are generated during the key generation phase. Three ranges are specified for key sizes small, large, and medium. In EECC, we added an extra parameter during transmission of information from the vehicle to RSU (Road Side Unit) for key generation. This additional parameter gives the information about vehicle ID and location of the vehicle to RSU and other vehicles.

Keywords VANET · AECC · EECC · Authentication · ITS · OBU

1 Introduction

Intelligent transportation system (ITS) is an important application of the vehicular ad hoc network. ITS helps to automate traffic monitoring and control system. In VANET, nodes are vehicles with each vehicle having OBU (On Board Unit) deployed on it. RSUs are deployed at fixed distances on the roadside. Figure 1

S. Godse (✉)
Department of Computer Engineering, SAOE, SPPU, Pune, India
e-mail: sachin.gds@gmail.com

P. Mahalle
Department of Computer Engineering, SKNCOE, SPPU, Pune, India
e-mail: parikshitmahalle@yahoo.com

© Springer Nature Singapore Pte Ltd. 2018
S. Bhalla et al. (eds.), *Intelligent Computing and Information and Communication*,
Advances in Intelligent Systems and Computing 673,
https://doi.org/10.1007/978-981-10-7245-1_57

shows the different components of VANET and communications in it. In VANET, nodes can communicate in three different ways: V2V (Vehicle to Vehicle), I2V (Infrastructure to Vehicle), and V2I (Vehicle to Infrastructure) communication. Due to open medium of VANET, security is a major issue with it. Malicious vehicle can enter into the network and hamper the performance of the network. Authentication is the need of vehicular network to avoid unauthorized vehicles and unauthorized messages. Previous research gives some solutions like ID-based, trust-based, cryptography-based, and signature-based authentication mechanisms, but overhead is a major issue in these solutions [1, 2]. The time required for authentication process is another important issue in VANET as nodes are moving fast compared to other ad hoc networks. The time span between nearest RSU and vehicles is very low. The presented schemes, improving the speed of authentication and improve the efficiency of RSU to serve more number of vehicles. These schemes are designed based on ECC. In Sect. 2, a survey of existing authentication schemes is discussed. Mathematical analysis of the time required and vehicle served by RSU with respective key sizes is explained in Sect. 3. Section 4 presents our authentication schemes AECC and EECC.

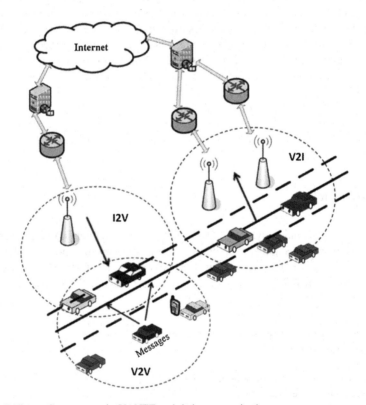

Fig. 1 Different Components in VANET and their communication

2 Existing Authentication Schemes in VANET

1. **PKI (Public-Key Infrastructure)**: It is used to verify the user. PKI is based on asymmetric key cryptography. It uses two keys: public and private key. CA issues certificates to the vehicles to decide the validity of the users. A vehicle manufacturer or government authority can act as a CA. Certificate revocation takes place at predefined time intervals, through this certificate revocation a list is generated.

2. **TESLA (Timed-Efficient Stream Loss-Tolerant Authentication)**: TESLA provides authentication for multicast and broadcast network communications. It uses symmetric key cryptography. The main flaw of TESLA is that it susceptible to attacks that happen due to memory-based denial of service.

3. **TESLA ++**: It is an advanced version of TESLA, which overcomes the drawbacks of TESLA.

4. **ECDSA (Elliptic Curve Digital Signature Algorithm)**: It includes three phases: key generation, signature generation, and signature verification. In ECDSA, a scalar multiplication of a given random point is used in signature generation and verification.

Table 1 summarizes existing authentication schemes and their merits and demerits.

3 Experimental Analysis of Authentication in VANET

There are two types of authentication in VANET: vehicle authentication and message authentication. The vehicle authentication verifies the identity of the vehicle. Vehicle authentication is further classified into 5 types: Node-level authentication, Group-level authentication, Unicast authentication, Multicast authentication, and Broadcast authentication [3]. Here, we studied and analyzed RSA- and ECC-based authentication scheme on the simulator.

3.1 Mathematical Analysis of Time Required for Authentication and Vehicles Serves by the RSU

We have given an analysis of authentication time required for RSA (Rivest–Shamir–Adleman) and ECC (Elliptic Curve Cryptography) algorithm. Consider the communication range of RSU is 10 km, i.e., 10,000 m. RSU requires 100 ms for authenticating one vehicle. The radius of RSU is 10 km. Figure 2 shows RSU communication range and vehicle "A" traveling from point S to D.

Table 1 Existing authentication scheme survey

References		A: Authentication Scheme; M: Merit; D: Demerit
[4]	A	A Hierarchical Privacy Preserving Pseudonymous Authentication Protocol for VANET
	M	1. No need of storage for storing a large pool of pseudonyms 2. Not need of Certificate Revocation List (CRL) 3. Valuable information is secured as compared to server
	D	1. If CA, RA, or RSU compromise scheme can fail
[5]	A	Security Enhancement in Group-Based Authentication for VANET
	M	1. A group-based V2 V communication framework is designed to secure VANET and preserve privacy 2. Eliminates the need of sign message in V2 V communication which leads to faster authentication
	D	1. Digital signature generation and verification of message in V2 V communication requires more time, which degrades performance of the network
[6]	A	Vehicular Authentication Security Scheme (VASS)
	M	1. The computational effort is much lower than other methods in the hash function 2. VASS provide security features such as privacy, authentication, and Sybil attack detection
	D	1. Vehicle to infrastructure communication not considered
[7]	A	Secure and distributed certification system architecture for safety message authentication in VANET
	M	1. False public-key certification is avoided 2. Provide secure and distributed certification system
	D	1. The storage required more for each vehicle to maintains a private key, a public key, an implicit certificate, a short-term key pair, and a public-key certificate delivered by the RCA 2. The high transmission range required to transmit various safety messages
[8]	A	A secure and efficient V2 V authentication method in heavy traffic environment
	M	1. Improve the speed of message processing by sending a low data volume in areas of heavy traffic 2. Detect replay attacks by checking timestamps of messages
	D	1. Vehicle to infrastructure communication not considered
[9]	A	PKI (Public-Key Infrastructure)
	M	1. The user ensures the integrity of the message by signing the encoded message using digital sign 2. Certificate authorities (CA) digitally sign the data and binds the public keys with private keys
	D	1. Cryptographic keys get compromised 2. Slower because of complexity
[10]	A	TESLA (Timed-Efficient Stream Loss-Tolerant Authentication)
	M	1. Lightweight broadcast authentication protocol 2. It uses symmetric key cryptography so it is faster
	D	1. TESLA is vulnerable to memory-based denial-of-service attack

(continued)

Table 1 (continued)

References		A: Authentication Scheme; M: Merit; D: Demerit
[11]	A	TESLA ++
	M	1. Protect against memory-based denial-of-service (DoS) attacks 2. Reduces the memory requirements at the receiver's end 3. Protect against computation-based denial-of-service (DoS) attacks with equal priority 4. Offers a more secure user authentication mechanism than TESLA
	D	1. In the lossy networks, performance of TESLA ++ gets degraded 2. TESLA ++ does not offer non-repudiation
[12]	A	ECDSA (Elliptic Curve Digital Signature Algorithm)
	M	1. It is mathematically derived form of Digital Signature Algorithm (DSA) 2. As an elliptic curve is used for key generations, strength per key bit is significantly greater
	D	1. Attacks on Elliptic Curve Discrete Logarithmic Problem (ECDLP) 2. Hash function can be hacked through attack

A literature survey of the existing authentication scheme shows that speed of authentication and security of authentication need to improve. Our research work is focused on how to improve speed of authentication and how to maintain security while authentication

Fig. 2 RSU Communication range

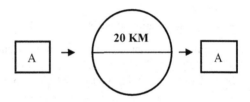

- Vehicle *A* required 20 s for traveling from *S* to *D*.
- Number of vehicles authenticated by RSU are 20,000 ms/100 ms = 200 vehicles.
- 200 vehicles serves by RSU in 20 s.

3.2 Analysis of Time Required for Authentication and Vehicles Serves by RSU with Respective Key Size (RSA Algorithm)

Table 2 shows the time required for authentication with respective key sizes. A number of vehicles served by RSU is calculated as follows:

- RSA with key size 512 bits required 461 ms for authentication.
- Number of vehicles authenticated by RSU are 20,000 ms/461 ms = 43 vehicles.
- 43 vehicles serves by RSU in 20 s.

Table 2 Vehicle served by RSU in 20 s for the RSA algorithm

Sr. no.	Key size	Authentication time (ms)	Vehicles serve by RSU in 20 s
1	256	64	312
2	512	128	156
3	1024	256	78
4	768	192	104
5	1536	384	52
6	2048	512	39

3.3 Analysis of Time Required for Authentication and Vehicles Serves by RSU with Respective Key Size (ECC Algorithm)

Table 3 shows the time required for authentication with respective key sizes. A number of vehicles served by RSU is calculated as follows:

- ECC with key size 112 bits require 301 ms for authentication.
- A number of vehicles authenticate by RSU are 20,000 ms/301 ms = 66 vehicles.
- 66 vehicles are served by RSU in 20 s.

Figure 3 summarizes time required for authentication with respective key sizes in RSA and ECC. We can see that the time required for authentication is less using ECC. Because of this, we took ECC for our authentication schemes.

4 Presented Authentication Schemes for VANET

The time required for authentication is an important factor for authentication. There is a need for efficient authentication protocol for VANET, which not only provides security but also take less time. Here, we proposed time-efficient authentication system based on ECC. Along with authentication, it provides the node location verification, replica attack, and Sybil attack detection. Figure 4 shows the

Table 3 Vehicle served by RSU in 20 s for ECC algorithm

Sr. no.	Key size	Authentication time (ms)	Vehicles serve by RSU in 20 s
1	40	12	1666
2	80	24	833
3	160	48	416
4	120	36	555
5	240	72	277
6	320	96	208

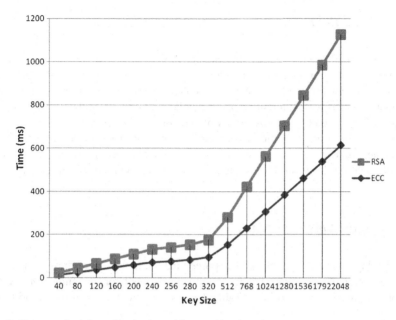

Fig. 3 Time required for authentication with respective key sizes

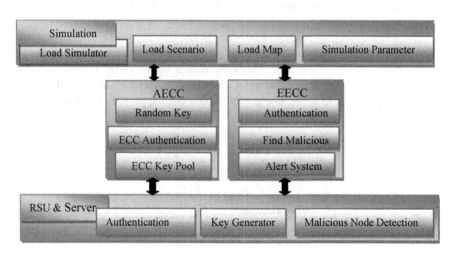

Fig. 4 Architecture for time-efficient attack-resistant authentication protocol

architecture of the proposed authentication protocol. The architecture shows three layers, simulation (front end), algorithms (processing part), and Server/RSU-side activities.

4.1 Simulation

We are using VSIM (VANET Simulator) for our protocol implementation and testing. VSIM is a Java-based simulator, which provides different classes to create an environment for VANET. We can implement our protocols and ideas in the simulator using Java. Different maps are available to test protocols, we can upload a map. Different road scenarios are available for each map; we can upload those scenarios in the simulator after uploading the map. The simulator has different input parameters like a number of vehicles, time stamp of each vehicle, road traffic density, event priority, time slot, etc., that we can provide to the simulator.

4.2 Algorithms/Processing

Middle layer in architecture shows processing part, i.e., algorithms which give authentication. Two types of algorithms are presented for authentication, AECC algorithm and EECC algorithm. Table 4 shows notations used in the algorithms.

4.2.1 AECC (Adaptive Elliptic Curve Cryptography)-Based Authentication

This algorithm uses random key size where no attacker can guess the key size at the current time and tries to break it. This system uses a cooperative system to decide the key size after every defined timeslot. When an attacker tries to guess the key to break the system as ECC is strong enough this not happens easily. But when an

Table 4 Terms/Notations Used in Algorithm

Terms/notations	Meaning
P	Key pool
Ts	Time slot. (Re-generate keys after every Ts seconds)
G	Key Generator
m, a, b	Unique parameters
K	Keys
Pu	Public key
Pr	Private key
Vc	Current vehicles
NR	Neighbor RSU
Ks	Key size
Kx	New Key
Us	Public-key server
Re	Verify—Sybil attack, replica attack

attacker succeeds to do so, because of our adaptive key size (AKS) algorithm that the key is no longer relevant to that attacker.

Algorithm/Pseudocode for AECC-based authentication:
Input: G, {Ts}, {Ks, P}, {V}
Output: Random_Keys, Access Granted/Rejected
Algorithm:

1. Sync {V,RSU,S} → Ts-Time Slot
2. Ser-Generated TimeSlots {Ts} and KeySizePool {Ks,P}
3. Generate ECC initial parameters G,PW
4. SessionKeyDistribution {Rc, Rs}

    a. Generate Random variable rA
    b. Compute Ra and Wa
    c. Get Ks → {Ks,P}
    d. Generate K → Ks size Client Side
    e. Generate K → Ks at Server Side

5. Session Key verify – H{K}

    a. Generate Hash{P}
    b. Verify

6. Session Granted/Rejected

End

4.2.2   EECC (Enhanced Elliptic Curve Cryptography)-Based Authentication

In enhanced ECC algorithm, we added an extra parameter during transmission of information from the vehicle to the RSU for key generation. This additional parameter gives the information about vehicle ID and location of the vehicle to RSU and other vehicles. This additional parameter also used in key generation. This algorithm provides replica and Sybil attack detection along with authentication.

Algorithm/Pseudo code for EECC-Based Authentication:
Input: G, {V}, {Ts}, {Ks, P}
Output: Detect Attack, Access Granted/Rejected
Algorithm:

1. Generate ECC initial parameters G,PW
2. SessionKeyDistribution {Rc, Rs}

    a. Generate Random variable rA
    b. Compute Ra and Wa

c. Get Ks → {Ks,P}
d. {ID, K,L,TS} → RSU
e. Verify V by RSU
f. If Verifed

   i. Generate K → Ks size Client Side
   ii. Generate K → Ks at Server Side

g. End IF
h. Else

   i. Start Re_verify

      1. Vehicle shares new {id, TS, L}
      2. Verify by RSU and Server

   ii. End

3. Session Key verify – H{K}

a. Generate Hash{P}
b. Verify

4. Session Granted/Rejected

End

4.3 RSU and Server

Key generation and authentication take place at server side and malicious node detection is take place at RSU.

5 Conclusion

In VANET, nodes are highly mobile and there is need to improve the speed of authentication mechanism. A cryptographic solution provided by the previous scheme need to enhance as per VANET requirement. Considering this we proposed a new effective protocol for VANET authentication. An ECC provides faster authentication than RSA, Diffie–Helman and other traditional cryptographic solutions with same key sizes. We proposed a modification in existing ECC algorithm to improve performances of VANET. The adaptive key size algorithm makes the system more secure against password cracking attack. Enhanced ECC algorithm provides security against Sybil attack and replica attack.

References

1. Saira Gillani, Farrukh Shahzad, Amir Qayyum et. al.: A Survey on Security in Vehicular Ad HocNetworks. In: *Springer, LNCS 7865*, pp. 59–74, 2013.
2. Jonathan Petit, Zoubir Mammeri.: Authentication and consensus overhead in vehicular ad hoc networks. In: *Springer Science*, Telecommun Syst (2013).
3. Khalid Haseeb, Dr. Muhammad Arshad, et. al.: A Survey of VANET's Authentication. In: *PG NET 2010. Liverpool. John Moores University UK*, June 2010.
4. Ubaidullah Rajput, Fizza Abbas, Heekuck Oh.: A Hierarchical Privacy Preserving Pseudonymous Authentication Protocol for VANET. In: IEEE Access, October 25, 2016.
5. Rajkumar Waghmode, Rupali Gonsalves et.al.: Security Enhancement in Group Based Authentication for VANET. In: IEEE International Conference on Recent Trends in Electronics Information Communication Technology, May 20–21, 2016, India.
6. Yongchan Kim, Jongkun Lee.: A secure analysis of vehicular authentication security scheme of RSUs in VANET. In: Springer-Verlag France 2016.22.
7. Tiziri Oulhaci, Mawloud Omar, Fatiha Harzine, et. al.: Secure and distributed certification system architecture for safety message authentication in VANET. In: Springer Science + Business Media New York 2016.
8. Myoung-Seok Han, Sang Jun Lee,Woo-Sik Bae.: A Secure and Efficient V2 V Authentication Method in Heavy Traffic Environment. In: Springer Science + Business Media New York 2016.
9. A. Rao, A. Sangwan, A. Kherani, A. Varghese, B. Bellur, and R. Shorey.: Secure V2 V Communication With Certificate Revocations. In: proceedings of the *IEEE Infocom 2007.*
10. Yih-Chun Hu and Kenneth P. Laberteaux.: Strong VANET security on a budget. In: Proceedings of the 4th Annual Conference on Embedded Security in Cars *(ESCAR 2006).*
11. Arzoo Dahiya, Mr. Vaibhav Sharma,: A survey on securing user authentication in vehicular adhoc networks. In: *6th EmbeddedSecurity in Cars Conference.*
12. Don Johnson, Alfred Menezes and Scott Vanstone.: The Elliptic Curve Digital Signature Algorithm (ECDSA). In: *Published in IJIS, Vol. 1 (2001)* pp. 36–63.

Inferring User Emotions from Keyboard and Mouse

Taranpreet Singh Saini and Mangesh Bedekar

Abstract This chapter emphasizes on retrieving user emotions from keyboard and mouse using different parameters. These parameters can be user keyboard typing style, mouse movements, and some physiological sensors are used. This field of retrieving emotions from machines comes under the field of affective computing.

Keywords Affective computing · Emotions · Keyboard · Mouse
Physiological sensors

1 Introduction

Emotions are the form of human feelings and sentiments that results in physical and psychological changes that affect human behaviors. The physiology of emotion is closely linked to arousal of the human nervous system with various emotional states and power of arousal relating, apparently, to particular emotions. Emotion is also linked to the behavioral tendency which changes over time. Emotions involve different components, such as subjective experience, cognitive processes, expressive behavior, psychophysiological changes, and instrumental behavior. The ability to recognize, interpret, and express emotions plays a key role in human communication and increasingly in HCI. Recent research could demonstrate that humans have an inherent tendency to interact with computers in a natural and humane way, mirroring interactions between humans in social situations. Computer systems currently are not able to recognize or respond to these complexities of a natural social interaction. Emotion recognition has become an important subject when it comes to human–machine interaction. Various methods have been used in the past to detect and evaluate human emotions. The most commonly used techniques

T. S. Saini (✉) · M. Bedekar
Department of Computer Engineering, Maharashtra Institute of Technology, Pune, India
e-mail: saini_13sf@yahoo.co.in

M. Bedekar
e-mail: mangesh.bedekar@mitpune.edu.in

© Springer Nature Singapore Pte Ltd. 2018
S. Bhalla et al. (eds.), *Intelligent Computing and Information and Communication*,
Advances in Intelligent Systems and Computing 673,
https://doi.org/10.1007/978-981-10-7245-1_58

include the use of textual information, facial expressions, speech, body gestures, and physiological signals. These signals are obtained from a skin temperature sensor, a heart rate sensor, and a force sensor.

The study of emotions from machines has developed into a new field known as affective computing. Affective computing is the process to learn, observe, and develop systems and applications that can sense, interpret, procedure, and pretend human-like effects. It is a field of work spanning computer science, psychology, and cognitive science. While the origins of the topic may be traced to far back in history as with humans early philosophical reasoning into emotion, the more quantifiable recent work based on computer science got shape with the path-breaking study of R. W. Picard's 1995 paper on affective computing.

2 Literature Survey

2.1 Affective Haptics

Affective haptics is the newest emerging area of study, which emphases on the study and plan of devices and structures that can incite, improve, or affect the emotional state of a human by means of intelligence of touch. The study area is highlighted in the research paper by Dzmitry Tsetserukou et al. Focused by the inspiration to enhance social communication and expressively immense understanding of users' real-time messaging, virtual, augmented realities, the idea of strengthening (intensifying) own moods, and replicating (simulating) the emotions felt by the communicating partner was scheduled [1].

Four simple haptic (tactile) networks leading our emotions can be demonstrated,

(1) Functional changes
(2) Bodily stimulation
(3) Social touch
(4) Emotional haptic design.

2.1.1 Applications of Affective Haptics

- Treating nervousness
- Monitoring and adjusting moods on the basis of physiological signals,
- Affective and combined games,
- Testing psychology,
- Communication systems for kids with autism (Fig. 1).

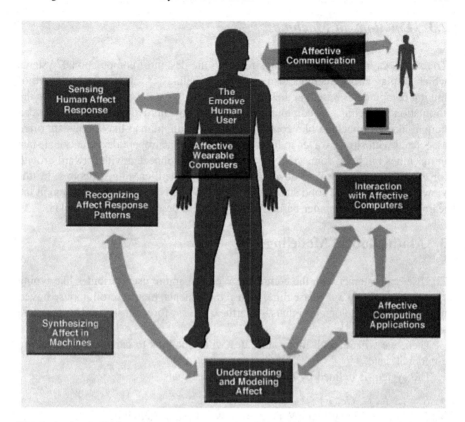

Fig. 1 Areas of affective computing

2.2 Detecting and Recognizing Emotional Information

Sensing strong feelings begins with information gathering from action-less sensors, which take in facts about the user's physical state or behavior. The facts gathered are of the form of human cues which are used to perceive feelings. For example, a camera might click photographs of the human face; recognize body position thus illuminating human feeling, while a microphone may transform human sound to electric signals based on verbal words. Many other explicit sensors can be used to determine specific feelings by directly calculating physiological facts, such as the human's skin, body temperature, etc. [1].

This information is treated to extract decisive information from the gathered facts. This is finalized using machine-learning methods that process unalike modalities, such as usage of words, natural language processing, or of the face words discovery, and create either tickets giving name-value pairs or orders in a valence-arousal space [1].

2.3 Emotions in Machines

The new area in computing feelings refers to the design of computational systems which display strong feelings or a part of the feelings. A more valuable move of linking computers is based on the current technology based on powers, which is the simulation of feelings in communicating agents to improve interactivity between human-like and machine-like communications. While human-like moods are often associated with strong waves in hormones and other neuropeptides, emotional state in machines might be connected with outline states connected with forward growth (or exist without of forward development) in self-ruled learning structures. In this viewpoint, feelings are solid states which are time derivatives (perturbations) in the learning turn of a computer-based learning system based on instructions [1].

3 Mathematical Modeling

The user can interact with the computer. We can capture user attributes like typing speed, key press—key release time, mouse movements, mouse scroll, mouse hover, mouse click, physiological signals—heartbeat sensor, force sensor, and temperature sensor.

Then, we will map all these attributes with emotions to capture user's current emotional state.

The system is defined by the tuple, "S", such that

- $S = \{I, Ic, O, Oc, F\}$
- I = Input to the system (Web pages = W and User Activities = U),
- Ic = Input Conditions (Mouse behaviors = MB and Keyboard behaviors = KB, time spent on web page, sensors),
- O = Output of the system (User Profile = $I + Ic + F$)
- Oc = Output Conditions (as behavior changes, user profile also changes),

  - F = Functions I to O, through Ic to Oc.
    (Table 1).

There can be four conditions,

- First condition where mouse, keyboard, and sensors work together.
- Second condition, only mouse will work.
- Third condition will be for keyboard only.
- Last condition where readings can be taken from sensors only.

4 Proposed Work

System flow describing the proposed work is as shown in Fig. 2.

Table 1 Mathematical modeling

V_M	Vertical mouse movement	H_{SM}	Mouse horizontal scroll
H_{OVM}	Mouse hover	B_K	Keyboard bookmark
H_{IM}	Mouse highlight	S_K	Keyboard save
H_M	Horizontal mouse movement	H_{SM}	Mouse horizontal scroll
B_M	Mouse bookmark	B_K	Keyboard bookmark
S_M	Mouse save	S_{LR}	Shift left/right
F/B	Forward and backward	C_A	Control all
S_E	Send email	C_K	Keyboard copy
P_M	Mouse print	P_{AK}	Keyboard paste
S_{TM}	Mouse select text	P_K	Keyboard print
C_M	Mouse copy	V_K	Keyboard vertical
P_{AM}	Mouse paste	H_K	Keyboard horizontal
V_{SM}	Mouse vertical scroll	T_K	Keyboard trace
S	Sensors	H_s	Heartbeat sensor
F_s	Force sensor	T_s	Temperature sensor

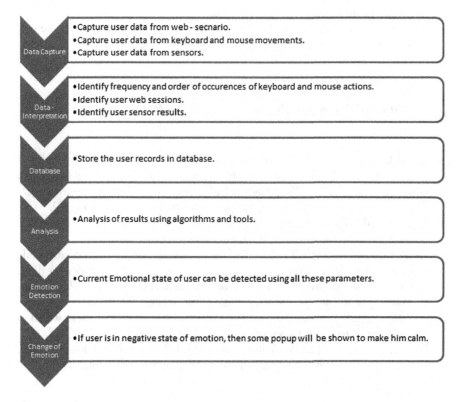

Fig. 2 System flow

5 Implementation

5.1 Detecting User Web Activities

- Time Spent on Web Page—User Engagement—Time was considered as one of the interest indicators. It consists of passive intervals and active intervals.
- User Click Behavior—Users' clicks on links is a factor for implicitly learning the user's interests.
- Copying Text from Web Page—The text copied indicates relevancy and that too in which parts of the web page the user is interested in.
- Mouse Movements on Web Page—Mouse movements can also determine the interest of the user.
- Amount and Speed of Scrolling on Web Page—several implicit measures related to the interests of the user and scrolling was one of the factors.
- Number of Time a Web Page is visited—greatest implicit indicators of interest were the time spent on the web page, number of link clicks, and arrow presses.
- User Behavior Based on Age and Gender—For understanding user's emotions, it is important to understand user's gender and age. It can be done by the process of gender identification and age detection.
- Most Number of Files Viewed—Types of files viewed (image, text) by the user also helps to understand the emotional state of the user.

5.2 Material Used for Physiological Signals

- ATMEGA 328—The ATMEGA 328 is a CMOS 8-bit microcontroller with low power based on the AVR gave greater value to RISC buildings and structure design. The ATMEGA 328 operates at 1 MIPS per MHz which allows the system designed to make the most out of the power consumed for processing [2].
- Bread Board—This component is used because it will be the base of our device with all the connections being connected with its help.
- Heart Beat Sensor TCRT1000—The use of this sensor simplifies the build process of the sensor part of the project as both the infrared light emitter diode and the detector are arranged side by side in a leaded package, thus blocking the surrounding ambient light, which could otherwise affect the sensor performance [3]. This sensor continuously monitors the pulse rate of the user, which is directly connected with the emotions.
- Temperature Sensor LM35—This component is selected because this is the only sensor, which is used to measure the body temperature accurately [4].

- Force Sensor MF01-N-221-A01—This sensor is mainly used on the keys, it can be a computer keyboard, mouse buttons, musical instruments like piano, harmonium, etc., or also in vehicles to detect force or the pressure [5].
- Bluetooth Module HC05—This Bluetooth module is used for transmitting data to the smartphone and to the computer also. This module directly sends data from the sensors and it is easily connected to any device [6]. We are using ArduDroid application on the android device, which directly gets data from this Bluetooth module.
(Fig. 3; Tables 2 and 3).

In Fig. 4, it is shown that human physiological signals from the heartbeat sensor, force sensor, and temperature sensor are the direct input into the microcontroller where all the processing takes place. The microcontroller used is the ATMEGA 328, where all the sensors are attached and LCD shows all the results. Heartbeat sensor continuously checks the pulse rate of the user. Force sensor senses the user pressure on the mouse button and keyboard keys. This will give result in the form of percentage value out of 100%. The temperature sensor senses the user temperature in the form of °C from users hand while working with a mouse. The microcontroller transmits data to Android and computer device through Bluetooth module which is wireless. Then, the data can be analyzed using different algorithms, techniques, and tools.

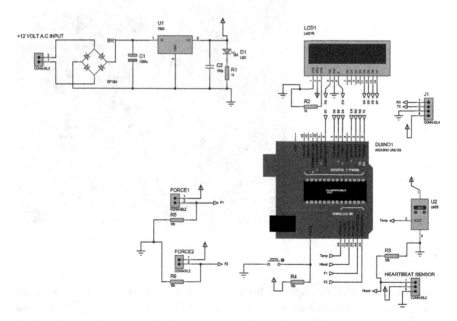

Fig. 3 Schematic diagram

Table 2 Heart rate analysis

Normal	Systolic rate is lesser than 120 mmHg and diastolic rate is lesser than 80 mmHg
At risk (pre-hypertension)	Systolic rate is higher than 120–139 mmHg and diastolic rate is higher than 80–89 mmHg
High	Systolic rate is 140 mmHg or higher diastolic rate is 90 mmHg or higher

Table 3 Temperature analysis

Normal	The average normal temperature is 98.6 °F (37 °C). But "normal" varies from person to person
Abnormal	Oral, temporal artery temperature: Fever: 100.4 °F (38 °C)–103.9 °F (39.9 °C), High fever: 104 °F (40 °C) and higher
	Armpit (auxiliary) temperature: Fever: 99.4 °F (37.4 °C)–102.9 °F (39.4 °C), High fever: 103 °F (39.5 °C) and higher
	A rectal or ear temperature of less than 97 °F (36.1 °C) means a low body

Fig. 4 Working of sensors

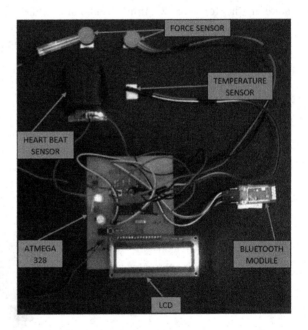

6 Results

Some of the results are captured from users using software and hardware. These results show the user data with respect to the website and their usage. The more the usage the more will be the engagement with the website. The heat maps show the user movements on the web page. This can help in capturing the user pattern (Figs. 5, 6, 7 and 8).

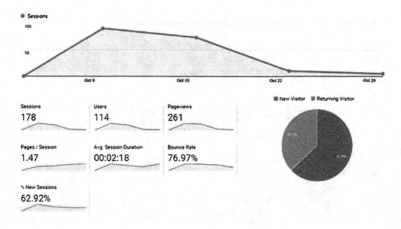

Fig. 5 User overview

Sessions
178
% of Total: 100.00% (178)

Pageviews
261
% of Total: 100.00% (261)

Session Duration	Sessions	Pageviews
0-10 seconds	138	139
11-30 seconds	2	4
31-60 seconds	2	16
61-180 seconds	4	9
181-600 seconds	16	44
601-1800 seconds	16	49

Fig. 6 User engagement data

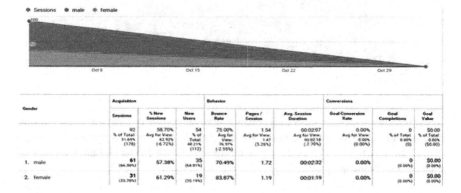

Gender	Acquisition			Behavior			Conversions		
	Sessions	% New Sessions	New Users	Bounce Rate	Pages / Session	Avg. Session Duration	Goal Conversion Rate	Goal Completions	Goal Value
	92 % of Total: 51.69% (178)	58.70% Avg for View: 62.92% (-6.72%)	54 % of Total: 48.21% (112)	75.00% Avg for View: 76.97% (-2.55%)	1.54 Avg for View: 1.47 (5.29%)	00:02:07 Avg for View: 00:02:18 (-7.70%)	0.00% Avg for View: 0.00% (0.00%)	0 % of Total: 0.00% (0)	$0.00 % of Total: 0.00% ($0.00)
1. male	61 (66.30%)	57.38%	35 (64.81%)	70.49%	1.72	00:02:32	0.00%	0 (0.00%)	$0.00 (0.00%)
2. female	31 (33.70%)	61.29%	19 (35.19%)	83.87%	1.19	00:01:19	0.00%	0 (0.00%)	$0.00 (0.00%)

Fig. 7 Gender data

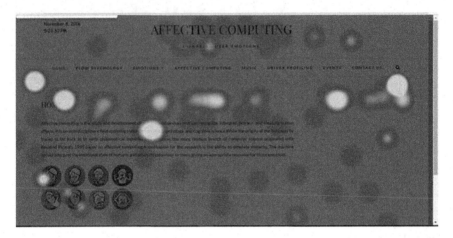

Fig. 8 Heat maps

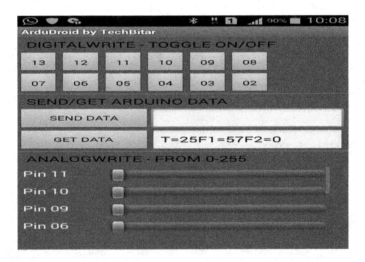

Fig. 9 Android results

Figure 9, shows the Android window where all results are shown in the get data column. These results are captured by the ArduDroid Android application in get data column, where T shows the temperature readings, F1 AND F2 are the readings for the force sensor.

7 Emotion Detection

Emotions play a very important role in problem solving, decision-making, communicating, and adapting to the unpredictable environment. These emotion detection methods have been used for improving the quality of human life. This model of emotion detection and recognition will help the users in early detection of any problem. The major agenda of this model is to recognize user emotions while working with keyboard and mouse. For the purpose of emotion detection, we will capture all the data from the sensors through Bluetooth module and map the values with different negative and positive emotions using the circumplex model.

8 Conclusion

The advantage of this system is that we are using keyboard and mouse, which are the natural type of biometrics, which does not require any special hardware. The sensors are small and are not very costly, so they can easily fit in the mouse and the keyboard for the continuous reading. These readings will help the computer users in evaluating and recognizing their emotions. This model can be used at any place where computers are used like universities, schools, offices, banks, hospitals, etc.

This system can also be used in the home for the old people, who are using different devices like television remote or computers, for continuously monitoring their health.

References

1. Taranpreet Singh Saini, Mangesh Bedekar, Saniya Zahoor. "Analysing human feelings by Affective Computing—survey", 2016 International Conference on Computing Communication Control and automation (ICCUBEA), 2016.
2. Russell, J. (1980). A circumplex model of affect. Journal of Personality and Social Psychology. http://doi.org/10.1037/h0077714.
3. Performance, L. Power, A. V. R. Microcontroller, A. R. Architecture, I. Programming, O. B. Program, S. P. W. M. Channels, O. A. Comparator, S. Sleep, M. Idle, and A. D. C. N. Reduction, "Features—32 × 8 General Purpose Working Registers—Optional Boot Code Section with Independent Lock Bits True Read-While-Write Operation—Programming Lock for Software Security—Internal Calibrated Oscillator—External and Internal Interrupt Sources and Extended Standby I/ O and Packages Operating Voltage: Temperature Range: Speed Grade: Microcontroller Bytes In-System Programmable Flash ATmega88PA ATmega168PA ATmega328P.".
4. "Heartbeat sensor information TCRT 1000.pdf".
5. Description, "Precision Centigrade Temperature Sensors LM35 LM35A LM35C LM35CA LM35D Precision Centigrade Temperature Sensors," no. December 1994, 1995.
6. Information, "μ A7800 Series Positive-Voltage Regulators Internal Thermal-Overload Protection μ A7800 SERIES," no. May 1976, 2003.

An Overview of Automatic Speaker Verification System

Ravika Naika

Abstract Biometrics is used as a form of identification in many access control systems. Some of them are fingerprint, iris, face, speech, and retina. Speech biometrics is used for speaker verification. Speech is the most convenient way to communicate with person and machine, so it plays a vital role in signal processing. Automatic speaker verification is the authentication of individuals by doing analysis on speech utterances. Speaker verification falls into pattern matching problem. Many technologies are used for processing and storing voice prints. Some of them are Frequency Estimation, Hidden Markov Models, Gaussian Mixture Models, Neural Networks, Vector Quantization, and Decision Trees. Mainly speaker verification depends upon speaker modeling and this paper represents a brief overview of the speaker verification system with feature extraction and speaker modeling. Bob spear toolkit is used for evaluation and experiment for the result and analysis. Bob spear is an open-source toolkit for speech processing. For evaluation purpose, three algorithms are proposed which are GMM, ISV, and JFA with the same preprocessing and feature extraction techniques.

Keywords ASV · MFCC · LPCC · LLR · PLDA · GMM · UBM
SVM

1 Introduction

Various methods and technologies are used for biometric authentication which include face, fingerprint, iris, palm print, retina, hand geometry, signature, and voice. Face recognition works with live capturing or digital image data. Fingerprint recognition works with the printed pattern of particular person's fingerprint. Likewise, speaker verification works on the basis of voice. As our primary need for

R. Naika (✉)
Institute of Technology, Computer Science and Engineering, Nirma University, S.G.
Highway, Ahmedabad 382481, India
e-mail: ravika.naika94@gmail.com

© Springer Nature Singapore Pte Ltd. 2018
S. Bhalla et al. (eds.), *Intelligent Computing and Information and Communication*,
Advances in Intelligent Systems and Computing 673,
https://doi.org/10.1007/978-981-10-7245-1_59

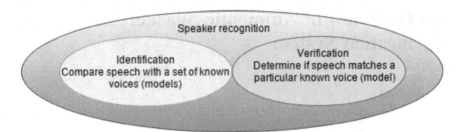

Fig. 1 Classification of speaker recognition [2]

communication is speech, we can say it is an appealing modality which includes individual differences in both physiological and behavioral manner [1]. For example, we can take vocal tract shape and intonation for use in automatic speaker verification.

The automatic speaker verification system is built on the bases of speech science, speaker characterization, and pattern recognition algorithm. It checks whether the person is who he/she claims to be. Speaker verification is the subpart of speaker recognition [2].

There are two parts of speaker recognition (1) speaker identification (2) speaker verification as described in Fig. 1.

Speaker verification is related to the physical shape of individual's vocal track with airways and soft tissues from where the sound is originated. All these components work in combination which includes physical movement of the jaw, larynx, and nasal passages. Speech acoustic patterns come from airways' physical characteristics and the motion of mouth and pronunciations. They are used as behavioral components of biometrics [3].

Several commercial applications are using speaker verification as a biometric for controlling or accessing important information or services and accounts. Speaker verification offers the replacement of PINs and passwords with something that cannot be stolen or forgotten.

2 Automatic Speaker Verification

Speaker verification is the task of determining whether a person is who he/she claims to be. Automatic speaker verification works on speech signals and the information which it conveys to the listener on different levels. First level conveys the message and the other levels convey the emotion, gender, and identity. This verification task is based on closed set and open set. In closed set, the input is matched with the stored samples and the closest to the input is returned, there is no rejection in closed set [4].

ASV can be classified in text-dependent and text-independent. In text-dependent, the speaker has to speak the predefined text while in text-independent there is no prior knowledge of any text that is why the text-independent is more difficult yet more flexible. Text-dependent mostly used with authentication systems while text-independent is more suitable for surveillance applications. Text-dependent technology is nowadays the most commercially viable and useful technology. Although the research has been done on both the task. The basic model for automatic speaker verification system is given here. As mentioned before, the result is given based on the accept or reject decision and its similarity is compared with a threshold. Here, the threshold is taken as likelihood ratio [5].

Figure 2 describes how automatic speaker verification works. First, input speech is given and then input speech is compared with the reference model and then it calculates the similarity score. Based on the threshold value, the decision of acceptance or rejection is taken. This is the basic flow of speaker verification system.

Mostly speaker verification is based on feature extraction and speaker modeling. Speaker verification is the pattern matching problem so different technologies like GMM (Gaussian Mixture Model), HMM (Hidden Markov Model), SVM (Support Vector Machine), VQ (Vector Quantization), ANN (Artificial Neural Network), JFA (Joint Factor Analysis), ISV (Inter-session Variability), etc., are used for performing the speaker verification. Different features like MFCC (Mel-Frequency Cepstral Coefficients), LPCC (Linear Predictive Cepstral Coefficients), PLP (Perceptual Linear Predictive), etc., and their fusions are used for better performance of the system.

Research and development techniques in this area have been undertaken for almost 3–4 decades, so it has been an active area for research. Different technologies and methods had been used for speaker verification. Its year wise representation is shown in the Fig. 3 [6–11]. From starting with pattern recognition technology to using GMM, HMM, and SVM.

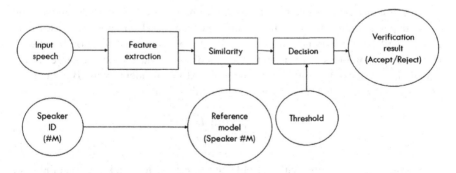

Fig. 2 Model of automatic speaker verification system [4]

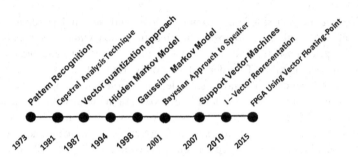

Fig. 3 Evaluation in speaker verification technologies [6–11]

2.1   Feature Extraction

Speech signal contains timber of voice, prosody, and content of language. Classification of speaker individuality can be done in three parts short-term spectral, prosodic, and high-level idiolectal features. MFCC, LPCC, and PLP are all spectral features.

Prosodic features are extracted from longer duration of time. Pitch, rhythm, energy, and duration can be included in prosodic features. However, these features are sparse in nature so it takes a large amount of training data to extract the features, also they are unreliable in a noisy environment.

The third one is high-level idiolectal features and they are extracted from lexicons, which represents speaker behavior or lexical cues [12]. They are less sensitive to the noise and channel effects. They require more complex front ends to extract the features.

2.2   Speaker Modeling and Classifiers

As mentioned above speaker verification is characterized in text-dependent and text-independent and both are using different datasets. Text-dependent uses relatively small size of data in comparison of text-independent verification because in text-independent the speech is not predefined so the data becomes large. Many classifiers are used for text-dependent and text-independent both. But in many classifier theoretical frameworks uses log-likelihood ratio [13].

$$1 = \log \frac{p(X|H_0)}{p(X|H_1)} \tag{1}$$

Its comparison is done with pre-obtained threshold to check the hypothesis of same speaker H_0 or different speaker H_1 where $p(X|H_0)$ and $p(X|H_1)$ are the likelihood for each. We can take $X = x_1, x_2 \ldots,$ and $p(X|H_1)$ is used to know the common effects which are not connected with speaker identity. In most approaches, X are considered as MFCCs and with that different classifier like GMM, PLDA, etc., is used.

2.3 System Fusion

It is a recent trend in ASV community. One of the examples of this fusion system we discussed in text-independent classifiers which is the combination of GMM, SVM, and UBM (Universal Background Model). It gives better result in comparison of single classifiers. Because it captures different aspects of speech together, this is not possible with the single classifier. Fusion system provides a platform for research on large-scale collaboration. Using the fusion of different classifier different feature extraction can be done together. But it uses the weighted summation of base classifiers and these weights are optimized.

3 Experiment Information

The experiment is done on bob spear toolkit. It is an open-source tool for signal processing and the machine learning algorithms. In the recent past few years, researchers are using this tool, so it has been a very popular tool in signal processing. It implements the whole chain from feature extraction to evaluation. It includes several techniques like GMM, SVM, JFA, i-vector with different databases like MOBIO, NISTSRE, VoxForge [14], etc. It also provides the system fusion for more accuracy in result and analysis. This tool provides the flexibility of using any data sets.

Speaker verification includes three things which are speech activity detection, feature extraction, and background data modeling with score computation. In order to do this, datasets are divided into background training set, DEV (Development) set, EVAL (Evaluation) set so the computation can be done easily [15]. Bob spear uses different techniques at each stage which are explained below.

Preprocessing
Mostly speech activity detection can be done in two ways. The first one is unsupervised and it computes energy values frame level wise. The other one computes the values based on the combination of energy and modulation 4 Hz. In both the techniques, a class with higher mean are considered as speech class.

Feature Extraction
This tool provides an efficient way to implement spectral and cepstral features. According to that, it computes its MFCC values. Cepstral mean and normalization of variance is also integrated. This tool provides different features for different datasets. For example, MFCC, LPCC, PLP are used with different number of Gaussians like MFCC60, MFCC20.

Modeling, Enrollment, and Score Computation

GMM, JFA, and ISV are some art modeling techniques which are integrated into the toolbox. GMM includes the background model of UBM, which uses maximum likelihood ratio. Enrollment is using Maximum-A Posteriori adaption. Matching is done using log-likelihood ratio. JFA works on the training set and used to estimate eigenvoice and eigenchannel and matrix which helps in modeling the residual noise. Matching is done based on linear score approximation. ISV works like JFA up to certain extents.

Generally, UBM model represents general person-independent feature characteristics in biometric verification system. So that they can be compared with the model of person-specific feature characteristic at the decision time. Here, UBM model is trained with speech samples of the specific enrolled speaker and then likelihood ratio test is done for a specific model with the UBM. It can also be used as a prior model, while training the speaker-specific model. That can be done with Maximum-A Posteriori adaption. Here, GMM-UBM works as the priority model for JFA and ISV model.

3.1 Information of Database

VoxForge is a collection of transcribed speech samples. It is free and can be easily used with speech recognition and verification engines. This database supports Windows, Linux, and Mac environment and it can be downloaded from www.voxforge.org. VoxForge collects these speech samples from volunteers and they make it available to the open-source speech recognition engine. Here, it is used with Bob spear toolkit. All these samples come from volunteers who visit this website and record their voice while reading written prompts provided by the website. These voice samples are recorded by volunteer's own microphone so the quality varies significantly. This corpus includes 6561 files and all are containing English audio files [16]. Audio files of 30 speakers are taken and all the speakers are selected randomly. This database is distributed in three parts: Training (10 speakers), Development (10 speakers), and Test (10 speakers). This database is of 1 GB size.

4 Experimental Results

The experiment is performed on VoxForge database with different algorithms and they are UBM-GMM, JFA, and ISV. According to different algorithms, it gives different values of EER for development and evaluation set. It also gives Cllr and Min Cllr values for both the development and evaluation set [17].

Table 1 Results of baseline ASV system for development set

Measure	UBM-GMM	ISV	JFA
Cllr	0.72080	0.68779	0.73322
Min Cllr	0.04816	0.024024	0.10287
EER	1.981%	1.037%	3.15%

Table 2 Results of baseline ASV system for evaluation set

Measure	UBM-GMM	ISV	JFA
Cllr	0.0633	0.03870	0.30423
Min Cllr	0.81157	0.076971	0.84141
HTER	2.074%	2.370%	8.944%

Cllr

It is the measure of log-likelihood ratio, which can properly evaluate the discrimination of all the log-likelihood ratio scores. It also evaluates the quality of the calibration.

Min Cllr

Min Cllr represents the minimum possible value of Cllr and that can be achieved by the optimally calibrated system. It ranges from zero to infinity but mostly the value is equal to one or below one. The smaller the Min Cllr better the indication of the classifier [18].

EER

It is an equal error rate, and it represents an overall measure of system performance. It corresponds to the threshold value where false acceptance rate and the false rejection rate is same.

With the algorithm GMM-UBM, it uses 25 training iterations and 256 Gaussian for the evaluation. Here, we have used the MFCC60 extractor and energy-2 Gauss pre-processor. The database has total 30 wave files so after completion of every iteration, it gives scores for both development and evaluation dataset and according to that it calculates the EER and HTER.

All the values of development and evaluation set is given in the below Tables 1 and 2.

5 Conclusion

In this paper, a brief summary is given about automatic speaker verification system. Also, we have discussed various techniques which were used in different years. Speaker modeling, feature extraction, and score computation are discussed in detail. Bob spear toolkit is a useful kit for speech and signal processing and it is helpful for implementing different algorithms. Results of evaluation and development dataset are given in the table of the algorithm UBM-GMM, JFA, and ISV. From the results, we can conclude that UBM-GMM algorithm gives the best result with the VoxForge database.

References

1. Campbell, Joseph P. "Speaker recognition: a tutorial." Proceedings of the IEEE 85.9 (1997): 1437–1462.
2. Tranter, Sue E., and Douglas A. Reynolds. "An overview of automatic speaker diarization systems." IEEE Transactions on Audio, Speech, and Language Processing 14.5 (2006): 1557–1565.
3. Voice Acoustics: an introduction, http://newt.phys.unsw.edu.au/jw/voice.html.
4. Rosenberg, A. E. "Evaluation of an Automatic Speaker Verification System Over Telephone Lines." Bell System Technical Journal 55.6 (1976): 723–744.
5. Reynolds, Douglas. "An overview of automatic speaker recognition." Proceedings of the International Conference on Acoustics, Speech and Signal Processing (ICASSP) (S. 4072–4075). 2002.
6. Sturim, Douglas E., et al. "Speaker indexing in large audio databases using anchor models." Acoustics, Speech, and Signal Processing, 2001. Proceedings. (ICASSP'01). 2001 IEEE International Conference on. Vol. 1. IEEE, 2001.
7. Sant'Ana, Ricardo, Rosngela Coelho, and Abraham Alcaim. "Text-independent speaker recognition based on the Hurst parameter and the multidimensional fractional Brownian motion model." IEEE Transactions on Audio, Speech, and Language Processing 14.3 (2006): 931–940.
8. Campbell, William M., et al. "Speaker verification using support vector machines and high-level features." IEEE Transactions on Audio, Speech, and Language Processing 15.7 (2007): 2085–2094.
9. Markov, Konstantin P., and Seiichi Nakagawa. "Text-independent speaker recognition using non-linear frame likelihood transformation." Speech Communication 24.3 (1998): 193–209.
10. Murthy, Hema A., et al. "Robust text-independent speaker identi cation over tele-phone channels." IEEE Transactions on Speech and Audio Processing 7.5 (1999): 554–568.
11. Moattar, Mohammad Hossein, and Mohammad Mehdi Homayounpour. "Text-independent speaker verification using variational Gaussian mixture model." ETRI Journal 33.6 (2011): 914–923.
12. Larcher, Anthony, et al. "Text-dependent speaker verification: Classifiers, databases and RSR2015." Speech Communication 60 (2014): 56–77.
13. Bimbot, Frdric, et al. "A tutorial on text-independent speaker verification." EURASIP journal on applied signal processing 2004 (2004): 430–451.
14. Khoury, Elie, Laurent El Shafey, and Sbastien Marcel. "Spear: An open source toolbox for speaker recognition based on Bob." 2014 IEEE International Conference on Acoustics, Speech and Signal Processing (ICASSP). IEEE, 2014.
15. https://pythonhosted.org/bob.bio.base/installation.html.
16. VoxForge, Free Speech Recognition http://www.voxforge.org/.
17. https://pythonhosted.org/bob.bio.spear/baselines.html.
18. N. Brummer and D. Van Leeuwen, On calibration of language recognition scores, in IEEE Odyssey Speaker and Language Recognition Workshop, 2006, pp. 18.

Topic Modeling on Online News Extraction

Aashka Sahni and Sushila Palwe

Abstract News media includes print media, broadcast news, and Internet (online newspapers, news blogs, etc.). The proposed system intends to collect news data from such diverse sources, capture the varied perceptions, summarize, and present the news. It involves identifying topic from real-time news extractions, then perform clustering of the news documents based on the topics. Previous approaches, like LDA, identify topics efficiently for long news texts, however, fail to do so in case of short news texts. In short news texts, the issues of acute sparsity and irregularity are prevalent. In this paper, we present a solution for topic modeling, i.e, a word co-occurrence network-based model named WNTM, which works for both long and short news by overcoming its shortcomings. It effectively works without wasting much time and space complexity. Further, we intend to create a news recommendation system, which would recommend news to the user according to user preference.

Keywords Data mining · Topic modeling · Document clustering
Online news · Recommendation

1 Introduction

Recently, a generative probabilistic model of textual corpora has been considered, to segregate representations of the news(information). It decreases the depiction length and discloses inter- and intra-document factual structure. Such models ordinarily will be classified into one of two arrangements that produces every word on the argument of some number of preceding words or word classes and those that

A. Sahni (✉) · S. Palwe
Department of Computer Engineering, MITCOE, Pune, India
e-mail: aas.sah91@gmail.com

S. Palwe
e-mail: sushila.aghav@mitcoe.edu.in

© Springer Nature Singapore Pte Ltd. 2018
S. Bhalla et al. (eds.), *Intelligent Computing and Information and Communication*,
Advances in Intelligent Systems and Computing 673,
https://doi.org/10.1007/978-981-10-7245-1_60

produce words based on latest variables induce from word correlations independent of the request in which the words show up.

Topic models are mainly applied to text in spite of a willful need of alertness of the hidden semantic structures that exist in natural language. In a topic model, the words of each document are accepted to be exchangeable; their probability is invariant to stage. This improvement is proven valuable for inferring proficient derivation procedures and quickly examining huge corpora.

However, exchangeable word models have limitations. For the classification or data recovery, these model are used, where a coarse statistical footprint of the topics of a document is adequate for achievement, exchangeable word models are weakly prepared for issues relying on all the more "fine-grained characteristics of language" [5]. For example, though a topic model can recommend documents important for a query, it cannot find particularly important phrases for question answering. Therefore, while a topic model could find a pattern, for example "eat" happening with "cheesecake", it does not have the illustration to represent selection preferences the procedure where certain words restrict the decision of the words that follow.

Due to the rapid growth of Internet and the development of different kinds of web applications, online news have become one of the most leading topic of Internet. Internet news covers topics ranging from politics, weather to sports, entertainment, and different important events, and, in this way, news captures, in a major way, the cultural, social, and political picture of society. Apart from the media networks, self-governing news sources also play a role in reporting the events which may get ignored in the midst of the major stories. Hence, accurately mining topics behind these online news texts are essential for a wide range of tasks, including content, query suggestion, document classification, and text clustering. Due to the increasing volume of short text data, it is essential to gather the latent topics from them, which would be useful for an extensive variety of substance examination applications, for example, content analysis, user interest formulation, and emerging topic detecting. The previously used topic models like LDA and its variations are successful in efficiently detecting topics from long texts (news), however, failing to do so in the case of short texts (news).

Short texts are common on the web, regardless of traditional websites, for example, titles of webpage, text endorsements, and image subtitles, or in social networking sites, e.g., status updates, tweets, and questions in Q and A websites. Dissimilar from traditional normal texts (for example, articles, scholarly paper publications). Short texts, as the name indicates, include a few words. As there is an increasing volume of short text datasets, gathering latest topics from them is essential for an extensive variety of substance examination applications, e.g., content analysis, user interest formulation, etc.

The content sparsity in short text data leads to some new difficulties in topic detection. "PLSA and LDA" [5] are an example of previously used topic models which explain that a document could contain a number of topics, where a topic conveys subject of the document by a set of related or connected words. To study the topic components (that is topic-word distributions) as well as mixture coefficients (that is topic proportions) of every document, statistical procedures are then used.

Traditional topic models detect topics within a text corpus by first capturing the document-level word co-occurrence patterns implicitly, and then, directly applying these models on short texts to tackle the acute data issues. In individual short text, the frequency of words assume less differentiate role than in case of long text, by making it difficult to collect the words which are more associated in every document.

Document clustering is grouping or categorizing documents or text. Document Clustering is an unsupervised learning technique; it is very useful for automatic topic detection and faster information retrieval. For example, as the volume of online data are increasing rapidly, users and information retrieval system needed to classify the desired document against a specific query. There are two types of clustering approaches used, top-down and bottom-up. In this paper, focus is on the performance of K-means clustering algorithm, which is a top-down clustering algorithm; every document is assigned to the cluster whose center also called centroid is nearest. Here, documents are presented as document vector and the center is the average of all the documents in the cluster, in vector space model.

In this paper, a study about the related work done in Sect. 2, the proposed approach modules description, mathematical modeling, algorithm, and experimental setup in Sect. 3 and finally, we provide a conclusion in Sect. 4.

2 Literature Survey

Cheng et al. a novel topic model, "biterm topic model (BTM)" [1], has been proposed for short text topic modeling. Are sparse and experience data ambiguity and imbalance due to lack of contextual information, which are not solved by traditional topic modeling approaches like LDA and its variants. In order to address these problems, in this paper, BTM has been proposed, which works on the corpus-level information (rather than document level) and is based on the word co-occurrence patterns. Also, online algorithms have been presented to manage large amount of incoming short content information.

Sahami et al. a "similarity kernel" [2] function has been described and its properties have been mathematically analyzed. Conventional document similarity measures (e.g., cosine) are not able to determine the similarity between two short text portions efficiently. In order to solve this problem, a new technique, a similarity kernel function, to measure the similarity between short text portions has been introduced in this paper. The author has also performed experiments using this kernel function in an expansive-level framework by suggesting associated queries to web.

In Phan et al. [3], a generic structure for building classifiers has been presented, which deals with short and sparse text and web portions by discovering hidden points found from large-scale information collections. Numerous examples of working with short portions of web, such as, forum messages, blog discussions, news headlines, image captions, etc., do not aim at gaining high precision because of the information sparsity. The author, hence, came up with an idea of gaining external learning, by collecting an external data compilation called "universal data set", to make the information more related and additionally extend the scope of classifiers to handle future information better.

Jin et al. [4] presents an unconventional approach to cluster short text messages via transfer learning from auxiliary long text data. To bridge the possible inconsistency between source and target data, we propose a unconventional topic model —Dual Latent Dirichlet Allocation (DLDA) model, which jointly learns two sets of topics on short and long texts and couples the topic parameters to cope with the potential inconsistency between data sets.

Ramage et al. present an adaptable execution of a partially "supervised learning model (Labeled LDA)" [5] that maps the substance of the Twitter channel into dimensions. These dimensions compare generally to substance, style, status, and social attributes of posts. The author characterizes users, tweets utilizing this model, and presents results on two data consumption oriented tasks.

Zhao et al. presents the comparison of the Twitter content and a customary news medium, New York Times, utilizing unsupervised topic modeling. The author utilizes a "Twitter-LDA model" [6] to find topics from a representative sample of the entire Twitter. The author then utilizes content mining strategies to compare these Twitter topics with themes from New York Times, taking into consideration topic classes and types. The author has also studied the relation among the extents of opinionated tweets and retweets and topic classifications.

Wang et al. a recent topic model, "Temporal-LDA" [7] or TM-LDA has been proposed, for proficiently modeling streams of social content such as a Twitter stream for a creator, by detecting the topic transitions that arise in these data. Existing models such as Latent Dirichlet Allocation (LDA) were developed for static corpora of relatively large documents. However, much of the textual content on web today, is temporarily sequenced and comes in short fragments, for example, microblog posts on sites such as Twitter, Weibo, etc.

Pramod Bide et al. an "improved document clustering calculation" [8] is taken it creates number of clusters for any content reports and to put similar documents in appropriate groups it uses cosine similarity measures.

Raihana Ferdous et al. proposed a "modified k-means algorithm" [9] which chooses distinct k documents as centroids for k clusters using a distance similarity measure. In experiments, it is proved that proposed K-means algorithm with Jaccard similarity distance measure to compute the centroid gives better results than traditional K-means algorithm.

3   System Architecture/System Overview

3.1   Problem Statement

The previous algorithms (LDA and its variants) work for long texts (news articles), however, fail to give efficient results for short texts (news articles) due to data sparsity problem. Our aim is to develop a novel way for topic modeling in online news by using two approaches named as, Biterm Topic Model (BTM) and Word Network Topic Model (WNTM), which give efficient results for long as well as short news articles, by tackling the sparsity and imbalance problems simultaneously. Also compare the results of both approaches in terms of time, accuracy, and memory. Further, apply Document Clustering algorithm to cluster the news related to a particular topic from different sources (news websites) together.

3.2 Proposed System Overview

In this paper, a lucid but generic explanation for topic modeling in online news has been presented. System presents a word co-occurrence network-based model named WNTM, which works for both long as well as short news articles by managing the sparsity and imbalance issues simultaneously. Dissimilar from existing methods, WNTM is modeled by assigning and reassigning (according to probability calculation) a topic to every word in the document rather than modeling topics for every document. It has better time and space efficiency, compared to previous approaches. Also, the context information saved in the word–word space likewise ensures to detect new and uncommon topics with convincing quality (Fig. 1).

In the proposed system, we retrieve real-time data from various news websites as an input to the system. The extracted data is forwarded to preprocessing where stemming and stop word removal takes place. As an output of preprocessing, we get documents represented as a bag of words which will be provided to the topic modeling algorithms (BTM and WNTM), in which topic of the news document is identified. In the case of BTM (Biterm Topic Model), [1] biterms are formed from the bag of words. These biterms go as input to the Gibbs Sampling algorithm which infers the topics of the documents.

In the case of WNTM (Word Network Topic Model), it takes lesser time and space as compared to BTM. This is because the input of BTM is in the form of biterms, while, input of WNTM is in the form of pseudo documents. In BTM approach, each time a word is scanned, it forms a new biterm with the word which occurs with it, irrespective of its previous occurences and then probability is re-evaluated or updated. However, in case of WNTM, a network or graph is formed containing nodes, representing words and edges with weights, representing relationships. In this case, each time a word is scanned, all the words which occurred with it in the entire document, along with their relationships are mapped or saved in the form of a graph or network or pseudo document. In this way, less number of tuples or rows are required, and, WNTM saves the time and space required for execution.

The output of the topic modeling step is given as input to the document clustering step, which groups or clusters the documents coming under specific topic together. Likewise, all the documents are clustered under corresponding topics.

Further, a summary is generated for the news documents, and, we intend to present the summarized news along with the respective topics to the user as an output.

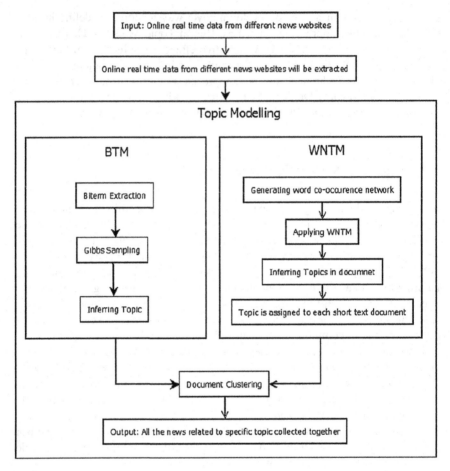

Fig. 1 System architecture with preprocessing, Topic modeling, document clustering, and summarization

Recommendation system: We also intend to create a news recommendation system, in which according to user preference we would try to prioritize the news for the user. User preference can be captured by the news which user browses more. We intend to create a newsbox (kind of a mailbox), in which the news related to topics which interest the user will be stored, the newsbox will be updated from time to time, so that whenever user wants to read the news, he/she can directly check the newsbox, and does not have to browse from all the available news. For example, suppose a particular user is interested in news related to sports and entertainment, then all the news associated with or in the field of sports and entertainment throughout the day would be stored in his/her newsbox, and he/she can directly check the newsbox, instead of browsing through the entire news.

3.3 Algorithm

Algorithm 1: WNTM Re-weighting Algorithm Required: the original word network G = (N, E, V), where E is the set of edges and V is the set of weights for edges.

Ensure: the re-weighted word network G' = (N, E, V')

1. compute degree D(n) and activity A(n) of each node n ε N
1. **for all** e = (n1, n2) ε **E do**
2. set V_e = [v_e/A(n$_i$)], argmin{D(n$_i$), i = 1, 2}
3. **end for**

This re-weighting is used to increase the time or space efficiency and improve the performance of WNTM effectively. After topic modeling step, we will get the set of topics, *T*, as an output. These topics will go as input along with the news data to the Document Clustering step.

Number of topics = Number of clusters

Algorithm 2: Document Clustering Algorithm Input: N: the number of clusters (number of topics)

Di: Size of documents (i.e. total no. of documents) in a database

Output: A set of n clusters which contain documents.

According to the output of Topic Modeling step, we group or cluster or categorize the documents under generalized topics such as Politics, Sports, Education, Entertainment, etc. In our system, instead of centroids, we consider the topics (from previous step), and calculate similarity from topics to documents. After this step, we will get clusters of documents, where each cluster consists of documents belonging to or related to the same topic. Cluster may contain documents from different newspapers, summarized together.

3.4 Mathematical Model

Let S be a system, Such that:

S = {Input Process, Output}

Input: Online Real-Time Data from News Websites

Output: All News Summarization Related to Specific Topic Together.

Process:

1. Preprocessing

   In preprocessing, system performs stop words removing and stemming.

   Step 1: Stop words removal

   Step 2: Stemming

2. Collection of Words into bags

After preprocessing, preprocessed words are collected into bag of words.

$$Bag = \{w1, w2, \ldots, wn\}$$

where, Bag is the collection on n number of words (w)

3. Topic Modeling

For topic modeling, BTM or WNTM is used. Common topics are identified in this step.

$$T = \{t1, t2, \ldots, tn\}$$

where T is the set of topics extracted form n number of documents.

4. Document Clustering

In this step, documents are clustered on the basis of detected topics. K-means clustering algorithm is used for clustering of documents.

$$DC = \{c1, c2, \ldots, cn\}$$

DC is the set of n number of documents clusters.

Each cluster c_n contains similar kind of documents based on relative topic. Cosine similarity measure is used to measure the similarity between topics and documents. It is given as:

$$\text{Sim}(C_i, C_j) = C_i \cdot C_j / |C_i||C_j|,$$

where C_i and C_j are the clusters.

4 System Analysis

4.1 Experimental Setup

The system built using Java framework on Windows platform. The Net beans IDE (version 8.1) are used as a development tool. The system does not require any specific hardware to run; this application can run on any standard machine.

4.2 Dataset

System extracts real-time news data from news websites (e.g., Indian Express, Hindustan times). We extract the data from the website links; links containing news data selected by passing appropriate attributes. On an average, at a time, a news website contains about 130 news documents, storylines, or articles, which are extracted. Every news extraction from website is stored in separate documents or files.

4.3 Performance Measures

The following are the performance measures of topic modeling algorithms (BTM or WNTM):

Time Efficiency: Amount of time taken for the execution of the algorithm.

Memory: Amount of space of memory used or required for the execution of the algorithm.

4.4 Result

Table 1 represents the comparison between existing system and proposed system on the basis of time efficiency and memory efficiency of topic modeling. Proposed system includes WNTM and Document clustering, it is more efficient than existing system which includes BTM and Document clustering, it finds out the topic modeling and document clustering.

Figure 2 shows the time efficiency comparison graph of the proposed system versus the base system. Time required to identify topics in existing system with

Table 1 Time and accuracy comparison of segmentation time

System	Time in ms	Memory in bytes
Existing system using BTM and document clustering	1294.7	898517444
Proposed system using WNTM and document clustering	338.2	484382868

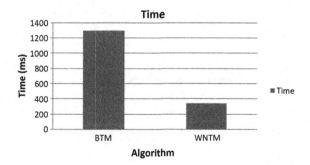

Fig. 2 Time efficiency comparison, X-axis = Topic modeling with BTM and WNTM, Y-axis = Time required in milliseconds

BTM is more than the time required in proposed system with WNTM. Here, we have shown the average result of about 20 sample cases (20 executions).

Figure 3 represents the memory comparison of BTM and WNTM for topic modeling in percentage. WNTM outperforms BTM. Here, we have shown the average result of about 20 sample cases (20 executions).

Figure 4 represents the variation in the results of time comparisons between WNTM and BTM of the 20 executions or 20 sample cases.

Figure 5 represents the variation in the results of memory comparisons between WNTM and BTM of the 20 executions or 20 sample cases.

Fig. 3 Memory Comparison, X-axis = System, Y-axis = Memory of topic modeling with BTM and WNTM

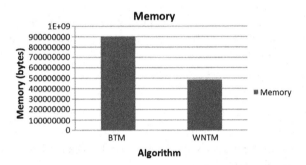

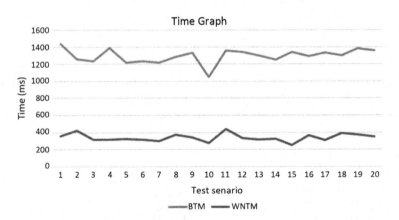

Fig. 4 Time efficiency comparison, X-axis = Topic modeling with BTM and WNTM, Y-axis = Time required in milliseconds

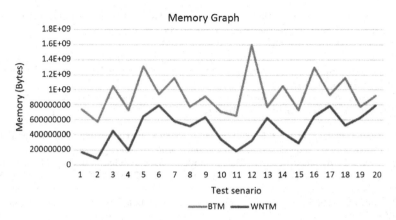

Fig. 5 Time efficiency comparison, *X*-axis = Topic modeling with BTM and WNTM, *Y*-axis = Time required in milliseconds

5 Conclusion and Future Scope

Topic modeling in online news is an increasingly essential job because of the existence of abundant unrecognized news data on the internet. When compared to normal texts, online short news data texts bring sparsity issue for existing topic models. In order to overcome these problems, in this paper, a topic model is proposed for general short texts, namely Word Network Topic Model, and apply it on online news data. Unlike the previous models, this model efficiently mines or detects topics for both long as well as short news data texts. WNTM can capture the topics in news texts by explicitly modeling word co-occurrence patterns in the entire corpus. Then, document clustering algorithm is used for clustering the news data related to specific topic together. Next, summarization of the news coming from various sources takes place, and, we intend to present the summarized news along with the topic to the user as the output. Further, we intend to create a news recommendation system according to user preference. In future, we can try to work with different languages as the news data may contain information in other languages as well.

References

1. Xueqi Cheng; Xiaohui Yan; Yanyan Lan; Jiafeng Guo. BTM: Topic Modeling over Short Texts. IEEE Transactions on Knowledge and Data Engineering, Year: 2014, Volume: 26, Issue: 12 Pages: 2928–2941.
2. M. Sahami and T. Heilman. A web-based kernel function for measuring the similarity of short text snippets. In *Proceedings of the 15th international conference on World Wide Web*, pages 377–386. ACM, 2006.

3. X. Phan, L. Nguyen, and S. Horiguchi. Learning to classify short and sparse text & web with hidden topics from large-scale data collections. In *Proceedings of the 17th international conference on World WideWeb*, pages 91–100. ACM, 2008.
4. O. Jin, N. Liu, K. Zhao, Y. Yu, and Q. Yang. Transferring topical knowledge from auxiliary long texts for short text clustering. In *Proceedings of the 20th ACM international conference on Information and knowledge management*, pages 775–784. ACM, 2011.
5. D. Ramage, S. Dumais, and D. Liebling. Characterizing microblogs with topic models. In *International AAAI Conference on Weblogs and Social Media*, volume 5, pages 130–137, 2010.
6. W. Zhao, J. Jiang, J. Weng, J. He, E. Lim, H. Yan, and X. Li. Comparing twitter and traditional media using topic models. *Advances in Information Retrieval*, pages 338–349, 2011.
7. Y. Wang, E. Agichtein, and M. Benzi. Tm-lda: efficient online modeling of latent topic transitions in social media. In *Proceedings of the 18th ACMSIGKDD*, pages 123–131, New York, NY, USA, 2012.ACM.
8. Pramod Bide, Rajashree Shedge, "Improved Document Clustering using K-means Algorithm" IEEE, 2015.
9. Mushfeq-Us-Saleheen Shameem, Raihana Ferdous, "An efficient Kmeans Algorithm integrated with Jaccard Distance Measure for Document Clustering", pp 1–6, IEEE, 2009.

Spam Mail Detection Using Classification Techniques and Global Training Set

Vishal Kumar Singh and Shweta Bhardwaj

Abstract Emails are Internet-based services for various purposes like sharing of data, sending notices, memos, and sharing data. Spam mail are emails that are sent in bulk to a large number of people simultaneously, while this can be useful for sending same data to a large number of people for useful purposes, but it is mostly used for advertising or scam. These spam mails are expensive for the companies and use a huge amount of resources. They are also inconvenient to the user as spam uses a lot of inbox space and makes it difficult to find useful and important emails when needed. To counter this problem, many solutions have come into effect, but the spammers are way ahead to find these solutions. This paper aims at discussing these solutions and identifies the strengths and shortcomings. It also covers a solution to these spam emails by combining classification techniques with knowledge engineering to get better spam filtering. It discusses classification techniques like Naïve Bayes, SVM, k-NN, and Artificial Neural Network and their respective dependencies on the training set. In the end of this paper, the global training set is mentioned which is a way to optimize these training sets and an algorithm has been proposed for the same.

Keywords Naïve Bayes · k-nearest neighbor · Support vector machine classifier method · Artificial Neural Network · Knowledge engineering · Global training set

V. K. Singh (✉) · S. Bhardwaj
Computer Science and Engineering Department, Amity University,
Noida, Uttar Pradesh, India
e-mail: vs48023@gmail.com

S. Bhardwaj
e-mail: shwetabhardwaj84@gmail.com

© Springer Nature Singapore Pte Ltd. 2018 623
S. Bhalla et al. (eds.), *Intelligent Computing and Information and Communication*,
Advances in Intelligent Systems and Computing 673,
https://doi.org/10.1007/978-981-10-7245-1_61

1 Introduction

Classification is based on the concept of data mining. Data mining is gathering meaningful data from a pool of information which is built using logical training sets and observation in related fields [1]. This helps in building a better system as we already have all the information of previous failures, success, and current problems [2]. We take the data from this pool, sort it according to our needs and then select the data which will help us in our current research. Data mining comprises of various methods of research and learning like artificial intelligence or machine learning. It is used in almost every fields like teaching, research, medical, defense, etc., to optimize the results.

If a spam mail is controlled, it can save a lot of resources and expenditure of the companies. Previously, a method called "Knowledge Engineering" was used to separate spam mail from important emails, but its success rate was not as much as that of spammers as they were able to find a way around it by changing a letter in the keywords which is used to differentiate spam mail from legitimate mail [3]. To overcome this, various methods for segregating spam mail using classification was introduced. These methods used machine learning, artificial intelligence, and various databases to develop a system to counter spam mail. These methods were more successful than knowledge bank as the data was updated more frequently, and were capable of learning on their own and had better algorithms than knowledge banking [4]. This paper will elaborate these techniques and compare them with previously used techniques, thus suggesting changes that can be made to spam mail filtering.

1.1 Why We Need Spam Filtering and Advantages of Classification Techniques for Spam Filtering

Why we need spam filtering	Advantages of classification techniques for spam filtering
• Spam mail use lots of resources • They contribute a lot to network traffic • They take inbox space which is a nuisance • It becomes hard to find useful mails among these spam emails • Service provider suffers loss because of these spam emails	• They are more efficient than knowledge engineering • Reduces consumer effort in spam filtering • Adaptive to change on their own, save time and other resources required for the spam filter

1.2   Type of Spam Mail and Their Composition

Type of spam mail	Percentage (%)
Leisure/Travel	2
Education	1
Business/Investment	18
Adult	18
Finance	17
Product/Services	19
Health	9
Computer services	7
Others	9

See Fig. 1.

2   Related Works

Mails were introduced to make communication easier and faster. With time its use for wrong purposes also increased and one of the major problems is sending bulk emails for advertising or other such purposes [5]. These bulk emails commonly known as "spam mail" were becoming a huge problem and had to be dealt with. First, it was dealt with the users manually by classifying or marking mail as spam mail and later when such similar mail came they were moved to spam folder. The biggest reason for the rise of spam mail is that it is faster and easier to use mail for advertising worldwide rather than using posters or TV advertisement for same [6].

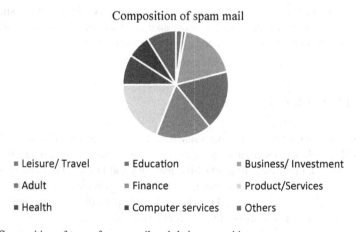

Fig. 1 Composition of type of spam mail and their composition

They provide a cheaper solution. According to a survey, several billion emails are exchange in a year and out of which a good amount approximately 50% turns out to be spam. But soon counters to this were found by spammers and new techniques had to be developed. In 1900s, classification techniques were introduced for spam filtering. These techniques were much better than the previous techniques as they were faster and more accurate in spam mail filtering. They could handle a large amount of mails, but the solution to these was also found [7]. The United State of America (USA) was the first country to take legal action against the spammers and passed a law for the same. Since then, several arrests have been made and many spammers were caught and punished for their crime. Some were sentenced to jail while some spammers were asked to compensate the ISP for the losses. These actions were a little successful in bringing down the number of spam mail but the problem still persists.

This constant battle between spammers and email service providers is still going on. Each comes up with a solution better than other [6]. But these classification techniques are very efficient and keep spammers at bay. New algorithm which uses classification was introduced to counter this problem. Each technique has its own advantages and disadvantages. Some of the techniques which were introduced are Naïve Bayes, SVM, Decision Tree, Neural Network, k-NN, etc. There are still being developed to be more effective and accurate.

3   Problem Statement

Whenever the email service provider comes up with a solution for spam filtering, the spammers come up with a way to get around that filter and it is a constant competition between these two. This is very costly for the companies, a huge amount of resources being used, and create a nuisance for the customer by filling up the inbox, wasting their time as it becomes hard to find useful mail among those spam emails. We need to find a way which is very effective against these spammers and can evolve with the attacks, which the spammer adapts. The training set which is used to train these techniques needs to be focused on more success.

4   Implementation

This paper suggests that we use both machine learning and data mining techniques along with previously used knowledge engineering approach to create a filter which is more robust and efficient than machine learning and knowledge bank techniques alone and make this learning technique global. Some providers provide this service while others either provide only knowledge-based spam filter or machine learning and are not very efficient in combining these two for a better filter.

5 Knowledge Engineering

In Knowledge Engineering, user input creates a set of rules for the incoming emails to be classified as a spam mail or uses binary value where 1 stands for spam and 0 stands for not spam. In this, the users define a specific email id and any incoming emails from this email id were automatically moved to the spam folder. They can also define text words, which can be used as a classification of spam mail. If the incoming mails contain these keywords, they were moved to spam filter [3]. Although this technique was successful for some time, the spammer found a way around it by changing the text. This was done by adding or omitting one or two alphabets in the email id or the main message and thus passing the filter settings set by the user.

6 Classification Techniques

Here, we discuss some of the classification techniques with their accuracy and precision obtained by passing a set of test spam data through the classifier. It can be depicted like as follows:

Training spam dataset ➡ Classifier ➡ Performance Analysis

6.1 Naïve Bayesian Classifier

It was one of the first classification techniques used for spam mail filtering. It works on Bayes Theorem of probability to check if the incoming mail is a spam or not. The filter in this classifier first has to be trained to check for spam mail. Training a data set means that the filter is given a set of words the user provides by manually identifying the mail as spam or not. Through the input from a user, the classifier is now trained and can check incoming emails. This classifier checks the probability of the words in the training set in incoming messages and with the result obtained it can filter spam mail. It creates a different folder for the spam and moves those emails directly to the folder. Although it is pretty old, it is still preferred over more sophisticated classifiers [8, 9].

6.2 k-Nearest Neighbor

In this technique of classification, a training set contains a sample of messages which can help in identifying whether the incoming mail is spam or not. The incoming mail is compared with the training set to find its k-nearest member found by comparing it with the training set and its k most similar document is found and then identified as spam, based on which group its k most similar document was found [10].

6.3 Support Vector Machine Classifier Method

In this method, a decision plane is formed to separate spam and legit emails. They are separated by a decision boundary which has certain conditions to separate these emails. A training set is formed for the sorting and incoming mail is compared with the training set. Like k-nearest, neighbor the incoming mail is compared with the training set to find similarity between the incoming mail and training set. A kernel function, K is used to determine the similarity and based on this the emails are sorted in decision plane [11].

6.4 Artificial Neural Network

Artificial Neural Network or commonly known as neural network acts like an artificial human brain. Artificial neurons are interconnected to form this network and data is passed through it for learning. Like the human brain, it learns by example and during training, the data is passed through the network so that it can learn and adapt according to the examples [12]. They change their structure based on information from the examples so that a better system of classification can be formed [13] (Fig. 2 and Table 1).

These comparisons may vary if the number of selected feature is changed, as in classification techniques each technique is dependent on how it is trained [14]. Among the discussed technique Naïve Bayes is the simplest technique which can be implemented and is comparatively faster too.

These were some of the classifier-based techniques for spam mail classification which used a training set for detection of spam mail. All techniques have a success rate of around 90%, neural network gives the best result. These techniques are

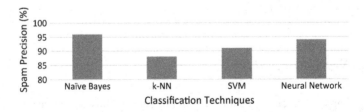

Fig. 2 Graphical representation of classification techniques and their precision

Table 1 Tabular comparison of classification techniques

Classification technique	Spam precision (%)
Naïve Bayes	96
k-nearest neighbor	88
Support vector machine	91
Neural network	94

dependent on the training set and examples and how they can be optimized so that the techniques can learn better and can give better results [15]. If we can combine knowledge engineering technique with these classification techniques, we can optimize it further.

7 Proposed Work

Data is collected from each user training set and then compared to find which mail is characterized as spam mail and then a global training set based on these inputs can be created to optimize the classification technique. Each technique is dependent on the training set for setting the criteria for spam mail identification. The user can add words to the training set of their personal devices or email accounts to extend the global training set and get better results. The service provider provides the user an option to choose the global training set or not. This global set can arrange the mails in the training according to their frequency. The spam mail which is sent to a large number of customers will be included in this training set while the spam mails which have a low-frequency rate, meaning the mails which are sent only to a few number of customers can be optional for the training set. The service provider can add a sensitivity controller to this training set with high sensitivity having all the spam mails for training the global training set irrespective of its frequency. Low sensitivity will only have spam mail in the training set with high frequency, thus making it easier for users to classify mail for the training set.

Proposed Algorithm

Keywords

Hf—high frequency
Mf—medium frequency
Lf—low frequency
Ef—email frequency
Gtb—global training set

For service provider

Create global training set → Gtb
{Set threshold for high, medium, and low frequency;

Get frequency >> hf, mf, lf
If (ef >= hf)
{Save email feature in all frequency training set}
else if (ef < hf && ef >= mf)
{Save email feature in medium- and low-frequency training set}
else
{Save email feature in low-frequency training set}
}

For the user

Provide option for global training set to user
{
If (Gtb == 1)
{Provide option between high-, medium-, and low-frequency training data set
if (user chooses high frequency training data set)
{(Combine local training set with high-frequency global training set)
else if (user chooses medium training data set)
(Combine local training set with medium-frequency training data set)
else
(Combine local training set with the low-frequency data set)}

8   Result

This algorithm is supposed to enhance and make the training set better for the classification techniques. With the increase in a number of data for training set which was submitted by users, it includes most of the keywords and phrases of recent spam mail as well. This helps in classification of spam mail better and increase the efficiency of the classification techniques. They can increase the precision of each technique by at least 2% and this can vary from technique to technique. A comparison of spam mail technique before using global training set and after using global training set is shown below which depicts the expected efficiency of proposed algorithm (Table 2).

Table 2 Comparison of precision of classification technique before and after using proposed algorithm for global training set

Classification technique	Spam mail precision before using algorithm for global training set (%)	Expected spam mail precision after using algorithm for global training set (%)
Naïve Bayes	96	97–98
k-nearest neighbor	88	91–94
Support vector machine	91	93–95
Neural network	94	95–97

9 Conclusion

All the classification techniques have to be trained first in separating spam emails from other emails before they are actually used. A data set called training set is used to train these techniques. Thousands of samples are used in these training set to make the classifier able to separate the spam mail. But even after this much work spam mail still persists. They persist because every day a new kind of spam mail is introduced. Thus, even if we get old spam mail sorted and marked, new one keep coming in. One of the solutions is to make the training set up-to-date by gathering information about the new kind of spam mail. The fastest way to do that is to make the user report the spam mail as soon as they encounter it and contribute to the global training set because it will take time if the service provided has to monitor each and every mailbox on their own to search for any new spam mail. This algorithm is expected to raise the efficiency of other techniques by some margin depending on the technique. If it is successful in doing so we will have the spam mail dealt with before it reaches our mailbox. This will also save our time and inbox will be less crowded thus making it easier to find useful emails.

References

1. Parhat Parveen, Prof. Gambhir Halse, "Spam mail detection using classification", International Journal of Advanced Research in Computer and Communication Engineering Vol. 5, Issue 6, June 2016.
2. Megha Rathi, Vikas Pareek, "Spam mail detection through data mining-A comparative performance analysis", I.J. Modern Education and Computer Science, May 2013.
3. W. A. Awad, S. M. ELseuofi, "Machine learning method for spam email classification", International Journal of Computer Science & Information Technology, Vol 3, No 1, Feb 2011.
4. R. Malarvizhi, K. Saraswathi, Research scholar, PG & Research, Department of Computer Science, Government Arts College, "Content-Based Spam Filtering and Detection Algorithms- An Efficient Analysis & Comparison", International Journal of Engineering Trends and Technology, Volume 4, Issue 9, Sep 2013.
5. Vinod Patidar, Divakar Singh, "A Survey on Machine Learning Methods in Spam Filtering", International Journal of Advanced Research in Computer Science and Software Engineering, Volume 3, Issue 10, October 2013.
6. Rekha, Sandeep Negi, "A review of spam mail detection approaches", International Journal of Engineering Trends and Technology (IJETT) – Volume 11 Number 6 - May 2014.
7. S. Roy, A. Patra, S. Sau, K. Mandal, S. Kunar, "An Efficient Spam Filtering Techniques for Email Account", [1,2,3]Computer Science & Technology & [4]Mechanical Engineering, NITTTR, Kolkata, American Journal of Engineering Research, Volume 02, Issue 10, pp-63–73.
8. Tianhao Sun, "Spam Filter Based on Naïve Bayes Classification", May 2009.
9. Rasim M. Alguliev, Ramiz M. Aliguliyev, and Saadat A. Nazirova, "Classification of Textual E-Mail Spam Using Data Mining Techniques", Institute of Information Technology of Azerbain National Academy of Sciences, September 2011.
10. Konstantin Tretyakov, "Machine Learning Techniques in Spam Filtering", Institute of Computer Science, University Of Tartu, May 2004.

11. Akshay Iyer, Akanksha Pandey, Dipti Pamnani, Karmanya Pathak and Prof. Mrs. Jayshree Hajgude, "Email Filtering and Analysis Using Classification Algorithms", IT Dept, VESIT, International Journal of Computer Science Issues, Vol. 11, Issue 4, No 1, July 2014.

12. Vinod Patidar, Divakar Singh, Anju Singh, "A Novel Technique of Email Classification for Spam Detection", International Journal of Applied Information Systems, Volume 5 – No. 10, August 2013.

13. Abhale Babasaheb Annasaheb, Vijay Kumar Verma, "Data Mining Classification Techniques: A Recent Survey", International Journal of Emerging Technologies in Engineering Research (IJETER) Volume 4, Issue 8, August (2016).

14. C. Neelavathi1, Dr. S. M. Jagatheesan, Research Scholar, Department of Computer Science, Gobi Arts & Science College, T.N., India, "Improving Spam Mail Filtering Using Classification Algorithms With Partition Membership Filter", Issue: 01, Jan 2016.

15. Ahmed Obied, "Bayesian Spam Filtering", Department of Computer Science, University of Calgary.

Smart Gesture Control for Home Automation Using Internet of Things

Sunil Kumar Khatri, Govind Sharma, Prashant Johri
and Sachit Mohan

Abstract Internet of Things (IoT) is a system where the machines involved in a system or an infrastructure can be monitored and certain activities can be controlled with the help of sensors and actuators. This research presents the model for controlling remote devices with the help of hand gestures. The gestures are recognized using template matching algorithms to identify the hand gestures and control remote things accordingly using the Internet of Things. The microcontrollers and the remote units which are involved in the architecture are connected to the Internet either via LAN or Wi-Fi module. The proposed system will help those who generally forget to switch off the power when not in use, as an example. It is a contribution to the body of knowledge in the field of home automation using a microcontroller.

Keywords IoT · Embedded microcontroller · Gesture control · Z-wave protocol
Template matching

S. K. Khatri (✉) · G. Sharma
Amity Institute of Information Technology, Amity University Uttar Pradesh,
Noida, UP, India
e-mail: skkhatri@amity.edu

G. Sharma
e-mail: govindsharma00709@gmail.com

P. Johri
Galgotias University, Greater Noida, India
e-mail: johri.prashant@gmail.com

S. Mohan
InventingThoughts Technologies, Ghaziabad, India
e-mail: er.sachit@gmail.com

© Springer Nature Singapore Pte Ltd. 2018
S. Bhalla et al. (eds.), *Intelligent Computing and Information and Communication*,
Advances in Intelligent Systems and Computing 673,
https://doi.org/10.1007/978-981-10-7245-1_62

1 Introduction

Internet of Things, more commonly referred to as IOT, is a system where we have a number of devices connected to the Internet that is made capable of interacting with each other through the use of sensors and actuators. The complete functioning of the devices can be controlled and monitored. The growth of the IoT has an influence on everyone and everything and will bring a great difference to lifestyle and environment in the next decades [1].

There have been a number of hand gesture-based systems using IoT that have been proposed by researchers globally. In our model, we have proposed the use of a camera to recognize the gestures. The previous methods have seen the use of accelerometers, which were fixed on the hand. This model is highly accurate but then cannot be used for controlling home appliances [2].

The use of gestures has emerged to be one of the most emerging ways to control devices. Wearable devices are the considered the best tool to carry out this practice [3].

Hand gestures tagged along with visual display is a popular choice in predicting sign language [4]. When it comes to the same concept for appliances, it is different when compared to the one for sign language. Controlling appliances has become a very complex and challenging task to execute.

Recently, hand gesture systems are becoming widely used in many products. Z-wave is used as a protocol in hand gesture system. It is a more effective, reliable mode of communication for the flow of small data packets. In this research paper, the main aim is to develop a hand gesture system with Z-wave protocol in order to make it cost effective and efficient in its working [5].

2 Literature Review

A real-time system developed by Madhuri and Kumar [6] was able to recognize different gestures made by an individual through a web camera. They used an algorithm called Haar Classifier. To get the most accurate data, a large amount of data and also images were required. The whole process took about a week to complete. This algorithm was deemed to be more flexible and applicable. The system had the ability to capture and track the gestures with high accuracy. The drawback to this system was that the distance between the camera and the user could not be too much.

Choondal and Sharavanabhavan [7] came up with a system which was for the natural user interface. This system depended upon how you place your fingers, and it kept track of the same by locating it frequently. Color markers were placed on different fingers. The whole system was dependant on the color recognition. The restriction to the system was when the same color appeared in the background too which would then end up confusing the system.

3 Analysis

During the initial stages of home automation, the whole system required the installation and updates to be done by a technician. The manual work that had to be put in during the initial stages was very high. As time passed by, the manual work got replaced by installation through Wi-Fi. A couple of examples would be Z-wave and ZigBee [4] These are wireless technologies that are used for monitoring and control. Even though their key features are similar, they differ in their specification and application. The use of such technologies is becoming more widespread as its ideal for home automating systems.

There are few aspects to differentiate between ZigBee and z-wave for home automation system using gesture control (Table 1).

4 Gesture Recognition

Gesture recognition refers to the interpretation of hand signals using a particular algorithm. It is something that is looked into as having the potential for the future market. The gestures that are interpreted are mainly from movements caused by the body, more commonly face or the hand. The person making gestures would be able

Table 1 Features of Z-wave and ZigBee [9]

	Z-wave	ZigBee
1. Compatibility	375 organizations and 1500 items comprise the Z-wave Alliance	A wide range of organizations make gadgets that speak with ZigBee's innovation
2. Security	Each Z-wave arrange, and the gadgets inside each system, are appointed extraordinary IDs that speak to your center point	ZigBee uses AES-128 encryption and make use of keys to secure it
3. Drawback	Z-wave items commonly go from $40 to $100 every, which can include rapidly on the off chance that you need to associate numerous gadgets	ZigBee is mainly renowned for its use in the retail and utility business. ZigBee was created for use in the utility and retail businesses
4. Advantage	Z-wave is easy to understand and gives a straightforward framework that clients can set up themselves	The main plus point to it is that it requires very less battery and so the gadgets can last up to 7 years on a single battery
5. Uses	Z-wave is for an essential comprehension of innovation that keeps home mechanization secure, proficient, easy to utilize, and simple to keep up	ZigBee is ideal for the innovation master who needs a framework they can tweak with their inclinations and introduce themselves

to control the whole system to which he is interacting with just by body movements. It allows an individual to interact with a system without the involvement of a mechanical instrument to connect the two. Human input-based interfaces can be developed that has great potential. This system, would propose a model for gesture recognition to control the various home appliances put in place. With the use of IOT, we will be able to monitor the performance of the device and also its efficiency.

1. Gestures for the templates

Few gestures which will be stored in the database for operating android devices using template matching technique are shown in the Fig. 1.

2. Template Matching

In the process of template matching, a template of the image is compared with the actual image to find the match. After receiving the input image, this method compares the received image with the template stored in the database. Template matching is generally done based on pixel or feature. Templates are frequently used for recognition of characters, numbers, objects, etc. The searching for the template depends upon the size of the image. More the pixels, more time it will take for appropriate template match.

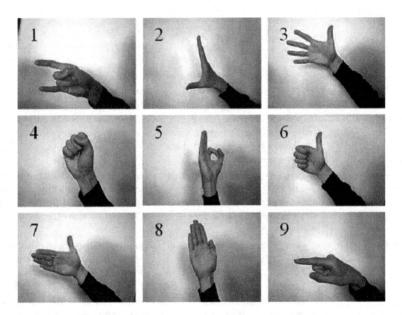

Fig. 1 Gestures depicting functions [7]

5 Proposed Work

5.1 Methodology Adopted

In order to apply the smart gesture control for home automation with IOT devices, a model is constructed as shown in the Fig. 2. The method that has been used is a pattern matching process, where in templates are set to understand the different gestures portrayed by an individual. The gestures are captured at 30 Frames per Second (fps) and are compared with the templates to derive a meaningful output. The end user can set up his\her own personalized gestures as a template to perform a particular action, but this can only be done during the initial stages when the system is being set up. Microcontrollers are responsible for handling the functional elements, and are developed using various programming languages. The model uses following devices to function:

Camera—Camera is situated on every android appliance so that it will easily capture our hand gesture and work as an input device to our gadget.

Microcontroller—The input received will be processed on the microcontroller, where each gesture is programmed with embedded c language using a particular function.

Templates—Templates are basically predefined gestures that will be used by the template matching algorithm.

Z-wave Transmitter—This is the IoT device that uses AES-128 encryption to send inputs over Wi-Fi. It has 900 MHz frequency to transmit data.

Z-wave Receiver—It receives the input from the z-wave transmitter process (and uses) the received input to perform a particular action for each device.

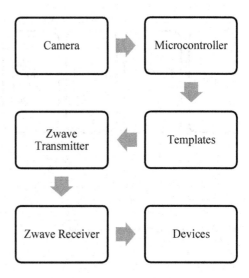

Fig. 2 Proposed operation flow

5.2 Proposed Model

For the application of gesture recognition in smart homes using IOT protocols, a prototype needs to be built. Users make various gestures to perform different actions [8]. To detect these gestures camera(s) are installed, it is used on the top of each Android device so that the input is taken from the user. The microcontroller is used to process the input that has been captured by the camera. The microcontrollers can be programmed using Embedded C. After the microcontroller has processed the input, the templates are used to map it to the in-built gestures, i.e., gesture's already present in the database. In Fig. 3, templates are basically the pre-set understanding of each action that the user performs. Each time a person performs an action (which is a template), it makes the system generate the desired output. If the gesture provided by the user is not present in the system or is unrecognizable, a beep sound can be heard, helping the user understand that he\she has provided a wrong gesture.

There are a few aspects that the user has to take care of while performing the gestures. The color pattern should be easily distinguishable from the system, which means that background color should not be the same as the skin color of the user. The gestures can only be detected in the presence of a certain amount of light, or else the system would not be able to capture the image. For example, gestures carried out in a dark room cannot be recorded by the system. Another key aspect that the user has to take care of is the speed at which the gesture is performed. If it is performed at a high speed, the camera would not be able to understand what action

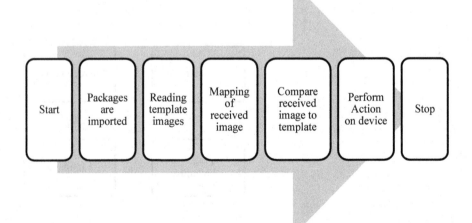

Fig. 3 Flow chart of proposed model

the user is trying to perform. Taking these points into consideration would help the user perform the various actions successfully.

The technique of finding a particular figure on a still image is called active shape models. Each template is stored with a number of frames in the database, on each frame. Active shape model is used to identify the figure as an initial approximation for the next frame. The number of templates to be used depends upon the number of gestures that is decided to perform on each device. As nine number of gestures are defined to use in the system and on the basis of the proposed model a certain time is shown for each gesture in Table 2.

The gestures are performed by a certain frame per second to analyze the performance time. The average time for performing these gestures is 1.34 s as shown in the graph Fig. 4, which is preferable time for performing such gesture in the system and also it is better than any other gesture recognition algorithms. To calculate the timing of each gesture, we conducted 15 trails to perform same gesture and took an average time of it.

Each gesture has its own priority for the availability and this priority is calculated according to the number of users for each device. As per every device has its On and Off state, thus its priority is high as it should be available for the users, that means high priority gesture should get recognized at any case for each device.

The quality of image should also be appropriate and should not below 480 pixels. For storing templates database is created which include 189 images, i.e., 21 images for each gesture (9). Each part of the template showing particular gesture is called as a model thus there are total 9 models have been proposed to store in the database. Each model is processed and compared with the gathered data and matching template is selected to perform the action on the device. Preprocessing of the data is needed before performing any gesture to the device. Preprocessing of data will remove noise, unwanted errors, make data effective and reliable.

Table 2 Number of templates performing gesture [10]

S. No.	Name of the gesture	Time of performing the gesture (Sec)	Availability (High/Med/Low)
1	Pause/Stop	1.49	High
2	Help	1.42	Medium
3	On	1.25	High
4	Off	1.55	High
5	Sleep	1.22	Low
6	Go/Play	1.40	High
7	Down	1.30	Medium
8	Up	1.19	Medium
9	Next	1.28	Medium

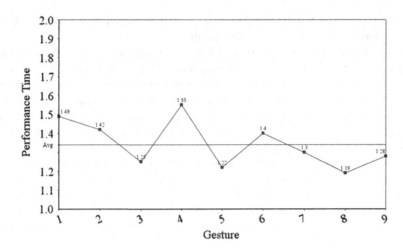

Fig. 4 Analysis of performance

6 Conclusion

This paper presents an effective and efficient home automation system, where devices can be controlled by hand gesture. It holds all the promises and has shown great signs to play a key role in home automation. The system is based on RF communication technology that shows low latency in forwarding data packets through devices. It enables the system to control up to 232 devices locally or remotely. Z-wave protocol of IoT is widely used for home automation. A couple of key advantages of this protocol are that it has low power consumption and supports mesh network. The work can be extended to control more machines by including particular Internet-empowered appliances station. This work proposed an approach to recognize nine kinds of gesture by counting the number of fingers. Each gesture is related to a particular task that will be performed on the device.

Acknowledgements The authors express their deep sense of gratitude to The Founder President of Amity Group, Dr. Ashok K. Chauhan, for his keen interest in promoting research in Amity University and have always been an inspiration for achieving great heights. I would also like to thank Mr. Sachin Mohan, who is IoT Architect at Inventingthoughts Technologies for providing vital information regarding various Android appliances. Mr. Sachit Mohan has provided his consent for the paper for which I am highly obliged. Inventingthoughts Technologies has been of great help in my research paper providing the base for my paper contents. Mr. Mohan's knowledge in the field of Android devices and their working has been of great help.

References

1. M A Rashid, X. Han, *"Gesture control of ZigBee connected smart home Internet of Things"*, 2016 5th International Conference on Informatics, Electronics and Vision (ICIEV), pp. 667–670.

2. J. Anderson and L. Rainie, *The Internet of Things Will Thrive by 2025*, 2014 Pew Research Center. Available at: http://www.pewinternet.org/2014/05/14/internet-of-things/.
3. G. R. S. Murthy, and R. S. Jadon, *A Review of Vision Based Hand Gestures Recognition*, International Journal of Information Technology and Knowledge Management, July-December 2009, pp. 405–410.
4. Shah J., Darzi A., "Simulation and Skills Assessment", IEEE Proc. Int. Workshop on Medical Imaging and Augmented Reality 2001, pp. 5–9.
5. Jun-Ki Min; Bongwhan Choe; Sung-Bae Cho, *A selective template matching algorithm for short and intuitive gesture UI of accelerometer-builtin mobile phones* 2010 Second World Congress on Nature and Biologically Inspired Computing, pp. 660–665.
6. Assad Mohammed, Amar Seeam, Xavier Bellekens, *Gesture based IoT Light Control for Smart Clothing*, 2016 Emerging Technologies and Innovative Business Practices for the Transformation of Societies (EmergiTech), IEEE International Conference, pp. 139–142.
7. J. J. Choondal and C. Sharavanabhavan, *Design and Implementation of a Natural User Interface Using Hand Gesture Recognition Method*, 2013 International Journal of Innovative Technology and Exploring Engineering, vol. 2, pp. 21–22.
8. K. Madhuri and L. P. Kumar, *Cursor Movements Controlled By Real Time Hand Gestures,* 2013 International Journal of Science and Research, vol.2, issue 2, pp. 4–5.
9. ZigBee Vs Zwave, Available: http://www.safewise.com/blog/zigbee-vs-zwave-review/.
10. Application of Template Matching Algorithm for Dynamic Gesture Recognition, Available: https://www.youtube.com/watch?v=fbeLUfTED00.

A Proposed Maturity Model for Himachal Pradesh Government e-Services

Alpana Kakkar, Seema Rawat, Piyush Gupta
and Sunil Kumar Khatri

Abstract With new leaps in the advancement of technology on a daily basis, it becomes quite essential that basic and necessary governmental services are provided through the internet as e-services. These are small steps to fulfil the dream of a "digitalized India". This paper is an aim to provide a maturity model for the existing e-services of Himachal Pradesh (HP) government. Its central focus is on how to make the services easily accessible for the masses and improve their functionality by giving a detailed and sophisticated maturity model. The paper also lists the major concerns of information security in the e-service portal of Himachal Pradesh (HP) government and an attempt has been made to provide a solution for the same. The proposed solution herewith will help the government to improve the level of its e-services and the security of the portal.

Keywords Information security · Maturity model · e-services · HP government

1 Introduction

Technology, today, has become the part and parcel of each individual's life. Most of the governments have gone online or are in the process of going online, thus providing the public services to its citizens through the Internet [1]. Both the state and central governments, in India, have taken the e-governance seriously and have been continuously endeavouring to provide citizen services in a better manner [2].

A. Kakkar (✉) · S. Rawat · P. Gupta · S. K. Khatri
Amity Institute of Information Technology, Amity University, Sec-125, Noida, UP, India
e-mail: akakkar@amity.edu

S. Rawat
e-mail: srawat@amity.edu

P. Gupta
e-mail: piyushhgg23@gmail.com

S. K. Khatri
e-mail: skkhatri@amity.edu

© Springer Nature Singapore Pte Ltd. 2018   643
S. Bhalla et al. (eds.), *Intelligent Computing and Information and Communication*,
Advances in Intelligent Systems and Computing 673,
https://doi.org/10.1007/978-981-10-7245-1_63

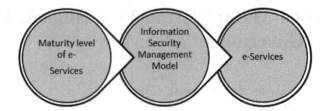

Fig. 1 Protection layers of e-services

Himachal Pradesh (HP) government have currently 16 existing e-services on their portal and this paper focuses on five major services of the government and suggests some improvements and changes that can be incorporated using a self-developed maturity model. These changes will make the experience of the user a simple and a straightforward process. The e-services currently offered by the Himachal Pradesh (HP) government are effort centric rather than result centric [3].

The information security maturity model offers a new approach to safeguard the individual's information over the Internet by improving the existing e-portal for the services of Himachal Pradesh (HP) government. The major lapse behind the security issue being that the Himachal Pradesh (HP) government site is on hypertext transfer protocol (http) rather being on hypertext transfer protocol secure (https), which allows any hacker an easy access to the sensitive information of the state government and its citizens. This issue needs to be considered by the state government IT department as a priority. To help resolve this issue, a multi-level authentication portal can be created for e-services, which has been described under information security model section.

Figure 1 displays the protection layers of e-services. The maturity level of e-services for an e-governance portal and proposed information security model is created and described in Sects. 4 and 5 respectively.

The e-services of Himachal Pradesh (HP) government included in the present paper are the ones that are providing essential services over the Internet to its residents. The e-services are (Fig. 2) as follows.

All these e-services need major improvements which can be achieved by the proposed maturity level in the Sect. 4.

2 Literature Survey

Governments have been slow in realizing the potential benefits of information technology and rather snail paced in providing e-services [4]. e-services are cost-effective services, which can drive the growth of economy and government productivity to new heights in brief period of time. Out of the existing 16 e-services of Himachal Pradesh (HP) government, the 5 e-services mentioned in the paper are the ones which provide essentials services to Himachal Pradesh (HP) citizens. It has

Employment Job Portal:

The portal provides information for the citizens looking for job opportunities in the HP state. Job seeker can apply for registration, view their details and update their details.

Lok Praman Patra:

This is the service which is essentially for the Department of HP Employees, but can be and should be extended to citizen's level as it provides the information of all the certificates needed to be a resident of a state.

MC, Shimla:

This E-Service provide the State Capital(Shimla) citizens to pay their Water and Property Tax bills online.

HRTC: This E-Service is related with booking tickets for State Deployed buses online

HPSEBL: This E-Service is related with Electricity online bill payment. There can be a separate portal for the bill payment as this site does not offer much to citizens rather than bill payment.

Fig. 2 e-services of HP government

been seen that since the emergence of e-services, the cost of data storage and hardware costs have also taken an upward ride. Fortunately, our technologists have discovered a solution to this as cloud computing. Cloud computing provides a new service consumption and delivery model inspired by consumer internet services. Cloud computing cuts down costs and increases the cost benefits [5]. Himachal Pradesh (HP) government is using a government initiative MeghRaj cloud for their e-services.

The existing e-services are currently not known to the citizens of the rural areas of the Himachal Pradesh (HP) state in comparison to the urban areas of the state. To do this an awareness campaign is needed to initiate by the government to educate the population of the remote areas of the state. There is also a major concern over trust, protection and safety of information which demands a high level of security within e-government organizations [6]. For this the state government must update its most critical security issue of not being a hypertext transfer protocol server (https) site, this is highly essential to popularize the use of e-services as it upgrades the security level of the portal to new level. The paper has the following sections which describe the work done and solutions proposed:

- Methodology Adopted
- Maturity model
- Information security model

The e-services portal is a great initiative by the Government of India and Himachal Pradesh government is concerned with making the initiative a success [7]. However, in Himachal Pradesh, there are challenges faced by government in effectively delivering the e-services to its citizens. These are majorly reaching out to its citizens and making them aware of the existence of these services. Challenges like this and others have been explained in different levels of the proposed model.

3 Methodology Adopted

- A set of questionnaires were created for NIC engineers working at the headquarters of IT department of Shimla, Himachal Pradesh (HP) to gather the information about how the e-services work and how well are they deployed.
- The questionnaires were instrumental in defining the working of IT department and their role in the maintenance of e-services of Himachal Pradesh (HP) government.
- NIC engineers, of IT department Himachal Pradesh (HP) government answered the questions effectively and imparted the information helping in the formation of the maturity model of the e-services.

These answers help to understand a lot about e-services and their working and maintenance over the internet. The government is using data centres to host all government applications and is also using a cloud service provider known as MeghRaj cloud to provide its services. The sites are using VPN to update the content and secure the information. The operating system used by Himachal Pradesh (HP) government for hosting their sites on the cloud is Red Hat Linux system. The technology currently in use by the government is MS platform and is working on other platforms such as Android, iOS, and windows applications. The sites' data is maintained and updated by NIC developer for any site and application. NIC has designated nodal officers for the purpose to keep the content up to date using CMS (Content Management System) available over VPN (Virtual Private Network).

The questionnaires answer helped me gain an understanding as to where the e-portal of Himachal Pradesh (HP) government can be enhanced. This made me develop a maturity model for the e-services portal listing the enhancement that can be made to the major e-services of the government. The model proposed is generic, i.e. it can be used to enhance any government e-services' portal which may need improvements.

4 Proposed Maturity Model

The maturity of e-services depends upon what kind of maturity level is used to describe their processes [8]. A maturity model is a method for judging the maturity of the processes of an organization and to identify the key practices that are required to increase the maturity of these processes.

A maturity model guides a government in selecting process improvement strategies by determining current process capability and its maturity and identifying the few issues that are most critical to e-governance quality and process improvement [9].

The paper proposes a new maturity model, first one for e-services of Himachal Pradesh (HP) government. It can provide the basis for developing a better e-services portal for the citizens to have access. Along with the proposed maturity model, the current level of e-services has also been presented.

Figure 3 describes the proposed levels for maturity model of e-services from which any state government can benefit to provide better and effective services to its citizens. These levels are explained as follows:

Level 1

- **Web Presence**: The websites for the e-services must be up and present on the Internet, it can be static or dynamic website.
- **Accessibility**: The e-service site and the services provided by site's data should be accessible 2×7.

Level 2

- **Usability**: The sites of e-services must be easy to use and user friendly.
- **Dynamic**: Site of services and their database should be updated regularly by an administrator.
- **Self-Service and Financial Transaction**: The e-services sites must have self-services with the possibility of electronic payments.
- **Transparency**: The information processing and data transparency should be there on the sites. Every recent update related to the service should be uploaded to the site.

Level 3

- **Customer Support**: The e-services sites need to provide a good customer support to attend queries of its citizens.

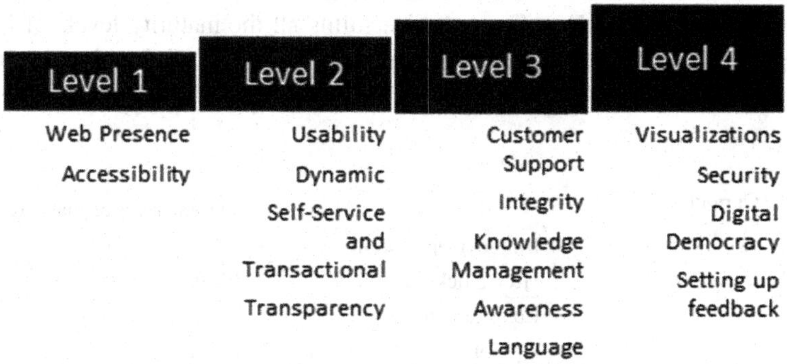

Level 1	Level 2	Level 3	Level 4
Web Presence	Usability	Customer Support	Visualizations
Accessibility	Dynamic	Integrity	Security
	Self-Service and Transactional	Knowledge Management	Digital Democracy
	Transparency	Awareness	Setting up feedback
		Language	

Fig. 3 Proposed levels of maturity

- **Knowledge Management**: A tab providing all the information related to the site and FAQs where all the questions should be answered related to all the sites.
- **Integrity**: Involves maintaining consistent accuracy and trustworthiness that data will not be misused over an entire life cycle. Data must not change in transit. Steps must be ensured that unauthorized people do not get access to the data.
- **Awareness**: The citizens must be made aware of the e-services and its content they have access to
- **Language**: The e-services are for all the sections of citizens, therefore it is mandatory that every e-service must be in English and Hindi.

Level 4

- *Visualisations*: They can be added to e-services to make the interaction with site simpler and impressive
- *Security*: The e-services portal needs to provide great security features to keep the content safe and secure.
- *Digital Democracy*: This maturity level specifies the option of online voting, public forums and opinion surveys to make the e-service delivery a more better prospect.
- *Setting up Feedback*: A tab where the user can give the feedback of the interaction and impart their valuable suggestions to make site and services more friendly for the user
- *Continuous Improvements*: The e-services must be looked upon as to what improvements can be made so to make the interaction with user more simpler and easy.

The maturity levels of current e-services of Himachal Pradesh (HP) government are tested under the proposed maturity model to define the present level of maturity these services have (Table 1).

Test Results
Employment Portal: The service only has web-presence and accessibility of data at all times is not available, so the e-services are not up to level 1.

HRTC Ticket Booking: The e-service fulfils all the maturity levels of level 1 and level 2 of the model, however, it can be matured to level 4 with most ease in comparison of other services.

Table 1 Current level of e-services

e-services	Current level of maturity
Employment portal	–
HRTC ticket booking	Level 2
Lok Praman	–
Hpsebl	Level 2
MC, Shimla	Level 1

Lok Praman: The e-service only has web presence, the data content of the site is not accessible to its citizens. Maharashtra government public e-services portal can serve as an example to improve the maturity level of this e-service.

HPSEBL: This e-service qualifies to be of level-2 service as its portal meets all the required levels.

MC, Shimla: The e-service fulfils all the maturity levels of level 1 of the model. It is a level 1 e-service, which can be easily moved up to level 2 by following Uttarakhand online water portal.

Following this model, the maturity of the services can be improved for any state government e-services.

The levels described in proposed maturity model are to classify the e-services of the portal as level 1, 2, 3 or 4 services. This model would help understand in enhancing the services and making them more developed. The test results describe the current level of maturity in the e-services.

5 Information Security Model

To ensure security, it is important to build-in security in both the planning and the design phases and adapts a security architecture which makes sure that regular and security-related tasks are deployed correctly [10]. Currently, the e-services of Himachal Pradesh (HP) government have a very basic level of security or if very honest no level of security. The major lapse behind the security issue is that their site is on hypertext transfer protocol (http) rather being on hypertext transfer protocol server (https), which allows any hacker and easy access to the sensitive information of the state government and its citizens. This issue needs to be considered by state government IT department as a priority.

Proposed e-Portal for the services
The paper proposes a multi-level authentication portal for all the existing e-services of Himachal Pradesh (HP) government. This portal will help the citizens to register once for all the e-services rather than registering individually for all e-services. This portal will partially help solve the issue of data security until the government decides to move from hypertext transfer protocol (http) to hypertext transfer protocol server (https). The following steps will result in creating the desired portal.

Registering on the portal

For simple access to e-services, a service portal can be created with a default page setup to register and login. Users can register with Aadhaar card no. or a working phone number. In case, the user does not possess an Aadhaar card, they can register in the portal by creating a username, password and working phone no. This will enable the generation of a unique ID of eight digits on the subsequent page. Once this unique ID is created, login tab shall appear.

Logging into services:

Users will be required to have two credentials to log in for to have access to all the services:

1. USER_ID
2. OTP (which will be sent to the registered phone.no)

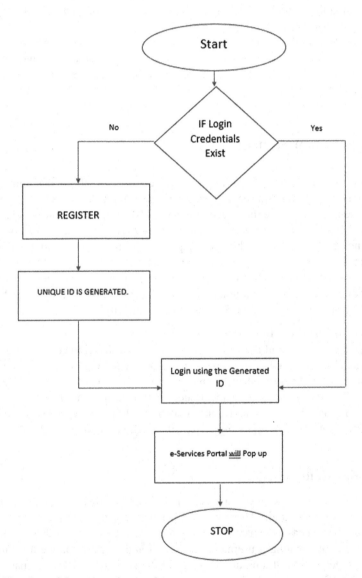

Fig. 4 Flowchart of the proposed e-portal

Once this is created, the user can have access to all the services provided by the portal and are saved from registering himself/herself at each service separately. This user id can also serve as the user's personal digital space where they can upload the needed documents and access all governmental documents, related to each service. This feature can be incorporated on the same page that indicates a list of all the services of the portal, but in a separate section and by the name of documents. This portal can be like a one-stop shop for the users as they will be able to access all the e-services at one go and will be saved from the pain of registering separately for each e-service (Fig. 4).

6 Conclusion

Through all the research done for the paper, it can be safely concluded that the proposed model can be of immense help to the Himachal Pradesh (HP) government in improving their e-services. Since this paper proposes the first ever maturity model for both information security and services of the e-services of Himachal Pradesh (HP) government, it can be appropriated as a standard base model for the improvement of the portal.

Acknowledgements I would like to express my gratitude to the Founder-President of Amity Group, Dr. Ashok K. Chauhan, for his keen interest in promoting research in Amity University. I have had the opportunity of interacting with several people whose inputs have been extremely vital. I would also like to thank the NIC engineers Mr. Sandeep Soodand for answering my questionnaires and providing help as to better understand the existing e-portal of the Himachal Pradesh (HP) government. Mr. Sandeep Sood has provided his consent for the paper for which I am highly obliged. NIC Engineer has been of great help in my research paper providing the base for my paper contents. Mr. Sood's consent in using the details provided for the paper have been of great help.

References

1. M. D. Waziri, Z. O. Yonah, "A Secure Maturity Model for Protecting e-Government Services: A Case of Tanzania", ACSIJ Advances in Computer Science: An International Journal, Vol. 3, Issue 5, No.11, pp. 98–106, 2014
2. M. R. Perumal, "E-governance in India: A Strategic Framework", International journal for Infonomics, Vol-1, pp. 1–12, 2014
3. S. Gupta, S. Kaushal, "E-Governance Infrastructure-Status and Challenges Case Study on Himachal Pradesh", National Informatics Centre Department of Electronics and Information Technology, Vol-1, pp. 1–6, 2013
4. V. Varma, "Cloud Computing for E-Governance", International Institute of Information Technology (White paper) Vol-1, pp. 10–23, 2010
5. R. Larbi, "A Proposed Cloud Security Framework for Service Providers in Ghana", International Journal of Computer Applications (0975 – 8887), Vol 158, pp. 17–22, 2017

6. V. A. Canal, "Information Security Management Maturity Model", www.isecom.org, Vol-2, pp. 11–24,2011 Lok Praman Patra, http://admis.hp.nic.in/epraman/

7. M.P. Gupta, "E-government evaluation: A framework and case study", Indian Institute of Technology, Department of Management Studies, Vol-1, pp. 365–387, 2010

8. T. N. Raja, "A Maturity Model Framework for e-Gov Applications – EGSARMM", International Journal of Computer Applications® (IJCA), Vol-1, pp. 20–25, 2012

9. A. F. Allah, L. Cheikhi, "E-Government Maturity models: A Comparative study", International Journal of Software Engineering & Applications (IJSEA), Vol.5, No.3, pp. 71–85, 2014

10. B. Stevanović, "Maturity Models in Information Security", International Journal of Information and Communication Technology Research, Vol 4, No. 2, pp. 15–19, 2011

Malaria Detection Using Improved Fuzzy Algorithm

**Mukul Sharma, Rajat Mittal, Tanupriya Choudhury,
Suresh Chand Satapathy and Praveen Kumar**

Abstract Malaria is one of the most life-threatening diseases, which need a serious attention in today's scenario. This disease is estimated to be having 3–6 billion infected cases worldwide annually having the mortality rate of 1–3 million people. Malaria should be diagnosed on time and treated precisely as it can lead to death of a person. The main objective of this paper is to design and describe an algorithm that can diagnose malaria in the early stage only so that a person cannot go up to the serious and hazardous stages or complications and also the mortality rate of malaria is reduced to an extent. Fuzzy logic is an approach to implement expert systems, which are portable and can diagnose malaria accurately as compared to the other systems.

Keywords Fuzzy sets · Fuzzy algorithm · Malaria · Fuzzy rules
Fuzzy values

M. Sharma (✉) · R. Mittal · T. Choudhury · P. Kumar
Amity School of Engineering & Technology, Amity University, Noida, Uttar Pradesh, India
e-mail: mukulrock38@gmail.com

R. Mittal
e-mail: rajatmittal100@gmail.com

T. Choudhury
e-mail: tchoudhury@amity.edu

P. Kumar
e-mail: pkumar3@amity.edu

S. C. Satapathy
P.V.P. Siddhartha Institute of Technology, Kanuru, Vijayawada, India
e-mail: sureshsatapathy@gmail.com

© Springer Nature Singapore Pte Ltd. 2018   653
S. Bhalla et al. (eds.), *Intelligent Computing and Information and Communication*,
Advances in Intelligent Systems and Computing 673,
https://doi.org/10.1007/978-981-10-7245-1_64

1   Introduction

Malaria is disease caused by the group of single-celled microorganism called parasitic protozoan. It is an infectious disease caused by liver parasites or spleen of genus plasmodium. Once the parasite is transmitted, they start getting multiplied in the liver before destroying red blood cell. Symptoms of malaria that are considered in this paper are fever, vomiting, headache, weakness, chills, nausea, sweating, diarrhoea and enlarged liver. Soft computing technique involves various methods like approximate reasoning and functional approximation. In approximate reasoning, collection of imprecise premises is used to deduce imprecise conclusion. Probabilistic reasoning and fuzzy logic are the methods involved in approximate reasoning. In functional approximation, a function among well-defined class is selected that closely matches a target function in a different way. Evolutionary computation, artificial neural networks and machine learning are the methods involved in functional approximation. There are several methods that exist for the malaria diagnosis. The methods can be classified on the basis of performance and cost. Polymerase chain reaction (PCR) and third harmonic generation (THG) are high cost methods. PCR method is used to detect nucleic acid sequences. According the studies these methods are best and are highly sensitive to malaria diagnosis. But due to their high cost, handling difficulties and infrastructures, they are rarely used in developing countries. Another method called RDT is relatively faster methods than other in diagnosis and can be performed by unskilled people but results are not reliable. Since RDT kits can be used for the single species, so four kits are used which make it expensive. Another method called conventional microscopy is used for the diagnosis but this time consuming and results are difficult to reproduce. From the above discussion, we found that more the sophisticated technique is, results are more reliable but very expensive and unaffordable in many countries. On the other hand, less sophisticated techniques are cheap and affordable but the results are not reliable.

2   Literature Survey

Chandra et al. [1] presents detection of malaria disease through soft computing. For this, image processing technique is used which is done by following some stages like image filtering, image pre-processing, image segmentation, feature extraction and artificial neural network (ANN). Mohamed et al. [2] presents a method to diagnose malaria parasite using the fuzzy logic systems. A proper methodology is mentioned for this approach of diagnosis of malaria. A certain number of data is taken from a database and then this data is received by the system through input technology. Fuzzification and defuzzification of the data are done and a desired output is obtained by the system. Djam et al. [3] describe (FESMM), i.e. fuzzy expert system for the management of malaria and giving a decisive as well as supportive platform to the scientists who are engaged in the research of the malaria disease, the specialist doctors and the healthcare experts. This FESMM comprises of 4 blocks which are (1) Data Collection (2) Fuzzification (3) Inference Engine and (4) Defuzzification. The basic idea behind the designing of the fuzzy expert system is based on the clinical perceptions, medical analysis and healthcare practitioners. Onuwa et al. [4], presents an expert system called fuzzy expert system for the detection of malaria disease. The paperwork also focuses on that the algorithm which is built to diagnose malaria rapidly and accurately is very simple to use, cheap and portable. These qualities of algorithm make it perfect to use in the diagnosis of malaria. This algorithm not only supports medical practitioners, but also helps malaria researchers to manage the imprecision and provides accuracy in the final output on the basis of given input. Sharma et al. [5], in this paper the authors have proposed a decision support system for malaria and dengue disease diagnosis for the remote (rural) areas where medical support is not available to the people easily. The decision support system for malaria diagnosis and treatment was designed, developed and simulated by using MATLAB, GUI function with the implementation of fuzzy logic. Duodu et al. [6] presents a brief description about malaria and its consequences, this also tells us about how fuzzy logic can be implemented in the various steps like collecting the data of the patients suffering from malaria disease, collected data is used as an input, fuzzification is done on the input, fuzzified data is matched with the pre-defined set of fuzzy logic in the inference engine and finally defuzzification is done which gives a representative output from the inference engine.

3 Fuzzy Logic System

In 1965, Prof. LotfiZadeh introduced a fuzzy logic system with the help of a fuzzy set of theory. In this theory, a group of objects having unclear boundaries can relate in which membership is just a matter of degree. By making the use of fuzzy approach, transition in terms can be binary, gradual or null; selection must be at the extreme end of the continual. Fuzzy logic system identifies and uses a real-world phenomenon. This method makes use of the concept of fuzziness in computationally better manner. It is basically associated with the reasoning that mainly uses approximate value rather being a fixed value.

- Fuzzy sets can be seen as an expansion and speculation of the basic concepts of the crisp set, which tells us that either an element is a member of set or not but on the other hand fuzzy sets gives partial memberships to the elements in a set with the degree of membership ranging from 0 to 1. 0 (not an element of a particular set) and 1 (element of a particular set). Different types functions such as triangular, trapezoidal, Gaussian and sigmoid membership functions are there, but in this paper, the membership function which is used is triangular membership function whose calculation can be done with the help of the below-mentioned functions (Fig. 1).

This Triangular Function is defined by a lower limit 'a', an upper limit 'c' and a value 'b' lying between 'a' and 'c'

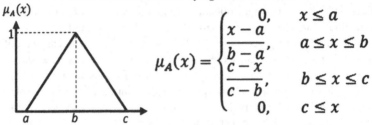

$$\mu_A(x) = \begin{cases} 0, & x \leq a \\ \dfrac{x-a}{b-a}, & a \leq x \leq b \\ \dfrac{c-x}{c-b}, & b \leq x \leq c \\ 0, & c \leq x \end{cases}$$

Fig. 1 Triangular membership function

A fuzzy set B in X can be expressed as a set of ordered pairs:

$$B = \{(x, \mu B())|x \in X\} \tag{1}$$

where μB = membership function,

$$\mu B : X \rightarrow P \tag{2}$$

where P is the membership space in which each element of X is mapped.

If $P = \{0, 1\}$, B is called crisp set.

However, if $\{0 \leq P \geq 1\}$, B is called fuzzy set.

Fuzzy logic has been applied in many fields such as artificial intelligence and control theory. It has various applications from customer end such as cameras, washing machines, ovens to various industrial processes.

4   Methodology

Structure of the Fuzzy System

Fuzzy system has an AMDFLT (algorithm for malaria diagnosis using fuzzy logic for treatment) structure shown in the figure, which consists of four major parts, i.e. fuzzification, knowledge base, fuzzy inference engine and defuzzification. Other components include input and output data.

4.1   Data Collection

In the literature survey, various observation and various other patient studies are used to collect the information to make a relation between input and output variables, create various functions and design fuzzy logic decision matrix.

Knowledge base and database are maintained in the data store. Malaria reference review and medical experts are major sources of data store. Knowledge base consists of decision matrices, which has a set of condition called IF and THEN which is formed using logic gate operator and forward chaining and storing information for references with the help of fuzzy inference engine.

4.2 Input Data

Mainly keyboard is used as input resource to enter various data into the system. Exact numbers are taken for patient's pulse rate, body temperature, weight and height from the different instruments in BPM, Celsius, kilogram and metres. Intensity values that are converted are determined in 100% and entered.

4.3 Fuzzification

Member functions (fuzzifier) are used to convert crisp input values into grades of membership degrees. In fuzzification, we can obtain a fuzzy value from real scalar value. This all can be obtained from different types of fuzzifiers. Three types are as follows: Are Singleton fuzzifier, Gaussian fuzzifier, Triangular fuzzifier.

Mathematically,

$$x = \text{fuzzifier}(x_0) \tag{3}$$

Where x is fuzzy set, x_0 is called crisp input value and fuzzifier represents fuzzification function. Mainly for changing the scalar values to fuzzy sets (range 0–1), triangular fuzzifier is used. Later crisp value is converted to a linguistic variable. To obtain degree of symptom, a formula is used, i.e.

$$X_i/X_n \tag{4}$$

X_i is no. of linguistic variable and X_n is total no. of linguistic variable. Through this formula, we can make the table for triangular fuzzifier. For a case, if a patient is complaining of very high fever then the degree of symptom will be 4/5 = 0.80 (Table 1).

4.4 Data Store

One of the main key factors responsible for the intelligent system is the knowledge. Knowledge consists of mainly database. Knowledge base acts as one of the major

Table 1 Ranges of fuzzy value

Linguistic variables	Fuzzy values
Very low	$0.00 \leq x < 0.20$
Low	$0.20 \leq x < 0.40$
Medium	$0.40 \leq x < 0.60$
High	$0.60 \leq x < 0.80$
Very high	$0.80 \leq x \leq 1.00$

source for the information to be searched collected, organised, shared and utilised. The knowledge base for the malaria diagnosis is composed of mainly structured and organised information. Knowledge base was taken from various books on malaria and medical experts. In knowledge base, forward chaining and logical AND operator is used to construct the if-then decision matrices and fuzzy inference engine is used to store data for references. This data is stored for the future references.

Representation of the above-mentioned rules from A to J is as follows:

A IF Fever = Very High, Vomiting = High, Headache = High, Weakness = High, Chills = Medium, Nausea = Medium, Sweating = Low, Diarrhoea = High and Enlarged Liver = Medium THEN Malaria = High

B IF Fever = Medium, Vomiting = Low, Headache = Very Low, Weakness = Low, Chills = Very Low, Nausea = Very Low, Sweating = Very Low, Diarrhoea = Very Low and Enlarged Liver = Very Low THEN Malaria = Very Low

C IF Fever = Low, Vomiting = Very Low, Headache = Very Low, Weakness = Low, Chills = Very Low, Nausea = Very Low, Sweating = Low, Diarrhoea = Very Low and Enlarged Liver = Very Low THEN Malaria = Very Low

D IF Fever = High, Vomiting = High, Headache = Low, Weakness = Medium, Chills = Low, Nausea = Very Low, Sweating = Medium, Diarrhoea = Low and Enlarged Liver = Low THEN Malaria = Medium

E IF Fever = Medium, Vomiting = Low, Headache = Very Low, Weakness = Medium, Chills = Very Low, Nausea = Very Low, Sweating = Low, Diarrhoea = Very Low and Enlarged Liver = Very Low THEN Malaria = Low

F IF Fever = Medium, Vomiting = High, Headache = High, Weakness = Medium, Chills = Low, Nausea = Very Low, Sweating = Medium, Diarrhoea = Medium and Enlarged Liver = Very Low THEN Malaria = Medium

G IF Fever = Very High, Vomiting = Very High, Headache = Very High, Weakness = Very High, Chills = Very High, Nausea = Very High, Sweating = High, Diarrhoea = High and Enlarged Liver = High THEN Malaria = Very High

H IF Fever = Very High, Vomiting = High, Headache = High, Weakness = Very High, Chills = Very High, Nausea = Very High, Sweating = Very High, Diarrhoea = Very High and Enlarged Liver = Very High THEN Malaria = Very High

I   IF   Fever = Very   High,   Vomiting = Very   High,   Headache = High, Weakness = High,   Chills = Medium,   Nausea = Medium,   Sweating = High, Diarrhoea = High and Enlarged Liver = High THEN Malaria = High

J   IF Fever = Low, Vomiting = Low, Headache = Very Low, Weakness = Low, Chills = Very   Low,   Nausea = Very   Low,   Sweating = Very   Low, Diarrhoea = Low and Enlarged Liver = Low THEN Malaria = Low

4.5 Fuzzy Inference Engine

Fuzzy inference system uses knowledge base to derive the answer. It is also sometimes referred as the brain of expert system that is used to reason the information in knowledge base and make new conclusions. In fuzzy inference engine, the degree of membership is determined by mapping the fuzzy inputs into their linguistic variables and weighting factors. The degree of the firing strength is determined by aggregation operator. And for this fuzzy logic AND is used. In fuzzy rule sets, combination many fuzzy logical operators such as AND, OR and NOT are used to have several predecessor. AND operator—used for minimum weight for all predecessor. OR operator—uses maximum value for all predecessors. NOT operator—used to find complementary function by subtracting membership function from 1. RSS also called as Root Sum Square is used in this research work whose formula is given by

$$\text{RSS} = \sqrt{\sum\nolimits_{i=1}^{n} \sigma_i^2} \tag{5}$$

RSS = value of firing rule and σ_i^2 are the values of different rules
Fuzzy inference engine has many uses.

Input data file and FIS structure file can be used to perform fuzzy inference. Fuzzy inference engine can be customised accordingly to include our own membership function. Executable codes can be embedded into other application. Fuzzy inference mainly consists of two c code source files, i.e. fis.c and fismain.c in fuzzy folder. In fisman.c file, it has only one main function which can be easily modified for our own application as it is ANSI C compatible.

4.6 Defuzzification

The process of converting a fuzzy output to a crisp value from inference engine is called defuzzification. Defuzzification is a method used to obtain a value that will represent the output obtained from the inference engine. That is, the level of the illness can be determined from the result which we get from the inference system using defuzzified root sum square. The technique developed by Takagi and Sugeno in 1985 called centroid of the area is also called centre of area or centre of gravity, which is used for defuzzification. Since it is very simple and accurate, it becomes one of the most commonly used techniques. The centroid defuzzification technique is

$$X^* = \frac{\int_{xmin}^{xmax} f(x)\mathrm{d}x}{\int_{xmin}^{xmax} f(x)\mathrm{d}x} \tag{6}$$

In the above equation, x^* refers to defuzzified output where $f(x)$ dx refers to aggregated membership function and the variable x indicates the output variable (Fig. 2).

5 Algorithm

The algorithm for the fuzzy logic system is as follows:

Step 1 Enter the details such as symptoms and indications of the patient into the system, where the variable m represents a number of symptoms and indications of the patient.

Step 2 Now, we have to rectify the signs and symptoms of the patient for which we have to explore the knowledge base for the disease which consists those symptoms.

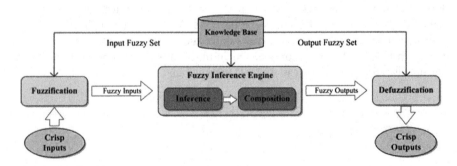

Fig. 2 Architectural structure of fuzzy system

Step 3 Now get the weighing factors (the associated degree of intensity) wf = 1, 2, 3, 4, 5 where 1 refers to very low, 2 refers to low, 3 refers to medium, 4 refers to high and 5 refers to very high.

Step 4 Now, fuzzy rules are applied.

Step 5 In the next step, the degree of membership is determined for which we have to map the fuzzy inputs into their particular weighing factors.

Step 6 Now the rule base is to be determined for which the non-minimum values can be evaluated.

Step 7 Evaluate the strength of firing (conclusion) for the rules R.

Step 8 The non-zero minimum values are evaluated to calculate the degree of truth R.

Step 9 Now the intensity of the disease is calculated.

Step 10 Now we get the output as fuzzy diagnosis.

6 Research Experimentation

In this research experiment, a sample of 5 patients who have already gone through medical doctor diagnosis and laboratory diagnosis has been taken consideration. Fuzzy system is used to diagnose the patient, based on the complaint made by him in order to get the knowledge of how the system works when implemented. Grades such as very low, low, medium, high and very high are used. A patient called Manmohan with patient ID AAS002 complains the doctor having symptoms of malaria as shown in Table 3. Fuzzy rules are shown in Table 4. About 10 rule bases are used for the patient in Table 2. Root sum square is used to compute the output membership function in the range of 0–1 for every linguistic variable. Now, the centroid of the area is used to defuzzify the outputs from the inference engine to obtain a crisp value.

7 Comparing Results of Malaria Diagnosis

The basic aim is to have a comparison between the algorithm's diagnosis with the doctor's diagnosis using laboratory diagnosis as accurate and standard diagnosis (Table 5).

Table 2 Fuzzy rule composition for the detection of malaria

Rule No	Fever	Vomiting	Headache	Weakness	Chills	Nausea	Sweating	Diarrhoea	Enlarged liver	Linguistic variables
A	0.9	0.7	0.7	0.7	0.5	0.5	0.3	0.7	0.5	High
B	0.5	0.3	0.1	0.3	0.1	0.1	0.1	0.1	0.1	Very low
C	0.3	0.1	0.1	0.3	0.1	0.1	0.3	0.1	0.1	Very low
D	0.7	0.7	0.3	0.5	0.3	0.1	0.5	0.3	0.3	Medium
E	0.5	0.3	0.1	0.5	0.1	0.1	0.3	0.1	0.1	Low
F	0.5	0.7	0.7	0.5	0.3	0.1	0.5	0.5	0.1	Medium
G	0.9	0.9	0.9	0.9	0.9	0.9	0.7	0.7	0.7	Very high
H	0.9	0.7	0.7	0.9	0.9	0.9	0.9	0.9	0.9	Very high
I	0.9	0.9	0.7	0.7	0.5	0.5	0.7	0.7	0.7	High
J	0.3	0.3	0.1	0.3	0.1	0.1	0.1	0.3	0.3	Low

Table 3 Symptoms of Malaria in MANMOHAN

Symptoms	Linguistic variables	Given values
Fever	Very high	0.9
Vomiting	High	0.7
Headache	Very high	0.9
Weakness	Very high	0.9
Chills	High	0.7
Nausea	Medium	0.5
Sweating	High	0.7
Diarrhoea	Medium	0.5
Enlarged liver	High	0.7

8 Results

By applying Eq 5, i.e. the formula of root sum square, we will get the actual fuzzy values corresponding to the respective linguistic values.

- Very Low—0.1414, Low—0.3162, Medium—0.4242, High—0.7071, Very high—0.9899

OUTPUT X*

$$= \frac{(0.1414 * 0.1) + (0.3162 * 0.3) + (0.4242 * 0.5) + (0.7071 * 0.7) + (0.9899 * 0.9)}{0.1411 + 0.3162 + 0.4242 + 0.7070 + 0.9899}$$

$$= 0.6619$$

Hence, the patient is having 66% malaria which is 'High' according to our linguistic variable which is at a complicated stage.

9 Conclusion

This research paper provides the algorithm of fuzzy logic for the diagnosis of the disease 'Malaria'. It also tells us that fuzzy logic is a very helpful technique which can deal precisely with the issues that occur in malaria. By following the proper methodology of fuzzy logic that is mentioned above, one can accurately detect malaria without going through the second opinion. Mamdani Fuzzy Inference System is a better system for the diagnosis of human disease. A sample of 5 patients was selected and the result of diagnosis by the doctor, the laboratory and the algorithm was compared with each other which proves that the fuzzy logic algorithm shows the best results amongst the doctor diagnosis and the laboratory

Table 4 Fuzzy rules for patient MANMOHAN

Rule No	Fever	Vomiting	Headache	Weakness	Chills	Nausea	Sweating	Diarrhoea	Enlarged liver	Linguistic variables	Non-Zero minimum value
A	0.9	0.7	–	0.7	0.5	–	–	–	0.5	High	0.5
B	–	–	0.1	0.3	–	0.1	0.1	–	0.1	Very low	0.1
C	0.3	–	–	–	–	–	–	0.1	0.1	Very low	0.1
D	–	0.7	–	–	0.3	–	–	–	–	Medium	0.3
E	–	–	0.1	0.5	–	–	0.3	0.1	–	Low	0.1
F	0.5	–	0.7	–	0.3	–	0.5	–	–	Medium	0.3
G	–	0.9	–	0.9	–	0.9	–	0.7	0.7	Very high	0.7
H	0.9	–	0.7	–	0.9	–	–	0.9	–	Very high	0.7
I	–	–	0.7	0.7	–	0.5	0.7	–	0.7	High	0.5
J	–	0.3	–	–	–	–	–	–	0.3	Low	0.3

Table 5 Compared results of malaria diagnosis

Patient id	Doctor diagnosis	Laboratory diagnosis	Algorithm diagnosis
AAS001	Uncomplicated malaria	Uncomplicated malaria	Uncomplicated malaria
AAS002	Free from malaria	Uncomplicated malaria	Complicated malaria
AAS003	Uncomplicated malaria	Free from malaria	Free from malaria
AAS004	Uncomplicated malaria	Uncomplicated malaria	Uncomplicated malaria
AAS005	Free from malaria	Complicated malaria	Uncomplicated malaria

diagnosis, some of the results were same in the laboratory and the algorithm diagnosis but the accurate results were shown by the algorithm diagnosis.

References

1. Upendra Kumar Chandra, Yogesh Bahendwar, Volume 5, Issue 10, October-2015,"Detection of Malaria Disease through Soft Computing" International Journal of Advanced Research in Computer Science and Software Engineering [ISSN: 2277 128X].
2. Khalid Alsir Mohamed, Eltahir Mohamed Hussein, Index Copernicus Value (2013): 6.14 | Impact Factor (2015): 6.391 Volume 5 Issue 6, June 2016, "Malaria Parasite Diagnosis using Fuzzy Logic" International Journal of Science and Research (IJSR) [ISSN (Online): 2319–7064].
3. Djam, X. Y, Wajiga, G. M, Kimbi Y. H, and Blamah, N. V (2011).A Fuzzy Expert System for the Management of Malaria. International Journal of Pure and Applied Sciences and Technology (5(2) (2011), pp. 84–108.
4. "Fuzzy Expert System for Malaria Diagnosis" by Ojeme Blessing Onuwa et. all, An International Open Free Access, Peer Reviewed Research Journal, Published By: Oriental Scientific Publishing Co., India. June2014, Vol.7, No. (2):Pgs. 273–284 [ISSN: 0974-6471].
5. "Decision Support System for Malaria and Dengue Disease Diagnosis (DSSMD)" by Priynka Sharma, DBV Singh, Manoj Kumar Bandil and Nidhi Mishra, International Journal of Information and Computation Technology.Volume 3, Number 7 (2013), pp. 633–640, [ISSN: 0974-2239].
6. Quashie Duodu, Joseph Kobina Panford, James Ben Hafron-Acquah, Volume 91 – No.17, April 2014, "Designing Algorithm for Malaria Diagnosis using Fuzzy Logic for Treatment (AMDFLT) in Ghana" International Journal of Computer Applications [ISSN: 0975 – 8887].

Feature Extraction Techniques Based on Human Auditory System

Sanya Jain and Divya Gupta

Abstract Feature extraction is a relevant method in the performance of the ASR system. A good technique not only removes irrelevant characteristics, but also represents important attributes of a speech signal. This paper intends to concentrate on the comparison between feature extractions techniques of speech signals based on human auditory system for better understanding and to enhance its further applications. In this review, we explain three different techniques for feature extraction. The main emphasis is to show how they are useful in processing signals and extracting features from unprocessed signals. The human auditory system is explained which combines this study altogether. The aim is to describe techniques like Zero Crossing with Peak Amplitude, Perceptual Linear Prediction and Mel Frequency Cepstral Coefficient. As each method has its own merits and demerits, we have discussed some of the most important features of these techniques.

Keywords Feature extraction · Perceptual linear prediction · Mel frequency cepstral coefficient · Zero crossing with peak amplitude

1 Introduction

Speech is the primary source of communication with people around us. If there is no speech, contacting people surely is difficult. With the emergence of digital technology, highly resourceful digital processors were invented with relatively high power, low cost and high speed. With their development, the researchers were able to transform analog speech signal to digital speech signal, which was undergone a study

S. Jain (✉) · D. Gupta
Department of Computer Science and Engineering, Amity School of Engineering and Technology, Amity University, Noida Uttar Pradesh, India
e-mail: sannyajain@gmail.com

D. Gupta
e-mail: dgupta1@amity.edu

© Springer Nature Singapore Pte Ltd. 2018
S. Bhalla et al. (eds.), *Intelligent Computing and Information and Communication*,
Advances in Intelligent Systems and Computing 673,
https://doi.org/10.1007/978-981-10-7245-1_65

for future applications. In the field of digital signal processing, speech recognition or processing has versatile relevance so it is an intensive field of research.

Automatic Speech Recognition (ASR) has emerged in digital speech signal processing applications [1]. In this process, the spoken words are automatically transformed into the written language by computer systems and smartphone also nowadays. The applications of speech processing are increasing day by day, for example in information searching on websites, data entry in fields and even in biometric recognition technologies [2].

Different data and algorithms are combined together such as statistical pattern recognition, linguistics, signal processing, etc., for speech recognition [3]. The front end of signal processing is feature extraction, which also has been proved to be of greater importance for parametric representation of signals.

A slight amount of data is extracted from the voice signal which is used to build a separate model for each speech utterance; this is known as the parametric representation [4]. It is then used for analysis for representation of specific speech utterance. Figure 1 depicts the speech recognition process. The successful techniques use the psychological process of human hearing in the analysis which is discussed in further sections in this paper.

There are many feature extraction techniques, which are as follows:

- Linear predictive analysis (LPC);
- Zero Crossing Peak Amplitude (ZCPA);
- Linear predictive cepstral coefficients (LPCC);
- Mel frequency cepstral coefficients (MFCC);
- Mel scale cepstral analysis (MEL);
- Perceptual linear predictive coefficients (PLP);
- Relative spectra filtering of log domain coefficients (RASTA);
- Power spectral analysis (FFT);
- First-order derivative (DELTA).

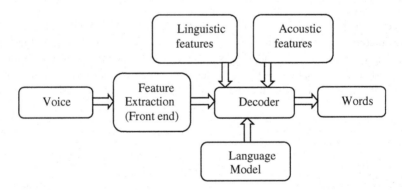

Fig. 1 Speech recognition

2 Human Auditory System

Technological development in speech recognition system presently focuses on providing trusted and robust interfaces for functional deployment. One of the greater challenges is achieving robust behavior of these techniques in noisy areas in the applications which use human–computer dialog system, dictation and voice-controlled devices [5]. Voice acquisition, its processing and its recognition are a complicated task, because of the presence of unwanted noises and variables. These additive noises from the environment in which speech is processed degrades the quality of recorded voice and makes it unfit for processing.

At the very first level of communication, the idea or thought of the message is developed in the mind of the speaker which is also known as the linguistic communication level. That thought is then converted to some significantly meaningful letters, words, phrases and finally sentences according to the grammatical rules of that particular language in which the speaker speaks in. The electric signal which activates muscles in vocal tract and vocal cord is first created by the brain that moves along the motor nerve [6]. This vocal tract movement creates pressure changes within vocal tract and lips, initiates sound waves that is to propagate in space. At last at the linguistic level brain performs speech recognition and understanding of words and sentences.

3 Issues in ASR System

There are many aspects of variability of the voice signal and these can be segregated with the help of a comparatively better ASR system. The system's robustness allows dealing with these variable characteristics.

For achieving a better overall result of voice recognition, it is relevantly important to train and test the system according to the environment it is going to be deployed in [7]. The differences between training and testing condition in a noisy environment often lead to the degradation in the ASR's system's recognition accuracy.

The issues in ASR system are shown in Table 1. When the data which is acquired in noisy condition is tested against the data recorded in clear environment, the quality and accuracy of the data get degraded over time, which is not possible in actual human hearing. To reduce the mismatch between the above-mentioned conditions and degradation in accuracy, various voice signal improvement, extraction and feature normalization techniques are applied one by one.

Table 1 Issues of ASR design

Conditions	Types of noises; Signal-to-noise ratio; training and testing environment conditions
Transducer	Microphone; mobile
Speech styles	Voice pitch; production of sound, e.g. continuous speech or isolated words; pace of speech
Speakers	Speaker dependence/independence; physical state
Channel	Amplitude; reverberation
Vocabulary	Aspects of available training and testing data; vocabulary

4 Overview of Feature Extraction in Speech Recognition

The front end of signal processing is feature extraction, which also has been proved to be of greater importance for parametric representation of signals. A relatively small detail of information is extracted from the signal that is used to build a separate model for each speech expression; this is known as the parametric representation. It is then used for analysis for the representation of specific speech utterance. It separates speech patterns from one another [8]. Figure 2 shows the feature extraction.

Speech signals are quasi-stationary as they are slow varying time signals. This is the reason it is better to perform feature extraction in short-term interval which would reduce these variabilities. Therefore, the examined period of time for the speech signals is 10–30 ms, where the characteristic of speech signal becomes stationary [9]. The features extracted from the speech signals should portray certain qualities such as it should not be subject to mimicry, it should be stable over a time period, does not deviate in different environments, etc. [3]. Parameters which can be determined through processing of speech signals are known as features.

Feature extraction aims to produce a perceptually significant representation of speech signal. Extraction of the feature vector is done through extraction methods

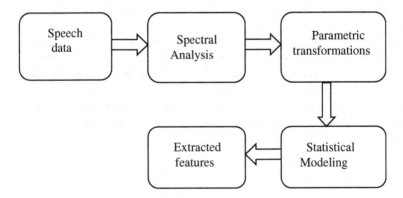

Fig. 2 Feature extraction process

like Mel Frequency Cepstral Coefficient (MFCC), Linear Predictive Coding (LPC), Perceptual Linear Prediction (PLP), Zero Crossing Peak Amplitude (ZCPA), Wavelet Transform, etc.

Feature extraction process can be divided into 3 operations:

- Spectral Analysis:

In time-varying signals, the characteristic properties of speech is represented by parameterization of spectral activity when speech is produced. Six classes in spectral analysis algorithms include Fourier transform, linear prediction, digital filter bank in speech processing system. From the mentioned classes, the linear prediction tends to give the best results.

- Parametric Transforms:

The two basic operations from which the parameters of the signal are generated are differentiation and concatenation. The resultant of both of these stages include the parameter vector of the raw signal.

Differentiation characterizes the temporal variations of speech signals and the signal model's higher order derivative. Concatenation is creating an individual parameter vector per frame that consists of all the signals which are desired.

- Statistical Modeling:

This is the third and last step of feature extraction. It is presumed here that some multivariate process generated the signal parameters. To understand this process, a model is imposed on the data, then the data is optimized for training and finally, the approximation quality of data is measured. The observed output (signal parameter) which we get from this process is the only information. Signal observation is another name given to the resultant parameter vector.

The main intention to perform statistical analysis is to find out if the spoken words are actually meaningful or just noise. Being the fundamental process of the speech recognition, it is very important to use a well-versed statistical model. Vector Quantization (VQ) is certainly the successful and sophisticated algorithm of statistical analysis model.

5 Feature Extraction Techniques

The various methods which are used for feature extraction based on the human auditory system are available, and some of them are discussed below:

5.1 Perceptual Linear Prediction (PLP)

The PLP representation was developed by Hermansky in 1990. As the name suggests, the main aim of the PLP approach is describing the psychophysical human hearing with accuracy throughout the whole feature extraction procedure [10]. Figure 3 depicts the steps to compute PLP coefficient. Using transformations relied on psychophysics; the spectrums are modified in PLP with contradiction to pure LPC analysis [11]. The PLP speech features are as follows: The PLP parameters of 18 filters cover the frequency of range 0–5000 Hz and its dependence are based on Bark spaced Filter bank [8]. The method which computes the PLP coefficients is:

- The $x(n)$ signal which is distinct in the time domain is subjected to the N—point DFT.
- The downsampled is applied with the pre-emphasis of equal loudness.
- Intensity-loudness compression is executed.
- An inverse DFT is performed on the result obtained so far to result in the equivalent functions which are correlated.
- At last, the coefficients of this PLP technique are calculated after autoregressive modeling and conversion of these autoregressive coefficients to cepstral coefficients.

The most often applied acoustic features in speech recognition are mel frequency cepstral coefficients (MFCC) and PLP features. As and when there is found a mismatch in training and test data, PLP features are found to be much more robust [12].

However, when the acoustic conditions are variable over whole data set in many applications, for example: in broadcast news transcription, segments with background music or noise are often intermixed with segments with clean speech [12].

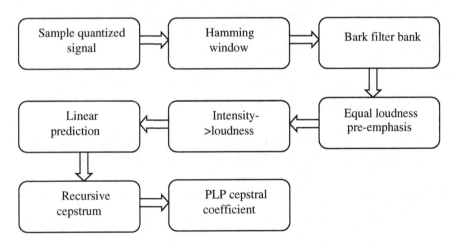

Fig. 3 Computational steps of PLP

The following changes in PLP are considered: (i) Mel filter Bank replaces the Bark filter-bank; (ii) The speech signal which is applied by the pre-emphasis substitutes the equal loudness weighting of spectrum; Finally, (iii), a very large number of filters are contained in new filter bank and there is no significant loss as they have the same bandwidth as the old ones.

5.2 Zero Crossing with Peak Amplitude (ZCPA)

The method ZCPA of extraction of features is based on an auditory system of humans. Zero crossing interval presents information of frequency signals and intensity information represented by amplitude value. Finally, the complete feature output is resulted by combining amplitude information and frequency information. This model is used in cochlear sound processing. Significant immunity is provided to white noise as compared to PLP and MFCC. Figure 4 portrays ZCPA block diagram for extraction of features. The diagram contains zero crossing detection block, Band Pass Filters, peak detection block, the frequency receiver and compression of nonlinear amplitude. The filters cover the range of 200–4000 Hz frequency and are based on the auditory system, which contains 16 BPF. The unprocessed speech information passes through the filter bank. These are then transformed to 16 different information processing paths.

Every signal is distributed in frames and each and every frame goes to detection of zero crossing interval and also peaks the detections in every gap. The equation which is a monotony function is used by nonlinear compression of peak values and is represented by x in peak upward going position in zero crossing intervals and when it is compressed logarithmically it results in $g(x)$.

$$g(x) = \log(1.0 + 20x). \tag{1}$$

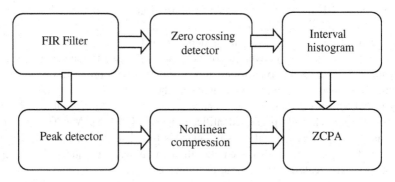

Fig. 4 Principal diagram of ZCPA

Each band called frequency bin forms a result when the frequency receiving block is divided into some sub-bands by frequency band. The ZCPA feature is the 16 path output together formed. The pitch and intensity features of speech are represented by the output feature vector as these are the basic features of speech.

5.3 Mel Frequency Cepstral Coefficient

The mel frequency cepstrum coefficient (MFCC) technique usually creates the basic fingerprint of the sound files. Human ear's critical bandwidth frequency is the basis of MFCC with filters which are usually spaced in linearity with typically less frequencies and logarithmically with more frequencies which captures the salient aspects of speech. To compute MFCC coefficients, the signal is divided into overlapping frames [12]. Every overlapped frame contains N samples and M samples separates adjacent frames. Hamming window is multiplied with every frame and is given by [13]:

$$W(n) = 0.54 - 0.46 \, \cos[(2\pi n)/(N-1)]. \tag{2}$$

Then the signal is gone through Fourier transform and converted to frequency domain from time domain by [13]:

$$X_k = \sum xi \, e^{-(j2\pi ki/N-1)}. \tag{3}$$

Now for a frequency, Mel is calculated by [13]:

$$M = 2595 \log_{10}(1 + (f/700)). \tag{4}$$

DCT converts log Mel scale cepstrum to the time domain by [13]:

$$X_k = \alpha \sum xi \cos\{(2i+1)\pi k/2N\}. \tag{5}$$

The result so calculated is mel frequency cepstral coefficient.

In MFCC, all input statements are converted in a series of acoustic vectors. A diagram of the MFCC procedure is depicted in Fig. 5.

The last step includes conversion of Mel spectrum scale to the standard scale of frequency. This gives a good presentation of some spectral properties which are the key to recognizing relevant characteristics of the speaker.

The MFCC is the most evident cepstral analysis based on feature extraction method for speech and speaker identification tasks [14, 15]. MFCC is popularly used as it is the most evident cepstral analysis based feature extraction technique and it portrays the human response better than any other technique. Table 2 depicts the summarized features of all these techniques.

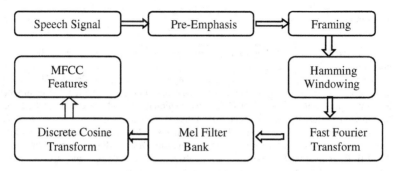

Fig. 5 Block diagram of MFCC

Table 2 Comparative analysis of feature extraction techniques

Techniques	Features
Perceptual linear prediction	By choosing a particular model, suppression of speaker-dependent information is allowed
	Robust to noise
	Uses engineering approximations for psychophysical hearing
Zero crossing peak amplitude	Used in cochlear implant sound processing
	Provides immunity to white noise
	Zero crossing of speech stimuli is the main basis of speech feature extraction
Mel frequency with peak amplitude	Good distinction
	Required phonetics data can be accumulated
	Less association among coefficient

6 Conclusion

The principal aim of this research is, to review basic feature extraction techniques which use the human auditory system for functioning. The three techniques discussed are Zero Crossing with Peak Amplitude, PLP and Mel Frequency Cepstral Coefficient. We compare the features of these three techniques and show that all the techniques have distinguished features. We come to a conclusion that PLP is entirely based on psychophysics of human hearing and is robust to noise as compared to the other two techniques. ZCPA provides immunity to white noise and its coefficient is calculated by zero crossing of speech signal whereas MFCC provides good distinction in the speech variables.

References

1. J. W. Picone, "Signal modelling technique in speech recognition," Proc. Of the IEEE, vol. 81, no.9, pp. 1215–1247, Sep. 1993.
2. Rabiner L.R. and Shafer R.W. Digital processing of speech signals, Prentice Hall, 19, 1978.
3. Vimala. C and Radha. V, "Suitable feature extraction and speech recognition techniques for isolated tamil spoken words" (IJCSIT) International Journal of computer Science and Information Technologies, Vol. 5(1), 2014378-383.
4. Hagen A., Connors D.A. & Pellm B.L.: The Analysis and Design of Architecture Systems for Speech Recognition on Modern Handheld-Computing Devices. Proceedings of the 1st IEEE/ACM/IFIP international conference on hardware/software design and system synthesis, pp. 65–70, 2003.
5. Thomas F. Quatieri, Discrete-Time Speech Signal Processing: Principles and Practice, Pearson Education, Inc. 2002.
6. Hermansky, H.: Should recognizers have ears? Speech Communication, Vol.1998, No. 25, pp. 2–27, 1998.
7. Lee Y. & Hwang K.-W.: Selecting Good speech Features for Recognition. ETRI Journal, Vol. 18, No. 1, 1996.
8. Mark D. Skowronski and John G. Harris, "Improving The Filter Bank Of A Classic Speech Feature Extraction Algorithm", IEEE Intl Symposium on Circuits and Systems, Bangkok, Thailand, vol IV, pp 281–284, May 25 - 28, 2003, ISBN: 0-7803-7761-31.
9. Urmila Shawankar and Dr. Vilas Thakare, Techniques for feature extraction in speech recognition system: a comparative study.
10. J. Mason and Y. Gu, "Perceptually-based features in ASR," in IEEE Colloquium on Speech Processing, 1988, pp. 7/1–7/4.
11. H. Hermansky, B. A. Hanson, and H. Wakita, "Perceptually based processing in automatic speech recognition," Proc. IEEE Int. Conf. on Acoustic, speech, and Signal Processing," pp. 1971–1974, Apr.1986.
12. H. Hermansky, "Perceptual Linear Predictive Analysis of Speech," The Journal of the Acoustical Society of America, vol. 87, no. 4, pp. 1738–1752, 1990.
13. Pratik K. Kurzekar, Ratnadeep R. Deshmukh, Vishal B. Waghmare, Pukhraj P. Shrishrimal, "A Comparative Study of Feature Extraction Techniques for Speech Recognition System", IJIRSET Department of Computer Science and Information Technology, Dr. Babasaheb Ambedkar Marathwada University, Vol. 3, Issue 12, December 2014.
14. Fang Zheng, Guoliang Zhang, and Zhanjiang Song "Comparison of Different Implementations of MFCC", The journal of Computer Science & Technology, pp. 582–589, Sept. 2001.
15. Hossan, M.A. "A Novel Approach for MFCC Feature Extraction", 4th International Conference on Signal Processing and Communication Systems (ICSPCS), pp. 1–5, Dec 2010.

Movie Recommendation System: Hybrid Information Filtering System

Kartik Narendra Jain, Vikrant Kumar, Praveen Kumar and Tanupriya Choudhury

Abstract The movie recommendation system is a hybrid filtering system that performs both collaborative and content-based filtering of data to provide recommendations to users regarding movies. The system conforms to a different approach where it seeks the similarity of users among others clustered around the various genres and utilize his preference of movies based on their content in terms of genres as the deciding factor of the recommendation of the movies to them. The system is based on the belief that a user rates movies in a similar fashion to other users that harbor the same state as the current user and is also affected by the other activities (in terms of rating) he performs with other movies. It follows the hypothesis that a user can be accurately recommended media on the basis others interests (collaborative filtering) and the movies themselves (content-based filtering).

Keywords Hybrid filtering system · Recommender · K-means Pearson correlation coefficient

1 Introduction

Movie recommendation systems are quite prevalent in today's market as people tend to spend a lot of money when they go to the movies or rent a movie, so they need to make an informed decision about it.

K. N. Jain (✉) · V. Kumar · P. Kumar · T. Choudhury
Amity School of Engineering and Technology, Amity University
Uttar Pradesh, Noida, India
e-mail: knjain1995@gmail.com

V. Kumar
e-mail: vikrantpositive@gmail.com

P. Kumar
e-mail: pkumar3@amity.edu

T. Choudhury
e-mail: tchoudhury@amity.edu

© Springer Nature Singapore Pte Ltd. 2018
S. Bhalla et al. (eds.), *Intelligent Computing and Information and Communication*,
Advances in Intelligent Systems and Computing 673,
https://doi.org/10.1007/978-981-10-7245-1_66

Over the past decade, a large number of recommendation systems for a variety of domains have been developed and are in use. These recommendation systems use a variety of methods such as content based approach, collaborative approach, knowledge based approach, utility based approach, hybrid approach, etc.

Lawrence et al. (2001) described a purchase behavior personalized product recommendation system that suggests new products to shoppers that shop in the supermarket. This system was adopted by IBM research as a PDA based remote shopping system called SmartPad. The recommendation was processed by content based and collaborative filtering for refinement.

Jung et al. (2004) improved people's search results by using previous ratings from the people to provide current users who are having similar interests, the proper recommendations. Their system works for a variety of items and collected their rating regarding the validity of their search results. This encouraged people to involve themselves in longer searches and use of more in-depth queries for their requirements and the system used these new rating to refine further ratings.

MovieLens is an online movie recommendation system created by the GroupLens research group that allows users to rate movies and utilizes rating from other sources along the user rating to collaboratively filter the input and generate ratings for recommendation to other people who have not seen said movies. They allow for account management services which allows people to rate movies and create profiles to receive recommendations based on their established tastes. This system utilizes multiple streams of data to provide comprehensive and personalized recommendations. This system can also be used over a mobile device online.

Websites and web services such as Netflix, IMdB, MovieLens etc. are some of the popular recommendation systems that harness the power of data mining and data analytics to analyze various factors associated with their users and the movies and make recommendations based on them.

2 Literature Review

Michael Fleischman et al. equated the problem of limited information about user preferences to one of content similarity measurement with the help of Natural Language Processing. They described a naïve word-space approach and a more sophisticated approach as their two algorithms and evaluated their performance compared to commercial, gold and baseline approaches [1]. A. Saranya et al. implemented a system based on probabilistic matrix factorization and provide users to solve cold start problems. Cosine similarity is used to judge similarity among users. They also implemented NB tree which is used to generate the user link formation. Their system considers various ranges of movie ratings which are given by review experts [2]. Hans Byström demonstrated a system in their paper by implementing k-means clustering and SoftMax regression classification. The baseline predictor showed an RMSE value as 0.90 and the system best achieved result was with RMSE value as 0.884 which is 1.81% improvement to baseline value [3]. Eyrun A. Eyjolfsdottir et al. introduced a system implemented using clustering and machine learning over a hybrid recommendation approach. Their system inputs user's personal information and with the help of the well-trained support vector machine (SVM) models predicts user's movies preferences. Using SVM prediction their system selects and clusters the movies from the dataset and generates questions. Then user answers those questions and their system refines its movie dataset and outputs movies recommendation for the users [4]. Arnab Kundu et al. developed a system by using collaborative filtering approach. In their system clustering is achieved among users and their preferences with the help of Expectation Maximization Algorithm [5]. Manoj Kumar et al. used collaborative filtering approach in their paper MOVREC. This approach makes use of the information provided by users. Their system then analyses the information and movies with highest rating using K-means algorithm is recommended to the user. In their system, user is provisioned to select attributes for the recommended movies [6]. Kaivan Wadia et al. implemented recommendation system based on self-organizing map (SOM). A SOM is defined as a neural network technique which comes in the domain of supervised learning. They used content based approach as well as SOM in their system for the categorization and retrieval of information [7]. Karzan Wakil et al. implemented their system which is based on human emotions. They used a hybrid approach which comprises of content based filtering, collaborative filtering and emotion detection algorithm. The output of their system provides better recommendation as it enables users to understand connection between their emotional states and the recommended movies [8]. Roberto Mirizzi et al. implemented a Facebook application that semantically recommends movies with the knowledge of user's profile. To detect similarity among movies Semantic version of Vector Space Model (SVM) is implemented by them [9]. Zan Wang et al. described improvements of the typical collaborative filtering methods in their paper. Their system uses collaborative filtering for extracting like-minded user's preferences. Then they used hybrid model of clustering which is a combination of Genetic Algorithms k-means (GA-KM) and Principal Component Analysis (PCA). They implemented PCA to reduce the sparseness in the feature space and GA-KM to reach to the global optimal solution [10]. Rupali Hande et al. implemented using hybrid filtering approach in their paper MOVIEMENDER. Various techniques like clustering, similarity and classification are implemented in their system to result in better recommendation. Their system resulted with high precision and accuracy as well as reduction in MAE score [11].

3 Proposed System

3.1 Architecture

The project is a web application that facilitates client server architecture, where the user of the application as a client through a web-browser on their system and the system resides on the server that responds to user's request with appropriate outputs.

The front end of the project is HTML 5 with CSS 3 with functioning and interactive bits driven by JavaScript and JQuery. Bootstrap served as the provider of the CSS and templates for the UI design. The Django web application framework was used to design the application along with Jinja performing the functions of dynamic HTML components.

The server is in charge of handling the client requests and authenticate clients by utilizing the CSRF token provided by the Django authentication services. Our application follows the Model-View-Template architecture wherein the model represents database schemas, View contains the user interface and the Templates are a combination of HTML and Jinja. The URL Dispatcher in Django handles the mapping of templates to appropriate URLs.

The recommendation engine is designed using Python language as python provides a plethora of libraries and data structures to make functioning of the system viable and efficient (Fig. 1).

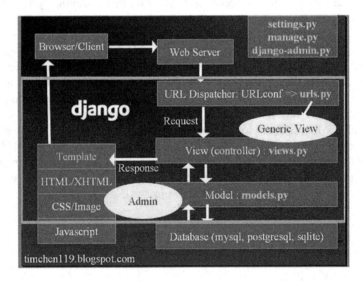

Fig. 1 System architecture

3.2 Working of the Hybrid Filtering System

The hybrid filtering system is the information filtering system that takes inputs of ratings on movies from users and recommends movies by application of a content based filter over a collaborative filter. The collaborative filter utilises the Pearson Coefficient Correlation and the content based filter is our proposed algorithm that utilises the genres associated with the active users' profile and those associated with the movies.

- The system first loads the movie-lens dataset into arrays for access by the various functions.
- Upon loading the data set, the utility matrix is created which contains all the ratings of each user to each movie. A similar test matrix is created for the test data.
- Post the creation of matrices, the movie genres are used to create clusters for clustering on items of the utility matrix the results of which are stored in matrix 'clustered utility matrix'. This matrix is dumped to a pickle file for access of the application.
- Following the clustering, user's average ratings are found and stored in user's object for accessing later.
- Collaborative filtering is performed in the form of Pearson Correlation coefficient which is calculated between two users. This is done for the full range of users and stored in the 'PCS matrix'.
- After the creation of the similarity matrix, the system proceeds to the testing part and employs guessing of rating of test movies given by users on a particular movie by incorporating a set of similar users:

  - First, the system access the similarity matrix to find the top certain number of users to the active user and normalise their scores by their average scores considering they have rated the movie in question.
  - The system computes the score for the movie by adding the average of the scores of the top users to the average of the current users if there were any similar users.
  - Following this computation, the proposed algorithm is applied and computations are made according the active user and movie to be rated to compute a new score.
  - The new score is compared with the actual score and after repeating process for the complete test set, the root mean square error is computed (Fig. 2).

3.3 Proposed Algorithm

The proposed algorithm is the calculation of score of movie for a particular user which utilizes the preference shown by users to movies of a range of genres and the

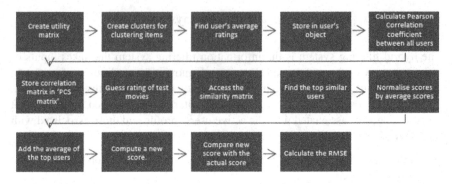

Fig. 2 Working of proposed system

association of those genres to the genres of the movie to be recommended to the user. The algorithm considers weights for different genres that all movies conform to for a user. Based on these observations a new score is computed for each movie.

$$\frac{x}{y} + \sum_{0}^{n}\left(\frac{z}{2*y^2}\right) - \sum_{0}^{n}\left(\frac{z'}{2*y'^2}\right) + \frac{n}{m} = x'. \qquad (1)$$

where,

x' the new calculated score of movies.
x the score derived by computing similar users and altering the movie average.
z weight of genre for the user that matches with his genre preference and movie.
z' weight of genre for the user that does not match with his genre preference.
y number of genres matched between user's profile and movie's profile.
y' number of genres not matched between user's profile and movie's profile.
n number of genres matched with preferred movies.
m number of genres that a user prefers based on his past rating.

The proposed algorithm employs a concept similar to genre correlations but does so between the user and the movie items instead of performing item-item collaboration approach.

4 Data

The system works on the movie-lens 100k dataset which consists of 9125 movies, 100,005 ratings provided by 671 users. The movies are classified with 20 genres.

The ratings file consists of the user-id, the movie-id's they are rating along with the rated score and timestamp of scoring.

The test and train data consists of 10 and 90% distributed data of the ratings file respectively.

The items file consists of the movies with their movie-ids, names along with the year of release, and the genres they are attached with which are represented as 1s for true and 0s for false. They correspond to a sequence of that is consistent throughout the data.

5 Results

The application results in a set of genres that are recommended to the user based on his ratings on the random movies. This factors in the genres attached to the random movies and weighted according to the rating the user gave it along with the preference of genres by users that have similar taste to the active user. The number along the genre so specifies the degree that the genre is recommended to the user. The more times a genre appear, the more it has been recommended to the active user.

The movies are the top movies that have been generated upon the computation of similarity to the other users and the proposed algorithm on the original ratings to generate new ratings that are specific for the user. The recommended movies need not be the highest rated as evidenced by the presence of 'Oliver Twist' which is a 4-starred movie but since the recommended genres according to the user is heavily inclined to 'adventure' and a bit of "children's" theme and 'Oliver Twist' happens to be a 'great adventure movie', hence it has been recommended to the user instead of a much higher rated movie (Fig. 3).

The recommender engine shows Mean Squared Errors in the range of 0.96–1 based on different sets of test data within the movie set. The resultant graph is based on an iteration with a mean squared error of 0.980763 and it shows the ratings originally given by user to movies and the ratings predicted by the system for the same (Fig. 4).

Fig. 3 Recommendation output page, genres and movies recommended

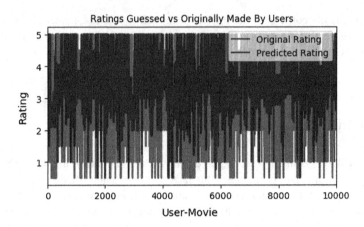

Fig. 4 Graph showing actual ratings (red) and predicted ratings (blue)

The graph below shows the rating provided by the users and the original ratings for the first 100 user-movie items of the test set (Fig. 5). Here we can see that considering rating 3 to be a central point (i.e. normalizing about 3), the ratings provided by the system are in the correct direction and differs by the magnitude only. This implicates the fact that since the system is predicting a new rating for every movie, although it does not match with the original rating, the variation between the movie set is consistent and hence it approximately makes the correct predictions. Another thing to note is that the initial movie-user sets, upon reflection on the data are heavily populated by ratings and well defined features (varied number of genres). In comparison to this the graph displaying the same data for the last 100 set (Fig. 6) shows predictions in the correct direction but with a more varied magnitude. This can be attributed to the fact that the last set of movies were

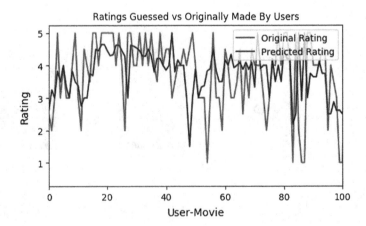

Fig. 5 Graph showing actual ratings (red) and predicted ratings (blue) for first 100 ratings

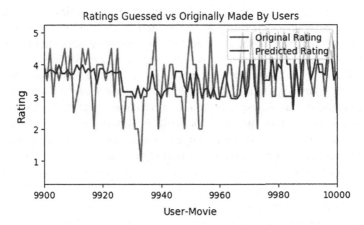

Fig. 6 Graph showing actual ratings (red) and predicted ratings (blue) for last 100 ratings

Table 1 Movielens 100 k database worked on using different techniques

Method	RMSE	MAE
UserItemBaseline	0.96754	0.74462
UserKNNCosine	0.937	0.737
ItemKNNCosine	0.924	0.727

less populated with ratings and also had a lack of refinement in terms of descriptors (lesser or no genres defined).

This observation leads us to believe that the system should work much better when a better quality of data in terms of sparsity would result in a better RSME value and better predictions (Table 1).

The table above specifies the mean squared error and root mean squared error on applying user-item filtering, cosine similarity between users and KNN and Item similarity using Cosine and KNN. The results show that our method having an average mean squared error of 0.980763 is comparable to these techniques considering they might have worked on a similar iteration of the dataset as similar to ours and that our system should work better with data having more user ratings.

6 Conclusion

The result shows that the hybrid system works within a respectable margin of error for recommending movies to users. The user can rate some movies to get recommendations of movies that suits the taste of the user instead of getting the most trending movies. The proposed algorithm accomplishes in this regard and couples over the similarity scores to add genre weighting in the mix. The fact that most of the predictions are at least in the correct direction are indicative towards the accuracy of the system. The results show that employing more dense data is bound to increase the accuracy of the recommender and increasing the number of genres will help in making more specialized recommendations. The system also establishes the viability of hybrid recommenders as good and effective recommender.

Movie recommendation has always been to a certain extent, a subjective opinion. Users tend to exhibit responses to recommendations that is at great odds from their past activities. This makes it considerably hard to evaluate recommenders mathematically and always involve a large margin of error. The system needs to be deployed in real-life scenario to receive manual responses of accuracy to judge the full extent of the system.

References

1. Michael Fleischman, Eduard Hovy. "Recommendations without User Preferences: A Natural Language Processing Approach". In proceedings of the 8th international conference on Intelligent user interfaces, pages 242–244. Miami, Florida, USA, 2003.
2. A. Saranya, A. Hussain. "User Genre Movie Recommendation System Using NB Tree". In proceedings of the International Journal of Innovative Research in Science, Engineering and Technology. Vol. 4, Issue 7, July 2015.
3. Hans Byström. "Movie Recommendations from User Ratings". Stanford University, 2013.
4. Eyrun A. Eyjolfsdottir, Gaurangi Tilak and Nan Li. MovieGEN: "A Movie Recommendation System". Technical report, Computer Science Department, University of California Santa Barbara, 2010.
5. Debadrita Roy, Arnab Kundu. "Design of movie recommendation system by means of collaborative filtering". Int. J. Emerg. Technol. and Adv. Eng, 2013.
6. Kumar, M., Yadav, D.K., Singh A., Gupta and V.K. "A movie recommender system: Movrec". In International Journal of Computer Applications, ACL 2002 Conference on Empirical Methods in Natural Language Processing, vol. 124, pp. 0975–8887, 2015.
7. Kaivan Wadia, Pulkit Gupta. "Movie Recommendation System based on Self-Organizing Maps". The University of Texas at Austin, Austin, Texas, 2011.
8. Karzan Wakil, Rebwar Bakhtyar, Karwan Ali and Kozhin Alaadin. "Improving Web Movie Recommender System Based on Emotions", International Journal of Advanced Computer Science and Applications, vol. 6, no. 2, 2015.
9. Roberto Mirizzi, Tommaso Di Noia, Azzurra Ragone. "Movie recommendation with dbpedia". In CEUR Workshop Proceedings, vol. 835, 2012.
10. Wang Z, Yu X, Feng N and Wang Z. "An improved collaborative movie recommendation system using computational intelligence". J Vis Lang Comput 25:667–675, 2014.
11. Rupali Hande, Ajinkya Gutti, Kevin Shah, Jeet Gandhi and Vrushal Kamtikar. "MOVIEMENDER - A MOVIE RECOMMENDER SYSTEM", International Journal of Engineering Sciences & Research Technology (IJESRT) (Vol. 5, No. 11), 2016.

Eco-Friendly Green Computing Approaches for Next-Generation Power Consumption

Seema Rawat, Richa Mishra and Praveen Kumar

Abstract The term Green Computing is combination of two words which basically means usage of electronic devices in an efficient way. Green computing is a trend/ study which came lately and is about the building, designing and operating electronic devices to maximize the efficiency in terms of energy. Green Computing includes the manufacturing and disposing of electronic devices including the computer resources such as the monitor, storage devices, CPU and associated subsystems etc. without causing any ill effect to the environment. Also, one of the important aspect of green computing is recycling of the devices which are used worldwide in IT industry. This research paper highlights the advantageous use of green computing and its associated approaches through a survey conducted over Power Consumption in Amity University.

Keywords Green computing · Eco-friendly · Resource utilization
Power consumption · Recycling · Benefits

1 Introduction

The idea of Green Computing was given by Environment Protection Agency in year 1992, when they launched a program called "Energy Star", which promoted hardware efficiency. All the operations which need sustainability of environment comes under the scope of green computing [1, 2]. Till date, the IT industry has delivered a lot of innovative products but on the other hand, it has also been a topic of major concern if we consider global issues such as global warming, climatic

S. Rawat (✉) · R. Mishra · P. Kumar
Amity University Uttar Pradesh, Noda, India
e-mail: srawat1@amity.edu

R. Mishra
e-mail: richa.m18@gmail.com

P. Kumar
e-mail: pkumar3@amity.edu

© Springer Nature Singapore Pte Ltd. 2018   687
S. Bhalla et al. (eds.), *Intelligent Computing and Information and Communication*,
Advances in Intelligent Systems and Computing 673,
https://doi.org/10.1007/978-981-10-7245-1_67

changes etc. A lot of factor are there to prove that Green Computing is an important solution to be followed as it minimizes the cost related to power management. Minimizing the electricity consumption which is risky for environment is the main aim of green IT. This model aims towards the reduced usage of electronic material that have negative impact on the environment. Increased knowledge and awareness about the environment has caused the industries all around world to recheck their environmental credentials.

2   Investigating the Power Consumption

This article focuses on calculating the total amount of power consumed by the desktops on an early basis [3]. Different costs including the cost of computing, cost of cooling process etc. has been provided in this research paper which also keep track on the amount of heat generated through the desktops. Further, this study throws light on the impact which the increased computing power has on the environment, it emits gases like Sulphur, CO_2 etc. The survey was conducted over the power consumption in Amity University, Noida which approximately has 10,000 desktops. With the help of meters, we were able to gauge the power consumption which was around $150,000, excluding the cooling process of these desktops. On the other hand, including the price of cooling process it would go around $220,000. To operate the desktop, we require power plant to generate power which in turn produces harmful gases like CO_2, mercury etc. This huge quantity of electricity initiates the downfall of environment. Here we are giving solutions to reduce the power consumption.

2.1   Advantage of Using Thin Clients

Currently organization have multiple servers and majority of them not even being used completely. Typically, server uses around 5–20% of their full capacity, even then they consume full energy. An efficient solution to the problem is using Thin Clients which consume only 8–20 W of energy in comparison to a typical PC which uses about 150 W. Thin clients maintain the environmental decorum by minimizing the consumption of energy. Disposing a PC is way too high as it contains heavy materials used in the hardware of system. A thin client uses a common or shred terminal which reduces the power consumption and provides high efficiency. Some of the pros which makes thin clients reliable [4].

- Low energy consumption.
- In comparison to PCs, they require lesser cost in transportation as they are made up of lightweight material.
- More availability.

- Little disposals cost.
- Cost of terminals are relatively low.
- Operational cost is low.

One of the examples of thin client is Sunray. It is a thin client which requires 4–8 watts, whereas intake for a video card is 12–15.

2.2 Consideration for General Changes

Green computing is not something which can be easily achieved. A collaborated attitude is needed to attain the same. The toxic substances must be urgently replaced to prevent environment from degradation [5, 6].

- Avoiding of plastic cases made up of brominated flame retardant (BFR).
- Using low capacity flash based drives instead of solid state drives.
- Controlling of landfills through periodic upgrading and repairing of resources.
- Organic light diode and Green LED should replace the CRTs displays.
- Substituting harmful chemicals like copper, lead etc.
- Avoiding the nonproductive mechanism to save energy resources.
- Instead of CPUs, GPUs can be considered as better option as it reduces power consumption while increasing the efficiency [7]. CPUs and GPUs data in idle state and 100% load state (Table 1).

2.3 Converting a Company into Green Company

Converting a company into sustainable business or green company can be achieved if a successful plan is followed. A company can be turned into a green company by

Table 1 Power consumed by CPUs and GPUs [7]

S. No.	CPU/GPU used	Idle state (W)	100% load state
1.	GPU ATI RADEON X1900XT	26	45
2.	CPU Pentium4XE@ 2.30 GHz	206	318
3.	GPU ATI RADEONHD 6990 CF(4)	43	0
4.	CPU Intelcorei3@ 2.40 GHz	32	65
5.	GPU ATI RADEON HD 6950 OC(900/1440)(32)	23	64
6.	CPU Intelcorei5@ 3.10 GHz	38	95

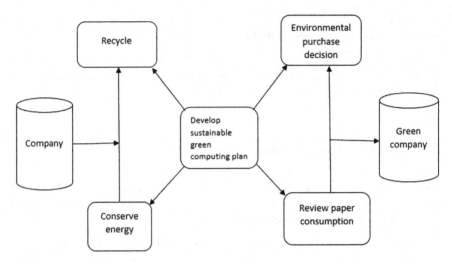

Fig. 1 Formation of green company [9]

some simple and easy guidelines [8]. A sustainable development procedure should of following processes—reusing, environmental purchase decisions, energy conservation and reviewing the paper consumption (Fig. 1). These processes cope up with each and in result form a green company which has silent and green infrastructure [9]. There are two major concerns attached to green computing—pollution control and reduction in energy consumption. While the first can be achieved through efficiently using electronic goods, later can be achieved by their reduced usage and proper recycling.

3 Feasible Green Computing Trends

Several researchers are trying to tend towards Green Computing. Research on Green computing leads to concepts that allow the usage of green concept in computing domain. Some practices adopted from green computing such as use of small screen size devices, using computer systems having ENERGY STAR label on them etc. [10]. Following topics are further discussed in the paper.

3.1 Virtualization

One of the main concept which fall under green computing is Virtualization. Virtualization is basically about the extraction of computer resources like using two or more logical systems on a single set of physical hardware. Virtualization is a green computing trend which offers both the management and the virtualized software. Virtualization is performed at different levels like storage, desktop and server level [10, 11]. In the virtualized environment, a single server can be used up-to its full utility. Through the virtualized datacenters, hardware costs can be reduced up to the half of original. But if we talk about virtualized servers, a server room is there which requires minimum maintenance services and minimizes real estate savings. In datacenters, the virtualized servers have been replaced by the blade servers which result in more than 70% reduction in consumption of energy [12]. It allows complete utilization of the resources and provides various benefits like

1. Reduced amount of hardware.
2. Reduced amount of space, rent and ventilation required etc.

3.2 PC Energy Usage

Before handing over the systems to the end users, companies should be wholly responsible to enforce recommended power settings. Allowing the monitor to sleep mode after it has been idle for long is a good practice in energy saving. Also, a lot of experiments have been performed as to check the performance of PC while in sleep mode and in idle mode (Graph 1).

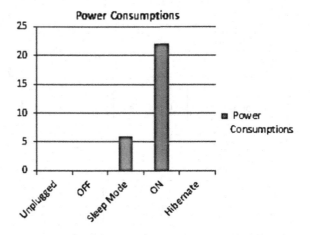

Graph 1 Power distribution of different modes [14]

3.3 Recycling Devices

Green computing should consider the product life cycle, from producing it to its operation and lastly recycling it. For sustaining environment recycling of cadmium, lead etc. is vital. The computer systems which have completed their lives can be re-purposed, reused or can be given to the needy one's (Fig. 2). Parts of systems which are no more in use or those which are outdated can be recycled. The computing supplies like batteries, paper, cartridges etc. can be recycled and could be converted from a total waste into useful objects. There is a constant requirement for new technologies and high tech products which results in huge amount of e-waste on a regular basis [1].

Such practices would help us in fewer costs, space saving low carbon emission, reduced power supplies etc. (Fig. 3). Some practices which promote green computing has been introduced in our education system as well and are discussed below [13]:

- Online learning methodology.
- Software based on intranet portals for attendance system, fee payment etc.
- E-admissions process.
- College e-brochures.

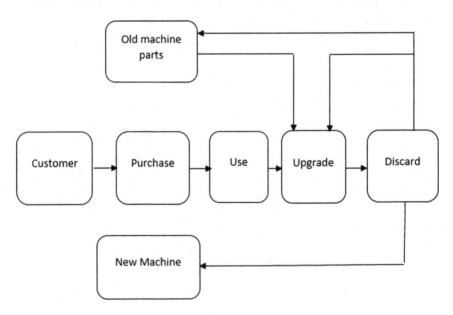

Fig. 2 Reusing of electrical components [15]

Fig. 3 Advantages of green computing [16]

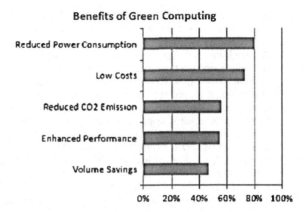

4 **Conclusion**

This paper highlights a lot of loopholes of the electronic industries which are highly hazardous to the environment as well as to human being. Technology isn't a passive observer, rather it is achieving green computing goals by being an active observer. Thin client cables, has reduced energy consumption up to half of the original consumption. Virtualization is another concept covered in this paper, it provides maximum utilization of the resources which results as a major advantage. Another aspect which has been covered here in this paper is about recycling devices. Recycling can improve the environment as well as the ecosystem on a huge basis. How reusing a system helps in saving a lot of energy and maintaining the ecology has been covered in this paper. There are enormous number of challenges to achieve the same. Also, the government is putting appreciable amount of effort. But it's a matter of huge concern as it cannot be achieved in a day or two. The overall approach of promoting Green Computing is to control environmental deterioration, enhanced life expectancy of devices etc. Implementing Green Computing trends is not the responsibility of a single person, instead it is something which can be achieved only when all of us work together in a positive way. The needs of present and future generation can only be fulfilled if we provide Eco-friendly environment. Therefore, I hope this paper works in support of development and enhancement of green computing based organizational infrastructure.

References

1. Gingichashvili, S., "Green Computing" 2007.
2. Tebbutt D., Atherton M. and Lock T., "Green IT for Dummies", John Wiley and Sons Ltd. 2009.
3. Mujtaba Talebi, "Computer Power Consumption Benchmarking for Green Computing", Villanova University, April 2008.

4. Aruna Prem Biazino, Claude Chaudet, Dario Rossi and Jean-Louis Rougier, "A Survey of Green Networking Research", IEEE Communications surveys and tutorials, vol. 14, no., First quarter 2012.

5. Praveen Kumar, Dr. Vijay S. Rathore "Improvising and Optimizing resource utilization in Big Data Processing "in the proceeding of 5th International Conference on Soft Computing for Problem Solving (SocProS 2015) organised by IIT Roorkee, INDIA (Published in Springer), Dec 18–20, 2015, pp. 586–589.

6. Seema Rawat, Praveen Kumar, Geetika, "Implementation of the principle of jamming for Hulk Gripper remotely controlled by Raspberry Pi "in the proceeding of 5th International Conference on Soft Computing for Problem Solving (SocProS 2015) organised by IIT Roorkee, INDIA, Dec 18–20, 2015, pp 199–208.

7. Y. Navneeth Krishnan, Vipin Dwivedi, Chandan N. Bhagwat, "Green Computing using GPU", IJCA, vol. 44, no. 19, April 2012.

8. http://www.berr.gov.uk/files/file35992.pdf.

9. Yan Chen, Shunqing Zhang and Shugong Xu, "Fundamentals Trade Offs on Green Wireless Networks", IEEE Communications Magazine June 2011.

10. Payal Mittal, Seema Rawat, Praveen Kumar "Proof Based Analytical Comparison of Energy Parameters between Android and iOS Platform for Heterogeneous Cloud Based Application in International conference on Computational Intelligence & Communication Technology (CICT), ABES Ghaziabad UP on 12–13 Feb. 2016.

11. S.K. Fayaz Ahamad, P.V. Ravikant, "Green Computing Future of liveliness", IJCER ISSN: 2250–3005.

12. Annapurna Patil, Abhinandan Goti, "Green Computing in Communications Networks", IJCA vol. 68, no. 23, April 2013.

13. Shalabh Agarwal, KaustuviBasu, Asoke Nath, "Green Computing and Technology Based Teaching and Learning Administration in Higher Education Institutions", IJACR vol. 3, no. 3, issue. 11, September 2013.

14. J.V. Mashalkar, S.K. Kasbr, M.B. Bhatade, "Green Computing: A Case Study of Rajarshi Shaher Mahavidyalaya, Latur, India", IJCA vol. 62, no. 2, January 2013.

15. Riyaz A. Sheikh, U.A. Lanjewar, "Green Computing- Embrace a Secure Future", IJCA vol. 10 no. 4, November 2010.

16. S. Murugesan, "Going Green With IT: Your Responsibility towards Environmental Sustainability" Cutter Business- IT Strategies Executive Report, vol. 10, no. 8, 2007.

Implementing Test Automation Framework Using Model-Based Testing Approach

Japneet Singh, Sanjib Kumar Sahu and Amit Prakash Singh

Abstract There are various stages present in the process of software testing life-cycle which starts mainly with the requirement review process, proceeds with test planning, test designing, test case development, and execution of test cases and ends with test reporting phase. This paper will emphasize mainly on the application of model based testing with test automation frameworks for automating the test designing and test development phases of software testing lifecycle. In the test development phase it mainly provides information about using model based testing in the activity related to the development of test automation scripts for execution of tests. This mainly tells about the steps involved in the implementation of test automation frame work using the approach of model based testing. Test development and test designing both can be automated to decrease the amount of human effort and cost for the same in the project. The development of test scripts which includes both automated and manual tests can be automated using the concept of model based testing. In this approach firstly the model is created for capturing the behaviour of the system which is under test. After this the model based testing tool parses the same to create the manual testing scripts. The model based testing tool then in addition generates the automated test scripts for automated test execution and can also be integrated with popular tools and test automation frameworks. So, using a model for automating the creation of both automated and manual test scripts not only saves cost and thereby effort, but also increases the amount of coverage and reduces significant amount of time for product to go to market.

Keywords Test automation framework · Model based testing · Software requirement specification · Quality assurance · Graphical user interface

J. Singh (✉) · S. K. Sahu · A. P. Singh
USICT, Guru Gobind Singh Indraprastha University, Dwarka, Delhi, India
e-mail: japneetheyer@yahoo.in

S. K. Sahu
e-mail: sahu_sanjib@rediffmail.com

A. P. Singh
e-mail: amit@ipu.ac.in

© Springer Nature Singapore Pte Ltd. 2018
S. Bhalla et al. (eds.), *Intelligent Computing and Information and Communication*,
Advances in Intelligent Systems and Computing 673,
https://doi.org/10.1007/978-981-10-7245-1_68

1   Introduction

The process of model based testing can be defined as an application of model based design used for execution of automation and manual based test cases to perform software testing and also used for test designing purposes. It can be stated that the models can be used for the representation of desired behaviour of the system which is under testing and also to represent the test environment and test strategies.

The general approach of model based testing has been discussed in Fig. 1. In this approach the model which is created is a partial description of the system which is under testing. The test cases which are derived from such a model will always be functional test cases. These test cases are also known as abstract test cases. These abstract test cases suite cannot be directly executed on the system which is under testing. So an executable test suite has to be derived from the corresponding abstract test suite cases. These tests then can directly be executed on the system under testing. This can be achieved by associating the abstract test cases to the specific test cases which will be suitable for execution.

This paper mainly focuses on the application of model based testing approach for automating the test development and test designing phases of software testing lifecycle. In the test development phase the model based testing is mainly implemented for the activity of developing test automation scripts. This approach mainly helps us in identifying steps to build the test automation framework using the model based testing.

Also, the test designing process can be called as the foundation phase for the software testing as this phase involves the analysis of the requirements or specification for the system and then generating the test cases using the same and validating them against the system that documented requirements or the specification have been met. This requires human effort so the model based testing approach can be used for automating the process of test designing as well to reduce the amount of effort required to write these test cases manually.

Fig. 1 Components of model based testing

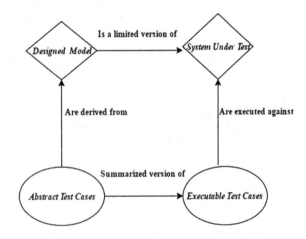

Figure 1 depicts the general approach of model based testing.

The roadmap for the remainder of the paper will be as follows. In Sect. 2 we have discussed about the related work being done in the area of implementing model based testing in software testing lifecycle process. Section 3 talks about the proposed solution and features of implementing test designing using model based testing. Section 4 talks about the proposed approach of test automation using model based testing and integration of test automation frameworks [1]. Finally the paper has been concluded and the future scope has been defined in Sect. 5 of the paper.

2 Related Work

There have been large number of advancements happening in the field of software testing. The process of capturing the requirements by creating models was introduced around twenty years back which is around two decades ago. But, the talk around using models for the process of software testing have been there around the last decade. There are good amount of tools like Test optimal, ConformIQ, Math Works and many more available today in the market which helps users in the creation of models and also further help in generating the test cases and test scripts by processing these models.

The test automation practises have also developed over the last ten years. As the linear scripts were developed earlier but now these have been replaced by the structured test scripts which use the test libraries, modifying and automating these test cases also require programming knowledge. There are new frameworks available today like keyword driven and data driven frameworks which have helped the testers to reduce the complexities in the test scripting [2]. All these advancements has helped in reducing the time, effort and cost required in the process of software testing. These advancements have also allowed the various software vendors to become aggressive and pretty much competitive in the market.

Model based testing usage is being done in various software processes, development organizations, application stacks and application domains. These includes commercial providers, transaction processing and open source providers. These also include implementation of programming languages like Python, C, C++, and Java and also include software development lifecycle models like V-model, Agile and Incremental model.

It is also true that testing itself is a noteworthy process of software development lifecycle in terms of both time and cost. All the features of a system need to be tested not only when the feature is introduced for the first time in the release but also whenever the next release comes up with the same set of features with some enhancements in it. So, the regression test cycles are very long and they do get repeated monthly or quarterly which increases the cost and effort required to do this testing manually.

So, the companies have been looking for an alternative solution in terms of performing automation testing to save the time and cost of the project. The model based testing process helps in the same way by making two specific advancements of automating the test design and test automation processes helping software vendors to automate the process of testing the software products [3].

3   Solution for Test Designing Using Model Based Testing

Test designing along with the process of test automation are very much specialized skills. These are always in less amount of supply. The test designing process involves the creation of test cases and also requires good amount of domain knowledge along with the general information related to software development.

The model based testing has now been accepted as a methodology for the purpose of test designing and test development as illustrated in Fig. 2. In the test development process the model based testing can be used for developing the manual test scripts as well as the automation based test scripts. From this statement it can be infer that model based testing can be used to build test automation frameworks.

The steps involved in the software test lifecycle can be represented in the form of a Pictorial representation implementing model based testing as illustrated in Fig. 2.

In a traditional test designing approach test cases are generated or written by going through the software requirement specification document which helps us in determining what all needs to be tested. Then tester design the test cases to test each of the listed requirements.

3.1   Features of the Solution

The user can start the process of implementing model based testing using the software requirement specification document by modelling the specified characteristics of the software under test. The model can be created in the form of a

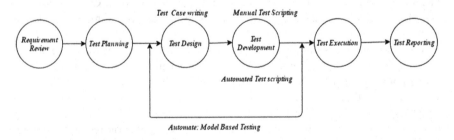

Fig. 2 Test automation across software test lifecycle using model based testing

markov chain or a state chart or in the form of finite state machine and are also represented using the UML diagrams.

The model can be used for representing the various requirements to a particular extent depending upon the focus of testing. It may also be very closer to the pictorial representation of requirements or a form of replica at a very small scale. Due to higher focus on the requirements it looks like a description of the requirements. The model helps in representing the key requirements. Then Model based testing tool is used for carrying out parsing on the model and helps in generating the test scripts and test cases for purpose of manual execution [4].

The model based testing tools additionally supports various coding languages like Python, Java and many more object oriented languages and model based languages such as Unified Modelling language which is used as a basis for the model based testing [5].

There are variety of model based testing tools available in the market today which can be used for implementing this type of testing which are listed as Conform IQ creator, JSXM, Model J unit and many more. These tools also support the code based model, helps in parsing of the code, creation of test scripts and test cases for the purpose of manual execution [6].

Figure 3 represents a model displaying the functionality of login page for web based application.

4 Proposed Approach for Test Automation Using Model Based Testing

The model based testing technique can also be used for developing the automation test scripts. The manual test scripts which have been generated using model based testing as a part of test designing process can easily be converted to automated test scripts for automatic execution. It is a type of fourth generation based test automation. It also supports requirement and defect identification, field defect prevention and automated generation of tests from the models which in turn helps in reducing the cost, effort and eliminates manual test designing [7].

Model based testing consist of majorly three steps in it which will be explained by using the model taken for the login functionality of web page in Fig. 3.

1. Creating a model suitable for execution: This step is required for the diagram based models generated in the form of finite state machine. Many model based testing tool can be used for the same purpose [8]. If the data is correctly associated and all the dependencies and constraints have been defined then it is an executable form of model. The coding is the best way to implement the association and it can easily be simulated as shown in Fig. 4.

Test data associated with the model:

Fig. 3 Model representing
the functionality of the login
page of web based application

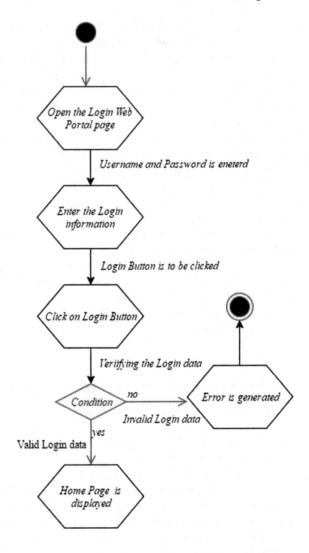

2. Furnishing the test cases from the model to create the test scripts for the purpose of test execution: Generally most of the model based testing tools support the process of furnishing or rendering using popular execution tools like Unified Functional Testing or Selenium and also supports unit testing based frameworks like JUnit.

The Model based testing tool also helps in creating the executable script with good amount of code for few of the steps and placeholder like providing additional graphical user interface details for the same. There is always a scope of further

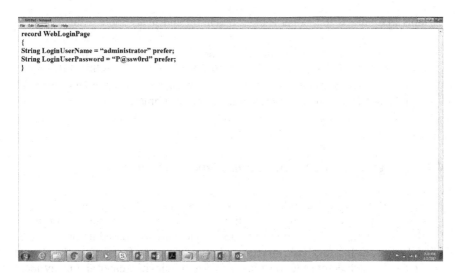

Fig. 4 Representation of the association of test data with the model

```
sendOpeningURL( new OpeningURL());
sendLoginUserCredential( new LoginUserCredential());
sendLoginClickButton(new LoginClickButton(/**ButtonClicked*/ "Login"));

public void sendOpenURL(OpeningURL argOpeningURL)
throws TestSuiteException {
try {
errorMesssage = "error encountered while opening the given URL";
basicURL1 =
objectsInstanceService.getApplicationProPropertyInstance().getProperty("Anoynymus_URL");
Log.warn(" --> Connecting server information is : " + basicURL1);
Log.info(" --> Entered URL is :" + basicURL1);
} catch (TimeoutException t) {
handleExceptionforFailure(t, errorMesssage);
} catch (NoSuchElementException t) {
handleExceptionforFailure(t, errorMesssage);
} catch (SeleniumException t) {
handleExceptionforFailure(t, errorMesssage);
} catch (Exception t) {
handleExceptionforFailure(t, errorMesssage);
}
}
```

Fig. 5 Representation for the stub written for test automation

modification in the stubs created before they can be executed as a complete automated test. This is further illustrated in Fig. 5.

Stub written for test automation:

3. Creating the scripts which are automatically executable and are complete tests:
   The small automation stubs which have been generated using the model based
   testing tool can be further developed for automated execution using automation

tools like unified functional testing, Selenium and many more available in the market today. This thing is achieved by using the record and play back feature of the right model based testing tool which is based up on the domain of the system or software under testing, also depends up on the capability of the tool to integrate with popular test automation frameworks and helps in generating the automated test scripts to build an automated test framework.

4.1 Integration with Test Automation Frameworks

1. Keyword driven framework: It is a type of software testing framework which separates the work of test automation involving test programming from the given test designs of a product. This technique is mainly based up on the usage of keywords for the basic functions or operations performed on the system which can include closing, opening, and button clicking, enter and many more [9]. Then the listed keywords are converted into automated test scripts for an automation tool to execute them. The process can be better understood by taking the example of web based login page functionality and determining the keywords for the same (Table 1).
2. Data driven framework: It is a type of framework which mainly revolves around data which is stored in a database or in the form of spreadsheets. It is a very useful approach as a single test can be tested multiple times with different test data which avoids huge number of test cases to be written for testing the same type of scenario which only varies with data [10]. Model based testing tool integrates with this framework by doing association of test data with different set of states and actions. It is really useful when the system under test needs to be tested with good number of input values (Table 2).

The data set represented above is then read in the script to test the similar scenarios by providing different set of values and leads to good coverage.

Table 1 Keywords or test directives generated for testing login page script

Graphical user interface window	Locator	Action based keyword	Parameters
Login web page	Browser	Open	Loginwebpage.com
Login web page	Browser	Enter	Admin, P@ssw0rd
Login web page	Login button	Click	Login Button
Login web page	Browser	Verify	Home Page

Table 2 Test data generated for login page script

LoginUsername	LoginUserPassword	Result
administrator	admin	False
administrator	P@ssw0rd	True
admin	admin	False
admin123	password	False

5 Conclusion and Future Scope

Model based testing can help in the establishment of test automation framework around it which begins with the process of test design automation and is followed by the manual test scripts development. This type of testing also allows us to integrate with different automation frameworks to achieve test script designing and also integrates with variety of tools for supporting automated execution of tests. Such a framework helps the organization to build there infrastructure for automated test designing and automated test execution [11]. There are many benefits of model based testing which mainly involves the exhaustive coverage of test scenarios in the form of a model, ease of maintenance of test cases and test scripts, expected shorter cycle time which helps in the faster execution of test cases and also helps in the elimination of need to modify the code which prevents the occurrence of human error in larger scripts.

It also helps the organizations in delivering the test automation at faster rate which can be utilized during the regression cycles and helps in saving time, effort and cost of the project. There are huge amount of benefits of model based testing which are derived by the organizations at smaller and larger level in terms of easy maintenance of automation and exhaustive coverage of test cases. The future work mainly lies around the area of analyzing the relative importance of model based testing and how it can be implemented more efficiently in the industry. Also, the future work can also involves the work of analysing the model based testing and its integration with more number of automation frameworks to efficiently test a software product.

References

1. Fei Wang and Wencai Du: A Test Automation Framework based on Web. IEEE/ACIS 11th International Conference on Computer and Information Science (2012)
2. Pekka Laukkanen: Data Driven and Keyword Driven Test Automation Frameworks (2006)
3. Rashmi and Neha Bajpai: A Keyword Driven Framework for Testing Web Applications. (IJACSA) International Journal of Advanced Computer Science and Applications, Vol. 3, No. 3 (2012)
4. Model-Based Testing, https://en.wikipedia.org/wiki/Model-based_testing
5. Yan Li and Li Jiang: The research on test case generation technology of UML sequence diagram. IEEE 9th International Conference on Computer Science & Education (2014)

6. Sebastian Wieczorek, Vitaly Kozyura and Matthias Schur: Practical model-based testing of user scenarios. IEEE International Conference on Industrial Technology (2012)
7. Test Automation, http://en.wikipedia.org/wiki/Test_automation
8. Naveen Jain: Model-Based Testing, HCL Whitepaper. http://www.hcltech.com/white-papers/engineering-services/model-based-testing (2013)
9. Anuja Jain, S. Prabu, Swarnalatha Muhammad Rukunuddin Ghalib: Web-Based Automation Testing Framework. International Journal of Computer Applications (0975 – 8887) Volume 45– No. 16 (2012)
10. Susanne Rösch, Sebastian Ulewicz, Julien Provost, Birgit Vogel-Heuser: Review of Model-Based Testing Approaches in Production Automation and Adjacent Domains Current Challenges and Research Gaps. Journal of Software Engineering and Applications (2015)
11. Ambica Jain: MBT, A superior alternative to traditional software testing (2012)

Improved Exemplar-Based Image Inpainting Approach

Hitesh Kumar, Shilpi Sharma and Tanupriya Choudhury

Abstract Image inpainting is a very common technique which is widely used to bring back or recover an image when it gets destroyed, or when there is an intention to perform a morphological operation on an image. It is a simple process to fill up the pixels of a particular region. The task is very similar to that of a skilled painter who has to draw or remove an object from its painting on its canvas. Inpainting's application may vary from its usage which includes object removal, object replacement etc. The aim of this process is to change the image properties by removing or adding objects. It has been also used to recover lost pixels or block of pixels of an image during a transmission of image through a noisy channel. It has been a useful method for red-eye removal and default stamped date from photographs clicked through primitive cameras. In this paper, we have built an exemplar-based image inpainting tool using basic functions of MATLAB. We have discussed the results by comparing the output generated by the image inpainting tool with Adobe Photoshop results to compare the results and check how much efficient the tool is.

Keywords Inpainting · Image restoration · Canny edge detection
Object replacement · Object removal · Exemplar-based inpainting

H. Kumar (✉) · S. Sharma · T. Choudhury
Computer Science and Engineering Department, ASET, Amity University,
Noida, Uttar Pradesh, India
e-mail: hitesh2194@gmail.com

S. Sharma
e-mail: ssharma22@amity.edu

T. Choudhury
e-mail: tchoudhury@amity.edu

© Springer Nature Singapore Pte Ltd. 2018
S. Bhalla et al. (eds.), *Intelligent Computing and Information and Communication*,
Advances in Intelligent Systems and Computing 673,
https://doi.org/10.1007/978-981-10-7245-1_69

1 Introduction

In computer science, this similar task is done by filling the damaged or corrupted pixels by the process of inpainting. There are multiple techniques and processes to do this task [1]. The techniques come from the area field of image inpainting and artificial intelligence. During taking a photograph, some problems are encountered:

i. Sometimes while sending or receiving an image, the pixels get corrupted or damaged
ii. While taking a photograph, there are sometimes unwanted objects that come in the frame which was not meant to be there.

To cope up with this problem, inpainting process comes into play and the desired image is obtained by removing the object and filling the cavity with desired pixel values. After the process, an image obtained must be plausible and the viewer must not able to detect any changes made to the output image [2].

In Fig. 1, the left image is corrupted and the right image is being inpainted to its original form. The red squares have been removed by the inpainting algorithm and the original image has been restored. Various Inpainting techniques include the following:

1.1 Texture Synthesis

Texture synthesis or TS inpainting takes into account the structural definition of the image. It tries to construct a large image from a small digitally defined image. It helps in expanding the image small image, filling cavities/holes for inpainting purpose and to fill the background of an image. Here, the texture can be either 'regular' or 'stochastic' [3]. Regular texture comprises similar patterns over an image. The example is as follows (Fig. 2).

Fig. 1 Inpainting example

Fig. 2 Regular and
near-regular texture [4]

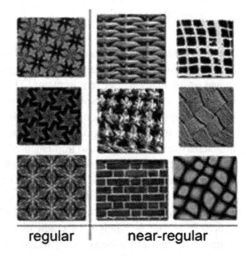

regular   near-regular

Following points must be taken care while using this technique:

(a) the output size of the image must be mentioned by the user
(b) the input image size should be similar to the output image size
(c) the image should be free from complex structures and artifacts.

1.2 Partial Differential Equation Based

PDE or Partial Differential Equation is an iterative algorithm that takes the information of the region by moving forward in those regions where the change is less. This algorithm is very effective where the corrupted regions are smaller in size [5]. When dealt with larger regions, it will take a longer time to process the image. Also, the results calculated will not be satisfactory. The total variation using the concept of differentiation is calculated for change in information in the image. This algorithm fails to restore those image which is partially degraded [6].

1.3 Exemplar and Search Based

Exemplar and search-based inpainting were initially introduced by Harrison [7]. The empty regions of the image were filled with the texture around the patch being created after the object removal. In other words, in exemplar and search based, the locality of the empty patch being created is taken into consideration and for the pursuit of information, the locality is being explored for the desired information. This is suitable for images having a linear structure. This technique works in the following manner.

2   Literature Review

Inpainting is a technique to regain or recover lost information in an image. This is done by gaining information related to the background of the image, surrounding of the image or manually as per requirements. The word 'inpainting' comes from ancient times where the skilled and specialist painters were used to re-paint a lost or degraded painting using some pre-requisites about the painting. Major researches have been done in this field. Many approaches have been implemented to do inpainting. Deep neural networks have been also implemented where the machine is trained to perform the desired results [8]. Some other fascinating works have been in done in [9], where the bifurcated image which is corrupted and separates out those corrupted into smaller sub-images. Then they feed those piece of images into a neural network using Euclidean loss function also calculated the MSE. The plus point for using this approach is that is efficient in eradicating small distortions but each corruption required a new and a separate neural layer. Criminisi et al. [10] have used the exemplar-based technique using texture synthesis. Their approach's result is quite impressive when it comes to region filling and background inpainting. The authors of this research paper [11] has proposed a new method to determine an enhanced 'priority factor' to fill the patches. The patch filling here is done from propagation from outer boundaries of the patch image to inwards of the image. The results then are compared with existing inpainting techniques by using PSNR as a metric to do the same. In this paper [12], the author has used a novel method by using exemplar-based inpainting as the base method. Here, initially the author has removed big objects from the image and has used both the combined benefits of inpainting and texture synthesis method. The main motto of the author here is to reduce the processing time and achieve better results. Their results showed that their computational time got reduced to 0.25 times to the existing inpainting techniques. In this paper [13], the target is a video on which the inpainting has been done. Author's aim here is to regenerate the part of the video that has been either deleted or destroyed. The workflow follows that the cross-correlation method to get the video frame by frame and the objects being analyzed in each frame. The source frame is obtained and inpainting technique is applied to these frames only. The order of filling the patch is done by determining the priority factor.

3   Methodology

The tool has been developed using MATLAB R2013a. The basic functions of MATLAB and those related to digital image processing have been used to get the desired result. The tool's working has been tested using various images. To check the quality of the image, Mean Square Error (MSE) and PSNR ratio have been used to compare the result generated by the tool and the Adobe Photoshop. The complete result has been shown in this paper by using three images and calculating their MSE and PSNR in tabular form and showing the result figures in graphical form (Fig. 3).

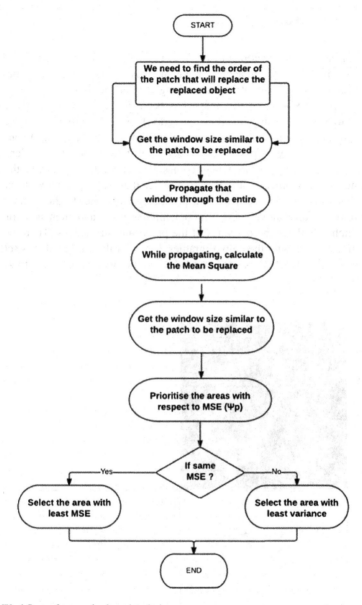

Fig. 3 Workflow of exemplar-based technique

4 About the Tool

The inpainting tool is developed using MATLAB 2013a. Basic functions of image processing have been used to achieve the goal. This tool is a simple GUI-based tool that fetches any image format in the working area of the tool.

Intially, preprocessing of the image has been done. The input RGB image is converted to grayscale using function 'rgb2gray()'. To differentiate the objects from the background, 'Canny edge detection' algorithm has been used. When there a change in the intensity of pixels and to deal with object boundaries, Canny edge detection is the best algorithm that comes into play. Canny labels the pixels depending on the gradient value larger or lesser than it [14]. Canny is one of the best methods to find the edges without producing any effect to the features of the image. Initially, to remove noise, smoothening is done and then the gradient is found which calculates the derivates of the pixel intensity values. Then the region highlighting is done with the help of gradient value. A threshold value is set, above which the pixel is labelled as edge and below it, the pixel value is set to zero [15] (Fig. 4).

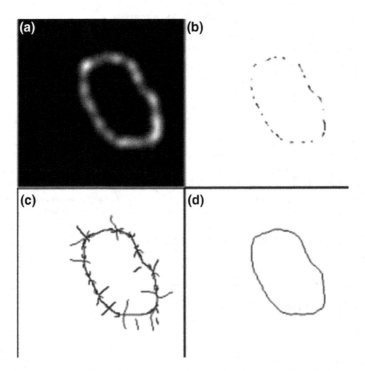

Fig. 4 **a** Synthetic image, **b** obtaining edge, **c** minor and major edges, **d** final edges [16]

To find the gradient:

$$G = \sqrt{G_x^2 + G_y^2}. \tag{1}$$

$$\theta = \arctan\left(\frac{G_y}{G_x}\right). \tag{2}$$

4.1 Structuring Element

A Structuring Element (SE) is used to define neighbourhood structures or to identify some kind of pattern. It is a binary image or also called as mask or kernel (Fig. 5).

In our coding, we have created a square-shaped structuring element. The code snap is as follows (Fig. 6).

4.2 Dilation

In morphological operation on the image, dilation is also known as growing of the image. It is usually applied to binary as well as grayscale image. It is called 'Growing' because it makes an object's pixel grow in size. The pixels present at the boundary tend to increase their size at the boundary. This method decreases the empty cavity/holes in the image. This is also used for region filling. Let M be a set of pixels for the input image. Let N be the structuring element matrix. The formula is as follows (Fig. 7):

$$M \oplus N = \{x | (N)_x \cap M \neq \phi\}$$

Fig. 5 Four-neighbourhood structuring element

$$H = \begin{array}{c} \bullet \\ \bullet \ \bullet \ \bullet \\ \bullet \end{array}$$ ▓ origin (hot spot)

Fig. 6 Use of strel() in MATLAB

```
se = strel('square',2);
BW2 = imdilate(BW,se);
axes(handles.axes1);
```

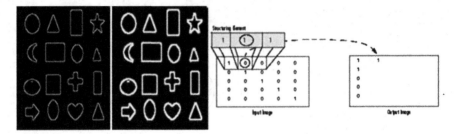

Fig. 7 Dilation results

5 Results

The inpainting tool has been developed on the MATLAB 2013a software. The working, GUI and the results have been described in this paper. The results have been compared with the same operation performed on ADOBE PHOTOSHOP software. The results of the tool have been compared by calculating PSNR and MSE values of each output. The tool's GUI is a follows (Fig. 8).

The above image shows the functioning of the tool. The image is fetched and the objects are detected using canny edge detection. The objects count is also mentioned by the tool.

The working of the tool has been represented in the following flowchart (Fig. 9).

In the below image, a middle face has been removed and has been replaced by the background. This inpainting technique is used to remove objects and replace with the background texture (Fig. 10).

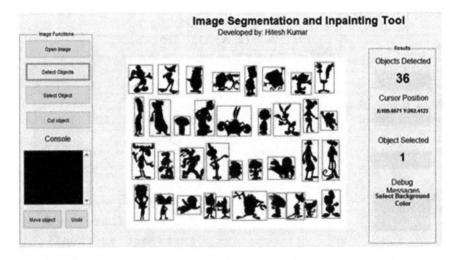

Fig. 8 Tool's GUI with image fetched and objects detected

Fig. 9 Working of the tool

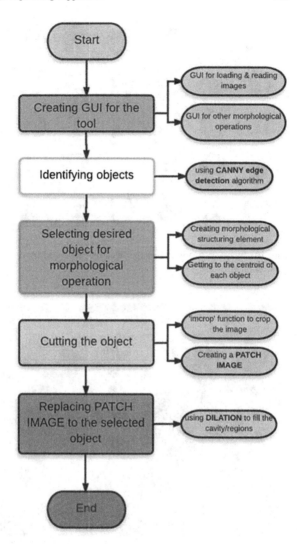

In the below image, the object has been replaced with the other object in the picture (Fig. 11).

In the below image, one face has been replaced with the other taken from the other image only manually by the user (Fig. 12).

The metric used to compare the output generated by the tool and the ADOBE PHOTOSHOP are MSE and PSNR. These two metrics are used to check the quality of the image generated by the tool with the same image being generated by ADOBE PHOTOSHOP software. The results are as follows (Fig. 13; Table 1).

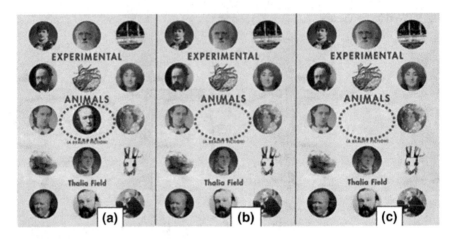

Fig. 10 Image 1: **a** original image, **b** object removed by our tool, **c** object removed by Adobe Photoshop software

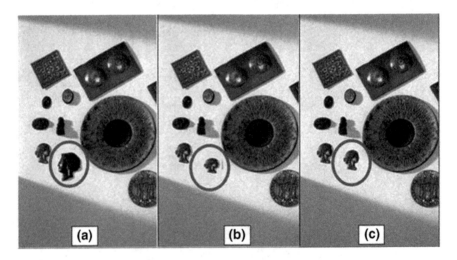

Fig. 11 Image 2: **a** original image, **b** object replacement by our tool, **c** object replacement by ADOBE PHOTOSHOP software

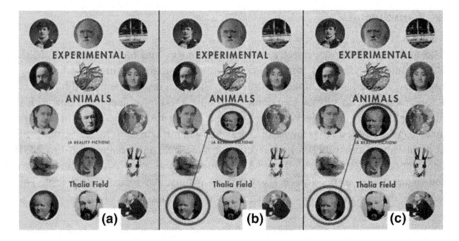

Fig. 12 Image 3: **a** original image, **b** face replacement by our tool, **c** face replacement by ADOBE PHOTOSHOP software

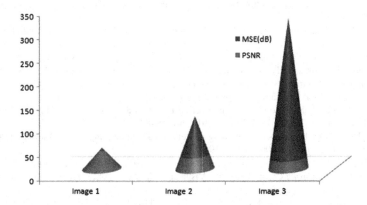

Fig. 13 Graph showing the comparison of MSE and PSNR of different operations on different images

Table 1 Table showing experiment results of three images

	PSNR	MSE (dB)
Image 1	37.9711	6.8229
Image 2	28.9338	83.1187
Image 3	23.4150	296.1951

6 Conclusion

The tool works almost perfectly with simple images and has a pretty low MSE. PSRN ratio calculated for the images are low except for the images which have object overlapped and complex in structure. The three results mentioned have good PSNR ratio. The developed tool comes out to be good for object replacement. It is also a helpful tool for object removal as well. As per results, the graph of image 1 and image 2 suggests that the MSE is less but when it comes to image 3, the MSE is high. Hence, this tool works perfectly for simple texture images with low MSE and a good PSNR value. Well, its MSE value increase when complex images are fed to the tool. As per Image 1 experiment results; the MSE is very less. This suggests that for images of type Image 1, the tool works comparatively well.

References

1. J. Electron., Image completion based on statistical texture analysis, Imaging. 24(1), 013032 (Feb 23, 2015)
2. Tao Zhou, Brian Johnson, Patch-based Texture Synthesis for Image Inpainting, 1605.0157 (May 5, 2016)
3. Survey on Image Inpainting Techniques: Texture Synthesis, Convolution and Exemplar Based Algorithms, IJIRST "International Journal for Innovative Research in Science and Technology| Volume 1 | Issue 7 | December 2014
4. https://upload.wikimedia.org/wikipedia/commons/0/02/Texture_spectrum
5. Rajul Suthar et al Int. Journal of Engineering Research and Applications, ISSN: 2248-9622, Vol. 4, Issue 2(Version 1), February 2014, pp. 85–88]
6. T. Chan and J. Shen, 'Local in painting models and TV in painting, SIAM Journal on Applied Mathematics, Vol. 62, 2001, pp. 1019–1043]
7. P. Harrison. A non-hierarchical procedure for re-synthesis of complex texture, In Proc. Int. Conf. Central Europe Comp. Graphics, Visua. and Comp. Vision, Plzen, Czech Republic, Feb 2001
8. Gupta, Kushagr, Suleman Kazi, and Terry Kong. "DeepPaint: A Tool for Image Inpainting"
9. Blind inpainting using the fully convolutional neural network, Nian Cai, et al., Springer 2015
10. Criminisi, Antonio, Patrick Prez, and Kentaro Toyama. 'Region filling and object removal by exemplar-based image inpainting.' Image Processing, IEEE Transactions on 13.9 (2004): 1200–1212
11. K. Sangeetha, P. Sengottuvelan and E. Balamurugan, "A Novel Exemplar based Image Inpainting Algorithm for Natural Scene Image Completion with Improved Patch Prioritizing", International Journal of Computer Applications (0975 – 8887), Volume 36– No. 4, December 2011
12. Waykule J.M., "Modified Image Exemplar-Based Inpainting", International Journal of Advanced Research in Computer and Communication Engineering, Vol. 2, Issue 9, September 2013
13. Sreelekshmi Das and Reeba R., "Robust Exemplar based Object Removal in Video", International Journal of Scientific Engineering and Research (IJSER), ISSN (Online): 2347–3878, Volume 1 Issue 2, October 2013

14. Lijun Ding and Ardeshir Goshtasby,On the Canny edge detector,Computer Science and Engineering Department
15. Nisha, Rajesh Mehra, Lalita Sharma, Comparative Analysis of Canny and Prewitt Edge Detection Techniques used in Image Processing, International Journal of Engineering Trends and Technology (IJETT), Volume 28 Number 1 - October 2015
16. On the Canny edge detector, Lijun Ding and Ardeshir Goshtasby Computer Science and Engineering Department

The Hidden Truth Anonymity in Cyberspace: Deep Web

Saksham Gulati, Shilpi Sharma and Garima Agarwal

Abstract The main objective of this paper is to study and illustrate the workings of the invisible Internet and practically determining the Onion Routing relays, bridges and exit nodes. A detailed research has been done on the working of exit nodes and privacy over the deep web through vulnerable nodes. This paper also illustrates a practical study of the depth of data available on the deep web. Emphasis is laid down towards the safe access to deep web without compromising one's privacy. A survey was conducted in the paper to show the awareness among the technically sound public and results were shown in pie charts.

Keywords Darknet · Onion router · Tor · Exit nodes · Privacy

1 Introduction

The deep web is that part of Internet which is not indexed by the usual search engines [1].

Although deep web has very common uses, it is most notoriously coined for illegal activities such as buying and selling of drugs, illegal armoury, anonymous chat platforms, child pornography and black hat hacking services. The websites which are generally accessed through deep web are not registered with regular search engines. The deep web is generally accessed through an anonymous network known as Tor, which completely shields an individual's identity. Although Google is known as the largest holder of indexed information, it is believed that information in deep web may be 2–3 times larger than that in the surface web [2]. In this paper,

S. Gulati (✉) · S. Sharma · G. Agarwal
Amity University, Noida, Uttar Pradesh, India
e-mail: saksham.gulati28@gmail.com

S. Sharma
e-mail: ssharma22@amity.edu

G. Agarwal
e-mail: garima.agarwal14@gmail.com

© Springer Nature Singapore Pte Ltd. 2018   719
S. Bhalla et al. (eds.), *Intelligent Computing and Information and Communication*,
Advances in Intelligent Systems and Computing 673,
https://doi.org/10.1007/978-981-10-7245-1_70

the invisible web has been categorized into three major subgroups which are deep web, dark web and the invisible Internet project.

In the methodology section of this paper, details are provided about the methods being used to collect the desired data and survey. The main emphasis is towards the dog command that is used to collect information about DNS [3, 4].

In the result section of this paper, the dataset was collected through systematic research. Former results contain a theoretical explanation of coexistence of deep web and surface web which has been explained practically using DIG command. Later, the section contains a survey which was conducted by over 300 IT experts. Graphical representation using pie charts were developed according to the awareness among the individuals about the deep web. The major category included General Public, IT Experts and Information Security Experts to survey user's knowledge and awareness of deep web and dark web.

1.1   Dark Web Services

Dark web generally deals with shady sides of Internet including the bitcoin mining and exchange, terrorism, hoaxes, illegal market [3]. Dark net can be considered as a subset of deep web. Often pages that are found in dark web are cached within a domain of the original web but can only be accessed by special browsing strategies known as onion routing or tor [5–7]. Hence, dark web is sometimes also considered as the onion world. Dark web is that part of deep web which is kept hidden, private and protected at source and destination. Examples are as follows:

a. **Fake Photo Identity**

Fake photo identity like U.S and U.K citizenships cards, driving license, etc., can be easily bought in through dark web [8].

The seller's promise for an original passport with information/picture. They guarantee that the information of the passport holder will be stored in the official passport database and thus make it possible to travels abroad [9].

b. **Renting Services**

Hiring a hacker over the web, who is willing to do practically anything for money [6, 7, 9].

Skilled hackers vary in ages even from 10 years onwards. These hackers are not identified because they are not able to find a decent job somewhere but because the pay is good and they get a thrill out of this job [10]. The amount depends on the type of service.

c. **Black Market**

No state or country allows buying a professional gun until and unless license and purpose of defence are clearly given to the required acting committee to do so.

Onion dark web is the largest place to buy guns and ammo's illegally. It happens to be one of the biggest markets for some of the most deadly and life-threatening psychedelics and drugs like LSD, COCAINE, METH, ECSTASY, etc. [10, 11].

The Dark net has a huge market for the buying and selling of guns. These sites do not check for licenses and they have every gun with little descriptions of them and cost. It is as easy as buying clothes from www.amazon.com [12–14].

d. Counterfeiting

The dark web also provides an easy market to counterfeit small bills in order to illegally launder money [12].

The money sold is said to be produced from cotton-based paper. It is often UVI incorporated hence it can pass the UV test. It can also pass the Pen test without any problems. They also have necessary security features to be spent at most retailers [12, 15].

1.2 Deep Web Services

Search engines—These search engines, unlike surface search engines—Google and Bing, are able to crawl much deeper into the peer-to-peer network [3]. Examples are:

a. CryptoCurrency

It is a virtual form of currency which is harvested using seeding and leaching methods similar to that of Torrents and holds a very high value in deep web. Almost anything and everything can be bought on deep web using cryptocurrency [11, 13, 15].

b. Drug Marketplace

One of the most common practices on dark web is the marketing of banned drugs and crack. Since the dark web is untraceable, drug dealers and their suppliers use this opportunity to connect to people who desire such substances [9, 14, 16].

c. Illegal Pornography

Dark Web is a major hub for child pornography. Due to its ability to maintain its user's anonymity, dark web offers a platform for child molesters, paedophiles and rapists alike to upload videos and photographs of their victims without the fear of authority [8]. These categories of porn receive many views from people with the same mindset and thus in turn encourage their uploaders to make more such videos [8–10].

2 Related Work

Deep Web and I2P have been around for several years now. Deep Web is that part of Internet which is generally not accessible by the standard search engines [1]. When talking about Deep Web, it is important to distinguish between traditional search engines and search engines capable of Deep Web harvesting [1–7]. Unlike traditional search technologies, like Google, that index links and allow to view the results, Deep Web engines take it a step further and harvest all of the results. The traditional web crawlers only focus on the surface web while the deep web keeps expanding behind the scene. The structure of the deep web pages makes it impossible for traditional web crawlers to access deep web contents [17]. Curiosity about Dark Web peaked when the FBI took down the Silk Road Marketplace in 2013 [2, 5, 11]. The media frenzy that took place made many people aware of the concept [2, 5].

Reference [1] introduces several different networks other than Tor that guarantee anonymity. It further analyzes malicious activities that take place over the Dark Web, which has now become like a safe haven for criminals by providing them a platform to conduct illegal activities without fear of authority. In the paper by Hardy [18], the process of utilizing a modelling technique, informed by trade auction theory is used to analyze the reputation of seller's in the encrypted online marketplace. This analysis of the seller's reputation helps in giving insights into the factors that affect and determine the prices of goods and services. The need for people to maintain their anonymity from government agencies is helping to expand the number of users on Darknet [8]. References [2, 19] state "Cybercriminals from every corner of the world take advantage of the anonymity of the Web, particularly the Deep Web, to hide from the authorities. Infrastructure and skill differences affect how far into the Deep Web each underground market has gone". Although mere access via TOR is not considered an illegal practice, it can arouse suspicion with the Law [1–7]. Terrorists use the deep web for their own advantage. Many terrorist organizations and their supporters use deep web, its anonymity and protection for recruitment and communication purposes. In their paper [21], H. Chen et al. developed a web-based knowledge portal, called dark web portal. This portal was developed to support the discovery and analysis of information discovered on the dark web. TOR has long been used by journalists, researchers, thrill seekers or in general curious people in heavily censored countries in order to hide their browsing histories and exchange information anonymously.

Dark net offers a variety of services [3, 4] including buying and selling of ammunition, access to child pornography and drugs. In this paper, we tried to discuss few services along with the awareness of people about the services of deep web.

3 Methodology

Dig command is a tool used to query DNS name servers for information about host addresses, mail exchanges, name servers and related information. The tool is widely used on Linux-based systems and Macintosh (Mac). This dig command is very useful in troubleshooting and educational purposes. It replaces older techniques like ns lookup and host programs.

The basic idea of dig is to pretend as a name server and working down from the root of the server by iterative queries. After it gets access to name servers it then requests for the specific details matching to the query.

We have implemented the following dig command to gather the information regarding domains which are part of the deep web but are operated at surface web along with deep web for maintaining high levels of anonymity.

Dig command connect to the domain while visiting all the intermediate nodes and tries to collect the information at each node. While looking up at the deep web domain the following nodes are used to route data.

a. Tor Relay

Tor relays are routers or nodes which are used to transmit data over peer networks. Since tor is a peer-to-peer service, a working user passes through three nodes before actually reaching its destination. This is done to mask the identity of the user but masking leads to slow and unreliable connection. There are three types of tor relays [22].

b. Middle Relays

These types of relays are used to mask the identity of the user entering the tor network. They are responsible for routing data from source to the destination while masking the identity. These are the middlemen in the network and store very little data. For security purposes, data travels through three nodes before reaching the actual destination, two of which are middle relays [22].

c. Exit Nodes

These nodes are the last nodes before which data reaches its destination and third node of the three. Generally, exit nodes are advertised all over the tor web or the deep web so that an uploader can easily setup a connection. Exit nodes are thoroughly protected using encryption so as to avoid eavesdropping and vulnerability extraction. But recently tor exit nodes have been attacked by various groups and defence forces.

d. Bridges

Bridges are Tor relays which are not publicly listed as part of the Tor network. Bridges are essential censorship-circumvention tools in countries that regularly block the IP addresses of all publicly listed Tor relays. A bridge is generally safe to run in homes, in conjunction with other services, or on a computer with personal files.

Other than dig, a survey has been conducted to gather the information regarding the awareness of the mass. The survey was conducted with over 500 individual and was divided into three major categories namely general, IT and Network Security.

The survey was kept anonymous. It contained no personal question like name, address or anything involving personal life except for work field which was relevant to the paper. The work field specifies the relativity of an individual to the hidden aspects of Internet, for example a network engineer is expected to be more aware of deep web than an individual in any other field of work.

The following set of questions was asked in survey.

1. Primary work field categorized into General, IT and Network Security
2. Have you heard about deep web?
3. If yes, have you ever used accessed deep web?
4. Do you access deep web on regular basis? If yes, no of days in week.
5. Have you ever come across any illegal activity on the Internet? Was the owner of this activity traceable?

4 Results

Interaction Between surface and deep web.

According to the current Internet protocols, the deep web can be considered to be hosted on current visible network with complete protection through overlay websites and hence it is the major reason why it is difficult to completely trace deep web. Some of the websites even use plain registered DNS, which is directly used to transfer files through wireless server [14, 22].

Case I—Hostmysite.com, the only single home page claims to upload the file anonymously but does not specify where it is to be uploaded. This DNS is accessible through resolved name but it does no information regarding the ownership of this page or relocation of the uploaded file can be found. Hence, it seems to be working on surface web but it is directly related to deep web [22].

Proof 1: It can be accessed without the use of http protocol or the https standard protocols.

This means that "hostmyshit.com" does not uses http or https before the DNS. It is unusual in web 3.0 and further to have such kind of anomaly since it was compulsory to use http or https tag to correctly determine the address of a website over the internet.

In case of deep web, this is an exception since deep web converts a normal DNS to the hash equivalent to reach to a desired website.

Proof 2: When this DNS was dug using an open-source software DIG (Figs. 1 and 2).

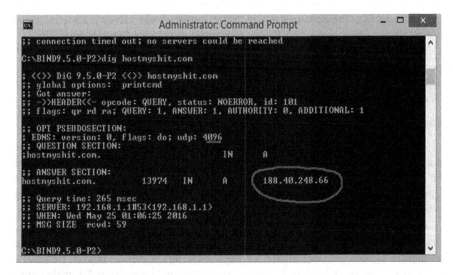

Fig. 1 DIG command result

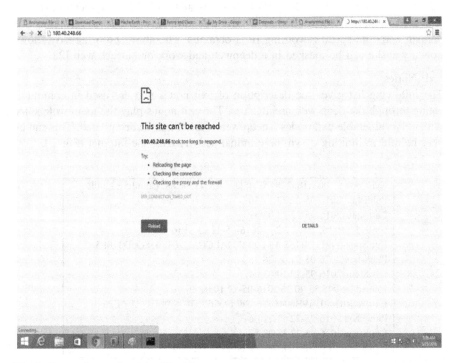

Fig. 2 Connection testing using IP address

But the DNS or the website is not traceable from original ip. Even if a host is behind a firewall, false hop proxy or VPN the Domain is searchable through proxy IP.

In such cases, an event where the host is not traceable from masked IP which positively replies on ICMP ping (ICMP ping is unreliable hence if it gives and active connection it means the connection is smooth.) means it has a constantly changing untraceable IP which is a property of Onion routing [1].

Proof 3: The website uses 4096 UDP port.

A website which is connectable without a packet drop cannot be relies on UDP port. Peer-to-peer connections generally work on UDP port. Hence this website runs on a peer-to-peer connection and since this uses FTP protocols (uploading) it has to spread connection and hence files. This is how the files are shared on peer-to-peer connection. Tor is entirely based on P2P connection reliability between the nodes.

Case II—In a recent chain of events which took place in the United States a major torrent site kickass torrent went down and was restricted to access. Torrent is a major example of coexistence of deep web and surface web. Torrents use peer-to-peer protocol for data download and upload with a registered domain name. Deep Web is completely based on peer based connection and Torrent search engines queries through deep web for accurate result because it is not possible to trace deep web log files. This makes piracy related crime much easier to execute. Kickass torrent can be reached easily through surface web but on resolving this domain name no specific working IP is obtained. The above proof already shows how a website can be masked in a deepweb and work on surface web [22].

Exit Nodes
The following list gives the description of exit nodes that are used in communicating through the deep web architecture. The exit nodes play the major role since with only vulnerable exit nodes a deep web website can be deciphered. This can be very helpful in limiting down the criminal activity over the internet (Fig. 3).

```
ExitNode      0011BD2485AD45D984EC4159C88FC066E5E3300E
Published 2016-05-22 22:17:50
LastStatus2016-05-22 23:02:51
ExitAddress162.247.72.201 2016-05-22 23:11:01
ExitNode 0091174DE56EFF09A7DAECCAC704F638D6D370F5
Published 2016-05-22 09:38:21
LastStatus2016-05-22 10:03:04
ExitAddress45.35.90.36 2016-05-22 10:07:30
ExitNode0111BA9B604669E636FFD5B503F382A4B7AD6E80
Published 2016-05-22 17:11:36
LastStatus2016-05-22 18:02:57
ExitAddress176.10.104.240 2016-05-22 18:05:00
```

Fig. 3 List of currently active exit nodes (as on 21 June 2016)

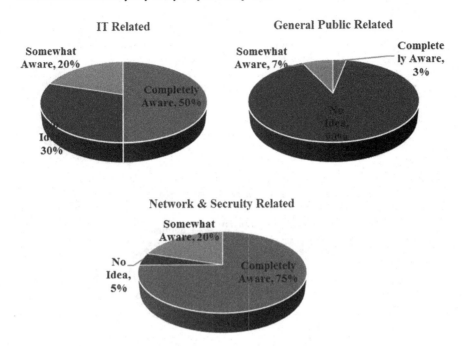

Fig. 4 Deep web awareness survey

These are the list of exit nodes which were accessed and verified as a part of deepweb data routing mechanism. The PGP signatures are mentioned along with their exit node addresses

A survey was conducted to check what percent of population amongst 300 data sets of users is aware of the deep web and black market. The survey was conducted and results were categorized into the general public, IT related and Network and Security Associated. The following results were obtained for deep web and dark web (Figs. 4 and 5).

IT Related   General Public Related

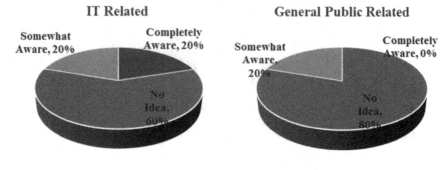

Network & Secruity Related

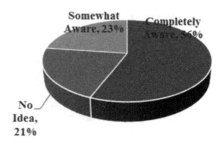

Fig. 5 Darkweb awareness survey

5 Conclusion

We conclude that there is very less number of people who actually know about the deep web and thus it raises a major concern. Also as 80% of the web is not visible and contains the most valuable data is results a great impact towards research area. The finer concern lies that the people have almost no idea about the dark web and illegal activities related to the Internet making them a potential victim of many crimes.

6 Future Work

This paper can be used to spread the awareness about the deep web, dark web and criminal related activities associated with the general public of internet. It not only prevents them from getting influenced by such activities but also helps to catch the criminals.

Internet is easily accessible by anyone at any place in the world and it takes a little effort to access deep web. It is very easy for anyone to indulge in the criminal market of deep web, which definitely leads to trouble. Through this chapter, we are aiming at spreading awareness among the masses to prevent themselves from getting involved in criminal activities.

This paper also explains the coexisting relationship between deep web and surface web hence further research studies can be done keeping this in mind that surface web is just subset of deep web and holds a relation which was not considered in many of the research work until now.

The exit nodes as listed in the paper can be monitored or attacked and information can be collected to perform a strategic way out of criminal activities in the coming upcoming researches.

References

1. Hamilton, Nigel. "The Mechanics of a Deep Net Metasearch Engine". CiteSeerX: 10.1.1.90.5847.
2. Owen, Gareth. "Dr Gareth Owen: Tor: Hidden Services and Deanonymisation". Retrieved 20 June 2015.
3. Devine, Jane; Egger-Sider, Francine (July 2004). "Beyond Google: the invisible web in the academic library". The Journal of Academic Librarianship 30 (4): 265–269.
4. http://www.scribd.com/doc/172768269/ Ulbricht-Criminal-Complaintulbr_mirror. Scribd. "Ulbricht Criminal Complaint." [Accessed on 10 June 2015].
5. Owen, Gareth. "Dr. Gareth Owen: Tor: Hidden Services and Deanonymisation". [Accessed 20 June 2015].
6. Vincenzo Ciancaglini. (8 November 2013). TrendLabs Security Intelligence Blog. "The Boys Are Back in Town: Deep Web Marketplaces Back Online." Last accessed on 10 June 2015, http://blog. trendmicro.com/trendlabs-security-intelligence/the-boysare-back-in-town-deep-web-marketplaces-back-online/.
7. Vincenzo Ciancaglini. (10 March 2015). TrendLabs Security Intelligence Blog. "The Deep Web: Shutdowns, New Sites, New Tools." Last accessed on 10 June 2015, http://blog. trendmicro.com/trendlabs-security-intelligence/the-deep-web-shutdowns-new-sitesnew-tools/.
8. Fake Photo Identity Services, 2ogmrlfzdthnwkez.onion [Accessed on 11 June 2016].
9. Illegal Pronography, Vfqnd6mieccqyiit,onion [Accessed on 11 June 2016].
10. Drug Market Place, S5q54hfww56ov2xc.onion [Accessed on 11 June 2016].
11. Illegal Services, Tuu66yxvrnn3of7l.onion [Accessed on 11 June 2016].
12. Illegal Services, Qkj4drtgvpm7eecl.onion [Accessed on 11 June 2016].
13. Robert McArdle and David Sancho. Trend Micro Security Intelligence. "Bitcoin Domains." [Accessed on 10 June 2015].
14. The Invisible Internet Project. [Accessed 11 June 2015] https://geti2p.net/en/.
15. FeikeHacquebord. (5 September 2015). TrendLabs Security Intelligence Blog. "The Mysterious MEVADE Malware." [Accessed on 10 June 2015], http://blog.trendmicro.com/trendlabs-security-intelligence/the-mysterious-mevade-malware/.
16. Trend Micro Incorporated. Trend Micro Security News. "Cybercriminal Underground Economy Series." [Accessed 11 June 2015], http://www.trendmicro.com/vinfo/us/security/special-report/cybercriminal-underground-economy-series/index.html.
17. R.Anita, R. "Deep iCrawl: An Intelligent Vision-Based Deep Web Crawler". World Academy of Science, Engineering and Technology, Internationals Science Index 50, International Journal of Computer, Electrical, Automation, Control and Information Engineering, (2011), 5 (2), 128–133.

18. HARDY, ROBERT AUGUSTUS, and JULIA R. NORGAARD. "Reputation in the Internet Black Market: an Empirical and Theoretical Analysis of the Deep Web." Journal of Institutional Economics, vol. 12, no. 3, 2016, pp. 515–539., doi:https://doi.org/10.1017/S1744137415000454.

19. http://blog.trendmicro.com/trendlabs-security-intelligence/cybercrime-in-the-deepweb/. Robert McArdle. (4 October 2013). TrendLabs Security Intelligence Blog. "Cybercrime in the Deep Web." [Accessed on 10 June 2015].

20. http://www.trendmicro.com/cloud-content/us/pdfs/security-intelligence/white-papers/wp-bitcoin-domains.pdf.

21. Qin, Jialun, et al. "The dark web portal project: collecting and analyzing the presence of terrorist groups on the web." *Proceedings of the 2005 IEEE international conference on Intelligence and Security Informatics.* Springer-Verlag, 2005.

22. https://www.torproject.org The Tor Project, Inc. Tor Project. [Accessed on 11 June 2015].

Test Case Optimization and Prioritization of Web Service Using Bacteriologic Algorithm

Gaurav Raj, Dheerendra Singh and Ishita Tyagi

Abstract Regression testing, testing is done on the changes made in existing software to check whether the existing software is working properly or not after the changes has been done. Therefore, retesting is performed to detect the new faults found. This type of testing is performed again and again after the changes have been made in the pre-existing software. Various methods are used for test case reduction and optimization for a web service. Regression testing creates a large number of test suites which consumes a lot of time in testing and many other problems are faced. Therefore, some technique or method should be used so that number of test cases are reduced and also test cases can be prioritized keeping in mind the time and budget constraints. The test case reduction and prioritization need to be achieved depending on various parameters such as branch coverage and also on basis of fault coverage etc. Therefore, this paper discusses about the analysis of the code of a web service and the technique used to analyze a web service based on branch or code coverage and also the fault detection for test case reduction and prioritization is bacteriologic algorithm (BA). The test cases generated and also other requirements are mapped with the branch coverage and fault coverage of the code of the web service.

Keywords Test case · Test suite · Bacteriologic algorithm · Web service
Regression testing · Test case reduction

G. Raj (✉)
Punjab Technical University, Kapurthala, Punjab, India
e-mail: graj@amity.edu

D. Singh
Computer Science and Engineering Department, Chandigarh College
of Engineering and Technology, Sector-26, Chandigarh, Punjab, India

I. Tyagi
Computer Science and Engineering Department, Amity University, Uttar Pradesh, India
e-mail: ishita27t@gmail.com

© Springer Nature Singapore Pte Ltd. 2018
S. Bhalla et al. (eds.), *Intelligent Computing and Information and Communication*,
Advances in Intelligent Systems and Computing 673,
https://doi.org/10.1007/978-981-10-7245-1_71

1 Introduction

Main goal of web service is to communicate and exchange information by using a common platform among different applications built in different programming languages and operating system. The confidentiality and integrity of a web service need to be maintained as attacks on web application have increased.

The term software engineering is defined as a branch of engineering which is related to software development and management with the help of properly defined methods, principles and steps. The aim to provide the customer a reliable and efficient product. The software engineering contains various models for the software development. The models are used according to the requirements and other parameters required for software development and management. Since, one of the most important phase of any model is testing phase, so the testing should be done properly and carefully to get reliable results. It is not practically possible to implement exhaustive testing. There can be both valid and invalid inputs because it is very difficult to perform testing on all the test cases available. There are many reasons like time constraints, budget etc. because of which exhaustive testing is not possible. It is not easy to completely test because of the design problems. In regression testing, testing is done on the changes made in existing software to check whether the existing software is working properly or not after the changes has been done. Therefore, retesting is performed to detect the new faults found. This type of testing is performed again and again after the changes has been made in the pre-existing software. Changes can be of different types such as adding extra features, changes in the configuration etc. [1]. But this testing increases the budget of the software or the product because of the features enhanced and also a lot of maintenance is required.

Various methods are used for test case reduction and optimization for a web service. But the technique should be selected in such a way that budget must be less. Regression testing also creates a large number of test suites which consumes a lot of time in testing and many other problems are faced [2, 3]. The test case reduction and prioritization need to be achieved depending on various parameters such as branch coverage and also on basis of fault coverage etc. so, [4]. The technique used to analyze the web service based on branch or code coverage and also the fault detection for reduction of test case and prioritisation is bacteriologic algorithm (BA). The test cases generated and also other requirements are mapped with the branch coverage and fault coverage of the code of the web service. This helps in calculating fitness of the code of the web service based on branch coverage and fault detection capability.

2　Terminologies

There are some terms that are explained under this section which will be used in the further sections of the paper.

2.1　Test Case

Test case is a collection the preconditions, post-conditions, test data and results which are documented for a specific test scenario so that compliance can be verified against the requirement [3]. For execution of the test, test case is the starting point. Some of the parameters of test case are: Test case ID, Test data, Expected results, Actual results etc.

2.2　Test Suite

A group of test cases is known as test suite. The test execution status can be reported by the testers with the help of test suites. Each test case can used in one or more test suites. Number of test cases are there in a test suite. Before creating test suites, a test plan is made. According to the scope and the cycle test suites are generated. There can be number of tests viz-functional or Non-Functional.

2.3　Test Case Minimization

It is defined as process to make lesser number of test cases with the help of making test suites from test cases. It is used for reducing the cost of resources, time taken in execution etc. The test suites should be created in such a way that satisfy all the requirements. The main aim of test suite is remove redundancy that is created because of the test cases. There are different techniques used for removing redundancy such as genetic algorithm etc.

2.4　Test Case Prioritization

There are number of test suites available in regression testing [2]. Since, it is not easy because of the time constraints to test all the test suites over the code, we use test case prioritization. It can be defined as ordering the test suites that follow one or more criteria. The aim is to find the faults by selecting least no of test cases. It is used so that the best results are found at the time of the result. There can be many be reasons for prioritizing the test suites: fault detection rate should be increased. Code coverage should me more and at a faster rate. Every system has a different architecture, Therefore, prioritization can also be done on basis depending on the system architecture. There are some parts of the architecture that have a major effect on the entire system, they should be tested properly. The components or parts of system can be tested individually also for test case prioritization.

2.5 Exhaustive Testing

The testing in which the testing is done for the all the test cases that are possible keeping in mind the quality [5, 6]. Every possible case is made but this type of testing is impractical. A product is considered to be perfect if the exhaustive testing of a software is done. It is actually not easy to pass exhaustive testing. There are strict deadlines because of which this type of testing becomes impossible to use. Some of the inputs become invalid after particular time period. Therefore, time constraint is also an issue. It is impossible if real time scenarios need to be tested like temperature etc. Also, it is impractical to test all the combinations of each user or many users [7]. All the outputs also cannot be checked.

2.6 Web Service

Goal of web service is to communicate and exchange information by using a common platform among different applications built in different programming languages and operating system. A web service contains UDDI, SOAP and WSDL.

2.7 Genetic Algorithm

The optimization problem that are related to natural selection are solved using the algorithm i.e. genetic algorithm. It changes each solution of the population again and again. In each step, the children are generated for the upcoming production from the parent test cases or test suites. And, as the generation passes, best or favourable results are seen. This type of algorithm is used in solving the issues that cannot use optimization algorithms etc. For knowing the nature of chromosome, fitness is calculated. There are various steps applied in this algorithm:

1. First step is selection in which test suites are selected and then crossover and also mutation are used over the test suites. The chromosomes or test suites are chosen generally which are having greater fitness.

2. Second step is crossover in which the successors are combined and then form new children.
3. The next operator used is mutation in which the current descendants are changed automatically.

2.8　Bacteriologic Algorithm

Mala et al. [8] this algorithm is the alternative form of the GA algorithm. It has slightly different steps because it follows the bacteria's nature. There is no operator like crossover present in it. It shows the properties of a bacteria. It only uses the operator like selection and mutation from the above algorithm referred. It is different because it contains a new function i.e. "memorization function" which uses the bacteria which is the most fit in each population. There is basic diagram showing the flow of the entire process.

1. First step is selection in which test suites are selected and then crossover and also mutation are used over the test suites. The chromosomes or test suites are chosen generally which are having greater fitness.
2. The next operator used is mutation in which the current descendants are changed automatically.
3. Next step is to find the "fitness" of the code using its formula.
4. Then, best test suite is memorized out of each population for addition to the next step/generation.

3　Related Works

Krishnamoorthi and Mary [1] in their paper *"Regression Test Suite Prioritization using Genetic Algorithms"* said that there are ways to use prioritization method with the help of genetic algorithm. The parameters checked are the effectiveness and time consumption. The faults are found which helps in knowing how effective is the proposed method. Time coverage plays an important role here. It helps in knowing the problems faced during time-aware prioritization. The paper also says also tells about the methods which help in reducing time above of prioritization.

Singh et al. [9] in their paper *"A Hybrid Approach for Regression Testing in Interprocedural Program"* mentioned that regression testing is performed again and again after the changes has been made in the pre-existing software. Changes can be of different types such as adding extra features, changes in the configuration etc. Paper contains a new algorithm related to variable which is applied on the variables with the help of hybrid method. The paper gives an algorithm for regression testing. Various codes have been analyzed to find a conclusion. The accuracy and performance according to the conclusion is much higher.

Jatain and Sharma [2] in their paper *"A Systematic Review of Techniques for Test Case Prioritization"* explained that there are many important activities that a maintenance phase faces like deletion, correction etc. For the retesting of the software retesting is done as modifications are done in the software time to time. But it Is expensive to retest again and again. There are many methods present for regression testing. The tester can choose the method according to their need and can do testing. Here, various methods used by different researchers for prioritization are presented for regression testing. Also different algorithms for the prioritization process is used. It helps in knowing scope of different methods. Also, explains the difficulties faced in prioritization based on requirements.

Joshi [5] in his paper *"Review of Genetic Algorithm: An Optimization Technique"* told about a technique which can be used for the optimization of the whole. After this there are many methods that come under genetic algorithm are used for optimization. The paper explains the basic of the algorithm and gives the proper flow of the "GA algorithm". Also, the operators of the algorithm are reviewed properly. It tells about the origin of the algorithm. There are many other optimization technique other than GA that are explained in the paper. Also it tells about different techniques of using the operators of the given algorithm.

Garg and Mittal [6] *"Optimization by Genetic Algorithm"* in their paper said the optimization problem that are related to natural selection are solved using the algorithm i.e. genetic algorithm. This research paper explains about the algorithm and the technique of optimization. The optimization problem that are related to natural selection are solved using the algorithm i.e. genetic algorithm. It changes each solution of the population again and again. In each step, the children are generated for the upcoming production from the parent test cases or test suites. The optimization method is related or present in every field like engineering etc. This paper explains about the "dejong function" and the experiments have been performed using this function in genetic algorithm. The paper with the help of this function gives better results. There is also the use "unimodal and multimodal benchmark functions" for the optimization. The results are represented using graphs and tables.

Ramesh and Manivannan [7] *"Test Suite Generation using Genetic Algorithm and Evolutionary Techniques with Dynamically Evolving Test Cases"* in their paper explained a new way for making test oracles with the help of evolutionary algorithm. This gave positive results which helped in finding bugs in various classes. This method is also easy to use. Test oracles take a lot of time and is difficult to make meaningful test cases if done manually on the data generated. In this paper coverage parameter is used for optimization of test suites. Also, branch coverage parameter has been used.

Shahid and Ibrahim [10] *"A New Code Based Test Case Prioritization Technique"* in their papers said that test cases and test suites play an important role in software testing. Validation is done of the software or product that is under inspection with the help of test cases and suites. It becomes difficult to test all the cases for all the code. Therefore, prioritization is used which can help in improving the effectiveness and also help in saving time. Here, a new algorithm or approach has been used for prioritization method and is done on basis of code covered. The algorithm is used on a case study for the results which are good and promising. The test case that cover most part are considered more important.

Mala et al. [8] *"A hybrid test optimization framework-coupling genetic algorithm with local search technique"* in their paper explained that there are three different algorithms that are used out of which "HGA algorithm" is used for making test cases of better quality. For this to happen path coverage and score of mutation is analyzed of every test case. The best test cases are selected having greater path coverage and score of mutation. It also takes less time as test cases get reduced. The test cases analyzed using this method are compared with the other algorithm on basis of efficiency. The other methods used are genetic and bacteriologic algorithm.

4 Architecture and Algorithm Used

4.1 Flow Diagram

The basic flow of the algorithm that is used to find the results is explained in Fig. 1.

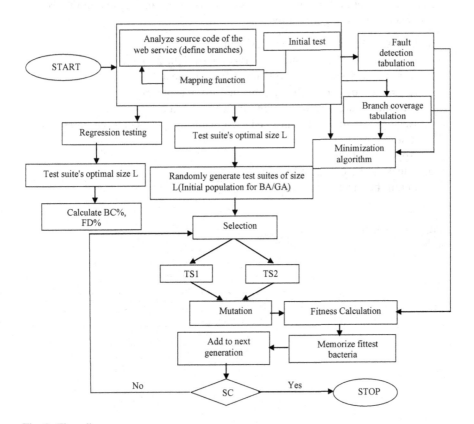

Fig. 1 Flow diagram

4.2 Algorithm Applied

Source code is the input i.e. P
There is a set/group of initial test cases (given by the user) Q
Reduced and prioritized test suites is the O/P.
Algorithm:

1. Find the branches in the code P which act as requirements from P.
   $R_b = \{B1, B2, B3 \ldots B_n\}$
2. Initial test cases(Q) are mapped with R_b
   BranchCoverage$_{table}$ = MappingFunction (Q, R_b)
3. In code P, search use of the variables used.
   $R_c = \{z_1(\text{line x, line y}) \ldots z_n(\text{line x, line y})\}$
4. Q is mapped with R_c, FaultDetectAbility$_{table}$ = MappingFunction(Q, R_c)
5. Search for test suite (T) having maximum branch coverage and fault detection ability.
   T = Search(FaultDetectAbility$_{table}$, BranchCoverage$_{table}$, Q)
6. Optimal length of test suites is to be searched,
   OL = SearchLength (T)
7. Make any random test suites $ts_i = \{ts_1, ts_2, ts_3 \ldots ts_n\}$ having length OL from Q
8. Use bacteriologic algorithm over ts_i

   a. Fitness calculation for every test suite.
      fitness = (weight1* branch coverage) + (weight 2* fault detection analysis)
      where branch coverage is obtained from Table 1.
      fda = fault detection ability (obtained from Table 2)
      w1 and w2 are weights ranging between 0 and 1.
   b. Arrange test suites present in tsi such that test suites with high fitness comes first.
   c. Apply mutation
   d. Fitness to be calculated again
   e. Learn the most fit test suite

Table 1 Value to be searched

Test cases	Search value
1	2
2	3
3	7
4	10
5	1
6	9
7	4

Table 2 Branch coverage

Test case	Search value	Branch b1	Branch b2	Branch b3	Branch b4
1	2	1	0	1	1
2	3	1	1	1	1
3	7	1	1	0	1
4	10	1	0	0	1
5	1	1	0	1	1
6	9	1	1	0	1
7	4	1	1	0	1

   f. Most fit test suite is must be added to next level/generation

   g. Repeat the steps from a and stop when stopping criteria is met. (stopping criteria here can be number of levels/generations)

   h. Rearrange all memorised bacteria in descending order of fitness, to prioritize them.

4.3 Formula Required

1. "Branch Coverage" is,
   (no. of branch covered)/(total no. of branches)*100.
2. "Fault Detection analysis" is, ((Variable(p,q)covered)/(TotalVariable(p/q)covered))*100
3. "Fitness" is,
   (Weight1*BranchCoverage) + (Weight2*Fault Detection analysis) and, the value of the weight can be in between or 0 or 1.

5 Case Study

To get a more practical idea the code of the web service of binary search is observed and is passed as the input into the algorithm. The code is given below:

```
int * BinarySearch (int value)
{
unsigned int A = 0, B = array_length(array), S;
while (A<B)
{
S = (A+B− 1)/2;
if (value = = array [S])
return array+S;
else if (value<array [S])
B = S;
```

```
else
A = S+1;
}
return null;
}
```

The variables used in the above code i.e. *A*, *B* and *S* are taken as input for creating test cases from the user.

6 Results and Discussion

This part of paper shows the result. First step is to take input from the user to generate the test cases. The code is divided into four branches to know the branch coverage. Each line of the code is assigned a number to find the fault detection coverage. The first step is to input the array for binary search code and set the number of test cases to be generated. Then the test cases are formed and values to be searched are given as input by the user. For example:

Array given: {2,3,4,7,9}

The value to be searched are given in Table 1. For the above array, the branch coverage is calculated for the values given by the user in Table 1.

The branch coverage analysis is given Table 2 where, 1 represents that branch has been covered to search the particular number entered by the user and 0 represents that branch has not been covered to search the particular number entered by the user.

Table 3 represents the fault coverage where variable(p, q) is represented as, L (3,6) means L is the variable used in the code and (3,6) is the line in which it initialized and used respectively. Also, 1 represents that test case is covering line p and line q for that particular test case.

After the fault coverage analysis the regression testing is performed on the test cases generated by the user in Table 1. All the possible suits are generated in Table 4 using regression testing (Fig. 2).

Table 3 Fault coverage table

Test case	L (3,6)	R (3,6)	M (6,8)	M (6,10)	M (6,12)
1	1	1	1	0	1
2	1	1	1	1	1
3	1	1	1	0	1
4	1	1	0	0	1
5	1	1	0	1	1
6	1	1	1	0	1
7	1	1	1	0	0

Table 4 Table showing the results of fit calculation for all possible test suite using Regression testing

Test suite dD	Test case	Bc	Fda	Fit
1	1,2	100	100	100
2	1,3	100	100	100
3	1,4	75	80	80
4	1,5	75	80	80
5	1,6	100	100	100
6	1,7	100	100	100
7	2,3	100	100	100
8	2,4	100	100	100
9	2,5	100	100	100
10	2,6	100	100	100
11	2,7	100	100	100
12	3,4	75	80	80
13	3,5	100	100	100
14	3,6	75	80	80
15	3,7	75	80	80
16	4,5	75	80	80
17	4,6	75	80	80
18	4,7	75	80	80
19	5,6	100	100	100
20	5,7	100	100	100
21	6,7	75	80	80

Also, branch coverage, fda(fault detection analysis) and fitness is calculated for all the possible test suites of pair 2 in Table 4. The formula used for:

1. "branch coverage" is,

   (no. of branch covered)/(total no. of branches)*100.

2. "fault detection analysis" is,

   ((variable(p, q)covered)/(total variable(p/q) covered))*100

3. "fitness" is,

   (weight1* branch coverage) + (weight 2* fault detection analysis)

4. and, the value of the weight can be in between or 0 or 1.
5. For example. In Table 4:

- For test suite id 18 i.e. 4,7, 3 out of total 4 branches are being covered. Therefore, (3/4)*100 = 75%
  Similarly, for fault detection analysis, (4/5)*100 = 80%
  Therefore, after putting the values in formula of fitness, answer is 80%.

Fig. 2 Graph showing bc, fda and fit calculated for regression testing

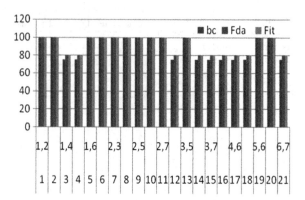

- For test suite id 19 i.e. 5,6, 4 out of total 4 branches are being covered. Therefore, (4/4)*100 = 100%

  Similarly, for fault detection analysis, (5/5)*100 = 100%

  Therefore, after putting the values in formula of fitness, answer is 100%.

Next step is to calculate fitness using BA algorithm in which minimization algorithm is applied and best possible test suites are calculated After the application of minimization algorithm, the bacteriologic algorithm is used to find the fittest bacteria and reduces the number of test suites in Table 5 using mutation operator. The final answers are shown in Table 5; (Fig. 3).

Table 5 Table showing results after applying BA algorithm

Test suites	Test cases	Bc	Fda	Fit
1	5,9	100	100	100
2	5,7	100	100	100
3	2,9	100	100	100
4	2,7	100	100	100
5	6,7	100	100	100
6	6,3	100	100	100
7	2,7	100	100	100
8	2,3	100	100	100
9	8,6	100	100	100
10	8,2	100	100	100
11	8,9	100	100	100
12	2,5	100	100	100

Fig. 3 Graph showing bc, fda, fit calculated using bacteriologic algorithm

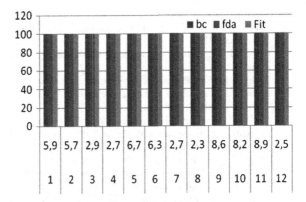

7 Conclusions and Future Work

The paper is about the web service being developed and its code being tested for fitness calculation using regression testing. The aim is to reduce and order the test suites in such a way that all the best results are obtained rather than creating and testing number of test cases as done in regression testing. The criteria to analyze the code is branch coverage and fault detection analysis. Therefore, bacteriologic algorithm is implemented on the code of the web service so that improved results are obtained rather than using regression testing for fitness calculation of the code. The usage of this approach is shown with the help of the example and the graphs showing the results generated using regression testing and then using bacteriologic algorithm on the same piece of code to get reduced and ordered results. The result have been shown in the result section of the paper. The results show the reduced set of test suites having the best fitness. Also, the terms that are used in the paper are explained in the terminologies section. Further analysis can be done in future by comparing the results of the various web service's code using the same algorithm. Also, the results of one web service can be compared with many other algorithm derived in future to make improvements and get best result for test case reduction and prioritization using code coverage and fault detection as parameters.

References

1. Krishnamoorthi, R. and Mary, S.A.S.A, *"Regression test suite prioritization using genetic algorithms"*, International Journal of Hybrid Information Technology, Vol. 2, No. 3, pp. 35–51, 2009.
2. Aman Jatain, Garima Sharma, *"A Systematic Review of Techniques for Test Case Prioritization"*, International Journal of Computer Applications (0975–8887), Volume 68, No. 2, April 2013.
3. Presitha Aarthi. M, Nandini. V, *"A Survey on Test Case Selection and Prioritization"*, International Journal of Advanced Research in Computer Science and Software Engineering, ISSN: 2277 128X, Volume 5, Issue 1, January 2015.

4. R. Beena, Dr. S. Sarala, *"Code coverage based test case selection and prioritization"*, International Journal of Software Engineering & Applications (IJSEA), Vol. 4, No. 6, November 2013.

5. Gopesh Joshi, *"Review of Genetic Algorithm: An Optimization Technique"*, Volume 4, Issue 4, April 2014.

6. Richa Garg, Saurabh mittal, *"Optimization by Genetic Algorithm"*, International Journal of Advanced Research in Computer Science and Software Engineering, ISSN: 2277 128X, Volume 4, Issue 4, April 2014.

7. K. Ramesh and P. Manivannan, *"Test Suite Generation using Genetic Algorithm and Evolutionary Techniques with Dynamically Evolving Test Cases"*, International Journal of Innovation and Scientific Research, ISSN 2351-8014 Vol. 2 No. 2 Jun. 2014.

8. Dharmalingam Jeya Mala, Elizabeth Ruby, Vasudev Mohan(2010), *"A Hybrid Test Optimization Framework-Coupling Genetic algorithm with local search technique"*, Computing and Informatics, Vol. 29, 2010.

9. Yogesh Singh, Arvinder Kaur and Bharti Suri, *"A Hybrid Approach for Regression Testing in Interprocedural Program"*, Journal of Information Processing Systems, Vol. 6, No. 1, 2010.

10. Muhammad Shahid and Suhaimi Ibrahim, *"A New Code Based Test Case Prioritization Technique"*, International Journal of Software Engineering and Its Applications, Vol. 8, No. 6, 2014.

Author Index

© Springer Nature Singapore Pte Ltd. 2018
S. Bhalla et al. (eds.), *Intelligent Computing and Information and Communication*,
Advances in Intelligent Systems and Computing 673,
https://doi.org/10.1007/978-981-10-7245-1

Printed in the United States
By Bookmasters